Artificial Intelligence in Complex Networks

Artificial Intelligence in Complex Networks

Editors

Xiaoyang Liu
Giacomo Fiumara
Pasquale De Meo
Annamaria Ficara

MDPI

Basel • Beijing • Wuhan • Barcelona • Belgrade • Novi Sad • Cluj • Manchester

Editors
Xiaoyang Liu
Chongqing University of Technology
Chongqing
China

Giacomo Fiumara
University of Messina
Messina
Italy

Pasquale De Meo
University of Messina
Messina
Italy

Annamaria Ficara
University of Palermo
Palermo
Italy

Editorial Office
MDPI
St. Alban-Anlage 66
4052 Basel, Switzerland

This is a reprint of articles from the Special Issue published online in the open access journal *Applied Sciences* (ISSN 2076-3417) (available at: https://www.mdpi.com/journal/applsci/special_issues/AI_Complex_Network).

For citation purposes, cite each article independently as indicated on the article page online and as indicated below:

Lastname, A.A.; Lastname, B.B. Article Title. *Journal Name* **Year**, *Volume Number*, Page Range.

ISBN 978-3-7258-0933-2 (Hbk)
ISBN 978-3-7258-0934-9 (PDF)
doi.org/10.3390/books978-3-7258-0934-9

Contents

About the Editors vii

Preface ix

Xiaoyang Liu
Special Issue "Artificial Intelligence in Complex Networks"
Reprinted from: *Appl. Sci.* **2024**, *14*, 2822, doi:10.3390/app14072822 1

Na Zhao, Qian Liu, Ming Jing, Jie Li, Zhidan Zhao and Jian Wang
DDMF: A Method for Mining Relatively Important Nodes Based on Distance Distribution and Multi-Index Fusion
Reprinted from: *Appl. Sci.* **2022**, *12*, 522, doi:10.3390/app12010522 5

Jie Li, Chunlin Yin, Hao Wang, Jian Wang and Na Zhao
Mining Algorithm of Relatively Important Nodes Based on Edge Importance Greedy Strategy
Reprinted from: *Appl. Sci.* **2022**, *12*, 6099, doi:10.3390/app12126099 19

Hairu Luo, Peng jia, Anmin Zhou, Yuying Liu and Ziheng He
Bridge Node Detection between Communities Based on GNN
Reprinted from: *Appl. Sci.* **2022**, *12*, 10337, doi:10.3390/app122010337 27

Suliman Aladhadh, Huda Alwabli, Tarek Moulahi and Muneerah Al Asqah
BChainGuard: A New Framework for Cyberthreats Detection in Blockchain Using Machine Learning
Reprinted from: *Appl. Sci.* **2022**, *12*, 12026, doi:10.3390/app122312026 41

Xiaoyang Liu, Nan Ding, Giacomo Fiumara, Pasquale De Meo and Annamaria Ficara
Dynamic Community Discovery Method Based on Phylogenetic Planted Partition in Temporal Networks
Reprinted from: *Appl. Sci.* **2022**, *12*, 3795, doi:10.3390/app12083795 58

Hyungsik Shin, Jeryang Park and Dongwoo Kang
A Graph-Cut-Based Approach to Community Detection in Networks
Reprinted from: *Appl. Sci.* **2022**, *12*, 6218, doi:10.3390/app12126218 77

Jian Huang and Yijun Gu
Unsupervised Community Detection Algorithm with Stochastic Competitive Learning Incorporating Local Node Similarity
Reprinted from: *Appl. Sci.* **2023**, *13*, 10496, doi:10.3390/app131810496 94

Min Ma, Qiong Cao and Xiaoyang Liu
A Graph Convolution Collaborative Filtering Integrating Social Relations Recommendation Method
Reprinted from: *Appl. Sci.* **2022**, *12*, 11653, doi:10.3390/app122211653 116

Jinlong Ma, Peng Wang and Huijia Li
Directed Network Disassembly Method Based on Non-Backtracking Matrix
Reprinted from: *Appl. Sci.* **2022**, *12*, 12047, doi:10.3390/app122312047 133

Yi Zuo, Shengzong Liu, Yun Zhou and Huanhua Liu
TRAL: A Tag-Aware Recommendation Algorithm Based on Attention Learning
Reprinted from: *Appl. Sci.* **2023**, *13*, 814, doi:10.3390/app13020814 148

Nan Xiang, Xiaoxia Ma, Huiling Liu, Xiao Tang and Lu Wang
Graph-Augmentation-Free Self-Supervised Learning for Social Recommendation
Reprinted from: *Appl. Sci.* **2023**, *13*, 3034, doi:10.3390/app13053034 . . . . **165**

Lifeng Yang, Shengzong Liu and Yiqi Zhao
Deep-Learning Based Algorithm for Detecting Targets in Infrared Images
Reprinted from: *Appl. Sci.* **2022**, *12*, 3322, doi:10.3390/app12073322 . . . . **183**

Yuhao Feng, Shenpeng Song, Wenzhe Xu and Huijia Li
5G Price Competition with Social Equilibrium Optimality for Social Networks
Reprinted from: *Appl. Sci.* **2022**, *12*, 8798, doi:10.3390/app12178798 . . . . **200**

Ganesh Sankaran, Marco A. Palomino, Martin Knahl and Guido Siestrup
A Modeling Approach for Measuring the Performance of a Human-AI Collaborative Process
Reprinted from: *Appl. Sci.* **2022**, *12*, 11642, doi:10.3390/app122211642 . . . . **214**

Abdulatif Alabdulatif, Muneerah Al Asqah, Tarek Moulahi and Salah Zidi
Leveraging Artificial Intelligence in Blockchain-Based E-Health for Safer Decision Making Framework
Reprinted from: *Appl. Sci.* **2023**, *13*, 1035, doi:10.3390/app13021035 . . . . **240**

Kun Wang, Defu Jiang, Lijun Yun and Xiaoyang Liu
Infrared Small and Moving Target Detection on Account of the Minimization of Non-Convex Spatial-Temporal Tensor Low-Rank Approximation under the Complex Background
Reprinted from: *Appl. Sci.* **2023**, *13*, 1196, doi:10.3390/app13021196 . . . . **257**

Yanchen Yang, Lijun Yun, Ruoyu Li, Feiyan Cheng and Kun Wang
Multi-View Gait Recognition Based on a Siamese Vision Transformer
Reprinted from: *Appl. Sci.* **2023**, *13*, 2273, doi:10.3390/app13042273 . . . . **276**

Kuei-Hu Chang
Integrating Spherical Fuzzy Sets and the Objective Weights Consideration of Risk Factors for Handling Risk-Ranking Issues
Reprinted from: *Appl. Sci.* **2023**, *13*, 4503, doi:10.3390/app13074503 . . . . **290**

José Maurício, Inês Domingues and Jorge Bernardino
Comparing Vision Transformers and Convolutional Neural Networks for Image Classification: A Literature Review
Reprinted from: *Appl. Sci.* **2023**, *13*, 5521, doi:10.3390/app13095521 . . . . **306**

Manuel A. López-Rourich and Francisco J. Rodríguez-Pérez
Efficient Data Transfer by Evaluating Closeness Centrality for Dynamic Social Complex Network-Inspired Routing
Reprinted from: *Appl. Sci.* **2023**, *13*, 10766, doi:10.3390/app131910766 . . . . **323**

About the Editors

Xiaoyang Liu

Xiaoyang Liu is a full professor at the Chongqing University of Technology, China. He received his B.S. degree in electronic information science and technology from Huainan Normal University, China, in 2006, his M.S. degree in communication and information systems from Kunming University of Science and Technology, China, in 2010, and his Ph.D. degree in communication and information systems from Northwestern Polytechnical University, China, in 2013. From 2015 to 2017, he completed his first post-doctoral fellowship at Chongqing University, China. Then, he completed his second post-doctoral fellowship in the Department of Electrical Computer Engineering, The University of Alabama, from 2017 to 2018, in Tuscaloosa, Alabama, USA. He has been a member of IEEE and ACM since 2018. His main research interests are in the areas of online social networks, misinformation diffusion, complex networks, data mining, and computer application. More information about his research can be found on his lab website.

Giacomo Fiumara

Giacomo Fiumara received his PhD degree in physics at the University of Messina. He has been an assistant professor of computer science with the University of Messina, Italy, since 2009. He is a member of the board of professors at the PhD school of Mathematics and Computational Sciences, University of Palermo. His research interests include network science, criminal networks, and simulations of model systems. He has published in top international journals such as *CACM, Expert Systems with Applications, Information Sciences* and *IEEE Transactions on Big Data*. He is an Academic Editor for *Complexity* and a member of various conference PCs.

Pasquale De Meo

Pasquale De Meo is an associate professor of computer science at the Department of Computer Science, University of Messina, Italy. His main research interests include social networks, recommender systems, and user profiling. His PhD thesis was selected as the Best Italian PhD thesis in artificial intelligence by the AI*IA (Italian Association for Artificial Intelligence). He has been the Marie Curie fellow at Vrije Universiteit Amsterdam. He has served as an Associate Editor for IEEE Transactions on Cybernetics.

Annamaria Ficara

Annamaria Ficara received her B.S. degree in Computer Science from the University of Messina, Italy, in 2016; her M.S. degree in Engineering and Computer Science from the University of Messina, Italy, in 2018; and her Ph.D. degree in Mathematics and Computational Sciences from the University of Palermo, Italy, in 2022. She has previously been an Assistant Professor at the University of Pisa, Italy. She is currently an Assistant Professor at the University of Messina, Italy. Her main research interests are in network science, social network analysis, criminal networks, and complex systems.

Preface

With the rapid development of artificial intelligence, big data, and self-media, computational thinking and analysis methods of network science have become prevalent in people's life and work, and a variety of theoretical calculation methods and analysis techniques for network science have emerged. This Special Issue will be rich in content, covering many disciplines such as computer science, mathematics, journalism and communication, sociology, and management. It is not only suitable for graduate students in computer science and artificial intelligence, but it is also suitable for researchers in related fields.

It focuses on the modeling and analysis of network science, including that of key node identification, community detection, personalized recommendation systems, image processing, object detection, and the optimization method of artificial intelligence technology.

These research results will not only provide decision references and theoretical guidance for relevant management departments, but will also help relevant enterprises and institutions to optimize products and services, and will provide accurate personalized recommendations for users according to the different characteristics of individuals in complex networks.

Xiaoyang Liu, Giacomo Fiumara, Pasquale De Meo, and Annamaria Ficara
Editors

applied sciences

Editorial

Special Issue "Artificial Intelligence in Complex Networks"

Xiaoyang Liu

School of Computer Science and Engineering, Chongqing University of Technology, Chongqing 400054, China; lxy3103@cqut.edu.cn

1. Introduction

Artificial intelligence (AI) in complex networks has made revolutionary breakthroughs in this century, and AI-driven methods are being increasingly integrated into different scientific research [1–3]. The scientific research of complex networks can be traced back to two aspects. Firstly, the main mathematical subjects of graph theory and statistical physics. One of the major breakthroughs in graph theory is the idea of random graph theory. Complex topologies arise from simple random rules. Random graph theory is often used in conjunction with percolation theory to describe random network modeling. Secondly, complex systems and statistical physics gave birth to a few important theoretical models, such as the Ising model [4–8], mean-field theory, nonequilibrium thermodynamics and dissipative structure theory, synergetic theory, and self-spinning glass model [8,9].

Moreover, AI plays a crucial role in improving the performance of network dynamics, key node mining, community detection, and recommendation behaviors in complex networks [10,11]. The social impact of artificial intelligence is becoming increasingly prominent. On the one hand, as the core force of a new round of scientific and technological revolution and industrial reform, artificial intelligence is promoting the upgrading of traditional industries, driving the rapid development of an "unmanned economy", and having a positive impact on people's livelihoods, such as intelligent transportation, smart homes, and intelligent medical care. On the other hand, issues such as personal information and privacy protection, intellectual property rights of AI-created content, possible discrimination and bias of AI systems, traffic regulations for driverless systems, and the scientific and technological ethics of brain–computer interfaces and human–machine symbiosis, have emerged and need to be urgently provided with solutions [12–15].

Despite the transformative potential of AI in complex networks, there are challenges such as data privacy, user privacy protection, data sample scarcity and diverse network structure, and so on. As these technologies continue to evolve, addressing these issues is paramount for ensuring their responsible and ethical implementation. In the future, further developments in AI technologies are expected to refine and expand their applications in social networks, social computing, transportation and finance networks, large model applications, etc.

Citation: Liu, X. Special Issue "Artificial Intelligence in Complex Networks". *Appl. Sci.* **2024**, *14*, 2822. https://doi.org/10.3390/app14072822

Received: 25 March 2024
Accepted: 26 March 2024
Published: 27 March 2024

2. An Overview of the Published Articles

Complex network theory is widely used in the field of artificial intelligence, and key node identification is the core technology of complex network theory research, which has been highly concerned by the academic community. Many scholars have conducted in-depth research on academic problems such as the identification of critical nodes or the ranking of node importance in complex networks, and have achieved a large number of research results (contributions 1, 4, 7, 10).

A community in a network is a set of nodes that are highly connected to each other, unlike other nodes in the network, which have relatively random and scattered relationships. A key role of community detection algorithms is that they can be used to extract useful information from the network. The biggest challenge for community detection is that the community structure is not universally defined (contributions 3, 5, 19).

As the most central component of the personalized recommendation system, the efficiency of the recommendation algorithm directly affects the performance of the entire recommendation system. More mature recommendation algorithms include content-based, collaborative filtering, and other algorithms. Although these algorithms have been widely used, there are still many areas to be improved. The recommendation algorithm based on complex network theory is a good attempt, and it is also one of the current research hotspots (contributions 9, 11, 12, 16).

The data modeling, image processing, object detection and optimization methods of artificial intelligence technology have all been widely used (contributions 2, 6, 8, 13, 14, 15, 17, 18, 20).

There are two main ways to predict information propagation in complex networks. One is feature-based methods; these methods rely on users to manually extract features, such as the content features of the information, the timing features of the current propagation, the structural features, and the user characteristics on the propagation path. Based on these features, a regression algorithm is used to predict the number of retweets. The effects of these kinds of methods depend heavily on the extraction of features. For different problems, users need to extract appropriate features according to their own experiences. The second category is the generative algorithm, which designs a model to simulate the mechanism of information diffusion, tries to retain the main characteristics of information diffusion in the model, and then uses the model to calculate the spread range of each piece of information in the future.

These papers were received from Europe and Asia, with a number combining the expertise of researchers from different countries and even different continents. Finally, we are particularly pleased with the breadth of authors, topics, techniques, and findings that can be found within this Special Issue, "Artificial Intelligence in Complex Networks".

3. Conclusions and Future Perspectives

With the publication of the present Special Issue, we hope to contribute to better links being formed between artificial intelligence and complex networks; as such, we have selected original works aimed at including key node identification, community detection, recommendation systems, object detection, data processing, and optimal decision algorithms in complex networks. Due to its interdisciplinary and complexity characteristics, the study of complex networks involves the knowledge and theoretical bases of many disciplines, especially those of system science, statistical physics, mathematics, computer and information science, etc. The commonly used analysis methods and tools include graph theory, combinatorics, matrix theory, probability theory, stochastic process, optimization theory, genetic algorithms, etc. The main research methods of complex networks are based on graph theory and its methods, and have achieved gratifying results.

Conflicts of Interest: The authors declare no conflicts of interest.

List of Contributions:

1. Zhao, N.; Liu, Q.; Jing, M.; Li, J.; Zhao, Z.; Wang, J. DDMF: A Method for Mining Relatively Important Nodes Based on Distance Distribution and Multi-Index Fusion. *Appl. Sci.* **2022**, *12*, 522. https://doi.org/10.3390/app12010522.
2. Yang, L.; Liu, S.; Zhao, Y. Deep-Learning Based Algorithm for Detecting Targets in Infrared Images. *Appl. Sci.* **2022**, *12*, 3322. https://doi.org/10.3390/app12073322.
3. Liu, X.; Ding, N.; Fiumara, G.; De Meo, P.; Ficara, A. Dynamic Community Discovery Method Based on Phylogenetic Planted Partition in Temporal Networks. *Appl. Sci.* **2022**, *12*, 3795. https://doi.org/10.3390/app12083795.
4. Li, J.; Yin, C.; Wang, H.; Wang, J.; Zhao, N. Mining Algorithm of Relatively Important Nodes Based on Edge Importance Greedy Strategy. *Appl. Sci.* **2022**, *12*, 6099. https://doi.org/10.3390/app12126099.
5. Shin, H.; Park, J.; Kang, D. A Graph-Cut-Based Approach to Community Detection in Networks. *Appl. Sci.* **2022**, *12*, 6218. https://doi.org/10.3390/app12126218.

6. Feng, Y.; Song, S.; Xu, W.; Li, H. 5G Price Competition with Social Equilibrium Optimality for Social Networks. *Appl. Sci.* **2022**, *12*, 8798. https://doi.org/10.3390/app12178798.
7. Luo, H.; Jia, P.; Zhou, A.; Liu, Y.; He, Z. Bridge Node Detection between Communities Based on GNN. *Appl. Sci.* **2022**, *12*, 10337. https://doi.org/10.3390/app122010337.
8. Sankaran, G.; Palomino, M.; Knahl, M.; Siestrup, G. A Modeling Approach for Measuring the Performance of a Human-AI Collaborative Process. *Appl. Sci.* **2022**, *12*, 11642. https://doi.org/10.3390/app122211642.
9. Ma, M.; Cao, Q.; Liu, X. A Graph Convolution Collaborative Filtering Integrating Social Relations Recommendation Method. *Appl. Sci.* **2022**, *12*, 11653. https://doi.org/10.3390/app122211653.
10. Aladhadh, S.; Alwabli, H.; Moulahi, T.; Al Asqah, M. BChainGuard: A New Framework for Cyberthreats Detection in Blockchain Using Machine Learning. *Appl. Sci.* **2022**, *12*, 12026. https://doi.org/10.3390/app122312026.
11. Ma, J.; Wang, P.; Li, H. Directed Network Disassembly Method Based on Non-Backtracking Matrix. *Appl. Sci.* **2022**, *12*, 12047. https://doi.org/10.3390/app122312047.
12. Zuo, Y.; Liu, S.; Zhou, Y.; Liu, H. TRAL: A Tag-Aware Recommendation Algorithm Based on Attention Learning. *Appl. Sci.* **2023**, *13*, 814. https://doi.org/10.3390/app13020814.
13. Alabdulatif, A.; Al Asqah, M.; Moulahi, T.; Zidi, S. Leveraging Artificial Intelligence in Blockchain-Based E-Health for Safer Decision Making Framework. *Appl. Sci.* **2023**, *13*, 1035. https://doi.org/10.3390/app13021035.
14. Wang, K.; Jiang, D.; Yun, L.; Liu, X. Infrared Small and Moving Target Detection on Account of the Minimization of Non-Convex Spatial-Temporal Tensor Low-Rank Approximation under the Complex Background. *Appl. Sci.* **2023**, *13*, 1196. https://doi.org/10.3390/app13021196.
15. Yang, Y.; Yun, L.; Li, R.; Cheng, F.; Wang, K. Multi-View Gait Recognition Based on a Siamese Vision Transformer. *Appl. Sci.* **2023**, *13*, 2273. https://doi.org/10.3390/app13042273.
16. Xiang, N.; Ma, X.; Liu, H.; Tang, X.; Wang, L. Graph-Augmentation-Free Self-Supervised Learning for Social Recommendation. *Appl. Sci.* **2023**, *13*, 3034. https://doi.org/10.3390/app13053034.
17. Chang, K. Integrating Spherical Fuzzy Sets and the Objective Weights Consideration of Risk Factors for Handling Risk-Ranking Issues. *Appl. Sci.* **2023**, *13*, 4503. https://doi.org/10.3390/app13074503.
18. Maurício, J.; Domingues, I.; Bernardino, J. Comparing Vision Transformers and Convolutional Neural Networks for Image Classification: A Literature Review. *Appl. Sci.* **2023**, *13*, 5521. https://doi.org/10.3390/app13095521.
19. Huang, J.; Gu, Y. Unsupervised Community Detection Algorithm with Stochastic Competitive Learning Incorporating Local Node Similarity. *Appl. Sci.* **2023**, *13*, 10496. https://doi.org/10.3390/app131810496.
20. López-Rourich, M.; Rodríguez-Pérez, F. Efficient Data Transfer by Evaluating Closeness Centrality for Dynamic Social Complex Network-Inspired Routing. *Appl. Sci.* **2023**, *13*, 10766. https://doi.org/10.3390/app131910766.

References

1. Bischof, R.; Milleret, C.; Dupont, P.; Chipperfield, J.; Tourani, M.; Ordiz, A.; de Valpine, P.; Turek, D.; Royle, J.A.; Gimenez, O.; et al. Estimating and forecasting spatial population dynamics of apexpredators using transnational genetic monitoring. *Proc. Natl. Acad. Sci. USA* **2020**, *117*, 30531–30538. [CrossRef] [PubMed]
2. Boccaletti, S.; De Lellis, P.; del Genio, C.I.; Alfaro-Bittner, K.; Criado, R.; Jalan, S.; Romance, M. The structure and dynamics of networks with higher order interactions. *Phys. Rep.* **2023**, *1018*, 1–64. [CrossRef]
3. Liu, X.; Ye, S.; Fiumara, G.; De Meo, P. Influence Nodes Identifying Method via Community-based Backward Generating Network Framework. *IEEE Trans. Netw. Sci. Eng.* **2024**, *11*, 236–253. [CrossRef]
4. Liu, X.; Miao, C.; Fiumara, G.; De Meo, P. Information Propagation Prediction Based on Spatial–Temporal Attention and Heterogeneous Graph Convolutional Networks. *IEEE Trans. Comput. Soc. Syst.* **2024**, *11*, 945–958. [CrossRef]
5. Asikis, T.; Böttcher, L.; Antulov-Fantulin, N. Neural ordinary differential equation control of dynamics on graphs. *Phys. Rev. Res.* **2022**, *4*, 013221. [CrossRef]
6. Baggio, G.; Bassett, D.S.; Pasqualetti, F. Data-driven control of complex networks. *Nat. Commun.* **2021**, *12*, 1429. [CrossRef] [PubMed]
7. Zhang, S.; Li, T.; Hui, S.; Li, G.; Liang, Y.; Yu, L.; Jin, D.; Li, Y. Deep transfer learning for city-scale cellular traffic generation through urban knowledge graph. In Proceedings of the 29th ACM SIGKDD Conference on Knowledge Discovery and Data Mining, Long Beach, CA, USA, 6–10 August 2023; pp. 4842–4851.

8. Zhao, Q.; Van den Brink, P.J.; Xu, C.; Wang, S.; Clark, A.T.; Karakoç, C.; Sugihara, G.; Widdicombe, C.E.; Atkinson, A.; Matsuzaki, S.-I.S.; et al. Relationships of temperature and biodiversity with stability of natural aquatic food webs. *Nat. Commun.* **2023**, *14*, 3507. [CrossRef] [PubMed]
9. Zhao, X.; Yu, H.; Huang, R.; Liu, S.; Hu, N.; Cao, X. A novel higher-order neural network framework based on motifs attention for identifying critical nodes. *Phys. A Stat. Mech. Its Appl.* **2023**, *629*, 129194. [CrossRef]
10. Zheng, Y.; Lin, Y.; Zhao, L.; Wu, T.; Jin, D.; Li, Y. Spatial planning of urban communities via deep reinforcement learning. *Nat. Comput. Sci.* **2023**, *3*, 748–762. [CrossRef] [PubMed]
11. Zheng, Y.; Su, H.; Ding, J.; Jin, D.; Li, Y. Road planning for slums via deep reinforcement learning. In Proceedings of the 29th ACM SIGKDD Conference on Knowledge Discovery and Data Mining, Long Beach, CA, USA, 6–10 August 2023; pp. 5695–5706.
12. Hu, W.; Xia, X.; Ding, X.; Zhang, X.; Zhong, K.; Zhang, H.F. SMPC-Ranking: A Privacy-Preserving Method on Identifying Influential Nodes in Multiple Private Networks. *IEEE Trans. Syst. Man Cybern. Syst.* **2022**, *53*, 2971–2982. [CrossRef]
13. Zhou, C.; Wang, X.; Zhang, M. Facilitating graph neural networks with random walk on simplicial complexes. *Adv. Neural Inf. Process. Syst.* **2024**, *36*, 1–35.
14. Zoller, L.; Bennett, J.; Knight, T.M. Plant–pollinator network change across a century in the subarctic. *Nat. Ecol. Evol.* **2023**, *7*, 102–112. [CrossRef] [PubMed]
15. Zou, W.; Senthilkumar, D.V.; Zhan, M.; Kurths, J. Quenching, aging, and reviving in coupled dynamical networks. *Phys. Rep.* **2021**, *931*, 1–72. [CrossRef]

applied sciences

MDPI

Article

DDMF: A Method for Mining Relatively Important Nodes Based on Distance Distribution and Multi-Index Fusion

Na Zhao [1,†], Qian Liu [1,†], Ming Jing [2], Jie Li [3], Zhidan Zhao [4] and Jian Wang [5,*]

1 Key Laboratory in Software Engineering of Yunnan Province, School of Software, Yunnan University, Kunming 650091, China; zhaona@ynu.edu.cn (N.Z.); liu_antoni0409@163.com (Q.L.)
2 School of Information Engineering, Kunming University, Kunming 650214, China; proofle@163.com
3 Electric Power Research Institute of Yunnan Power Grid Co., Ltd., Kunming 650217, China; lj1226645407@163.com
4 Department of Computer Science, School of Engineering, Shantou University, Shantou 515063, China; zzhidanzhao@gmail.com
5 College of Information Engineering and Automation, Kunming University of Science and Technology, Kunming 650504, China
* Correspondence: jianwang@kust.edu.cn
† These authors contributed equally to this work.

Abstract: In research on complex networks, mining relatively important nodes is a challenging and practical work. However, little research has been done on mining relatively important nodes in complex networks, and the existing relatively important node mining algorithms cannot take into account the indicators of both precision and applicability. Aiming at the scarcity of relatively important node mining algorithms and the limitations of existing algorithms, this paper proposes a relatively important node mining method based on distance distribution and multi-index fusion (DDMF). First, the distance distribution of each node is generated according to the shortest path between nodes in the network; then, the cosine similarity, Euclidean distance and relative entropy are fused, and the entropy weight method is used to calculate the weights of different indexes; Finally, by calculating the relative importance score of nodes in the network, the relatively important nodes are mined. Through verification and analysis on real network datasets in different fields, the results show that the DDMF method outperforms other relatively important node mining algorithms in precision, recall, and AUC value.

Keywords: complex network; distance distribution; multi-index fusion; relatively important node

Citation: Zhao, N.; Liu, Q.; Jing, M.; Li, J.; Zhao, Z.; Wang, J. DDMF: A Method for Mining Relatively Important Nodes Based on Distance Distribution and Multi-Index Fusion. *Appl. Sci.* **2022**, *12*, 522. https://doi.org/10.3390/app12010522

Academic Editors: Giacomo Fiumara, Pasquale De Meo, Xiaoyang Liu and Annamaria Ficara

Received: 1 December 2021
Accepted: 3 January 2022
Published: 5 January 2022

1. Introduction

With the vigorous growth of network and information technology represented by the Internet, human society has entered a new and complex era of networks. Information mining in complex networks is important in theoretical research and offers great application and socioeconomic values [1–4]. For example, if users can unearth important nodes or edges in the spread network of a virus, then they can curb the spread of the virus in a short time by isolating or cutting off the important nodes or edges in the virus network at the beginning of the virus spread and thereby eliminate unnecessary economic losses [5]. Efficient information mining in complex networks has naturally become a key topic that continues to attract the attention of many scholars.

The existing studies on complex network information mining are generally ranked on the basis of the importance of all nodes and edges in the network [6–10]. However, determining which nodes are the most important in the network relative to one or one group of specific nodes presents an issue. This problem reminds us about the practical significance of mining relatively important information in networks, especially very large-scale ones.

The relative importance of nodes refers to the importance of nodes relative to known important nodes. It is also called proximity or similarity [11]. According to the key idea of relative importance, information mining in a complex network can be described as a process in which the importance of a node in a network relative to a known important node is quantified and the importance of a node relative to a known important node set is calculated to identify the relatively important nodes in the network.

The central idea of relative importance can be widely used in many fields. For example, potential criminals can be found using known criminal data in the field of criminal networks, and terrorists in hiding can be captured on the basis of known terrorist data [12,13]. In the bionetwork field, people susceptible to diseases can be identified for timely treatment and isolation on the basis of relevant information on populations infected with known infectious diseases. Unknown pathogenic genes may be determined according to known pathogenic gene information in protein networks [14]. In the field of power grids, on the premise that the information on important power generation units or circuit breakers is known, finding relatively important power generation units, circuit breakers, etc. is prioritized for protection, in order to effectively avoid large-area power outages caused by successive faults. Mining relatively important nodes in complex networks obviously offers great research significance and application value [15].

The node distance distribution in a complex network quantifies many types of topological information in the network, including the degree of nodes, average degree of the network, diameter of the network, closeness centrality of nodes, and average path length of the network [16]. Therefore, the study on the relative importance of nodes in a network based on node distance distribution in the network will contribute to the accurate mining of relatively important nodes in networks. In the current study, the distance distribution of all nodes in a network is calculated. On the basis of known important node information, the differences in distance distribution between known important nodes and target nodes are measured from three dimensions, namely, direction, distance, and distribution. A relatively important node mining method based on distance distribution and multi-index fusion (DDMF) is proposed.

The DDMF method involves two main steps: First, the distance distribution of all nodes (including known important nodes and target nodes) is calculated on the basis of the shortest distance between nodes in the network. Then, the calculated results are converted into vector form. Second, multi-index fusion is made for cosine similarity, Euclidean distance, and relative entropy. The weights corresponding to different indexes are calculated using the entropy weight method to obtain the relative importance scores of the nodes. The nodes with high scores are regarded as a relatively important nodes in the network.

Our key contribution is in proposing a novel method based on network topology to find relatively important nodes in the network. The DDMF method not only fills the gap of relatively important node algorithms in the scientific field of complex network theory, but also provides a new idea for community detection and link prediction. Since the network in real life exists in different kinds of fields, we also conduct some experiments on different types of real network datasets to verify whether the method has practical application value in real life. Experiments demonstrate that DDMF method outperforms other relatively important node mining algorithms in terms of precision and applicability.

The remainder of this paper is organized as follows. In Section 2, works related to the proposed method are given. Section 3 deals with detailed descriptions of the proposed algorithm. The experimental results and analysis are presented in Section 4. Finally, we summarize in Section 5.

2. Related Work

At present, many researchers in the field of complex networks focus on the mining of important nodes in networks; that is, ranking the importance of all nodes in a network as a whole. Existing research has primarily aimed to develop an identification algorithm for influential nodes. Inspired by the heuristic scheme, Wang et al. [17] proposed the price-performance-ratio PPRank method, selecting nodes in a given range and aiming to improve the performance of the diffusion. Yang et al. [18] proposed a method of ranking node importance based on multi-criteria decision-making (MCDM). The weight of each criterion is calculated by an entropy weighting method, which overcomes the impact of the subjective factor. Li et al. [19] proposed a method of calculating the importance degree of urban rail transit network nodes based on h-index, which considers the topology, passenger volume, and passenger flow correlation of the urban rail network. Luo et al. [20] proposed a relationship matrix resolving model to identify vital nodes based on community (IVNC), as an attempt to identify influential nodes in OSNs.

However, the study on node mining based on relative importance remains limited. The earliest study on relative importance in networks is that on a personalized variant HITS algorithm [21]. Haveliwala [22] and Jennifier et al. [23] later proposed their own variant PageRank algorithms, which consider the relative importance of nodes in a network. Alzaabi et al. [24] defined the universal framework of mining algorithms for relatively important nodes and proposed that the relative importance of nodes in a network relates to one node set or one group of specified node sets. Wang et al. [25] proposed a path probabilistic summation method, which defines the importance of any node relative to the nearest neighbor node as the probability of jumping from the node to the nearest neighbor node in the random walk process. Rodriguez et al. [26] proposed a cluster particle propagation method, which is used to evaluate the relative importance of nodes. Magalingam et al. [27] used shortest distance as a measurement indicator of relative importance. Langohr et al. [28] used the reciprocal of the P norm of the shortest distance as a measurement indicator of relative importance. In addition, some researchers have considered mining deep network information by using network embedded learning methods [29–35]. For example, some classical network-embedded learning algorithms have been used to mine relatively important nodes in networks.

Although some algorithms have been employed to mine relatively important nodes in networks, they suffer from problems that require immediate resolution, such as low accuracy and narrow use range. Therefore, novel and efficient methods for mining relatively important nodes need to be developed.

In the study of complex networks, the most classic and most widely used relative importance calculation indicators include the Ksmar index [11], PPR index [21], and Katz index [36]. Zhao et al. [37] proposed a relatively important node mining algorithm based on neighbor layer diffuse (NLD) in 2021, which is the latest relatively important node algorithm. In Section 4, we empirically compare our method with these methods using various real world networks.

3. Relative Importance Measure Based on Distance Distribution and Multi-index Fusion

To fully measure the impact of network structure information on the relative importance of nodes, this study proposes a relatively important node mining method based on distance distribution and multi-index fusion, i.e., the DDMF method. In this section, we first introduce the problem definition in complex networks and use a specific example to explain what is the distance distribution. Then three indicators of cosine similarity, Euclidean distance, and relative entropy are described in detail. Finally, we discuss how to calculate the relative importance score of a node based on multi-index fusion.

3.1. Problem Definition

Under normal conditions, a complex network can be represented by $G(V, E)$. Here, V refers to the node in the network G and E refers to the edge of the network G. The network G comprises n nodes. Among them, n nodes can be divided into important node set V_1 and unimportant node set V_2. The important node set V_1 has n_1 nodes, while the unimportant node set V_2 has n_2 nodes. The important node set V_1 includes known important node set R and unknown important node set U. The unimportant node set V_2 and unknown important node set U constitute target node set T, i.e., $T = V_2 \cup U$.

The key to finding the relatively important nodes in the target node set T is to first calculate the importance of a node in the target node set T relative to a known important node, and then calculate the importance of a node relative to all nodes in the known important node set R.

The main contents of this work include the following: For the information of known important node set R, the importance of any node in the target node set T relative to the node in the known important node set R is analyzed and calculated. The expectation is to find $top-k$ relatively important nodes in the target node set T. The final results are analyzed and evaluated on the basis of three evaluation indicators, namely, precision, recall, and area under the curve (AUC).

3.2. Distance Distribution

Distance distribution in complex networks is usually represented by the shortest path distribution between nodes. The node distance distribution in the network mainly considers the number of nodes with different shortest path lengths to the current node; thus, it can intuitively obtain the shortest path information of nodes in the network and reflect many important topological information in the network [38].

The distance distribution of each node v_i in the complex network can be represented as $P_i = \{p_i(j)\}$; the calculation formula of $p_i(j)$ is

$$p_i(j) = \frac{N_i(j)}{n} \tag{1}$$

where j represents the shortest path length with a value in the range of $0 \le j \le D(G)$. $D(G)$ refers to the diameter of the network G, and its value is the maximum distance between any two nodes in the network G. $N_i(j)$ represents the number of nodes with j of the shortest path length to node v_i in the network G; n represents the number of nodes in the network G.

Take a network $G_{example}$ as an example. The detailed calculation process of node distance distribution in $G_{example}$ is introduced as follows. In Figure 1, the red nodes are the nodes in the current study while the yellow, light green, blue, and pink nodes represent the nodes that can be reached by taking one, two, three, and four steps consecutively, starting from the nodes studied currently.

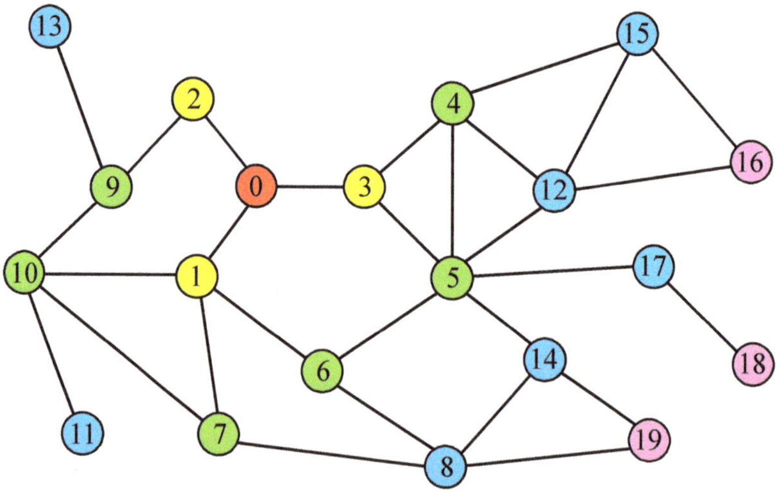

Figure 1. The topology of the example network.

The number of nodes n in the sample network $G_{example}$ is 20, and the diameter $D(G)$ is 7. The distance distribution dimension mainly depends on the diameter of the network. Its value range is from 0 to $D(G)$, with a total of $D(G)+1$ cases. Therefore, the distance distribution dimension d of each node in $G_{example}$ is 8. Provided that node 0 is used as the starting research node, the set of nodes $N(i) = \{N_i(j)|0 \le j \le D(G)\}$ that can be reached by node 0 in turn can be obtained by calculating the shortest path length between this node and other nodes in $G_{example}$, that is, according to Formula (1) and $N(i)$ obtained through the above analysis, the distance distribution P_0 of node 0 can be obtained as: $P_0 = \{0.05, 0.15, 0.30, 0.35, 0.15, 0, 0, 0\}$.

Similarly, the distance distribution of any node in the sample network $G_{example}$ can be obtained. For a network G with n nodes, if the distance distribution of n nodes is known $P = \{P_0, P_1, \cdots, P_{n-1}\}$, then much important topology information in the network G can be obtained on the basis of the distance distribution of nodes. For example, the degree k_i of any node v_i in G, the average degree $\bar{k}$ of G, the average path length APL of G, and the closeness centrality CC_i corresponding to node v_i.

For the network G with n nodes, a distance distribution matrix $X = [x_{ij}] \in R^{n\times d}$ is established on the basis of the distance distribution information of all nodes in G; n refers to the total number of nodes in the network G.

(1) Degree k_i of node v_i

$$k_i = nx_{i1} \tag{2}$$

(2) Average degree $\bar{k}$ of network G

$$\bar{k} = \frac{1}{n}\sum_{i=0}^{n-1} nx_{i1} \tag{3}$$

(3) Average path length APL of network G

$$APL = \frac{2}{n(n-1)}\sum_{i=0}^{n-1}\sum_{j=1}^{D(G)} j \times nx_{ij} \tag{4}$$

(4) Closeness centrality CC_i of node v_i

$$CC_i = \frac{n}{\sum_{j=1}^{D(G)} j \times nx_{ij}} \tag{5}$$

The analysis indicates that the distance distribution of nodes contains abundant network topology information. Therefore, taking the distance distribution P_i of each node v_i in the network G as the main subject investigated and converting it into vector form, the difference in the distance distribution between nodes in the known important node set R and the target node set T is analyzed to find the relatively important nodes in the network G.

3.3. Introduction to Indicators

Cosine similarity is a measurement method for the difference between two individuals and involves calculating the cosine value of the angle between two vectors in the vector space, mainly focusing on the measurement of the difference between two individuals from the dimension of direction. The basic idea is to covert the individual's index data into the vector space and then measure the difference between individuals by comparing the cosine values of the angle in the inner product space between different individual vectors [39].

In a M-dimensional space, assuming that A and B are M-dimensional vectors, namely $A = [a_1, a_2, \cdots, a_M]$, $B = [b_1, b_2, \cdots, b_M]$, then the cosine similarity Cos_{AB} can be expressed as:

$$Cos_{AB} = \frac{\sum_{i=1}^{M} (A_i \times B_i)}{\sqrt{\sum_{i=1}^{M} (A_i)^2} \times \sqrt{\sum_{i=1}^{M} (B_i)^2}} = \frac{A \cdot B}{|A| \times |B|} \tag{6}$$

where the value range of Cos_{AB} is [–1, 1], that is, $Cos_{AB} \in [-1, 1]$.

In this work, the distance distribution of nodes in the network is converted into vector form; that is, in the network G with n nodes, the vectors of distance distribution of any node x and node y can be expressed as P_x and P_y, respectively. Then, the formula for the cosine similarity between nodes can be represented as:

$$D_{Cos}(P_x \| P_y) = \frac{P_x \cdot P_y}{|P_x| \times |P_y|} \tag{7}$$

$$C_{xy} = \frac{1 + D_{Cos}(P_x \| P_y)}{2} \tag{8}$$

Normalization is performed for the cosine similarity between nodes $D_{Cos}(P_x \| P_y)$ based on Equation (8), and C_{xy} is obtained. Among them, $C_{xy} \in [0, 1]$.

Euclidean distance, also called Euclidean metric, originates from the distance formula between two points in Euclidean geometry [40]. It is mainly used to measure the real distance between two points in M-dimension space; that is, focusing on the numerical difference between individuals.

In a M-dimensional space, assuming that A and B are M-dimensional vectors, namely $A = [a_1, a_2, \cdots, a_M]$, $B = [b_1, b_2, \cdots, b_M]$, then the Euclidean distance Euc_{AB} can be expressed as:

$$Euc_{AB} = \sqrt{\sum_{i=1}^{M} (a_i - b_i)^2} \tag{9}$$

Similarly, the distance distribution of nodes in the network G is first converted into vector form. Then, the Euclidean distance between any node x and node y can be represented as Euc_{xy}:

$$E_{xy} = \frac{Euc_{xy}}{Euc_{\max}} \tag{10}$$

Normalization is performed for Euclidean distance Euc_{xy} between node x and node y based on Equation (10), and E_{xy} is obtained. Among them, $E_{xy} \in [0, 1]$.

From information theory, relative entropy, also called KL divergence or information divergence, is generally used to measure the difference between two probability distributions [41]. In this work, the difference in the distance distribution between different nodes in the network is calculated from the dimension of distribution to effectively find relatively important nodes in the network.

For the network G with n nodes, the distance distributions of node x and node y are P_x and P_y, respectively. Then, relative entropy can be defined as the difference in the distance distribution between the two nodes. The formula is as follows:

$$D_{KL}(P_x \| P_y) = \sum_{j=0}^{D(G)} p_x(j) \ln \frac{p_x(j)}{p_y(j)} \tag{11}$$

If relative entropy $D_{KL}(P_x \| P_y)$ is small, then the difference in the distance distribution between node x and node y is small. The denominator of the logarithmic function cannot be 0. Therefore, in $p_x(j) = 0$ or $p_y(j) = 0$, the values of $\ln \frac{p_x(j)}{p_y(j)}$ are uniformly set to 0.

In addition, relative entropy is an asymmetric measure. Therefore, this study symmetrically converts the relative entropy between node distance distributions. The specific formula is as follows:

$$Q_{xy} = \frac{D_{KL}(P_x \| P_y) + D_{KL}(P_y \| P_x)}{2} \tag{12}$$

$$R_{xy} = 1 - \frac{Q_{xy}}{Q_{max}} = \frac{Q_{max} - Q_{xy}}{Q_{max}} \tag{13}$$

The relative entropy in asymmetric form is converted into symmetric form Q_{xy} in Equation (12). On the basis of Equation (13), normalization processing is implemented for the relative entropy in symmetric form, then R_{xy} is obtained. Among them, $R_{xy} \in [0,1]$.

This study aims to find relatively important nodes in the network G by calculating the relative entropy of the distance distribution of nodes in the known important node set R and target node set T. If the relative entropy is small, then the difference in the distance distribution between different nodes is small. That is, the nodes with a smaller relative entropy in the target node set T compared to the known important node set R are more likely to be relatively important nodes in the network G.

3.4. *Relative Importance Score Based on Multi-index Fusion*

To fully integrate the advantages of cosine similarity, Euclidean distance, and relative entropy in the direction, distance, and distribution dimensions, this study performs the multi-index fusion of cosine similarity, Euclidean distance, and relative entropy and calculates the weights of the different indexes by using the entropy weight method [42] to maximize the advantages of the different indexes. The entropy weight method is an objective weighting method that is widely used and often depends on the discreteness of data. It mainly weighs different indexes according to the amount of information of different evaluation indexes.

Cosine similarity, Euclidean distance, and relative entropy are mainly considered in this work. Thus, weight allocation becomes necessary. A relative importance score matrix, $Z = [z_{tg}] \in R^{|T| \times 3}$, is defined herein.

$$z_{t1} = \frac{\sum_{r=1}^{|R|} C_{tr}}{|R|}, t = 1, 2, \cdots, |T| \tag{14}$$

$$z_{t2} = \frac{\sum_{r=1}^{|R|} E_{tr}}{|R|}, t = 1, 2, \cdots, |T| \tag{15}$$

$$z_{t3} = \frac{\sum_{r=1}^{|R|} R_{tr}}{|R|}, t = 1, 2, \cdots, |T| \tag{16}$$

where z_{t1}, z_{t2} and z_{t3} represent the arithmetic mean of cosine similarity, Euclidean distance and relative entropy between the $t-th$ node in the target node set T and all nodes in the known important node set R respectively. $|T|$ refers to the number of nodes in the target node set T, $|R|$ refers to the number of nodes in the known important node set R, and g refers to the number of indexes, $g = 1, 2, 3$.

Based on the relative importance score matrix, the entropy corresponding to cosine similarity, Euclidean distance, and relative entropy can be further calculated. The formulas are as follows:

$$e_g = -\frac{1}{\ln|T|} \sum_{t=1}^{|T|} p_{tg} \ln(p_{tg}) \tag{17}$$

$$p_{tg} = \frac{z_{tg}}{\sum_{t=1}^{|T|} z_{tg}} \tag{18}$$

where e_g represents the entropy of the index in the g column and p_{tg} represents the proportion of the index in the g column of the $t-th$ node in the target node set T in this column of indexes.

After the entropies of different indexes are obtained, the weight coefficient ω_g of each index can be further calculated. The weights corresponding to different indicators determine the relative importance scores of the target nodes in the network. The specific formula is:

$$\omega_g = \frac{1 - e_g}{\sum_{g=1}^{3} (1 - e_g)} \tag{19}$$

where $1 - e_g$ refers to information entropy redundancy. At the same time, ω_g should meet the restrictive conditions of $\sum \omega_g = 1, g = 1,2,3$.

Therefore, the relative importance score of $t-th$ node in the target node set T can be expressed as:

$$s_t = \omega_1 z_{t1} + \omega_2 z_{t2} + \omega_3 z_{t3} \tag{20}$$

Finally, the relative importance scores of all nodes in the target node set T are sorted in descending order, and the nodes with high scores can be regarded as relatively important nodes.

The calculation of the relative importance scores of the nodes in a network by using the DDMF method consists of the following steps:

First, on the basis of the information of the shortest distance between nodes in the network G, the distance distribution vectors of all nodes in the network G are calculated, along with all the nodes of the known important node set R and target node set T.

Second, the differences in the distance distribution of the nodes between the known important node set R and the target node set T are determined. The cosine similarity, Euclidean distance, and relative entropy of the distance distribution of the two node sets are then calculated and normalized.

Finally, multi-index fusion is made for cosine similarity, Euclidean distance, and relative entropy, and the weights corresponding to different indexes are calculated using the entropy weight method. The relative importance scores of all the nodes in the target node set T are further obtained. The nodes with high scores are regarded as relatively important nodes.

4. Experimental Results and Analysis

The data of four real networks are used to analyze and verify the accuracy of the DDMF method. The Node2vec algorithm [43] is a network-embedded learning algorithm that cannot be directly used to calculate the relative importance scores of nodes. Therefore, the NMF index is obtained on the basis of the improvement of the Node2vec algorithm. The basic idea of the NMF index is as follows: first, the Node2vec algorithm is adopted to generate the embedded vector of the network. Second, the multi-index fusion is made for the obtained vectors so as to calculate the relative importance scores of the nodes. The multi-index fusion method of the NMF index is consistent with proposed DDMF method.

The comparative algorithms included the Ksmar index, PPR index, Katz index, NLD algorithm, and NMF index obtained on the basis of the Node2vec algorithm improvement.

4.1. Datasets

Experimental analysis is performed for the selected algorithms by using four classical real network datasets. The selected datasets are of different sizes and come from different network fields as much as possible, including virus networks, gene networks, and protein networks. The weight and direction of each network linking edge are ignored in this experiment.

(1) The international aviation network where the SARS virus spread [44] comprises 224 nodes and 2247 edges. The nodes represent the countries where flights arrived while the edges represent the routes between two countries. The important node set of the network is defined as the countries where the SARS virus spread at the early stage.

(2) The Genepath human gene signaling network [45] comprises 6306 nodes and 57,340 edges. Nodes represent genes while edges represent the relationship between nodes. The important node set of the network is defined as the Alzheimer's disease gene.

(3) The mouse protein interaction network [46] comprises 1187 nodes and 1557 edges. Nodes represent mouse proteins while edges represent the interaction between proteins. The important node set of the network is defined as mouse protein kinase.

(4) The yeast protein network [47] comprises 5093 nodes and 24,743 edges. The nodes represent proteins while edges represent the relationship between proteins. The important node set of the network is defined as the important protein of the yeast network.

The basic topology characteristics of the four real networks used in this work are shown in Table 1.

Table 1. Basic topological characteristics of real networks.

Dataset	n	m	n_1	$\bar{k}$	C
SARS	224	2247	18	20.06	0.65
Genepath	6306	57,340	51	18.19	0.32
Mouse	1187	1557	67	2.62	0.09
Yeast	5093	24,743	1167	9.72	0.1

Here, n refers to the number of nodes in the network, m refers to the number of edges in the network, n_1 refers to the number of important nodes in the network, $\bar{k}$ refers to the average degree of the network, and C refers to the average clustering coefficient of the network.

4.2. Evaluation Indexes

Precision, recall, and AUC are the three evaluation indexes used to quantify the relatively important nodes obtained by several algorithms in this work.

Precision is mainly used to measure whether the $top - L$ nodes in the results by the algorithm are predicted correctly. It is specifically defined as the proportion of correct predictions in $top - L$ nodes among the predicted results. The formula is defined as:

$$\text{precision} = \frac{N_r}{L} \tag{21}$$

where N_r refers to the frequency at which the $top - L$ nodes predicted by the algorithm occurred in the unknown important node set U.

Recall is mainly used to measure how many of the $top - L$ nodes predicted by the algorithm are correctly predicted. It is specifically defined as the proportion of the number of unknown important nodes n_r found in the $top - L$ nodes in the prediction results relative to all nodes in the unknown important node set U. The formula is defined as:

$$\text{recall} = \frac{n_r}{|U|} \tag{22}$$

AUC is mainly used to measure the precision of the algorithm as a whole. The formula is defined as:

$$\text{AUC} = \frac{0.5N_1 + N_2}{N} \tag{23}$$

The specific calculation process for AUC is as follows: one node is selected from the unknown important node set U, and another is selected from the unimportant node set V_2 in each experiment, and the relative importance scores of the two nodes are compared. If the two nodes receive the same score, then the score is recorded as 0.5 point; if the relative importance score of the node selected from the unknown important node set U is greater than that from the unimportant node set V_2, then the score is recorded as 1 point. N represents the number of all node combinations from the two sets U and V_2. After N independent experiments, the final AUC value is the sum of the scores of N experiments. Among them, the frequencies of getting 0.5 point and 1 point are N_1 and N_2, respectively.

4.3. Experimental Analysis

The core goal of this work is to find relatively important nodes from the target node set T. Therefore, the major subjects investigated from the four real networks selected, that is, all nodes of target node set T, need to be determined. From the important node set V_1, 10%, 20%, 30%, 40%, 50%, 60%, 70%, 80%, and 90% of the nodes are selected and used as known important nodes. The experiment in this paper treats the proportion of nodes equally; that is, the number of experiments corresponding to different proportion of nodes is the same. Different algorithms are used to find the relatively important nodes in the network. At the same time, precision, recall, and AUC values corresponding to different algorithms are calculated, and their values obtained from the experiments are averaged. Finally, the proposed DDMF method is used and compared with other comparative algorithms in terms of the three evaluation indexes.

The parameters of five other comparative algorithms are adjusted to be close to the optimal ones in the four networks. The specific values are as follows: $K = 3$ is taken from the Ksmar indexes, $S = 0.75$ is taken from the PPR indexes, and $\varphi = 0.0001$ is taken from the Katz indexes. In the NMF algorithm, random walk length $walk\_length$ is valued as 10, embedded vector length $size$ is set to 128, and hyperparameters $p, q \in \{0.25, 0.50, 1, 2, 4\}$. In the NLD algorithm, the selection method of known important nodes hub is the same as that of the DDMF method. The experimental results of the three evaluation indexes are shown in Figures 2 and 3 and Table 2.

In this study, different proportions of nodes are selected from the important node set V_1 as the known important nodes R. The precision, recall, and AUC values are calculated by six relatively important node mining algorithms on the basis of experiments. The average value of 50 times in the experimental results is used as the final experimental result. Figure 2 shows the precision values of six relative importance node mining methods in the four networks. The X axis represents the proportion of nodes in the target node set T while the Y axis represents the precision of different node proportions. Figure 3 shows the recall rates of the six relative importance node mining algorithms in the four networks. The X axis represents the proportion of nodes in the target node set T while the Y axis represents the recall rates of different node proportions. Table 2 shows the AUC values obtained by the six relative importance node mining algorithms in the four networks.

The experimental results show that with the increase in the number of nodes in the target node set T, the precision of the algorithm decreases gradually while the recall rate increases gradually. In order to better simulate the actual situation of different real-world networks and to reduce accidental error, the important nodes of different batches are selected in different proportions from the important node set. Then the relatively important nodes corresponding to the important nodes of these different batches are calculated and mined. By calculating the arithmetic average of the relatively important nodes of different batches, the final relatively important nodes are obtained. In terms of precision, the proposed DDMF method is obviously better than the other five comparative algorithms in the SARS and Genepath networks, and all of them perform well in the mouse and yeast networks. In terms of recall, the DDMF method performs well in the SARS and mouse networks. Specifically, its recall, under multiple node proportions, is better than those of the comparative algorithms. The DDMF method ranks second for the Genepath and yeast networks. In terms of the AUC, the DDMF method outperforms the others in the SARS, Genepath, and mouse networks and ranks second in the yeast network. In sum, the proposed DDMF method performs well in terms of all the evaluation indexes in the SARS, Genepath, and mouse networks and comes in second place in the yeast network. Specifically, the proportion of the important nodes in the yeast network is relatively large. Therefore, some errors may occur in calculating the distance distribution of important nodes.

In general, the proposed DDMF method achieves excellent performance in real and complex network datasets, especially in terms of the evaluation of precision and AUC. It is obviously better than several comparative algorithms. At the same time, the selected datasets come from different fields. The results indicate that the DDMF method is characterized by high precision and wide applicability in mining relatively important nodes in networks.

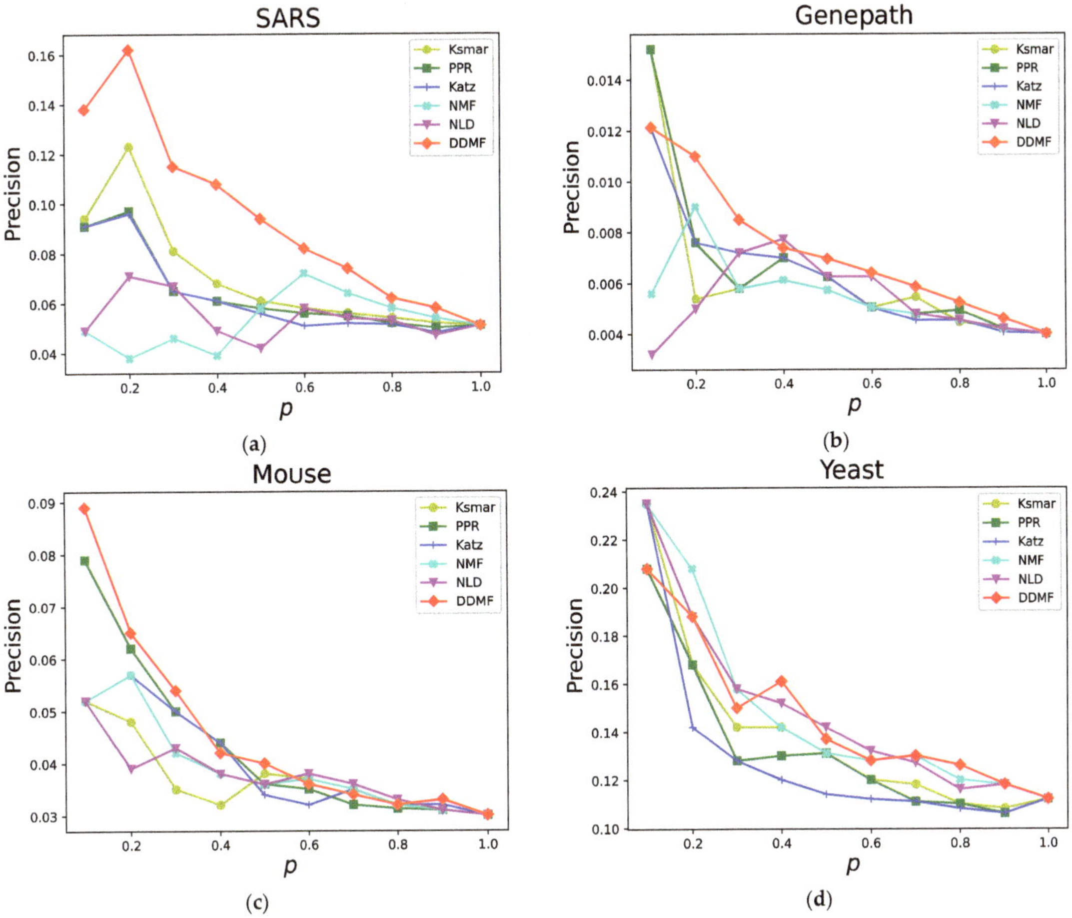

Figure 2. Precision rate results in four networks: (**a**) SARS network; (**b**) Genepath network; (**c**) Mouse network; (**d**) Yeast network.

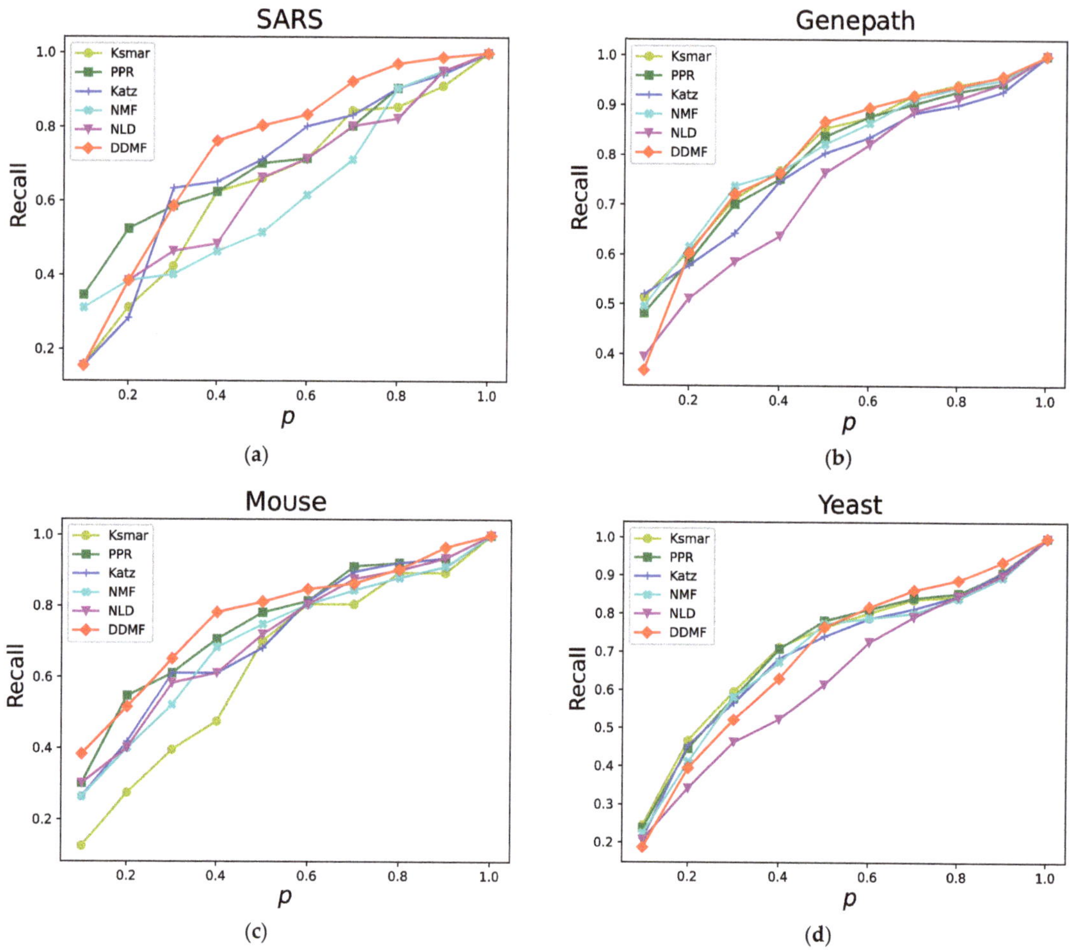

Figure 3. Recall rate results in four networks: (**a**) SARS network; (**b**) Genepath network; (**c**) Mouse network; (**d**) Yeast network.

Table 2. AUC results in four networks.

Dataset	Ksmar	PPR	Katz	NMF	NLD	DDMF
SARS	0.686	0.683	0.650	0.635	0.667	0.692
Genepath	0.545	0.526	0.482	0.568	0.565	0.675
Mouse	0.696	0.693	0.685	0.654	0.669	0.737
Yeast	0.596	0.582	0.564	0.686	0.665	0.669

5. Conclusions

A relatively important node mining method based on DDMF is proposed in this work. The DDMF method is mainly based on the distance distribution information of nodes. Starting from known important nodes, it aims to find relatively important nodes in a network. The detailed comparative experiments with five other algorithms for mining relatively important nodes in four real networks reveal that the DDMF method performs well in terms of precision and applicability. Moreover, the DDMF method can not only be used to mine the relatively important nodes in a network, but also be considered as a new idea for community detection and link prediction.

Mining relatively important nodes in complex networks is a challenging task with practical value. The DDMF method can be effectively used to find relatively important nodes in networks and provides a new idea and direction for the related work of network information mining in the future. With that being said, the limitation of the DDMF method can be summarized as something that it only considers mining relatively important nodes in single-layer networks. In the future, our relatively important nodes mining method can be applied to complex and diversified multilayer networks. Random walk could also be considered as a direction in future research.

Author Contributions: Conceptualization, N.Z. and Q.L.; methodology, Q.L.; software, J.W.; validation, Q.L., M.J. and J.L.; formal analysis, N.Z.; investigation, J.W.; resources, Z.Z.; data curation, M.J.; writing—original draft preparation, Q.L.; writing—review and editing, Q.L.; visualization, N.Z.; supervision, J.W.; project administration, Z.Z.; funding acquisition, J.L. All authors have read and agreed to the published version of the manuscript.

Funding: This research was funded by the Special Plan of Yunnan Province Major Science and Technology Plan (202102AA100021), the National Natural Science Foundation of China (62066048), the Yunnan Natural Science Foundation Project (202101AT070167) and the Open Foundation of Key Laboratory in Software Engineering of Yunnan Province (2020SE311).

Institutional Review Board Statement: Not applicable.

Informed Consent Statement: Not applicable.

Data Availability Statement: The data presented in this study are available on request from the corresponding author.

Conflicts of Interest: The authors declare no conflict of interest.

References

1. Ren, T.; Li, Z.; Qi, Y.; Zhang, Y.X.; Liu, S.M.; Xu, Y.J.; Zhou, T. Identifying vital nodes based on reverse greedy method. *Sci. Rep.* **2020**, *10*, 18. [CrossRef]
2. Li, A.W.; Xiao, J.; Xu, X.K. The Family of Assortativity Coefficients in Signed Social Networks. *IEEE Trans. Comput. Soc. Syst.* **2020**, *7*, 1460–1468. [CrossRef]
3. Liao, H.; Shen, J.; Wu, X.T.; Zhou, M.Y. Empirical topological investigation of practical supply chains based on complex networks. *Chin. Phys. B* **2017**, *26*, 144–150. [CrossRef]
4. Li, J.; Peng, X.Y.; Wang, J.; Zhao, N. A Method for Improving the Accuracy of Link Prediction Algorithms. *Complexity* **2021**, *2021*, 8889441. [CrossRef]
5. Paduraru, C.; Dimitrakopoulos, R. Responding to new information in a mining complex: Fast mechanisms using machine learning. *Min. Technol.* **2019**, *2019*, 1577596. [CrossRef]
6. Wang, T.; Chen, S.S.; Wang, X.X.; Wang, J.F. Label propagation algorithm based on node importance. *Phys. A: Stat. Mech. Its Appl.* **2020**, *551*, 124137. [CrossRef]
7. Meng, Y.Y.; Tian, X.L.; Li, Z.W.; Zhou, W.; Zhou, Z.J.; Zhong, M.H. Exploring node importance evolution of weighted complex networks in urban rail transit. *Phys. A: Stat. Mech. Its Appl.* **2020**, *558*, 124925. [CrossRef]
8. Liu, F.; Wang, Z.; Deng, Y. GMM: A generalized mechanics model for identifying the importance of nodes in complex networks. *Knowl.-Based Syst.* **2020**, *193*, 105464. [CrossRef]
9. Wen, T.; Jiang, W. Identifying influential nodes based on fuzzy local dimension in complex networks. *Chaos Solitons Fractals* **2019**, *119*, 332–342. [CrossRef]
10. Zhao, G.H.; Jia, P.; Zhou, A.M.; Zhang, B. InfGCN: Identifying influential nodes in complex networks with graph convolutional networks. *Neurocomputing* **2020**, *414*, 18–26. [CrossRef]
11. White, S.; Smyth, P. Algorithms for estimating relative importance in networks. In Proceedings of the 3th ACM SIGKDD International Conference on Knowledge Discovery and Data Mining, Washington, DC, USA, 24 August 2003; pp. 266–275.
12. Magalingam, P.; Davis, S.; Rao, A. Ranking the importance level of intermediaries to a criminal using a reliance measure. *arXiv Preprint*, 2015; arXiv:1506.06221.
13. Magalingam, P. Complex network tools to enable identification of a criminal community. *Bull. Aust. Math. Soc.* **2016**, *94*, 350–352. [CrossRef]
14. Zhao, J.; Lin, L.M. A survey of disease gene prediction methods based on molecular networks. *J. Univ. Electron. Sci. Technol. China* **2017**, *46*, 755–765.
15. Zhu, J.F.; Chen, D.B.; Zhou, T.; Zhang, Q.M.; Luo, Y.J. A survey on mining relatively important nodes in network science. *J. Univ. Electron. Sci. Technol. China* **2019**, *48*, 595–603.

16. Schieber, T.A.; Carpi, L.; Díaz-Guilera, A.; Pardalos, P.M.; Masoller, C.; Ravetti, M.G. Quantification of network structural dissimilarities. *Nat. Commun.* **2017**, *8*, 110. [CrossRef]
17. Wang, Y.F.; Vasilakos, A.V.; Jin, Q.; Ma, J.H. PPRank: Economically Selecting Initial Users for Influence Maximization in Social Networks. *IEEE Syst. J.* **2017**, *11*, 2279–2290. [CrossRef]
18. Yang, Y.Z.; Yu, L.; Wang, X.; Zhou, Z.L.; Chen, Y.; Kou, T. A novel method to evaluate node importance in complex networks. *Phys. A: Stat. Mech. Its Appl.* **2019**, *526*, 121118. [CrossRef]
19. Li, X.L.; Zhang, P.; Zhu, G.Y. Measuring method of node importance of urban rail network based on h index. *Appl. Sci.* **2019**, *9*, 5189. [CrossRef]
20. Luo, J.W.; Wu, J.; Yang, W.Y. A relationship matrix resolving model for identifying vital nodes based on community in opportunistic social networks. *Trans. Emerg. Telecommun. Technol.* **2021**, *12*, e4389. [CrossRef]
21. Chang, H.; Cohn, D.; McCallum, A.K. Learning to create customized authority lists. In Proceedings of the Seventeenth International Conference on Machine Learning, San Francisco, CA, USA, 29 June–2 July 2000; pp. 127–134.
22. Haveliwala, T.H. Topic-sensitive pagerank: A context-sensitive ranking algorithm for web search. *IEEE Trans. Knowl. Data Eng.* **2003**, *15*, 784–796. [CrossRef]
23. Jennifier, G.; Widom, J. Scaling personalized web search. In Proceedings of the 12th international conference on World Wide Web, New York, NY, USA, 20–24 May 2003; pp. 271–279.
24. Alzaabi, M.; Taha, K.; Martin, T.A. CISRI: A crime investigation system using the relative importance of information spreaders in networks depicting criminals communications. *IEEE Trans. Inf. Forensics Secur.* **2015**, *10*, 2196–2211. [CrossRef]
25. Wang, H.; Chang, C.K.; Yang, H.I.; Chen, Y. Estimating the relative importance of nodes in social networks. *J. Inf. Processing* **2013**, *21*, 414–422. [CrossRef]
26. Rodriguez, M.A.; Bollen, J. An algorithm to determine peer-reviewers. In Proceedings of the 17th ACM conference on Information and knowledge management, New York, NY, USA, 26–30 October 2008; pp. 319–328.
27. Magalingam, P.; Davis, S.; Rao, A. Using shortest path to discover criminal community. *Digit. Investig.* **2015**, *15*, 117. [CrossRef]
28. Langohr, L. Methods for finding interesting nodes in weighted graphs. *Hels. Yliop.* **2014**, *11*, 145.
29. Cui, P.; Wang, X.; Pei, J.; Zhu, W.W. A survey on network embedding. *IEEE Trans. Knowl. Data Eng.* **2018**, *31*, 833–852. [CrossRef]
30. Zhang, P.Y.; Yao, H.P.; Li, M.Z.; Liu, Y.J. Virtual network embedding based on modified genetic algorithm. *Peer-Peer Netw. Appl.* **2019**, *12*, 481–492. [CrossRef]
31. Nelson, W.; Zitnik, M.; Wang, B.; Leskovec, J.; Goldenberg, A.; Sharan, R. To embed or not: Network embedding as a paradigm in computational biology. *Front. Genet.* **2019**, *10*, 381. [CrossRef]
32. Su, C.; Tong, J.; Zhu, Y.J.; Cui, P.; Wang, F. Network embedding in biomedical data science. *Brief. Bioinform.* **2020**, *21*, 182–197. [CrossRef]
33. Yao, H.P.; Ma, S.; Wang, J.J.; Zhang, P.Y.; Jiang, C.X.; Guo, S. A continuous-decision virtual network embedding scheme relying on reinforcement learning. *IEEE Trans. Netw. Serv. Manag.* **2020**, *17*, 864–875. [CrossRef]
34. Li, B.T.; Pi, D.C.; Lin, Y.X.; Cui, L. DNC: A Deep Neural Network-based Clustering-oriented Network Embedding Algorithm. *J. Netw. Comput. Appl.* **2021**, *173*, 102854. [CrossRef]
35. Song, G.J.; Wang, Y.; Du, L.; Li, Y.; Wang, J.S. Network Embedding on Hierarchical Community Structure Network. *ACM Trans. Knowl. Discov. Data* **2021**, *15*, 123. [CrossRef]
36. Zhao, J.; Yang, T.H.; Huang, Y.; Holme, P. Ranking candidate disease genes from gene expression and protein interaction: A Katz-centrality based approach. *PLoS ONE* **2011**, *6*, e0024306. [CrossRef]
37. Zhao, N.; Li, J.; Wang, J.; Peng, X.Y.; Jing, M.; Nie, Y.J.; Yu, Y. Relatively important nodes mining method based on neighbor layer diffuse. *J. Univ. Electron. Sci. Technol. China* **2021**, *50*, 121–126.
38. Mu, J.F.; Liang, J.Y.; Zheng, W.P.; Liu, S.Q.; Wang, J. Node similarity measure for complex networks. *J. Front. Comput. Sci. Technol.* **2019**, *14*, 749–759.
39. Liu, D.; Chen, X.; Peng, D. Some cosine similarity measures and distance measures between q-rung orthopair fuzzy sets. *Int. J. Intell. Syst.* **2019**, *34*, 1572–1587. [CrossRef]
40. Balaji, R.; Bapat, R.B.; Goel, S. Generalized Euclidean distance matrices. *arXiv Preprint*, 2021; arXiv:2103.03603. [CrossRef]
41. Gour, G.; Tomamichel, M. Entropy and relative entropy from information-theoretic principles. *IEEE Trans. Inf. Theory* **2021**, *67*, 6313–6327. [CrossRef]
42. Li, Y. Scheduling analysis of intelligent machining system based on combined weights. In Proceedings of the 2nd International Conference on Frontiers of Materials Synthesis and Processing, Sanya, China, 10–11 November 2018; Volume 493, p. 012146.
43. Grover, A.; Leskovec, J. node2vec: Scalable feature learning for networks. In Proceedings of the 22nd ACM SIGKDD International Conference on Knowledge Discovery and Data Mining, San Francisco, CA, USA, 13–17 August 2016; pp. 855–864.
44. Jani, P. Airport, Airline and Route Data. Available online: https://openflights.org/data.html (accessed on 28 December 2021).
45. Krauthammer, M.; Kaufmann, C.A.; Gilliam, T.C.; Rzhetsky, A. Molecular triangulation: Bridging linkage and molecular-network information for identifying candidate genes in Alzheimer's disease. *Proc. Natl. Acad. Sci. USA* **2004**, *101*, 15148–15153. [CrossRef]
46. Xenarios, I.; Rice, D.W.; Salwinski, L.; Baron, M.K.; Marcotte, E.M.; Eisenberg, D. DIP: The database of interacting proteins. *Nucleic Acids Res.* **2000**, *28*, 289–291. [CrossRef]
47. Li, M.; Zhang, H.H.; Wang, J.X.; Pan, Y. A new essential protein discovery method based on the integration of protein-protein interaction and gene expression data. *BMC Syst. Biol.* **2012**, *6*, 15. [CrossRef]

applied sciences

MDPI

Article

Mining Algorithm of Relatively Important Nodes Based on Edge Importance Greedy Strategy

Jie Li [1,†], Chunlin Yin [1,†], Hao Wang [2], Jian Wang [3] and Na Zhao [2,*]

1 Electric Power Research Institute of Yunnan Power Grid Co., Ltd., Kunming 650217, China; lj1226645407@163.com (J.L.); 18487125168@139.com (C.Y.)
2 Key Laboratory in Software Engineering of Yunnan Province, School of Software, Yunnan University, Kunming 650091, China; wang3706@mail.ynu.edu.cn
3 Faculty of Information Engineering and Automation, Kunming University of Science and Technology, Kunming 650504, China; jianwang@kust.edu.cn
* Correspondence: zhaona@ynu.edu.cn
† These authors contributed equally to this work.

Abstract: Relatively important node mining has always been an essential research topic in complex networks. Existing relatively important node mining algorithms suffer from high time complexity and poor accuracy. Therefore, this paper proposes an algorithm for mining relatively important nodes based on the edge importance greedy strategy (EG). This method considers the importance of the edge to represent the degree of association between two connected nodes. Therefore, the greater the value of the connection between a node and a known important node, the more likely it is to be an important node. If the importance of the edges in an undirected network is measured, a greedy strategy can find important nodes. Compared with other relatively important node mining methods on real network data sets, such as SARS and 9/11, the experimental results show that the EG algorithm excels in both accuracy and applicability, which makes it a competitive algorithm in the mining of important nodes in a network.

Keywords: complex network; important nodes; relative importance; important edge

Citation: Li, J.; Yin, C.; Wang, H.; Wang, J.; Zhao, N. Mining Algorithm of Relatively Important Nodes Based on Edge Importance Greedy Strategy. *Appl. Sci.* **2022**, *12*, 6099. https://doi.org/10.3390/app12126099

Academic Editors: Pasquale De Meo, Giacomo Fiumara, Xiaoyang Liu and Annamaria Ficara

Received: 19 April 2022
Accepted: 25 May 2022
Published: 15 June 2022

1. Introduction

With the advances in human scientific cognition and information technology, network science has become a hot topic in academia. As the primary research object of network science, complex networks are gradually emerging in the eyes of scholars [1]. A complex network refers to a network with some or all of the properties of self-organization, self-similarity, attractor, small world, and scale-free. Complex networks can model all aspects of real-life human society, and through the study of these networks abstracted from reality, people can explore the laws of the real world. Therefore, analysis of complex networks and their applications is a crucial issue.

Many research papers on complex networks have been published [2–5]. Early research covered the traditional statistical properties of networks (e.g., the two papers that laid the foundation of complex networks—scale-free networks [6] and small-world networks [7]). Later works described the structural properties of networks (e.g., the exploration of "community phenomena" [8] and "network modalities" [9]). Even later, papers addressed the deeper study of points and lines. The examination of complex networks has undergone tremendous evolution, and the study of the importance of the nodes or edges of complex networks is one of the most important topics.

Nodes and edges are the basic elements of network structure. Studying important nodes or edges helps us protect the system better, but also helps us understand the system better. For example, a disease transmission network can search for known infected people. Then, susceptible people can be searched, treated, and isolated, in order to prevent further spread of the virus. For another example, Fan et al. collected 20 years of trade data from 232 countries and regions around the world, and then constructed a trade network. In this network, a novel node importance ranking and analysis method was proposed by comprehensively considering factors such as generalized degree, DHC theorem and weight [10]. This method helps the formulation of trade policies in countries around the world, and deepens our understanding of the history of world trade. Xu et al. also pointed out that for traditional information retrieval evaluation metrics based on citation network structure, it is difficult to accurately assess the impact of a particular piece of literature. They experimentally argued that the adapted PageRank and LeaderRank methods are still the most accurate evaluation criteria available [11].

As an interdisciplinary subject, there are examples of applying various computer technologies to the research of complex networks. For example, Liu et al. proposed an artificial neural network-based model for information dissemination and opinion evolution, IPNN (Information propagation and public opinion evolution model based on artificial neural network, IPNN) [12]. Fan et al. proposed a reinforcement learning-based algorithm for node importance identification in complex networks, FINDER. It first learns the exhaustive method in a simulated BA network, continues training the previously trained model in the real network, and evaluates the performance of the model based on the order of node removal [13]. In addition, many excellent improvements to the traditional node importance recognition algorithm have also been proposed. For example, Fan et al. proposed a new node importance ranking metric, the circle ratio, beginning from a circular structure in the network [14]. Traditional methods judge the importance of a node by the contributions of neighboring nodes. However, the circle ratio judges the importance of the current node by the amount of information it brings to its neighbors, which inspires a new research idea [14]. Lu and Liao et al. summarized and sorted out the current node importance identification and ranking methods in various existing networks [15,16].

As we can see, most of the current research on nodes focuses on the mining and discovery of important nodes in a network, but little research exists on the mining of relatively important nodes. The idea of relative node importance considers questions like "which node in a network is the most important relative to a specific node or a specific group of nodes?" Compared with other research fields, the relative or local importance of nodes also has practical implications, especially when the scale of the network grows larger. Some research results on relative node importance mining are available today, but these methods still need improvement. Areas of improvement include whether the time complexity and space complexity can be further reduced, how to further improve the accuracy of exiting method, which kind of method performs the best on a specific type of network, and the parameter selection and optimization method, etc.

In this paper, we consider the connection role of important edges in an unweighted network, where the edge importance represents the degree of association between two nodes. The connections between important nodes should be closer, thus we propose a metric to measure edge closeness for important nodes in an unweighted network. It is based on the idea of "the node with the largest edge closeness to a known important node is likely to be an important node" for which an edge importance greedy strategy (EG) is proposed to mine relatively important nodes. Through the comparison experiments with the NN [17] and the RD [18], which originated from protein networks, and the Katz [19], which is based on random wandering, it can be proved that the EG strategy achieves ideal experimental results and shows its application value in identifying the importance level of unprivileged network nodes.

2. Greedy Strategy Based on Edge Importance

The EG algorithm uses greedy strategy that requires an importance measure for the edges in a network before each use. It works by adding known important nodes to a set C that includes all their neighbors, after which one can assign an importance score to the connected edges of the known important nodes based on the topological information of the network. The importance score can be calculated as follows:

$$SV = \frac{k_j}{k_i} \cdot \frac{1}{d^2} + \frac{1}{cn+1} \tag{1}$$

where $(k_j/k_i) \cdot (1/d^2)$ is an important component called NP value, which measures the importance of a certain node; k_i is the degree of known important node i; k_j is the degree of known important neighbor node j; d is the shortest distance between node j and the set of known important nodes; and cn is the number of common neighbors of node j and known important nodes.

The core idea of Equation (1) is actually very straightforward: the larger the k_j is, the greater the importance of node j. On the other hand, since d is the shortest distance between node j and the set of known important nodes, the larger the d is, the smaller the value of $1/d^2$. A small k_j or a large d results in a smaller NP, and thus a lower importance score for node j. In other words, nodes with low degree and large distance from known important nodes tend to have lower importance, and vice versa.

In particular, there will be cases where two nodes have the same NP value. In order to solve this problem, a bias value $1/(cn+1)$ is added to Equation (1) to distinguish their importance.

Once the importance score is obtained, the greedy strategy is used to find the node whose edge has the largest score from an edge set that corresponds to known important nodes. The found node will then be added to set C. The edge scores corresponding to each node in set C will be again calculated, and the node will be found by using the same strategy stated previously. This process is repeated until all nodes are added. The order in which nodes are added to set C is exactly the order of their importance.

The pseudo-code for the EG algorithm, also known as Algorithm 1, is as follows:

Algorithm 1: EG Algorithm

Begin
Input Network $G = (V, E)$, the set of known important nodes R;
Initialization $C = R$; $S = 0$;
1. **While** the number of elements in C is less than $|V|$ **do:**
2. **for** i in C **do:**
3. **for** j in i neighbors **do:**
4. $SV(i,j) = k\_j/k\_i \cdot 1/d2 + 1/(cn+1)$;
5. **end for**
6. $t = \{j|\max(SV)\}$;
7. Place the node t into set C;
8. **end for**
9. **end while**
10. **return** C
End

The EG algorithm is divided into three parts, which are as follows: computation of the shortest path of a single source; computation of the node importance score; and selection of the greedy policy. The EG algorithm chooses to compute the neighboring nodes of known important nodes. Lines 1–2 of the algorithm are the traversal of the set C, and the nodes in it are computed and analyzed. Lines 3–5 of the algorithm calculate the node importance score by first selecting a node i from among the set of known important nodes, then traversing its neighbor nodes and calculating the importance score of each neighboring node. Lines 6–8 of the algorithm apply a greedy strategy to select nodes, and the node with the highest SV is added to set C. The above process is repeated until all nodes are added to set C.

It can be found that the time complexity of the EG algorithm depends mainly on the calculation of the shortest path of a single source and the importance score. It is easy to see that the calculation of the node importance score is a cumulative process, and depends on the degree of the important nodes. When there is only one element in the set of important nodes, its time complexity is $\bar{k}$, and $\bar{k}$ is the average degree of the network; when there are two elements, its time complexity is $2\bar{k}$, and so on. The time complexity of this part can be written as $\bar{k} + 2\bar{k} + 3\bar{k} + \ldots + n\bar{k}$, which equals to $O\left(n^2\bar{k}\right)$.

One can also notice that the previously calculated node importance scores are fully reusable when new nodes are added to the set of important nodes. In this case, each time an important node is added, only the neighboring node scores of the newly added node need to be calculated. Based on above analysis, one can conclude that the time complexity actually depends on the number and average degree of nodes in the network, thus the overall time complexity would be $O\left(n\bar{k}\right)$.

The following section will use the network shown in Figure 1 to illustrate the computation process of the edge greedy strategy.

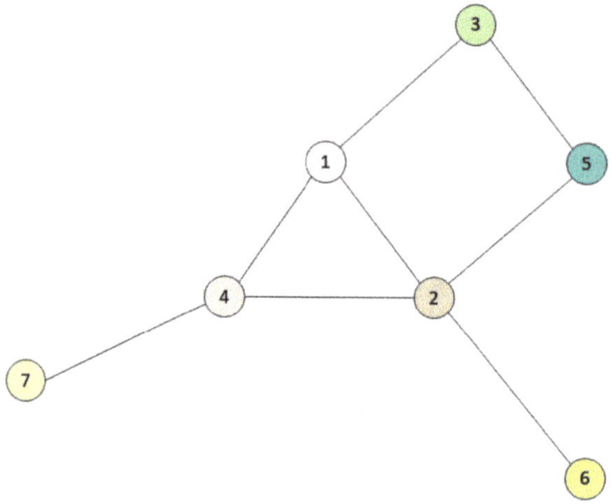

Figure 1. Example network of the edge importance greedy strategy.

Step 1, number the nodes in Figure 1 sequentially are 1, 2, 3, 4, 5, 6 and 7. Let node 1 be a known important node, and add node 1 to set C. Calculate the edge scores first according to the known important node. The edge scores here are (1,2):2.33, (1,3):1.65 and (1,4):2.0. Node 2, which corresponds to edge (1,2), has the largest value of 2.33, and it is added to set C.

Step 2, calculate the edge importance score for nodes 1 and 2 in set C, and one obtains (1,3):1.65, (2,4):1.75, (2,5):1.125, and (2,6):1.0625. Node 4, which corresponds to edge (2,4), has the largest value and is added to set C.

Step 3, calculate the edge importance score for nodes 1, 2 and 4 in set C, and one obtains (1,3):1.65, (2,5):1.125, (2,6):1.0625 and (4,7):1.083. Node 3 has the largest value and is added to set C.

Step 4, calculate the edge importance score for nodes 1, 2, 4 and 3 in set C and obtain (2,5):1.125, (2,6):1.0625 and (4,7):1.083. Node 5 is selected and added to set C.

Step 5, calculate the edge importance score for nodes 1, 2, 4, 3 and 5 in set C and obtain (2,6):1.0625 and (4,7):1.083, and node 7 is added to set C.

Step 6, calculate the edge importance score for nodes 1, 2, 4, 3, 5 and 7 in set C. The edge importance score is (2,6):1.0625 and node 6 is added to set C.

At this point, all nodes are added and the loop ends. For node 1, the order of relatively important nodes possibilities would be 2, 4, 3, 5, 7 and 6.

3. Experiment

The experiments in this paper use four real network data sets.

(1) The 9/11 criminal relationship network [20]. The nodes represent the terrorists who hijacked the planes and those who had contact with them; the edges represent the interpersonal relationships between them; and the set of important nodes represents the group of terrorists who hijacked the planes on 9/11.
(2) SARS international aviation network [21]. Each node represents a country; the edge represents the existence of routes between two countries; and the earliest group of countries with the SARS virus is the set of important nodes.
(3) Mouse protein interaction network [22]. The nodes represent mouse proteins; edges represent the existence of interactions between proteins; and the group of mouse protein kinases is the set of the important nodes.
(4) Human protein interaction network [22]. The nodes represent human proteins; edges represent the existence of interactions between proteins; and the group of human protein kinases is the set of the important nodes.

The topological information of these network data sets is listed in the following Table 1.

Table 1. Network topology information.

Network	N	N'	M	K	C
9/11	37	19	85	4.59	0.52
SARS	224	18	2247	20.06	0.65
Human	3574	186	6002	3.36	0.15
Mouse	1187	67	1557	2.62	0.09

The table header indicates the topological attributes of a network. N is the number of nodes in the network; N' is the number of important node sets; M is the number of edges; and K and C are the average degree and clustering coefficient of the entire network, respectively.

In this paper, we use AUC (Area Under Curve) to evaluate the overall results of this algorithm for mining relatively important nodes. The AUC is calculated as follows.

$$AUC = \frac{0.5n_1 + n_2}{n} \quad (2)$$

where n_1 represents the number of times that the importance score of a node selected from the unknown important node set equals that of a node selected from the unimportant node set; n_2 represents the number of times that the importance score of a node selected from the unknown important node set is larger than that of a node selected from the unimportant node set. n represents the number of comparisons, which is the product of the size of the unknown important node set and the size of the unimportant node set.

In this paper, we conduct nine rounds of experiments for each network, and the ratios (p) of known important nodes for each round are set to 10%, 20%, 30%, 40%, 50%, 60%, 70%, 80% and 90%. Each round uses a different algorithm to calculate the relative importance of the nodes, according to which these nodes are ranked. Twenty independent experiments are conducted, and the AUC values are calculated for the ranking results.

This paper uses NN [17], RD [18] and Katz [19] for comparison, and the results are as follows.

In Figure 2, the horizontal axis represents the percentage of known important nodes of the set of important nodes, and the vertical axis represents the average AUC value. A higher AUC value represents a higher effectiveness. It is easy to see that the AUC values of most algorithms show an increasing trend as the ratio of important nodes in the network increases. The effect of the EG algorithm is most significant for the 9/11 network, and its AUC value remains stable at around 0.8, which is far ahead of other algorithms. In the SARS network and the Human network, our algorithm also achieved satisfactory results.

The *AUC* index maintains its leading position against comparison algorithms, although the margins are not as prominent as those for 911net. One can also note that when the proportion of important nodes exceeds 30%, the *AUC* value of the EG algorithm in the Mouse network is slightly lower than that of the Katz algorithm, which may be due to the fact that the EG algorithm depends on adjacent nodes for judgment.

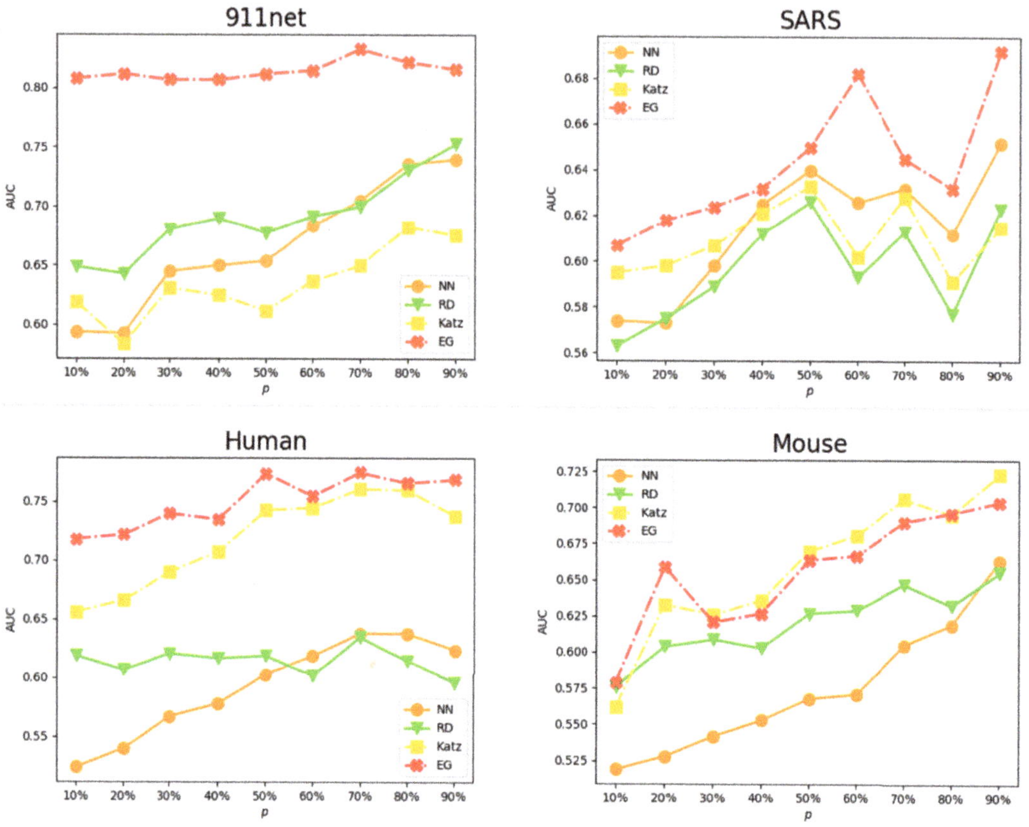

Figure 2. Comparison results from *AUC*s of the edge importance greedy strategy algorithm.

The experiments proved that the EG method performs best on the 9/11, SARS, and Human networks based on the evaluation of the *AUC*, regardless of the ratio of known important nodes. It performs second-best on the Mouse network. From the perspective of network topology, the SARS network is different from other networks and possesses the highest average degree (20). At the same time, both the 911net and SARS networks are with high node aggregation. It can be inferred that the EG algorithm performs the best in those networks that are similar to SARS and 911net. The fluctuation amplitudes of the *AUC*s for the four algorithms at different important node ratios reside in the scope of 0.08~0.12. This indicates that the ratio value in dense graphs has limited effect on these four algorithms. However, the clustering coefficients for both the Human and Mouse networks are smaller compared with those of the SARS and 9/11 networks, and we believe it is this attribute that reduces the advantage of EG over Katz, which performs second best for the Human network.

Further analysis finds that the NN algorithm considers the nodes with more connections to known important nodes as relatively important nodes, so its accuracy decreases when the known important nodes are fewer. On the other hand, when the known important nodes are in large amounts, it is also very challenging for an algorithm like NN to distinguish important nodes when the number of connected nodes equals that of the known important nodes. The RD method uses the inverse sum of the shortest path lengths with known important nodes to measure relative importance. However, in real networks, the nodes close to known important nodes are not necessarily the nodes we are looking for, and those far from the known important node may actually be the ones we need. The Katz method adopts a random wandering strategy that can better complete the mining task of relatively important nodes in some networks, but it still needs to improve its accuracy for other networks.

4. Discussion and Conclusions

In this paper, we propose the EG method for mining relatively important nodes based on the greedy strategy of edge importance. This method measures the importance of the edges of known important nodes, and uses the most important edges to find nodes that are closely related to the current important nodes. The EG method does not calculate or consider network paths, and thus avoids the limitations of some methods that consider path information only. By comparing with the existing methods of mining relatively important nodes, such as NN, RD and Katz, based on the indicator of *AUC*, we proved the performance and the feasibility of the EG algorithm.

Although the adaptability of our proposed algorithm to different networks has been discussed, we still consider it necessary to further analyze the impact network topology has on an algorithm in a more detailed and systematic way, in order to obtain better guidance for the use of the EG algorithm. The EG method is designed for undirected networks, and does not consider applications in directed or weighted networks. Extending the EG method to directed graphs, weighted graphs, or even time series networks could be another possible future direction. Lastly, it seems that the EG method does not possess a rigorous physical or mathematical meaning; some theoretical work can be carried out surrounding this issue in the future.

Author Contributions: Conceptualization, J.L. and C.Y.; methodology, J.L.; software, J.W.; validation, J.L., C.Y. and H.W.; formal analysis, J.L.; investigation, J.W.; resources, N.Z.; data curation, C.Y.; writing—original draft preparation, J.L. and H.W.; writing—review and editing, J.L. and H.W.; visualization, J.W.; supervision, N.Z.; project administration, N.Z.; funding acquisition, J.L. All authors have read and agreed to the published version of the manuscript.

Funding: This research was funded by the Special Plan of Yunnan Province Major Science and Technology Plan (202102AA100021), the National Natural Science Foundation of China (62066048), the Yunnan Natural Science Foundation Project (202101AT070167) and the Open Foundation of Key Laboratory in Software Engineering of Yunnan Province (2020SE311).

Institutional Review Board Statement: Not applicable.

Informed Consent Statement: Not applicable.

Data Availability Statement: The data presented in this study are available on request from the corresponding author.

Conflicts of Interest: The authors declare no conflict of interest.

References

1. Zhou, T.; Bai, W.J.; Wang, B.H.; Liu, Z.J.; Yan, G. Overview of complex network research. *Physics* **2005**, *34*, 6.
2. Xuan, Q.; Wang, J.; Zhao, M.; Yuan, J.; Fu, C.; Ruan, Z.; Chen, G. Subgraph networks with application to structural feature space expansion. *IEEE Trans. Knowl. Data Eng.* **2019**, *33*, 2776–2789. [CrossRef]
3. Yue, P.; Fan, Y.; Batten, J.A.; Zhou, W.X. Information transfer between stock market sectors: A comparison between the USA and China. *Entropy* **2020**, *22*, 194. [CrossRef] [PubMed]
4. Ran, Y.; Liu, T.; Jia, T.; Xu, X.K. A novel similarity measure for mining missing links in long-path networks. *Chin. Phys. B* **2021**, preprint. [CrossRef]
5. Li, R.; Wang, W.; Di, Z. Effects of human dynamics on epidemic spreading in Côte d'Ivoire. *Phys. A Stat. Mech. Its Appl.* **2017**, *467*, 30–40. [CrossRef]
6. Barabasi, A.L.; Albert, R. Emergence of scaling in random networks. *Science* **1999**, *286*, 509–512. [CrossRef] [PubMed]

7. Watts, D.; Strogatz, S. Collective dynamics of 'small-world' networks. *Nature* **1998**, *393*, 440–442. [CrossRef] [PubMed]
8. Newman, M.; Girvan, M. Finding and evaluating community structure in networks. *Phys. Rev. E* **2004**, *69*, 026113. [CrossRef] [PubMed]
9. Alon, U. Network motifs: Theory and experimental approaches. *Nat. Rev. Genet.* **2007**, *8*, 450–461. [CrossRef]
10. Fan, T.; Li, H.; Ren, X.L.; Xu, S.; Gou, Y.; Lü, L. The rise and fall of countries on world trade web: A network perspective. *Int. J. Mod. Phys. C* **2021**, *32*, 2150121. [CrossRef]
11. Shuqi, X.; Msma, B.; Linyuan, L.; Medo, M. Unbiased evaluation of ranking metrics reveals consistent performance in science and technology citation data. *J. Informetr.* **2020**, *14*, 101005.
12. Liu, X.; He, D. Information propagation and public opinion evolution model based on artificial neural network in online social network. *Comput. J.* **2019**, *63*, 1689–1703. [CrossRef]
13. Fan, C.; Zeng, L.; Sun, Y.; Liu, Y.Y. Finding key players in complex networks through deep reinforcement learning. *Nat. Mach. Intell.* **2020**, *2*, 317–324. [CrossRef] [PubMed]
14. Fan, T.; Lü, L.; Shi, D.; Zhou, T. Characterizing cycle structure in complex networks. *Commun. Phys.* **2020**, *4*, 272. [CrossRef]
15. Lü, L.; Chen, D.; Ren, X.L.; Zhang, Q.M.; Zhang, Y.C.; Zhou, T. Vital nodes identification in complex networks. *Phys. Rep.* **2016**, *650*, 1–63. [CrossRef]
16. Liao, H.; Mariani, M.S.; Medo, M.; Zhang, Y.C.; Zhou, M.Y. Ranking in evolving complex networks. *Phys. Rep.* **2017**, *689*, 1–54. [CrossRef]
17. Biagioni, R.; Vandenbussche, P.Y.; Novacek, V. Finding Explanations of Entity Relatedness in Graphs: A Survey. *arXiv* **2018**, arXiv:1809.07685.
18. Wu, X.; Jiang, R.; Zhang, M.Q.; Li, S. Network-based global inference of human disease genes. *Mol. Syst. Biol.* **2008**, *4*, 189. [CrossRef] [PubMed]
19. Gori, M.; Pucci, A. Research paper recommender systems: A random-walk based approach. In Proceedings of the 2006 IEEE/WIC/ACM International Conference on Web Intelligence (WI 2006 Main Conference Proceedings) (WI'06), Washington, DC, USA, 18–22 December 2006; pp. 778–781.
20. Krebs, V.E. Mapping networks of terrorist cells. *Connections* **2002**, *24*, 43–52.
21. Airport, Airline and Route Data. Available online: https://openflights.org/data.html (accessed on 16 May 2021).
22. Xenarios, I.; Rice, D.W.; Salwinski, L.; Baron, M.K.; Marcotte, E.M.; Eisenberg, D. Dip: The database of interacting proteins. *Nucleic Acids Res.* **2000**, *24*, 289–291. [CrossRef] [PubMed]

applied sciences

MDPI

Article

Bridge Node Detection between Communities Based on GNN

Hairu Luo, Peng Jia *, Anmin Zhou, Yuying Liu and Ziheng He

School of Cyber Science and Engineering, Sichuan University, Chengdu 610065, China
* Correspondence: pengjia@scu.edu.cn

Abstract: In a complex network, some nodes are relatively concentrated in topological structure, thus forming a relatively independent node group, which we call a community. Usually, there are multiple communities on a network, and these communities are interconnected and exchange information with each other. A node that plays an important role in the process of information exchange between communities is called an inter-community bridge node. Traditional methods of defining and detecting bridge nodes mostly quantify the bridging effect of nodes by collecting local structural information of nodes and defining index operations. However, on the one hand, it is often difficult to capture the deep topological information in complex networks based on a single indicator, resulting in inaccurate evaluation results; on the other hand, for networks without community structure, such methods may rely on community partitioning algorithms, which require significant computing power. In this paper, considering the multi-dimensional attributes and structural characteristics of nodes, a deep learning-based framework named BND is designed to quickly and accurately detect bridge nodes. Considering that the bridging function of nodes between communities is abstract and complex, and may be related to the multi-dimensional information of nodes, we construct an attribute graph on the basis of the original graph according to the features of the five dimensions of the node to meet our needs for extracting bridging-related attributes. In the deep learning model, we overlay graph neural network layers to process the input attribute graph and add fully connected layers to improve the final classification effect of the model. Graph neural network algorithms including GCN, GAT, and GraphSAGE are compatible with our proposed framework. To the best of our knowledge, our work is the first application of graph neural network techniques in the field of bridge node detection. Experiments show that our designed framework can effectively capture network topology information and accurately detect bridge nodes in the network. In the overall model effect evaluation results based on indicators such as Accuracy and F1 score, our proposed graph neural network model is generally better than baseline methods. In the best case, our model has an Accuracy of 0.9050 and an F1 score of 0.8728.

Keywords: social network analysis; bridge node detection; graph neural network; community

Citation: Luo, H.; Jia, P.; Zhou, A.; Liu, Y.; He, Z. Bridge Node Detection between Communities Based on GNN. *Appl. Sci.* **2022**, *12*, 10337. https://doi.org/10.3390/app122010337

Academic Editors: Giacomo Fiumara, Xiaoyang Liu, Annamaria Ficara and Pasquale De Meo

Received: 14 August 2022
Accepted: 9 October 2022
Published: 13 October 2022

1. Introduction

At present, various complex network structures have been integrated into our life, such as transportation networks, computer networks, citation networks, and so on. Thus, emerging network science has become an important research field. In a complex network, there is a type of node that plays a key role in the information dissemination between local network structures, which is called a bridge or bridge node. In this paper, we use bridge node to refer to such a node. Due to their special topological position in the network, when these bridging nodes are activated, they can effectively promote the information flow between local structures in the network; on the contrary, when immune to them, they can effectively prevent the information flow between local structures. Accurately discovering bridge nodes in the network is an important research topic in network science, and its results have important application value in scenarios such as community immunization [1,2] and drug analysis [3].

So far, researchers have proposed many detection methods for bridge nodes, which are mainly divided into two categories: methods based on community structure and methods that do not consider community structure. The method based on community structure focuses on the detection of bridge nodes as important information transmission media between communities, which is more in line with the definition of "bridge"; the method of bridge node detection that does not depend on community structure focuses on the bridging role of nodes in the global network; this "bridging effect" is more similar to the definition of node "influence". On the other hand, the method based on community structure is more suitable for some real-world application scenarios. Suppose a large computer cluster network is infected by a virus, and it is necessary to immunize the computers in key locations to prevent the further spread of the virus. If the location of the bridge nodes is determined based on the network community structure, the original network topology will be protected to the greatest extent from damage after the node computer is removed, and most functions of the network will be maintained normally. Therefore, in recent years, more and more bridge node-related work is based on the community structure.

As far as we know, the current community-based algorithms for detecting bridge nodes are based on the local structure information of nodes and perform index calculations to define the bridging role of nodes. The calculation of a single index means that such methods cannot comprehensively consider the multi-dimensional information of nodes to characterize the bridging effect of nodes, thus affecting the accuracy of evaluation results. In addition, community-based structures may rely on community detection algorithms, resulting in the consumption of additional computing resources.

This paper systematically studies bridge nodes in complex networks and proposes a deep learning-based method for detecting bridge nodes between communities. The paper contains the following three main contributions:

- A deep learning-based framework named BND is proposed to detect bridge nodes, through which we can avoid expensive community detection algorithms;
- On this basis, we applied graph learning technology and constructed a GNN model, BND-GCN, for bridge node detection on complex networks;
- We test our model on sex real social networks and compare it with other baseline methods. Experiments show that BND-GCN performs well on bridge node detection tasks, and is generally better than the baselines.

The rest of this paper is organized as follows. In Section 2, we systematically introduce a series of related works on bridge node detection and graph representation learning. Section 3 details our proposed bridge node detection framework, BND. In Section 4, a method for constructing the training dataset is presented. In Section 5, extensive experiments are designed and conducted to verify the effectiveness of the framework. Finally, in Section 6, we summarize our research work.

2. Related Works

Our research is related to the following works.

2.1. Bridge Nodes Detection Methods

Regarding the definition of node bridging, there are many definitions given by researchers due to differences in network types and research ideas. Before conducting research on bridge nodes, it is necessary to establish a standard and universal definition. According to the research of Meghanathan [4], the existing approaches can be roughly divided into two categories: community-unaware approaches and community-aware approaches.

2.1.1. Community-Unaware Approach

The community-unaware approach does not rely on the community partition algorithm, but uses the local or global topology information of nodes to define the bridging of nodes. The research ideas are not limited to the definition and calculation of indicators [3–6], but also including random walk-based methods [7], heuristic algorithms [8], and so on.

Since the general definition of bridging is more based on node neighborhood, network local information is used more in the decision method of bridge nodes.

Hwang et al. [3] first proposed bridging centrality in 2008 to evaluate drug targets. The definition of this indicator combines the calculation of random betweenness centrality and the bridging coefficient.

Through this similar idea of combining global and local indicators, Liu et al. [5] proposed a bridge node quantification indicator named BNC, which combines route-betweenness and the bridgeness-coefficient. For each node, its route-betweenness is the sum of routing weights through that node, and its bridgeness-coefficient is defined as the reciprocal of the sum of distances from that node to all its neighbors and indirect neighbors. Similarly, after the dispersion standardization of the above two indicators, the product is used as the final BNC score for the node.

The community bridge finder algorithm (CBF) proposed by Salathe et al. [7] is a random walk-based bridge node detection algorithm, which attempts to detect bridging nodes between communities without relying on the community structure of the network. The basic idea is that the first node that is not connected back to the current random walk that has already been visited is more likely to belong to a different community. The algorithm selects a random node at the beginning and then follows a random path until a node is found not connected to multiple previously visited nodes during the random walk; then, this node is identified as a potential community bridge. It then randomly selects two of its neighbor nodes, and if neither of them is connected to a previously visited node, then the community bridge is an effective bridge.

Meghanathan et al. [4] summarized the research results of bridge nodes in recent years in detail, and designed a neighbor-based bridge node centrality triplet NBNC to more comprehensively evaluate the bridging of nodes, which has the following form:

$$(NG_i^{\#comp}, NG_i^{ACR}, |NG_i|) \tag{1}$$

where $NG_i^{\#comp}$ is the number of components in the neighborhood graph NG_i of node i, $|NG_i|$ is the number of nodes in NG_i, and NG_i^{ACR} is the ratio of the algebraic connectivity of NG_i and $|NG_i|$. In this paper, the neighborhood graph of a node i is defined as a graph composed of its neighbor nodes and all edges connecting the neighbor nodes, and a component is defined as a subgraph composed of nodes and edges in the graph and is not connected to the outside world. Obviously, by describing the state of the neighborhood graph after removing nodes, $NG_i^{\#comp}$ intuitively defines the bridging role that nodes play in their neighborhoods. NG_i^{ACR} improves algebraic connectivity [9] and describes the bridging of nodes from another aspect. When using NBNC tuples to determine the bridging rank of nodes, the priority of each element is decreasing; that is, when the previous element cannot determine the bridging rank of two nodes, the next-level element index is used.

Although the community-unaware approach eliminates the limitation of the community, it can also be directly applied to the network without community structure, but due to the limitation of the available information, it can only mine the properties related to the bridging effect from the local or global topology information of the nodes. Additionally, the goal of this type of algorithm is to find influential nodes from the perspective of bridging, and we think it is inappropriate to define bridge nodes in the community-unaware way.

2.1.2. Community-Aware Approach

The community-aware approach focuses on detecting nodes that play a bridging role in the process of information exchange between communities, so it must be executed under the premise of divided communities. Due to the different partitioning algorithms, the community structure is further divided into overlapping communities and non-overlapping communities, so there are correspondingly two different types of detection algorithms.

Approach based on overlapping community

Overlapping communities means that there are some shared nodes between two different communities on the same network. This kind of situation is common in the real world (for example, a social network user participates in multiple groups at the same time), and some existing community partitioning algorithms are also able to reveal the existence of overlapping communities. Due to the special topological location of such shared nodes, some researchers started to define the bridging of nodes by using overlapping nodes in overlapping communities [1,2,10].

Nepusz et al. [10] proposed an extended overlapping community detection method and, based on this, they proposed a way to define node bridging. However, in some special cases, this indicator will identify nodes outside the community as bridge nodes. Therefore, the author proposes a method of correcting bridgeness with indicators such as degree centrality, which is called degree-corrected bridgeness.

For networks that have divided overlapping communities, Taghavian et al. [1] proposed a random walk-based sorting algorithm for bridge nodes to enforce network immunity strategies. First, extract overlapping nodes according to the community division results; then perform random walk RWOS from random nodes in the network and set the overlapping nodes visited in this process as immune targets until enough targets are collected.

On the basis of the above results, Kumar et al. [2] conducted another similar work and proposed the overlapping neighborhood-based immune strategy, overlap neighborhood. The basic idea of this strategy is that among the neighboring nodes of overlapping nodes, there is a high probability of nodes with high degree, and immunizing these nodes can effectively control the spread of epidemics. Based on this, the process of the algorithm is as follows. First, use the overlapping community division algorithm to find overlapping nodes; then, find the neighbor nodes of all overlapping node and arrange them in descending order of node degree; then, the immune priority sequence of the network can be obtained.

In general, the method of defining bridge nodes based on overlapping community is simpler and more intuitive, but it relies too much on overlapping community, so that such algorithms can only focus on the overlapping parts of the communities in the network and cannot measure the bridging effect of the vast majority of nodes, which cannot give enough target objects in tasks such as network immunity.

Approach based on non-overlapping community

This type of method is based on non-overlapping community division, focusing on two attributes of nodes, that is, the connection of nodes within the community and the outside world.

Gupta et al. [11] first proposed the Commn centrality index, which considers the in-degree and out-degree of a node in the community. In-degree and out-degree respectively represent the number of edges directly connected to the node in the community and out of the community. For the nodes in the community C, the calculation process of Commn centrality is as follows:

$$CC(i) = (1 + \mu_C) * (\frac{k_i^{in}}{max(k_j^{in} \forall j \in C)} * R) + (1 - \mu_C) * (\frac{k_i^{out}}{max(k_j^{out} \forall j \in C)} * R)^2 \quad (2)$$

where R is an arbitrary positive integer and its function is to make the obtained in-degree and out-degree values within the same value range, i.e., $[0, R]$. The author proposes R to take the largest in-degree in the community C, i.e., $max(k_j^{in} \forall j \in C)$. $\mu_C$is the ratio of outgoing connections to the total number of connections in the community C and can be calculated as:

$$\mu_C = \frac{\sum_{i \in C} \frac{k_i^{out}}{k_i}}{size(C)} \quad (3)$$

Another metric, gateway local rank (GLR) [12], is based on the idea that nodes that have shortest path to core nodes of community can spread information more efficiently. First, a local critical node and a gateway node are defined In each community. The GLR value of node v is calculated as follows:

$$GLR(v) = [\alpha_1 \sum_{u \in \Gamma_k} d(v,u) + \alpha_2 \sum_{p \in \Gamma_G} d(v,u)]^{-1} \tag{4}$$

where Γ_k is the set of local cores, Γ_G is the set of gateway nodes, and $d(v,u)$ is the shortest path between node v and u. Parameters α_1 and α_2 are set to weight the different parts.

In terms of centrality based on community structure, there are some other related works such as community hub-bridge centrality [13], modular centrality [14], etc. Another feasible idea is the method proposed by Magelinski et al. [15], which is based on the modularity of the network about the community, and measures the importance of the node in maintaining the community structure by calculating the change of the corresponding modularity after removing the node.

The algorithms based on non-overlapping community structure focus on the connection between communities, and most of them have the same idea, that is, to comprehensively consider the connection status of nodes inside and outside the community and then make an evaluation. Therefore, some related indicators, such as in-degree, out-degree, community size, etc., are used in the intermediate process of calculating such indicators. As a cost, such methods ignore the topology features of nodes in the global network, and rely on computationally expensive community detection algorithms.

2.2. Graphical Representation Learning

In recent years, due to its powerful performance and wide application range, the research on graph representation learning [16] has received more and more attention, and it has been widely used in social network analysis such as community detection, node classification, link prediction, and behavior analysis. There is significant related work in this field, and researchers divide these into dimensionality reduction-based, random walk-based, graph decomposition-based, and neural network-based methods according to different graph embedding methods. The methods based on dimensionality reduction include PCA [17], LDA [18], MDS [19], etc. Representative works based on random walk include DeepWalk [20] and node2vec [21]. The method based on graph decomposition realizes graph embedding by decomposing the adjacency matrix of the graph; representative works include graph Laplacian eigenmaps [22], GraRep [23], and so on. Recently, inspired by RNNs and CNNs, researchers have begun to generalize these to graphs, resulting in a new class of neural network-based methods. This class of methods integrates semi-supervised information into graph representation learning and has strong performance; representative works include GCN [24], GraphSAGE [25], GAT [26], etc.

Before graph representation learning, there were also many studies applying traditional machine learning methods, such as logistic regression, SVM, etc., to social network analysis. Due to the high degree of fit between the graph data structure of social networks and graph neural networks, some researchers have begun to try to apply graph representation learning to social network influence recognition, such as with Deepinf [27], RCNN [28], and InfGCN [29]. These research works realize the embedding process of the network by extracting the relevant attributes of the node network topology structure; further, graph convolution is used to realize the aggregation of node features and generate a low-dimensional representation of the node. The trained and tuned graph neural network model can finally predict the influential nodes in the network. Experiments show that although transforming the network influence prediction task into a classification problem [28] produces more severe application scenarios, this method is better than some other methods in terms of efficiency and accuracy.

3. Main Framework

In this section, we formally propose the inter-community bridge node detection framework BND to detect bridge nodes between complex network communities. Its basic structure and process are shown in Figure 1. For a target social network that does not define bridge nodes, the framework first uses the Louvain algorithm and three bridge node detection algorithms to define bridge node labels (see Section 3.1 for details); then, it performs feature extraction of nodes (see Section 3.2 for details) to generate the corresponding attribute graph. The neural network model takes the attribute graph as input and outputs the result after propagation through the multi-layer neural network (see Section 3.3 for details); finally, the model output is compared with the ground truth value to calculate the loss.

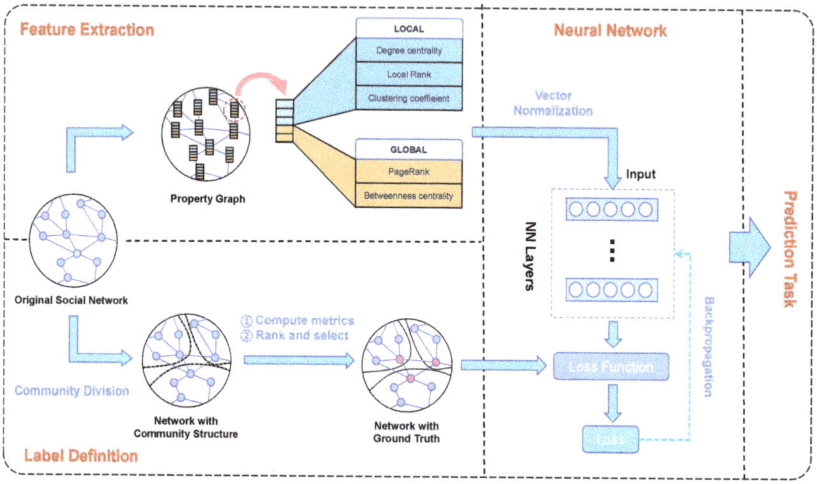

Figure 1. The structure of bridge nodes detection framework.

3.1. Label Definition

In a framework based on graph deep learning, high-quality label data is an important condition to ensure the effect of the model. For network datasets with real labels of bridge nodes, our framework can directly receive these as input; for complex networks without labels, we need to define the bridge nodes in the network ourselves.

In this paper, we combine three bridge-related algorithms—Commn, GLR, and NBNC—to define the bridging role of nodes in complex networks. The three algorithms are proven to be precise and efficient. Commn and GLR are based on community structures and can be combined to detect bridge nodes efficiently. As a community-unaware approach, NBNC is used to correct the contingency of the above two methods.

First, we use the classic Louvain algorithm to divide the community structure of the network. Then, the Commn value of all nodes can be calculated based on community information, as can GLR value and NBNC tuple. Finally, we can obtain three node bridging ranks according to the result, and nodes that obtain high rank in all of the three ranked lists are defined as bridge nodes.

3.2. Feature Extraction

We extract node features from two dimensions, namely local features and global features. The purpose is to ensure that the extracted features can more comprehensively describe the topology information of each dimension of the node. At the same time, considering the need to avoid over-reliance on feature engineering as much as possible, we only selected five types of node topology-related indicators as node features, and most of these indicators are simple and easy to calculate. Among them, local features include degree centrality, LocalRank [30], and clustering coefficient, and global features include PageRank and betweenness centrality. They are detailed in Table 1.

Table 1. Descriptions of the selected features.

Type	Feature	Description
Local	Degree	Measure the number of the neighbors of a node.
	LocalRank	Aggregate the information contained in the fourth-order neighbors of each node.
	Clustering coefficient	Describe the degree of interconnection between the neighbors of a node.
Global	PageRank	Measure the importance of a particular webpage relative to other webpages.
	Betweenness	Measure the degree of interaction between the node and other nodes based on the shortest path.

In order to improve the generalization ability and convergence speed of the model, we normalize the selected features. For each feature X:

$$X^{norm} = \frac{X}{max(X)} \tag{5}$$

3.3. Graph Neural Network

In order to detect bridge nodes in the network more accurately, the model we build must have the ability to extract the deep topological information from the complex network. Therefore, based on the principle of graph representation learning, we propose a deep neural network model BND-GCN (bridge node detection-GCN) for detecting bridge nodes.

As shown in Figure 2, we first superimpose two GCN layers as the main part of the model; with the graph structures and feature vectors they can learn representation vectors of nodes. Each GCN layer has propagation rules defined as follows:

$$H^{i+1} = \sigma(AH^iW^i + b^i) \tag{6}$$

where H^i is the representations of nodes at the ith GCN layer and W^i and b^i are its trainable parameters. A is the symmetric normalized Laplacian of the network and σ denotes the nonlinear activation function.

After propagation through the multi-layer graph neural network, we can obtain a low-dimensional node representation that aggregates the topological information of the complex network. At the end of the neural network, we add three fully connected layers to process the node representation and give the final output as a classifier. The activation function between fully connected layers is LeakyReLu, and the activation function of the output layer is Sigmoid. Finally, we choose the cross-entropy function as the loss function of the model. According to the classifier output and the known ground-truth labels of the nodes, the loss function can calculate the loss of the model and then adjust the relevant parameters of the neural network layer through backpropagation. In addition, we adopt the dropedge [31] strategy on GCN layers and apply the dropout technique on fully connected layers to prevent the model from overfitting. The skip connection technique is also used in our model.

On the other hand, we also try to replace GCN layers in the model with GraphSAGE and GAT layers. Accordingly, we modified some model parameters to adapt to different network layer structures. Experiments show that the variants of BND-GCN, which we named BND-GraphSAGE and BND-GAT, also have good results on the bridge node detection task.

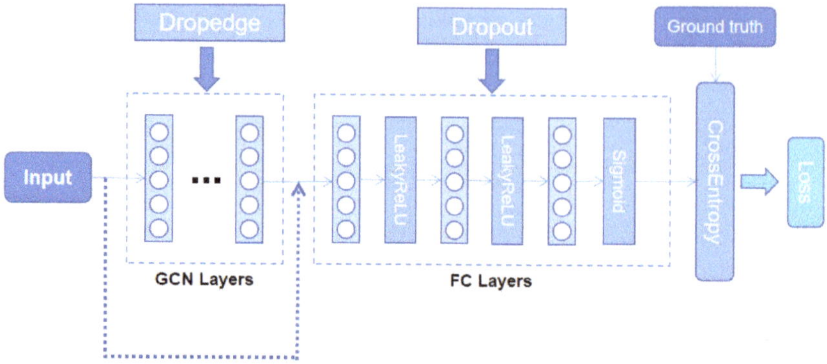

Figure 2. BND-GCN model architecture.

4. Experiment

In this section, we will elaborate on our experimental framework process and the reasons for designing the experiments in this way. In addition, we also give a detailed description of the parameter settings in the experiment in this section.

4.1. Dataset Construction

We selected six real networks as our dataset sources: (1) collaboration network of Arxiv General Relativity category CA-GrQc [32], (2) collaboration network of Arxiv High Energy Physics Theory category CA-HepTh [32], and (3) a series of snapshots of the peer-to-peer file sharing network of Gnutella, among which three networks, p2p-Gnutella04, p2p-Gnutella08, and p2p-Gnutella25, were selected [32]. Related information about these networks is shown in Table 2.

Table 2. Information of the five selected networks

Network	Nodes	Edges
CA-GrQc	5242	14,496
CA-HepTh	9877	25,998
p2p-Gnutella04	10,876	39,994
p2p-Gnutella08	6301	20,777
p2p-Gnutella25	22,687	54,705

For the convenience of experiments, we ignored other properties of these networks, treated them as undirected and unweighted networks, and removed self-loops in the network. After obtaining the processed network, we divided each network into a community structure according to the Louvain algorithm, and obtained defined labels as described above. It should be noted that we may encounter two special types of nodes when dealing with the network: isolated nodes with degree 0 and small community nodes (the number of community nodes is less than 3). We directly defined these as non-bridging nodes, because such nodes have little bridging effect.

After ranking nodes with three bridging algorithms, we counted the top 30% of ranked nodes for each rank and took their intersection as the bridge nodes of the network. In contrast, other nodes in the network are non-bridge nodes. We used labels "1" and "0" to identify bridge and non-bridge nodes in the network, respectively. In the machine learning-based classification task, when the proportion of labels of a certain category is too large, it will easily lead to a long tail problem, resulting in poor model performance. Therefore, in order to avoid this problem, we randomly selected some non-bridge nodes and all of the bridge nodes to form the dataset, so that the ratio of the number of bridge nodes and non-bridge nodes was 1:2.

During experiments, we randomly divided the training set and testing set according to the ratio of 7:3 and, at the same time, ensured that the ratio of positive and negative labels in training and testing remained the original 1:2. We built the model to train and test on five network datasets and the obtained test results can measure the effect of the model on bridge node detection. In the experiments, we selected Accuracy, Precision, Recall, and F1 score as the evaluation metrics for our proposed deep learning model.

4.2. Parameter Settings

We used two hyperparameters when building the dataset, Bridge-Percentage—the proportion of nodes we select from the three ranks, and Label-Ratio—the ratio of negative labels to positive labels. In the experiment, we set Bridge-Percentage to 30% and Label-Ratio to 2 by default. Additionally, we conducted experiments to explore the settings of these two hyperparameters; the results are presented in detail in Section 5.

Our BND-GCN model consists of two GCN layers and three fully connected layers, which contain 16, 16, 16, 8, and 2 neurons, respectively. The learning rate in the model is uniformly set to 0.01 and the dropout probability is set to 0.2. The dropedge rate is the percentage of the dropped edges to the total during each round of the training, which we set to 0.1. When it came to BND-GraphSAGE and BND-GAT, we maintained most model parameters including the number of layers and neurons. The number of hidden layer neurons of GAT is set to 8. We use the GCN aggregator [25] in the BND-GraphSAGE model with the following:

$$\mathbf{h}_v^k \leftarrow \sigma(\mathbf{W} \cdot MEAN(\{\mathbf{h}_v^{k-1}\} \cup \{\mathbf{h}_u^{k-1}, \forall u \in N(v)\})) \quad (7)$$

where $\mathbf{h}_v^{k-1}$ is the node's previous layer representation and $\mathbf{h}_{N(v)}^k$ is the aggregated neighborhood vector.

In order to ensure that the model can converge well on networks of different sizes, we set the number of model training rounds to 200.

4.3. Baseline Methods

In this part of the work, we have performed many comparative experiments to verify the advantages of our model. The following baselines are compared to our work.

Logistic regression (LR)

Logistic regression (LR) is a classic machine learning model. We trained an LR model with the features mentioned earlier to predict bridge nodes.

Support vector machine (SVM)

Support vector machine (SVM) is a generalized linear classifier based on supervised learning. We also used support vector machine (SVM) with a linear kernel as the classification model. The model uses the same features as the logistic regression model.

Multilayer perceptron (MLP)

An MLP consists of multiple fully connected layers. We built a five-layer multilayer perceptron as a representative to test its performance.

Inf-GCN

Inf-GCN [29] is a GCN-based method proposed to find influential nodes in complex networks, which performs very well in its area. The author claims that they can reach an F1 score of 90.7 in the best case. We used a Inf-GCN model on the bridge node detection task to compare it with our work.

Variants of BND-GCN

We implemented two variants of BND-GCN, denoted by BND-GAT and BND-GraphSAGE. Related parameters of these models have been introduced in the previous section.

5. Experimental Results and Analyses

5.1. Bridge Node Prediction Experiment

First, we conducted bridge node detection experiments on five real networks. In this part of the experiment, we used selected features to generate the feature vector of the node, and trained and tested the models on this basis. The experiments included our three proposed models—BND-GCN, BND-GraphSAGE, BND-GAT—and four baselines. Table 3 shows the test results of these models on various networks.

Table 3. Bridge node detection experimental results

Model	Score	CA-GrQc		CA-HepTh		p2p-Gnutella04		p2p-Gnutella08		p2p-Gnutella25	
LR	Accuracy Precision	0.6667	0.5000	0.6860	0.7857	0.7731	**0.9043**	0.5971	**1.0000**	0.8007	0.8591
	Recall F1 score	0.0141	0.0274	0.0866	0.1560	0.3574	0.5123	0.0276	0.0537	0.5903	0.6998
SVM	Accuracy Precision	0.6230	0.3333	0.6887	0.6667	0.8293	0.8777	0.6728	0.8000	0.8832	0.8013
	Recall F1 score	0.0141	0.0270	0.1417	0.2338	0.5670	0.6889	0.0276	0.0533	0.8641	0.8315
MLP	Accuracy Precision	0.7789	**0.7647**	0.5955	0.4661	0.7292	0.5925	0.7604	0.8361	0.7676	0.6105
	Recall F1 score	0.5493	0.6393	0.9213	0.6190	0.8694	0.7047	0.3517	0.4951	0.9777	0.7516
Inf-GCN	Accuracy Precision	0.7418	0.5889	**0.8285**	**0.6802**	0.8832	0.7723	0.7742	0.7238	0.8683	0.7199
	Recall F1 score	0.7465	0.5684	0.9213	**0.7826**	0.9210	0.8401	0.5241	0.6080	0.9907	0.8339
BND-GraphSAGE	Accuracy Precision	0.7559	0.6044	0.8047	0.6448	0.8305	0.7527	0.7604	0.5945	0.7789	0.6189
	Recall F1 score	0.7746	0.6790	0.9291	0.7613	0.7320	0.7422	0.8897	0.7127	0.8771	0.7257
BND-GAT	Accuracy Precision	0.7277	0.6585	0.7784	0.6443	0.8706	0.7947	0.8041	0.8409	**0.9050**	**0.7883**
	Recall F1 score	0.3803	0.4821	0.7559	0.6957	0.8247	0.8094	0.5103	0.6352	0.9777	**0.8728**
BND-GCN	Accuracy Precision	**0.7934**	0.6364	0.7995	0.6269	**0.8877**	0.7629	**0.8779**	0.7347	0.8925	0.7563
	Recall F1 score	**0.8873**	**0.7412**	**0.9921**	0.7683	**0.9622**	**0.8511**	**0.9931**	**0.8446**	**1.0000**	0.8613

For each metric, we bolded the highest score obtained in each network. Experimental results on five networks of different sizes show that our proposed BND-GCN model exhibits outstanding advantages compared with baseline models on the task of identifying bridge nodes between communities with known communities. In addition, the two variants of BND-GCN, BND-GraphSAGE and BND-GAT, also perform well. Compared to traditional methods, GNN-based methods show that the attribute graph constructed by selecting node features can describe the network structure well, so as to achieve a good embedding effect. On the other hand, the BND-GCN model is more robust compared to the baseline method Inf-GCN according to the results. This may be due to improvements we made to the model.

5.2. Hyperparameter Analysis

5.2.1. Label-Ratio

The ratio of negative to positive labels in the dataset, Label-Ratio, will have a certain impact on the effect of the model, because a change of Label-Ratio may lead to too little label data or long tail problems. In order to verify the influence of Label-Ratio on the effect of our proposed model and to help us set the appropriate Label-Ratio parameter value in the experiment, we designed multiple sets of experiments based on the BND-GCN model; the experimental results are shown in Figure 3.

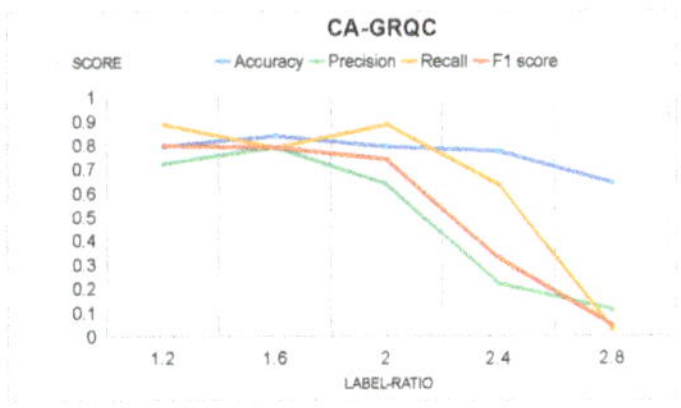

(a) CA-GrQc

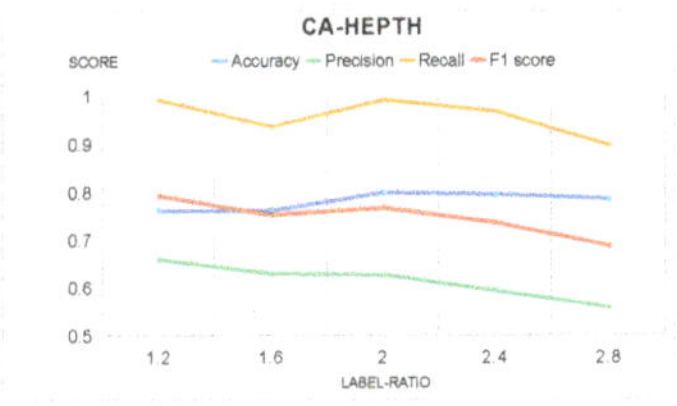

(b) CA-HepTh

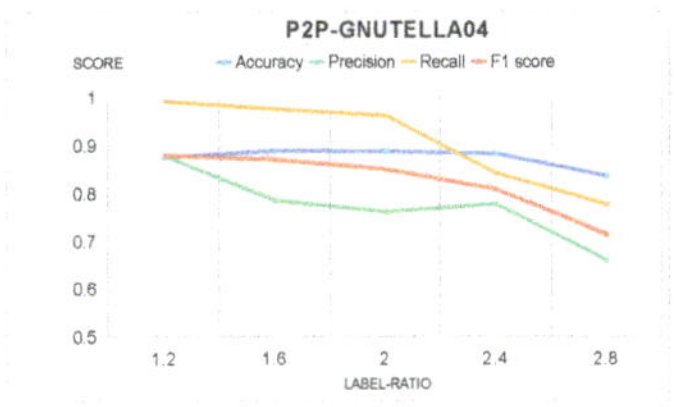

(c) p2p-Gnutella04

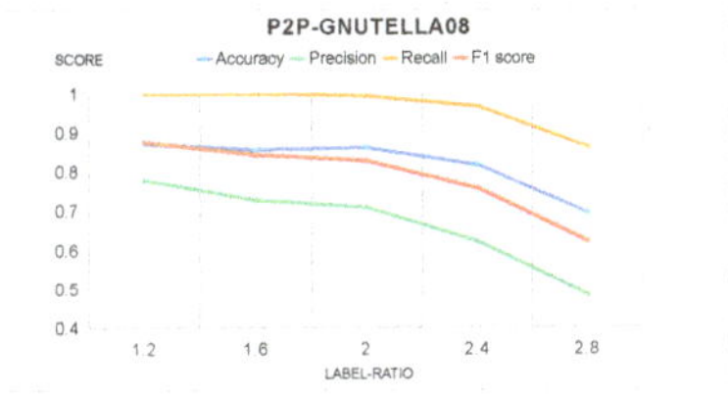

(d) p2p-Gnutella08

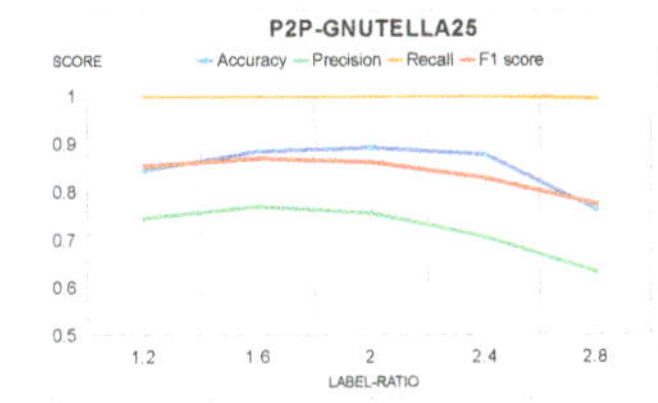

(e) p2p-Gnutella25

Figure 3. Effects of different Label-Ratio on five social networks.

Through the experimental results, we can see that with the increase of Label-Ratio, the overall effect of the model generally tends to decline. This is due to the increase in the number of negative labels, and the long-tail problem becomes more and more significant, resulting in a poorer model classification effect. A relatively balanced Label-Ratio can improve model performance; this may have good guiding significance for us when choosing data to build a training set.

5.2.2. Bridge-Percentage

According to the work in [29], the proportion of bridge nodes in the network also affects the model test results. Therefore, we set different Bridge-Percentage for repeated experiments to change the proportion indirectly. For example, if Bridge-Percentage is set to 10%, the top 10% of nodes by Commn are marked, and we do the same for nodes ranked with NBNC and GLR. Then, the intersections of the three parts are defined as bridge nodes. It is clear that Bridge-Percentage directly determines the proportion of bridge nodes. Therefore, we set different value of Bridge-Percentage to verify the effect of the parameter on the effect of the BND-GCN model. The results are shown in Figure 4.

The experimental results in this part show that as the Bridge-Percentage increases, the number of nodes defined as bridge nodes in the network increases, and the overall effect of the model decreases. Based on the experimental results, we speculate that when the Bridge-Percentage value is smaller, the bridging value of the selected node is more concentrated, and the characteristics of bridge nodes that are different from other nodes are more prominent and easier to capture. This may mean that our model will be more applicable in networks where there are few bridge nodes. When Bridge-Percentage is too small, it will lead to the problem of insufficient labels and data waste. This is another aspect that we should take into account when setting parameters.

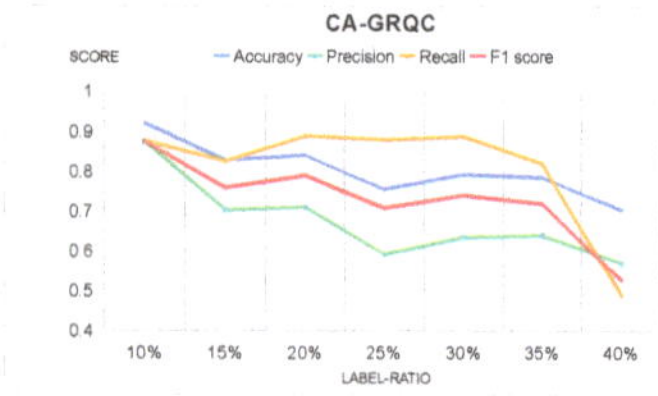

(a) CA-GrQc

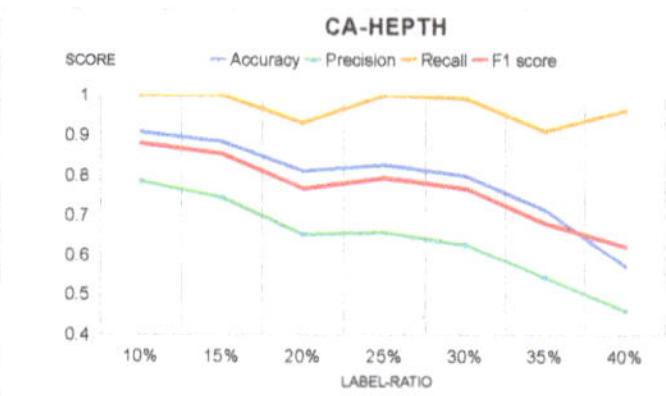

(b) CA-HepTh

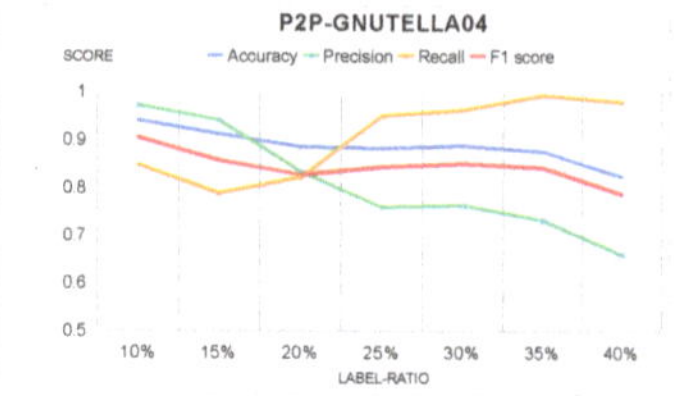

(c) p2p-Gnutella04

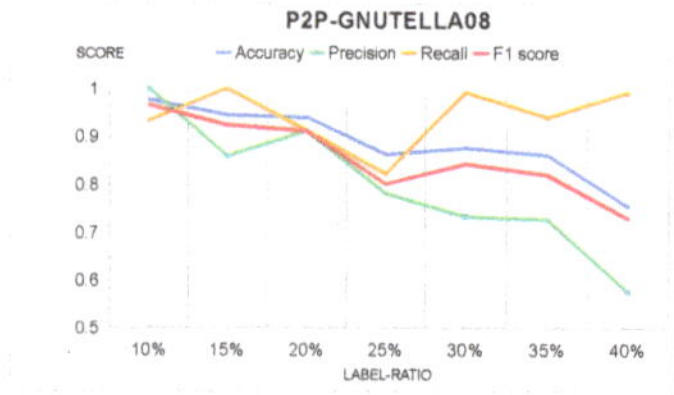

(d) p2p-Gnutella08

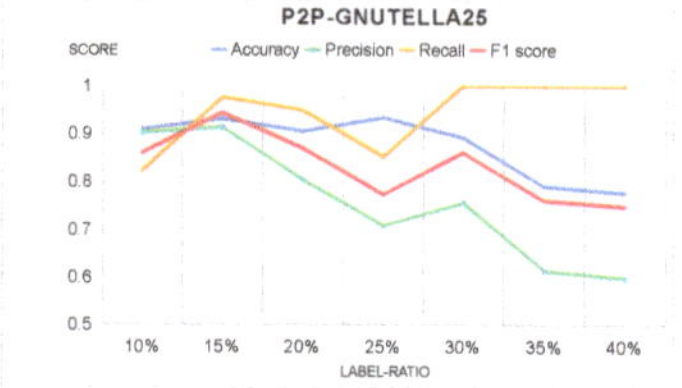

(e) p2p-Gnutella25

Figure 4. Effects of different Bridge-Percentage on five social networks.

5.3. Pre-Training Experiments

The idea of transfer learning [33] has been widely used in various fields. Using transfer learning, the transfer and application of knowledge between different fields can be realized, helping people solve problems such as insufficient label data. In our scenario, due to the inconsistency of the node embedding vector space between different networks, the node features of another network cannot be directly used in the training process of this network. However, this barrier can be broken down by transfer learning; a similar graph neural network pre-training process has been applied in [29] with remarkable results. Therefore, we conducted similar pre-training experiments to verify whether the bridge node features learned in the network can help the training process in other networks. We trained the BND-GCN model on network CA-HepTh and fine-tuned the model on the other four networks. With the fine-tuned model, we can successfully implement bridge node detection tasks on different networks. Figure 5 shows the changes in Loss, Accuracy, and F1 score of the model during training when using the pre-trained model and the directly trained model.

By comparing the loss curve of the model before and after pre-training, it is obvious that our pre-training process accelerates the convergence rate in the early stage of model training and plays a good role in initializing the model. Through transfer-learning technology, we can transfer the bridge node feature information learned on another network to this network for use. This is very meaningful where there are few label resources available.

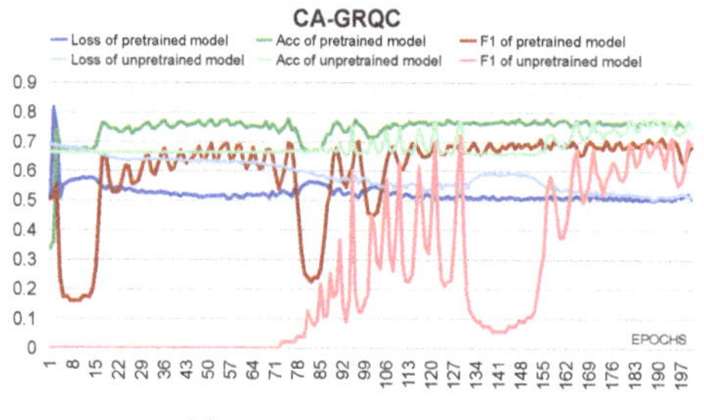

(a) performance in CA-GrQc

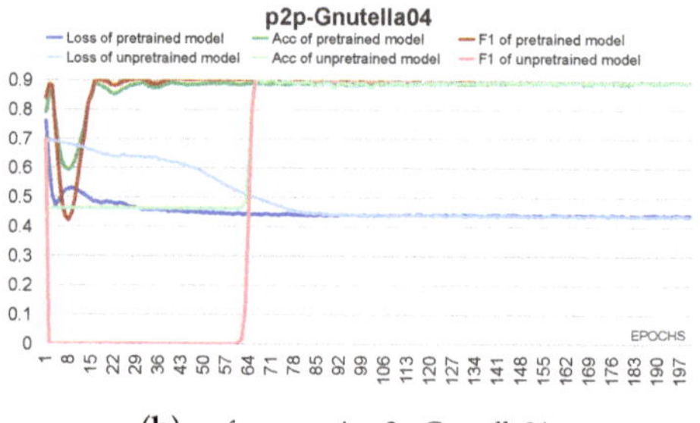

(b) performance in p2p-Gnutella04

Figure 5. *Cont.*

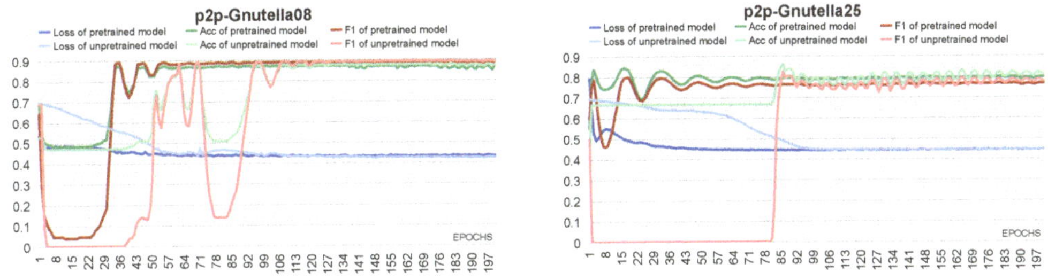

(**c**) performance in p2p-Gnutella08

(**d**) performance in p2p-Gnutella25

Figure 5. Comparison of the performance between the pretrained model and unpretrained model.

6. Conclusions

In this paper, we first propose a new inter-community bridge node detection framework, BND, which transforms the bridge node detection task into a classification problem and uses deep learning techniques to detect bridge nodes in the network, thus overcoming the traditional method's dependence on the community division algorithm. At the same time, BND makes judgments on bridge nodes after synthesizing network topology information and node characteristics, so it has high accuracy. On this basis, we propose a graph neural network model BND-GCN and two variants, BND-GAT and BND-GraphSAGE, for bridge node detection. Our extensive experiments on five real social networks show that they outperform other deep learning models and some traditional machine learning models in bridge node detection. As far as we know, this is also the first application of graph neural network technology in the field of bridge node detection, and there is no similar application work in this field.

In future work, we will continue to study how to mine community structure information in the graph embedding or node feature stage, improve the accuracy of inter-community bridge node detection, and further improve our framework.

Author Contributions: Conceptualization, P.J.; methodology, H.L.; software, H.L.; validation, Z.H.; formal analysis, H.L.; investigation, Y.L.; resources, P.J.; data curation, Z.H.; writing—original draft preparation, H.L.; writing—review and editing, H.L.; visualization, Z.H.; supervision, A.Z.; project administration, A.Z.; funding acquisition, P.J. All authors have read and agreed to the published version of the manuscript.

Funding: This research was funded by the Key Research and Development Program of Sichuan Province under Grant 2021YFG0156.

Institutional Review Board Statement: Not appliable.

Informed Consent Statement: Not appliable.

Data Availability Statement: Not applicable.

Acknowledgments: The authors would like to thank the Key Research and Development Program of Sichuan Province for its funding and support in this research.

Conflicts of Interest: The authors declare no conflict of interest.

References

1. Taghavian, F.; Salehi, M.; Teimouri, M. A local immunization strategy for networks with overlapping community structure. *Phys. A Stat. Mech. Appl.* **2017**, *467*, 148–156. [CrossRef]
2. Kumar, M.; Singh, A.; Cherifi, H. An efficient immunization strategy using overlapping nodes and its neighborhoods. In Proceedings of the Companion Proceedings of the Web Conference 2018, Lyon, France, 23–27 April 2018; pp. 1269–1275.
3. Hwang, W.C.; Zhang, A.; Ramanathan, M. Identification of information flow-modulating drug targets: A novel bridging paradigm for drug discovery. *Clin. Pharmacol. Ther.* **2008**, *84*, 563–572. [CrossRef] [PubMed]

4. Meghanathan, N. Neighborhood-based bridge node centrality tuple for complex network analysis. *Appl. Netw. Sci.* **2021**, *6*, 47. [CrossRef]
5. Liu, W.; Pellegrini, M.; Wu, A. Identification of bridging centrality in complex networks. *IEEE Access* **2019**, *7*, 93123–93130. [CrossRef]
6. Jiang, L.; Jing, Y.; Hu, S.; Ge, B.; Xiao, W. Identifying node importance in a complex network based on node bridging feature. *Appl. Sci.* **2018**, *8*, 1914. [CrossRef]
7. Salathé, M.; Jones, J.H. Dynamics and control of diseases in networks with community structure. *PLoS Comput. Biol.* **2010**, *6*, e1000736. [CrossRef]
8. Morone, F.; Makse, H.A. Influence maximization in complex networks through optimal percolation. *Nature* **2015**, *524*, 65–68. [CrossRef]
9. Fiedler, M. Algebraic connectivity of graphs. *Czechoslov. Math. J.* **1973**, *23*, 298–305. [CrossRef]
10. Nepusz, T.; Petróczi, A.; Négyessy, L.; Bazsó, F. Fuzzy communities and the concept of bridgeness in complex networks. *Phys. Rev. E* **2008**, *77*, 016107. [CrossRef]
11. Gupta, N.; Singh, A.; Cherifi, H. Centrality measures for networks with community structure. *Phys. A Stat. Mech. Appl.* **2016**, *452*, 46–59. [CrossRef]
12. Salavati, C.; Abdollahpouri, A.; Manbari, Z. Ranking nodes in complex networks based on local structure and improving closeness centrality. *Neurocomputing* **2019**, *336*, 36–45. [CrossRef]
13. Ghalmane, Z.; Hassouni, M.E.; Cherifi, H. Immunization of networks with non-overlapping community structure. *Soc. Netw. Anal. Min.* **2019**, *9*, 45. [CrossRef]
14. Ghalmane, Z.; El Hassouni, M.; Cherifi, C.; Cherifi, H. Centrality in modular networks. *EPJ Data Sci.* **2019**, *8*, 15. [CrossRef]
15. Magelinski, T.; Bartulovic, M.; Carley, K.M. Measuring node contribution to community structure with modularity vitality. *IEEE Trans. Netw. Sci. Eng.* **2021**, *8*, 707–723. [CrossRef]
16. Hamilton, W.L. Graph representation learning. *Synth. Lect. Artifical Intell. Mach. Learn.* **2020**, *14*, 1–159.
17. Jolliffe, I.T.; Cadima, J. Principal component analysis: A review and recent developments. *Philos. Trans. R. Soc. A Math. Phys. Eng. Sci.* **2016**, *374*, 20150202. [CrossRef]
18. Ye, J.; Janardan, R.; Li, Q. Two-dimensional linear discriminant analysis. *Adv. Neural Inf. Process. Syst.* **2004**, *17*.
19. Robinson, S.L.; Bennett, R.J. A typology of deviant workplace behaviors: A multidimensional scaling study. *Acad. Manag. J.* **1995**, *38*, 555–572. [CrossRef]
20. Perozzi, B.; Al-Rfou, R.; Skiena, S. Deepwalk: Online learning of social representations. In Proceedings of the 20th ACM SIGKDD International Conference on Knowledge Discovery and Data Mining, New York, NY, USA, 24–27 August 2014; pp. 701–710.
21. Grover, A.; Leskovec, J. node2vec: Scalable feature learning for networks. In Proceedings of the 22nd ACM SIGKDD International Conference on Knowledge Discovery and Data Mining, San Francisco, CA, USA, 13–17 August 2016; pp. 855–864.
22. Ahmed, A.; Shervashidze, N.; Narayanamurthy, S.; Josifovski, V.; Smola, A.J. Distributed large-scale natural graph factorization. In Proceedings of the 22nd International Conference on World Wide Web, Rio de Janeiro, Brazil, 13–17 May 2013; pp. 37–48.
23. Cao, S.; Lu, W.; Xu, Q. Grarep: Learning graph representations with global structural information. In Proceedings of the 24th ACM International on Conference on Information and Knowledge Management, Melbourne, Australia, 19–23 October 2015; pp. 891–900.
24. Kipf, T.N.; Welling, M. Semi-supervised classification with graph convolutional networks. *arXiv* **2016**, arXiv:1609.02907.
25. Hamilton, W.; Ying, Z.; Leskovec, J. Inductive representation learning on large graphs. *Adv. Neural Inf. Process. Syst.* **2017**, *30*.
26. Veličković, P.; Cucurull, G.; Casanova, A.; Romero, A.; Lio, P.; Bengio, Y. Graph attention networks. *arXiv* **2017**, arXiv:1710.10903.
27. Qiu, J.; Tang, J.; Ma, H.; Dong, Y.; Wang, K.; Tang, J. Deepinf: Social influence prediction with deep learning. In Proceedings of the 24th ACM SIGKDD International Conference on Knowledge Discovery & Data Mining, London, UK, 19–23 August 2018; pp. 2110–2119.
28. Yu, E.Y.; Wang, Y.P.; Fu, Y.; Chen, D.B.; Xie, M. Identifying critical nodes in complex networks via graph convolutional networks. *Knowl.-Based Syst.* **2020**, *198*, 105893. [CrossRef]
29. Zhao, G.; Jia, P.; Zhou, A.; Zhang, B. InfGCN: Identifying influential nodes in complex networks with graph convolutional networks. *Neurocomputing* **2020**, *414*, 18–26. [CrossRef]
30. Chen, D.; Lü, L.; Shang, M.S.; Zhang, Y.C.; Zhou, T. Identifying influential nodes in complex networks. *Phys. A Stat. Mech. Appl.* **2012**, *391*, 1777–1787. [CrossRef]
31. Rong, Y.; Huang, W.; Xu, T.; Huang, J. Dropedge: Towards deep graph convolutional networks on node classification. *arXiv* **2019**, arXiv:1907.10903.
32. Leskovec, J.; Kleinberg, J.; Faloutsos, C. Graph evolution: Densification and shrinking diameters. *ACM Trans. Knowl. Discov. Data (TKDD)* **2007**, *1*, 2-es. [CrossRef]
33. Weiss, K.; Khoshgoftaar, T.M.; Wang, D. A survey of transfer learning. *J. Big Data* **2016**, 3, 9. [CrossRef]

applied sciences

MDPI

Article

BChainGuard: A New Framework for Cyberthreats Detection in Blockchain Using Machine Learning

Suliman Aladhadh *, Huda Alwabli, Tarek Moulahi * and Muneerah Al Asqah

Department of Information Technology, College of Computer, Qassim University, Buraidah 52571, Saudi Arabia
* Correspondence: s.aladhadh@qu.edu.sa (S.A.); t.moulahi@qu.edu.sa (T.M.)

Abstract: Recently, blockchain technology has appeared as a powerful decentralized tool for data integrity protection. The use of smart contracts in blockchain helped to provide a secure environment for developing peer-to-peer applications. Blockchain has been used by the research community as a tool for protection against attacks. The blockchain itself can be the objective of many cyberthreats. In the literature, there are few research works aimed to protect the blockchain against cyberthreats adopting, in most cases, statistical schemes based on smart contracts and causing deployment and runtime overheads. Although, the power of machine learning tools there is insufficient use of these techniques to protect blockchain against attacks. For that reason, we aim, in this paper, to propose a new framework called BChainGuard for cyberthreat detection in blockchain. Our framework's main goal is to distinguish between normal and abnormal behavior of the traffic linked to the blockchain network. In BChainGuard, the execution of the classification technique will be local. Next, we embed only the decision function as a smart contract. The experimental result shows encouraging results with an accuracy of detection of around 95% using SVM and 98.02% using MLP with a low runtime and overhead in terms of consumed gas.

Keywords: blockchain; Ethereum; cyberthreats; machine learning; MLP; SVM; smart contract

Citation: Aladhadh, S.; Alwabli, H.; Moulahi, T.; Al Asqah, M. BChainGuard: A New Framework for Cyberthreats Detection in Blockchain Using Machine Learning. *Appl. Sci.* **2022**, *12*, 12026. https://doi.org/10.3390/app122312026

Academic Editor: Xiaoyang Liu

Received: 13 October 2022
Accepted: 18 November 2022
Published: 24 November 2022

1. Introduction

The concept of blockchain was firstly proposed by Satoshi Nakamoto in [1] as a decentralized system for money transfer.

Blockchain is a system that consists of a network of multiple blocks where each block contains several transactions, and each block connects to its previous block through a hash-based procedure. This hashing procedure and the utilization of other cryptography methods have given blockchain technology its property of protecting block content integrity [1]. Figure 1 shows the list of blockchain characteristics [2], which are:

- Block: contains a list of information such as the block number, nonce, time stamp, data, the hash of the previous block, and the hash of block itself.
- Ledger: a list of block forms a ledger.
- Distribution: is an important characteristic of blockchain, as blockchain architecture is based on P2P network, and each miner contains the whole blockchain.
- Transaction: the data in the block are a list of transaction
- Confirmation: is needed by at least 51% of miners to validate the list of transactions in the block.
- Proof of work: is the right value of nonce giving a block hash starting by a list of zeros.
- Result: add a new block to the blockchain.

1.1. Motivation and Problem Statement

Blockchain has attracted the research community, as well as the industry. Actually, blockchain is used in serval domains such as the Internet of Things [3,4] and also in healthcare [5,6], in addition to other fields such as finance [7].

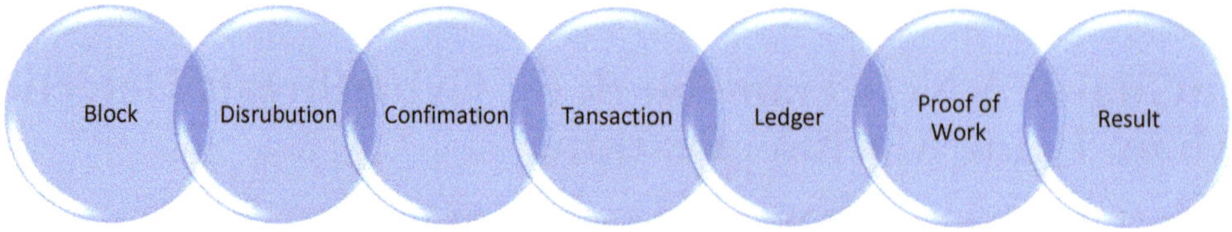

Figure 1. Blockchain characteristics.

Blockchain characteristics ensure content and operations safety, making it a suitable choice for system security, although its characteristics are the subject of many cyberthreats such as DDoS, eclipse, 51% attack, and so on. Therefore, developing a cyberthreats detection system on the blockchain is highly needed.

Although many efforts have been made by the research community to tackle blockchain threats, this domain needs more investigation. Most of the proposed framework in this context is using statistical methods or simple algorithms, but machine learning techniques are lowly used. Furthermore, there are many limitations in existing related works such as the overheads due to deployment and runtime. More details will be given later in this paper.

1.2. Objectives

The main objective of this paper is to develop a framework called BChainGuard for cyberthreats detection in Blockchain. BChainGuard will be based on smart contracts in order to assure safe execution and on machine learning tools in order to provide high-accuracy detection. This will help to distinguish between normal and abnormal behavior of the traffic linked to the blockchain network. This choice will decrease the deployment and the runtime overhead. In the proposed framework, the execution of the classification technique will be in a local machine. Next, we embed only the decision function as a smart contract. This will create a new layer in blockchain for cyberthreat detection.

The contributions of this research work can be summarized as follows:

- Build a novel framework to detect cyberthreats on blockchain by merging machine learning and deep learning to increase the accuracy of detection and the blockchain itself.
- Propose a new layer in the blockchain-based on smart contracts for cyberthreats detection in order to protect the blockchain.
- Measure the effectiveness of the proposed technique by comparing it with an existing solution [8] in terms of accuracy, deployment, and execution overhead.

1.3. Paper Organization

The rest of this paper is organized as follows: Section 2 outlines the literature review including a study of recently proposed techniques that deal with the issue of blockchain attacks and protection. Section 3 describes the BChainGuard contribution. Section 4 presents the experimental study. Finally, the conclusion is given in Section 5.

2. Literature Review

This section is divided into three subsections. First, a study on the most relevant attacks on the blockchain is given. Next, it goes through a literature review to outline the framework and system for intrusion detection in the blockchain. Finally, we discuss and compare these frameworks.

2.1. Review of Blockchain Cyberthreats

Many cyberthreats and their countermeasures have been studied by research communities. In what follows, we outline the most relevant works in this context.

In [9], the authors studied (Distributed Denial of Service) DDoS attacks. This attack can cause an increase in mining fees, and it can have an effect on the cryptocurrency systems' memory pools. The authors use Bitcoin mempools to study DDoS and to observe its effect. To tackle DDoS attacks, the authors propose fee-based and age-based designs. The proposed technique is validated through simulation studies with different kinds of attack conditions.

In [10], the authors discussed a DAO attack. They proposed a framework called VeriSolid to be used to formally verify smart contracts, which are using a transition system-based model. This will help developers to check, at a high level of abstraction, the contract behavior. The proposed framework helps in generating the Solidity code from the verified models. This allows the development of smart contracts based on correct-by-design concept.

Another type of BC attack was studied in [11]. The authors showed that the most widely used Ethereum implementation called Go Ethereum (Geth) is exposed to eclipse attacks. For that reason, they study the fundamental properties of Geth's node discovery, which can be the cause of false friends' attacks, and they propose countermeasures to avoid the eclipse attack.

In [12], the authors proposed a countermeasure for a malleability attack. This attack happens when the malicious tries to change some byte in the signature, while it may still be effective, so that the attacker can control bitcoin transactions. To tackle this, attack the authors suggest modifying the specification of bitcoin by calculating the transaction hash deprived of its input script.

The paper of [13] dealt with time hijacking attacks that happened due to the vulnerability of the time stamp process in bitcoin. The attacker alters the bitcoin time counter, as well as the node's time. This can lead to a perturbation of times. To avoid this attack, the authors propose to not accept time ranges or not utilize the system time of nodes.

Sybil attack was discussed in [14]. This kind of attack occurs when the malicious try to control the blockchain network by creating a great number of pseudonymous identities and manipulating blockchain redundancy and anonymity. The authors proposed to avoid this attack by proving identities using a trusted agency. This credible agency is responsible for certifying identities. Before being part of a Peer-to-Peer network, a third party will be used to authenticate it.

In [15], the authors outlined the phishing attack which happens when the attacker tries to steal user credentials through websites or using untrue emails or both of them simultaneously. To defend against this attack, the authors suggest adopting a strategy based on excluding the unreasonable behavior of the user in addition to detecting and filtering phishing resources. This will help to avoid the stolen of phishing infrastructure elements such as passwords.

Border Gateway Protocol (BGP) attacks can target blockchains [16]. This attack is linked to routing protocols. It happens when a malicious system creates and broadcast fake advertisements to its neighbors. This can result in the redirection of traffic to specified destinations. In this case, the attacker may control the node traffic destroying the consensus mechanism. To defend against BGP attacks, the authors propose a new Bitcoin relay network with great extensibility and safety denoted as SABRE, where blocks are relayed using a list of connections that can resist attacks on routing.

In [17], the authors discussed selfish mining attacks. This attack occurs when selfish miners use selfish techniques for getting non-deserved incomes. As an example, the malicious miner pool decides to not disseminate the block after finding the next block and to continue the hashing process in order to create a new valid chain and neglect the right one. To tackle this attack, the authors proposed to neglect blocks that are not achieved in time and give awards to blocks merging links into their previous competing blocks.

Another serious attack targeting blockchain is called the integer overflow attack [18]. This attack is a serious problem linked to Ethereum smart contracts, which are program codes. The integer in these codes has an upper and lower limit. The integer overflow attack can happen essentially in the value-type conversion, which can cause a big loss. To tackle this kind of attack, the authors propose the Osiris framework in order to control integer overflow in a smart contract by combining taint analysis and symbolic execution.

2.2. Intrusion Detection System in Blockchain

Although blockchain has been widely used as a tool for IDS in other environments, there is an extreme lack of IDS solutions on the blockchain.

In [19], the author proposed to use a GPU solution based on TRS (Target Rooted Subgraph) to detect anomalies in transactions by using a part of the data. Through an experimental study, the proposed techniques archive 195 times faster than the existing method. Furthermore, it outperforms the existing method in terms of accuracy and true positive rate, due to the use of subgraphs to detect local anomalies in the case of transactions on a small scale. This rate is near the existing method in the case of transactions on a large scale.

The authors In [20] proposed a method to detect anomalies in bitcoin transactions. They use a dataset including a list of bitcoin transactions with normal behavior. The authors created another list of transactions with abnormal behavior by including three types of attacks, which are: DDoS, 51% attack, and Double spending attack. For the detection, they used SVM (support vector machine) and K-means. Through experimental study, they show that both techniques give good accuracy.

Since smart contracts are vulnerable to attacks, in [21], the authors discussed tackling these vulnerabilities by proposing ContractGuard as the first intrusion detection system for smart contracts in Ethereum. After deploying the proposed system with real smart contracts in Ethereum, the authors showed the effectiveness of their solution in terms of protection with low execution overheads (only add 28.27%) and with light deployment (only add 36.14%). The vulnerabilities decreased by a rate of 83%.

In [22], the authors proposed SODA as a new framework for the online detection of many attacks. The authors showed that SODA outperforms existing solutions in terms of efficiency, compatibility, and capability. The aim of SODA is to let users rapidly develop a new application for cyberthreat detection. To show the effectiveness of SODA, the authors developed 8 applications, including new detection methods for the detection of attacks focusing on smart contract vulnerabilities. SODA is also embedded into the EVM-based blockchain. Through experiments, SODA shows effective attack detection with low overhead.

Another proposed IDS was denoted BAD: Blockchain Anomaly Detection [23]. BAD can detect only two types of attack: eclipse attack and zero-day attack. The main goal of BAD is to detect anomalies in transactions in addition to preventing their spreading. The prevention is based on the collection of malicious activities in addition to building a distributed database of threats. The authors make an analysis of BAD overhead, its implementation, and in order to show its effectiveness in the detection of eclipse attacks, as well as zero-day attacks.

Recently, an important technique called SolGuard was proposed in [24] in order to prevent all issues linked to smart contracts. The proposed work is based on using multi-agent robotic systems. The work is based on studying Ethereum smart contracts to show its vulnerabilities due to several programming problems—in particular, the use of low-level calls to malicious resources. The proposed technique is performed by implementing SolGuard, aiming to prevent three serious issues linked to low-level calls to malicious resources done by smart contracts written in solidity. Through an empirical study, based on efficiency and accuracy, the proposed technique is outperforming existing tools in the same context.

In [25], DefectChecker was proposed aiming to detect defects in smart contracts by using symbolic execution to analyze the bytecode of the smart contract. In fact, DefectChecker is using various rules in order to detect eight vulnerabilities in contracts. After running it on a previous work dataset, it accomplishes encouraging results. DefectChecker outperforms some existing tools for defect detection in smart contracts. The experiment shows that 15.89% of Ethereum smart contracts include at least one example of the eight vulnerabilities in the contract.

2.3. Discussion

In this subsection, we make a discussion of the previously outlined system and framework for intrusion and threats detection in the blockchain. Table 1 presents a general comparison of them.

Table 1. Comparison of the existing frameworks for blockchain intrusion detection.

Tools	Technique of Detection	Cyberthreat Coverage	Limitations
TRS [17]	Using graph theory to detect anomalies in smart contract	Smart contract vulnerabilities	In the case of a transaction having a large scale this rate of anomalies detection is similar to the existing method
ADM [18]	Using one-class SVM and K-means for transaction anomalies detection	DDoS, 51% attack, and Double spending attack	The anomalies are executed by the authors themselves so they can be nonrealistic
ContractGuard [19]	A mechanism for intrusion detection	Attempts of malicious intrusion	Needs more experience with real smart contract vulnerabilities
SODA [20]	On-chain applications to detect anomalies	DAO attack, time hijacking, smart contract vulnerabilities	Used only for online detection of a specific attack list
BAD [21]	Blockchain Anomaly Detection solution	Eclipse attack, zero-day attack	Used only for online detection of 2 kinds of attacks
SolGuard [22]	A solution to detect anomalies in smart contract	Smart contract vulnerabilities and DoS attack	High overheads due to deployment and runtime
DefectChecker [23]	Analyzing byte code to detect anomalies in smart contracts	Detecting if a contract is controlled by an attacker	functions call issues cannot be detected

Most of the proposed techniques to deal with attacks on BC are using statistical methods or simple algorithms, but machine learning techniques are lowly used. There are many limitations in existing related work such as the overheads due to deployment and runtime. There is a need for more simulations to measure the effectiveness of some proposed techniques. Some attacks are difficult to be detected, such as 51% attack, because it is linked to the consensus protocols themselves. Some other solutions are designed only for the Ethereum platform by using smart contracts. Consequently, the development of new techniques to detect abnormal behavior in the blockchain is still needing investigation. For that reason, it is important to develop a framework for cyberthreat detection in Blockchain. This framework will be based on smart contracts (to guarantee safe execution) and on machine learning tools (to ensure high-accuracy detection). BChainGuard's main goal is to distinguish between normal and abnormal behavior of the traffic linked to the BC network. To significantly decrease the deployment and the runtime overhead, our framework will be a hybrid. The execution of the classification technique will be done locally. Next, we embed only the decision functions as smart contracts. This will create a new layer in blockchain for cyberthreat detection.

3. Contribution: BChainGuard

In this section, we describe the proposed framework in Figure 2 in addition to the list of used algorithms to detect cyberthreats.

3.1. BChainGuard Phases

Phase 1: Selection of the dataset and artificial intelligence tool

In this phase, the dataset of attack on the blockchain is selected which is located in [26]. In addition, we plan to use SVM (support vector machine) as a machine learning tool and MLP (multi-layer perceptron) as a deep learning tool.

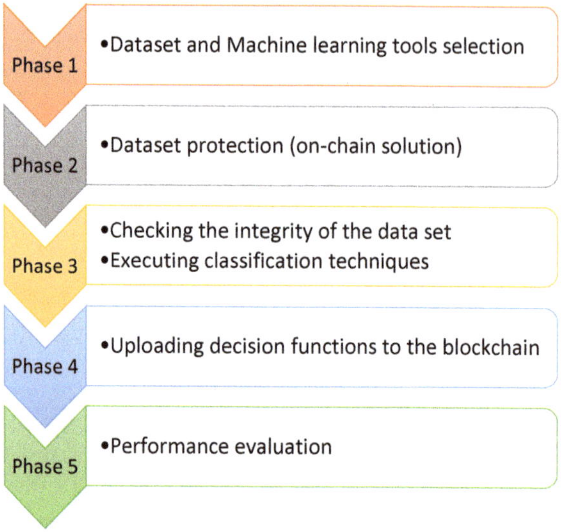

Figure 2. Overview of the proposed work.

Phase 2: Linking the dataset to the Ethereum blockchain

To link the dataset from the client to the blockchain, there are two strategies: (1) on-chain solution, which means uploading the whole dataset in the blockchain, or (2) off-chain solution, which means uploading only the hash of the dataset to the blockchain. Both strategies help to protect the integrity of the dataset.

Phase 3: Checking the integrity of the data set and executing classification techniques

Before executing the classification techniques on the client side using Python, we can either: (1) download the dataset from the blockchain in case we choose on-chain strategies or (2) hash the dataset and compare the hash with that of the dataset in the blockchain in case of off-chain strategies.

Phase 4: Uploading decision functions to the blockchain

During this phase, classification techniques are executed in the local machine using Python. Next, we extract the parameters of the decision function for both SVM and MLP. Finally, upload these parameters to be used by smart contracts on blockchain to segregate between normal behavior and abnormal behavior. This helps to protect the blockchain against attack.

Phase 5: Performance evaluation

The performance evaluation of the proposed techniques can be measured following many parameters, such as the accuracy, the recall, the f1-score, and the time of execution. The proposed technique will also be compared with an existing technique using the same dataset [27] to validate the contribution and to show its effectiveness.

3.2. SVM Training and Parameters Extraction

To train SVM, we used the Radial Basis Kernel Function (RBF). This function calculates the Euclidean distance between vectors. After the training process was completed, the training parameters of the following model were extracted to be kept on-chain, as shown in Table 2.

Table 2. Extracted SVM training parameters.

Name	Description	Size	Data Type
support_vectors_	Datapoints defining hyperplane decision boundaries placements	18 × 395	Decimal matrix
_dual_coef_	Weights of data points	1 × 395	Decimal array
intercept	The bias	1	Decimal
_gamma	The parameter that handles non-linear classification	1	Decimal

Algorithm 1 describes the steps of SVM execution and the decision parameter extractions. Those parameters will be sent next to the blockchain to be used for the decision function.

Algorithm 1: SVM execution

Input: *dataset*
Output: *decision parameters*
1: Read dataset
2: Read dataset hash from the blockchain
3: Check dataset authenticity
4: Execute SVM on the dataset
5: Print SVM performances Accuracy, Precision, F1-score, Recall
6: Extract decision parameters
7: Send decision parameters to smart contract
8: Print Consumed Gaz, Time

3.3. MLP Training and Parameters Extraction

The MLP training process included two hidden layers of sizes 5 and 2 over 1000 epochs. The feedforward deep learning NN relied on a nonlinear activation function, known as the Rectifier Linear Unit activation function (ReLu). Indeed, Table 3 shows the MLP parameters stored on-chain after the training is complete.

Table 3. MLP parameter details.

Name	Description	Size	Data Type
coefs_	Weights of neuron's inputs in three layers, two input layers, and the output layer	5 × 18 2 × 5 1 × 2	Decimal matrices
intercepts_	Biases of each neuron in three layers	1 × 5 1 × 2 1 × 1	Decimal arrays

Algorithm 2 describes the steps of MLP execution and the decision parameter extractions. Those parameters will be sent next to the blockchain to be used for the decision function.

Algorithm 2: MLP execution

Input: *dataset*
Output: *decision parameters*
1: Read dataset
2: Read dataset hash from the blockchain
3: Check dataset authenticity
4: Execute MLP on dataset
5: Print MLP performances Accuracy, Precision, F1-score, Recall
6: Extract decision parameters
7: Send decision parameters to smart contract
8: Print Consumed Gaz, Time

3.4. SVM Decision Function

The values returned by the decision function of an RBF kernel SVM vary between −1 and 1. This function is described in the mathematical representation below.

$$h^*(\mathrm{x}) = (\mathrm{w}^{\bigstar}\phi(\mathrm{x})) + w_0^{\bigstar} = \sum_{i \in \mathcal{P}_S} \alpha_i^{\bigstar} u_i \cdot \mathcal{K}(\mathrm{x}_i - \mathrm{x}) + w_0^{\bigstar}$$

where:

- h^* is the decision function;
- $\alpha_i{}^{\bigstar}$ is the value of the coefficients;
- u_i is the support vector output of the kernel function $\mathcal{K}$;
- x is the new data point;
- x_i is the support vector;
- $w_0^{\bigstar}$ is the bias or the intercept of each vector.

$$\mathcal{K}(\mathrm{x}, \mathrm{x}') = \exp(-\gamma \|\mathrm{x} - \mathrm{x}'\|^2)$$

The core function of this SVM implementation was the RBF function. Its mathematical representation is described above.

- The RBF kernel function returns the product of negative gamma with the Frobenius norm of two input vectors.
- The exp() function is the exponent of Euler's number, e.

$$\|\mathrm{x} - \mathrm{x}'\|_F = \sqrt{\sum_{i=1}^{m} \sum_{j=1}^{n} |a_{i,j}|^2}$$

The above math representation shows that the Frobenius norm is the square root of the summation of two input vectors' squared difference. In this study's implementation case, the two input vectors are the new data point and the support vector.

Algorithm 3 shows the steps taken to compute the decision function of an RBF kernel SVM using the previous procedures.

Algorithm 3: SVM decision function for attack detection

Input: *V: input vector including list of features*
Output: *attack or normal behavior*
Read SVM decision parameters from Python
Create decision function based on RBF kernel and using decision parameters
Execute decision function based on the input vector and get the result
If the result is near −1 **Then**
    Normal behavior
Else
    Attack
End If

3.5. MLP Decision Function

MLP is a supervised classifier that takes an n-dimensional input to return an m-dimensional output.

In the implementation, the input layer of MLP is 9-dimensional, and the two hidden layers 5 and 2-dimensional. Additionally, there is a 1-dimensional output layer. Figure 3 shows a visualization of the implemented MLP layers.

Where:

- x denotes the input characteristics;
- a designates the neurons of the first hidden layer;

- p designates the neurons of the second hidden layer;
- b denotes the bias;
- y denotes the output neuron.

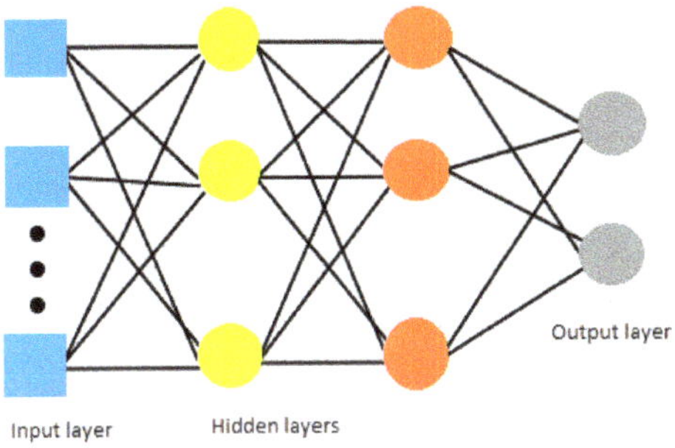

Figure 3. MLP layers visualization.

MLP's decision function concludes a series of additions and multiplications to classify an input. In this calculation, the value of each hidden neuron is equal to the linear sum of all the neuron values of the previous layer multiplied by their coefficients, knowing that the weights between the neuron's layer and the last layer.

An additional intercept value, or bias, is added to this summation.

where:

- x is the value of the neuron;
- w is the weight of the neuron.

Note that the bias, b, is denoted by x_0 with a value of +1, and the w_0 is the intercept value in each layer's bias table.

$$h_i^{(1)} = \phi\left(\sum_j x_j w_{ij}^{(1)} + b_i^{(1)}\right)$$

$$h_i^{(2)} = \phi\left(\sum_j h_j^{(1)} w_{ij}^{(2)} + b_i^{(2)}\right)$$

$$y_i = \phi\left(\sum_j h_j^{(2)} w_{i,j}^{(3)} + b_i^{(3)}\right)$$

In the above representation, h in is the neuron, i, value in the nth layer. This implementation includes two layers and a final output layer with one neuron,

- y_i, gives the final summation value.
- $\phi()$ is the non-linear activation function that calculates neuron's value by a weighted sum.
- x_j is the input features vector.
- $h_j{}^n$ is the neurons' values at layer $i-1$.
- $w_{i,j}$ is the weight.
- $b_i{}^n$ is the intercept of neuron i at the nth layer.

$$f(x) = \max(0, x)$$

This study employed the ReLu nonlinear activation function. The above annotation shows that the ReLu function returns the maximum between x and 0. Algorithm 4 shows steps of implementing the MLP decision function.

Algorithm 4: MLP decision function for attack detection

```
Input: V: input vector including list of features
Output: attack or normal behavior
Read MLP decision parameters from Python
Create decision function using decision parameters
Execute decision function based on the input vector and get the result
If the result is −1 Then
        Normal behavior
Else
        Attack
End If
```

4. Experimental Results

In this section, we give the description for the dataset and the environment of execution. We also discuss the experimental results.

4.1. Dataset Description

This study implementation used the dataset stored in Github [27]. It is a website that makes it easy for programmers and developers to work together to improve application code. It relies on the principle of version control, which uses branching and merging to ensure seamless collaboration without affecting the integrity of the original project [26]. In Table 4, we present the used dataset that contain a list of attack on the Ethereum blockchain.

Table 4. Dataset features.

	Features	Features Details
Old features	hash	Transaction hash
	nonce	How many transactions did the sender's account make?
	transaction indicator	Transaction index in block
	From the address	Origin account
	To the address	destination account
	The value	The transferred value in Wei which is the smallest Ether unit
	gas	Quantity of gas per source
	gas_price	The price of gas (Wei) which is provided by the source
	input	The data sent during the transaction
	cumulative gas reception used	How much gas was used by this transaction while executing a block
	receiving gas used	Total gas has been used by this single given transaction
	timestamp_block	Block timestamp was used by this transaction
	block_number	Operation block number
	block hash	Hashing the block used during the transaction
New added features	Of fraud	- 1 indicates that the return address is the result of forgery - 0 indicates that the sender's address is correct
	to sheat	- 1 indicates that the return address is the result of forgery - 0 indicates that the sender's address is correct
	from_category	Determine if the abnormal activity that occurred from the sender address is phishing or scamming, and (null) for normal operation
	to_category	Determine if the abnormal activity that occurred from the sender address is phishing or scamming, and (null) for normal operation

To evaluate the detection systems that use labeled data (i.e., transactions), the proposed data set should be labeled. Therefore, the transactions included in the proposed data set were classified as normal or harmful transactions, so the number "0" indicates that the transaction is valid., while the number "1" means that the transaction is an attack. As a reminder, each transaction has two addresses: (1) for the sender and (2) for the recipient.

The address account has been passed to the Etherscamdb API through the Python programming language to find out the source of the fraud. Is it the sender or the receiver? Then, four new columns were added to the transaction table: "of_fraud", "to sheat ", "from_category", and "to_category"; the description of these extensions is found in Table 4.

Additionally, two types of attacks were recorded in relation to the proposed technique when API response, as there are two main categories, namely phishing and scamming, as shown in Table 5 and Figure 4. The percentages of abnormal transactions are 22% and 80%, respectively.

Table 5. Ethereum transactions in the dataset.

Type	Transaction	Ratio
Threat transaction	14,250	20%
Normal transaction	57,000	80%
Total	71,250	100%

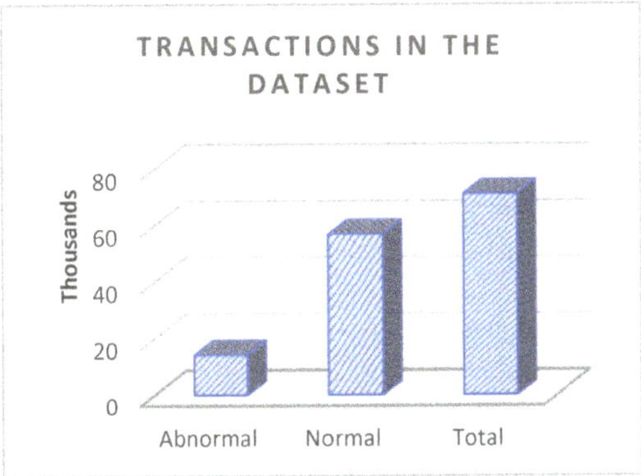

Figure 4. Distribution of transactions in the dataset.

4.2. Environment Description

Table 6 shows the development computer system and the specifications of the development tools.

Table 6. System specifications.

Item	Specifications
Computer OS	Windows 10
CPU	AMD64 3.20 GHz
RAM	8 GB
Ganache	v2.5.4
Solidity	v0.5.16
Vyper	v0.3.0
MetaMask	v10.2.2
Python	v3.9.7

The experimentation is done as follows:

- We use a smart contract to upload the dataset line by line to Ethereum (Ganache) for integrity protection.

- SVM and MLP are executed locally on Python on the same dataset. Next, the decision parameters are extracted and sent to Ethereum (Ganache).
- We use smart contracts to receive the decision parameters on Ethereum (Ganache).
- We re-write the decision MLP decision function and SVM decision function as a smart contract on Ethereum (Ganache).
- Finally, we test the performance of MLP and SVM decision functions.

The aim of BChainGurad is to embed the protection on Ethereum (Ganache). BChainGuard is using smart contracts which work as a defender by checking the traffic of transactions to decide if abnormal behavior is detected or not based on the training already done.

In the experiment, we use Ganache as a local Ethereum platform. Solidity and Vyper are used to write the smart contract for decision function and to protect the dataset. MetaMask is a gateway allowing the communication between web applications and blockchain using 'web3.js'. The local training and testing of SVM and MLP are executed using Python. The right values of each item used in the experiment are indicated in Table 6.

4.3. Evaluation Parameters

The evaluation parameters are described by the following equation: The accuracy is in Equation (1)

$$\text{Accuracy} = (\text{TP} + \text{TN})/(\text{TP} + \text{FP} + \text{FN} + \text{TN}) \quad (1)$$

The equation of precision is given in the following:

$$\text{Precision} = \text{TP}/(\text{TP} + \text{FP}) \quad (2)$$

Equation (3) describes the recall:

$$\text{Recall} = \text{TP}/(\text{TP} + \text{FN}) \quad (3)$$

Finally, the F1-score is described using the equation below:

$$\text{F1 Score} = 2 \times (\text{Recall} \times \text{Precision})/(\text{Recall} + \text{Precision}) \quad (4)$$

4.4. Machine Learning Result Analyses

We compared both models on the same dataset. Table 7 and Figure 5 show the performance of the MLP and SVM models. We concluded from this comparison that the two models have distinct performances. In addition, our analysis also showed that the MLP belonging to the deep learning family outperforms SVM in all metrics with a small difference (3.02% for accuracy, 3.27% for precision, 5.03% for F1-score and 3% for the recall metric). In summary, after conducting several experiments, the MLP and SVM networks have confirmed that they have very distinct solutions for regression, classification and prediction tasks.

Table 7. SVM vs. MLP performance.

Classifier	Accuracy	Precision	F1-Score	Recall
SVM	95	95.03	93	95
MLP	98.02	98.5	98.03	97

Although MLP has the best generalizability, the observed performance differences are negligible in most cases. The main difference lies in the complexity of the networks. An MLP network that implements a global approximation strategy typically uses a very small number of hidden neurons.

4.5. Blockchain Decision Function Result Analyses

Gas fees are payments made by users to pay validators and miners for the computational energy consumed to process and validate transactions on the Ethereum blockchain.

Figure 6 shows a graphical representation of the performance of MLP and SVM models in term of consumed gas in gwei. Figure 7 shows a graphical representation of the performance of MLP and SVM models in term of decision function elapsed time in seconds.

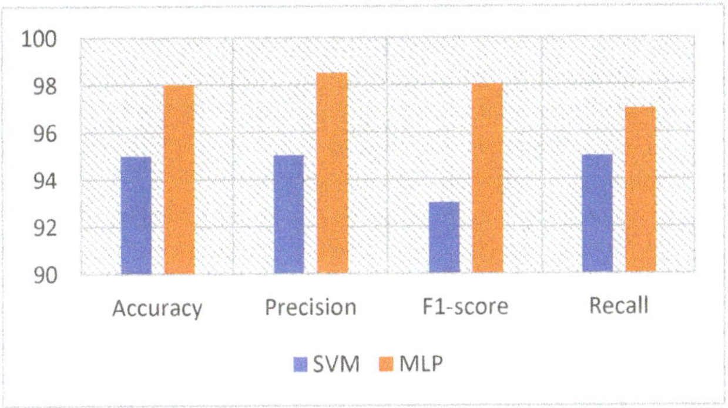

Figure 5. SVM vs. MLP performance.

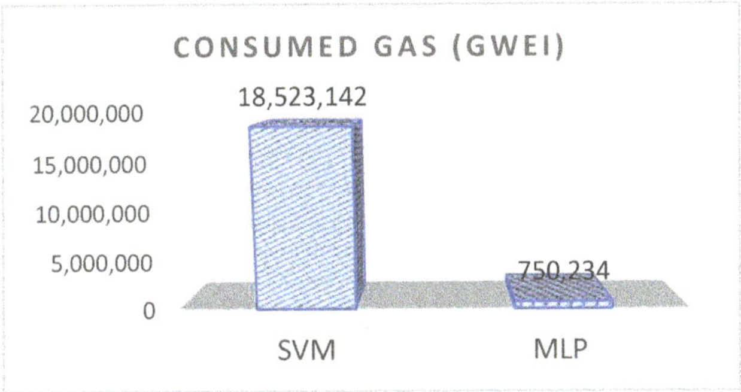

Figure 6. Consumed gas in gwei for SVM decision function and MLP decision function.

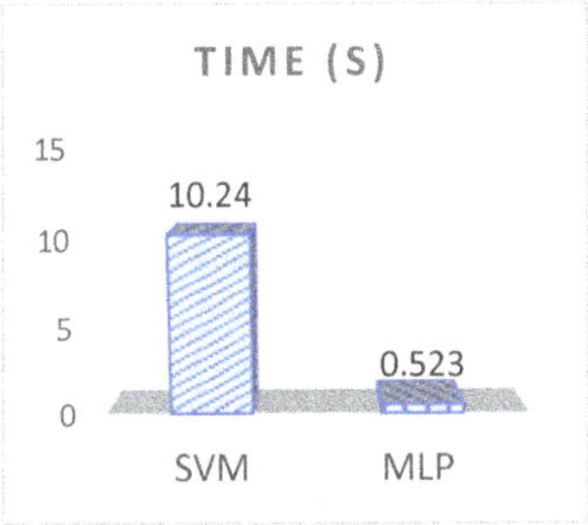

Figure 7. Elapsed time in seconds for SVM decision function and MLP decision function.

This comparison analysis is useful to verify the effectiveness of both models. The consumed gas in gwei for SVM decision function is equal to 0.1914, while this consumption is multiplied more than 44 times for SVM and reaches 8.4415. This big difference can be explained in two ways.

First explanation: the implementation of the MLP approximative strategy usually employs a very small number of hidden neurons. On the other side, the SVM is based on the local approximation strategy and uses a large number of hidden units. This large number can be the cause of differences in consumption and also need more time for execution.

Second explanation: If a gas price is set too low, the transaction could be ignored, missed or the wallet could become stuck, freezing transactions from that wallet. Therefore, the wallet can remain blocked until the transaction is resolved.

4.6. Overeall Result Discussion

An overall result comparison shows that MLP is providing the best results. Indeed, the SVM decision function is more expensive in terms of consumed gas and time. This can be explained by the number of parameters and operations in SVM decision function. In our case, we adopt only two layers in MLP. For that reason, the MLP decision function is cheaper than that of SVM.

Furthermore, MLP outperforms SVM in terms of accuracy and precision; this can be rendered to the data itself. In addition, these results are linked to the implementation of the MLP as an approximative strategy that usually employs a very small number of hidden neurons while SVM is based on the local approximation.

The provided performances' favorite MLP is used in this case. However, this is always linked to the type of the used transaction themselves. An enlargement of the dataset by considering more types of attacks may favorite another technique to be used for cyberthreat detection.

4.7. Comparison between BChainGuard and Works on the Same Dataset

In this section, we make a comparison between our framework result and the result of work on the same dataset. In BChainGuard, the accuracy of MLP in is 98.2% and 95% for SVM. However, in [28], the best accuracy is performed at 98.8% using Random Forest, while the smallest one is 82% for logistic regression. In the same context, the accuracy of detection abnormal transaction using SVM and KNN is 95%. Our contribution is not only improving the accuracy of detection but also securing the dataset against poisonous attack, making the detection safe by embedding the decision function as a smart contract. In addition, we chose to embed only the decision function on the blockchain to minimize the overhead of runtime and deployment, which was discussed in the previous section. In what follows, we make a comparison between BChainGuard and other frameworks.

Table 8 shows that our framework is the first one that uses machine learning, embedded in the blockchain itself, and with the lowest deployment and runtime overhead compared to others. In future works, we plan to make a secure analysis of BChainGuard by injecting attacks on the dataset, as well as on Ethereum, to see the performance of our framework.

Table 8. Frameworks comparison.

Tools	Technique	Place	Overhead
BChainGuard	Machine learning	On-chain	Low
TRS [17]	Graph theory	On-chain	High
ADM [18]	Machine learning	off-chain	-
ContractGuard [19]	Statistical analysis	On-chain	High
SODA [20]	Statistical analysis	On-chain	High
BAD [21]	Statistical analysis	On-chain	High
SolGuard [22]	Statistical analysis	On-chain	High
DefectChecker [23]	Statistical analysis	On-chain	High

Recently, many research efforts have been conducted to show the great impact of applying blockchain in Industry 4.0 [28]. This generation of industry investigates the application of the latest technology innovation in Artificial Intelligence, Big Data, the Internet of Things and Blockchain for supply chain and manufacturing improvement. On one hand, Blockchain with its great potential can raise many opportunities for Industry 4.0. On the other hand, although Blockchain is one of the most secure peer-to-peer systems, it can be the target of attacks and cyberthreats. Consequently, securing service-based blockchain has high importance. In this context, we believe that BChainGuard is a successful key to supporting the application of Blockchain in Industry 4.0. Our framework can be improved by considering more types of attacks and next integrated into Ethereum as a new protection layer against smart contract vulnerabilities. This can improve Ethereum's trustworthiness to be more attractive to the industry.

4.8. BChainGuard Limitations and Possible Future Improvements

BChainGuard is based on executing SVM and MLP locally and next embedding only the decisions function in blockchain in order to detect attacks. Despite the advantages of our contribution, it can be improved in the future by:

- Considering more realistic datasets linked to blockchain smart contract and transactions.
- Using different scenarios of transactions and smart contracts that help to convert more situations that can be the target of attacks.
- Applying other machine learning and deep learning techniques that may be offered the best results.
- Adopting federated learning instead of machine learning in the case of the nonavailability of the dataset, since federated learning helps to protect privacy.

5. Conclusions

The blockchain, as any system, and despite of its power protection, can be prone to many attacks. For that reason, many efforts have been made by the research community to protect it. The majority of efforts have been based on analyzing statistically the smart contracts. To the best of our knowledge, we propose, for first time BChainGuard, a new layer in Ethereum for blockchain protection. Our idea is based on executing SVM and MLP locally and next embedding only the decisions function in the blockchain in order to detect attacks on blockchain. Our contribution is not only improving the accuracy of detection but also securing the dataset against poisonous attack. In addition, we choose to embed only the decision function on the blockchain to minimize the overhead of runtime and deployment.

Smart contracts can contain numerous security vulnerabilities, such as reentrancy, unhandled exceptions, Integer Overflow and unrestricted action. The aim of BChainGuard is to ensure the safe execution of smart contracts by avoiding vulnerabilities. As an open system, Ethereum can be the target of many attacks embedded in smart contracts themselves. New types of attacks can always appear. To tackle them, a strategy of BChainGuard continuous improvement must take place. This strategy must be defined to take into consideration the integration of new attack detection.

The limitations of BChainGuard are: (1) detecting only if an attack is happened or not without preventing the type of attack, (2) dealing only with two types of attacks in the used dataset, which are phishing and scamming attacks, and (3) using only MLP and SVM techniques, when other techniques may provide best detection accuracy.

Despite the advantages of our contribution, it can be improved in the future by considering more realistic datasets linked to blockchain smart contracts and transactions. Additionally, using different scenarios of transactions and smart contracts helps to convert more situations that can be the target of attacks. The use of other machine learning and deep learning techniques may offer the best results. As a future work, we also plan to make a secure performance analysis of BChainGuard. This can be done by injecting poisonous attacks on the dataset to see how blockchain can protect its integrity. In addition, we can run some malicious smart contracts on Ethereum to measure the reaction of our framework.

Author Contributions: Conceptualization, T.M.; methodology, H.A.; software, M.A.A.; validation, T.M.; formal analysis, S.A.; investigation, H.A.; resources, H.A.; data curation, M.A.A.; writing—original draft preparation, T. M.; writing—review and editing, S.A.; visualization, S.A.; supervision, T.M. and project administration, T.M. All authors have read and agreed to the published version of the manuscript.

Funding: This research was funded by the Deputyship for Research& Innovation, Ministry of Education, Saudi Arabia, through project number QU-IF-4-4-1-31851.

Data Availability Statement: The used dataset was downloaded from (https://github.com/salam-ammari/Labeled-Transactions-based-Dataset-of-Ethereum-Network (accessed on 31 May 2022)). No ethics approval was required for this dataset.

Acknowledgments: The authors extend their appreciation to the Deputyship for Research& Innovation, Ministry of Education, Saudi Arabia for funding this research work through the project number (QU-IF-4-4-1-31851). The authors also thank to Qassim University for technical support.

Conflicts of Interest: The authors declare no conflict of interest.

References

1. Nakamoto, S. Bitcoin: A Peer-to-Peer Electronic Cash System. Available online: https://bitcoin.org/bitcoin.pdf (accessed on 15 April 2022).
2. Alsayegh, M.; Moulahi, T.; Alabdulatif, A.; Lorenz, P. Towards Secure Searchable Electronic Health Records Using Consortium Blockchain. *Network* **2022**, *2*, 239–256. [CrossRef]
3. Samaniego, M.; Deters, R. Blockchain as a Service for IoT. In Proceedings of the IEEE International Conference on Internet of Things, Honolulu, HI, USA, 25–30 June 2017.
4. Alfrhan, A.; Moulahi, T.; Alabdulatif, A. Comparative study on hash functions for lightweight blockchain in Internet of Things (IoT). *Blockchain Res. Appl.* **2021**, *2*, 100036. [CrossRef]
5. AlAsqah, M.; Moulahi, T.; Zidi, S.; Alabdulatif, A. Leveraging Artificial Intelligence in Blockchain-Based E-Health for Safer Decision Making Framework. 2022. Available online: https://europepmc.org/article/ppr/ppr501665 (accessed on 15 April 2022).
6. Dubovitskaya, A.; Xu, Z.; Ryu, S.; Schumacher, M.; Wang, F. Secure and Trustable Electronic Medical Records Sharing using Blockchain. In Proceedings of the AMIA 2017 Annual Symposium Proceedings, Washington, DC, USA, 4–8 November 2017; pp. 650–659.
7. Eyal, I. Blockchain Technology: Transforming Libertarian Cryptocurrency Dreams to Finance and Banking Realities. *Computer* **2017**, *50*, 38–49. [CrossRef]
8. Al-E'mari, S.; Anbar, M.; Sanjalawe, Y.; Manickam, S. A Labeled Transactions-Based Dataset on the Ethereum Network. In *Advances in Cyber Security. ACeS 2020. Communications in Computer and Information Science*; Anbar, M., Abdullah, N., Manickam, S., Eds.; Springer: Singapore, 2021; Volume 1347. [CrossRef]
9. Saad, M.; Thai, M.T.; Mohaisen, A. POSTER: Deterring ddos attacks on blockchain-based cryptocurrencies through mempool optimization. In Proceedings of the 2018 on Asia Conference on Computer and Communications Security, Incheon, Republic of Korea, 4–8 June 2018; pp. 809–811.
10. Mavridou, A.; Laszka, A.; Stachtiari, E.; Dubey, A. VeriSolid: Correct-by-design smart contracts for Ethereum. In Proceedings of the International Conference on Financial Cryptography and Data Security, Frigate Bay, St. Kitts and Nevis, 18–22 February 2019; Springer: Cham, Switzerland, 2019; pp. 446–465.
11. Henningsen, S.; Teunis, D.; Florian, M.; Scheuermann, B. Eclipsing Ethereum Peers with False Friends. *arXiv* **2019**, arXiv:1908.10141. [CrossRef]
12. Andrychowicz, M.; Dziembowski, S.; Malinowski, D.; Mazurek, Ł. Fair two-party computations via bitcoin deposits. In *Financial Cryptography and Data Security*; Böhme, R., Brenner, M., Moore, T., Smith, M., Eds.; Springer: Berlin/Heidelberg, Germany, 2014; pp. 105–121.
13. Apostolaki, M.; Zohar, A.; Vanbever, L. Hijacking bitcoin: Routing attacks on cryptocurrencies. In Proceedings of the 2017 IEEE Symposium on Security and Privacy (SP), San Jose, CA, USA, 22–26 May 2017; pp. 375–392.
14. Swathi, P.; Modi, C.; Patel, D. Preventing Sybil Attack in Blockchain using Distributed Behavior Monitoring of Miners. In Proceedings of the 10th International Conference on Computing, Communication and Networking Technologies (ICCCNT), Kanpur, India, 6–8 July 2019; pp. 1–6. [CrossRef]
15. Andryukhin, A.A. Phishing attacks and preventions in blockchain based projects. In Proceedings of the 2019 International Conference on Engineering Technologies and Computer Science, EnT, Moscow, Russia, 26–27 March 2019; pp. 15–19.
16. Apostolaki, M.; Zohar, A.; Vanbever, L. Hijacking bitcoin: Large-scale network attacks on cryptocurrencies. *arXiv* **2016**, arXiv:1605.07524.
17. Zhang, R.; Preneel, B. Publish or Perish: A Backward-Compatible Defense Against Selfish Mining in Bitcoin. In *Topics in Cryptology—CT-RSA 2017*; Handschuh, H., Ed.; Springer International Publishing: Cham, Switzerland, 2017; pp. 277–292. [CrossRef]

18. Torres, C.F.; Schütte, J.; State, R. Osiris: Hunting for integer bugs in Ethereum smart contracts. In Proceedings of the 34th Annual Computer Security Applications Conference, ACSAC'18, San Juan, Puerto Rico, 3–7 December 2018; ACM: New York, NY, USA, 2018; pp. 664–676.
19. Morishima, S. Scalable anomaly detection method for blockchain transactions using GPU. In Proceedings of the 20th International Conference on Parallel and Distributed Computing, Applications and Technologies (PDCAT), Gold Coast, Australia, 5–7 December 2019; pp. 160–165.
20. Sayadi, S.; Rejeb, S.B.; Choukair, Z. Anomaly detection model over blockchain electronic transactions. In Proceedings of the 15th International Wireless Communications & Mobile Computing Conference (IWCMC), Tangier, Morocco, 24–28 June 2019; pp. 895–900.
21. Wang, X.; He, J.; Xie, Z.; Zhao, G.; Cheung, S. ContractGuard: Defend Ethereum Smart Contracts with Embedded Intrusion Detection. *IEEE Trans. Serv. Comput.* **2019**, *13*, 314–328. [CrossRef]
22. Chen, T.; Cao, R.; Li, T.; Luo, X.; Gu, G.; Zhang, Y.; Liao, Z.; Zhu, H.; Chen, G.; He, Z.; et al. SODA: A Generic Online Detection Framework for Smart Contracts. In Proceedings of the 27th Network and Distributed System Security Symposium, NDSS, San Diego, CA, USA, 23–26 February 2020. [CrossRef]
23. Signorini, M.; Pontecorvi, M.; Kanoun, W.; Di Pietro, R. BAD: A Blockchain Anomaly Detection Solution. *IEEE Access* **2020**, *8*, 173481–173490. [CrossRef]
24. Praitheeshan, P.; Pan, L.; Zheng, X.; Jolfaei, A.; Doss, R. SolGuard: Preventing external call issues in smart contract-based multi-agent robotic systems. *Inf. Sci.* **2021**, *579*, 150–166. [CrossRef]
25. Chen, J.; Xia, X.; Lo, D.; Grundy, J.; Luo, X.; Chen, T. DefectChecker: Automated Smart Contract Defect Detection by Analyzing EVM Bytecode. *IEEE Trans. Softw. Eng.* **2021**, *48*, 2189–2207. [CrossRef]
26. Chacon, S.; Straub, B. *Pro Git: Everything You Need to Know About Git*, 2nd ed.; Apress: New York, NY, USA, 2014.
27. Dataset. Available online: https://github.com/salam-ammari/Labeled-Transactions-based-Dataset-of-Ethereum-Network (accessed on 31 May 2022).
28. Javaid, M.; Haleem, A.; Singh, R.P.; Khan, S.; Suman, R. Blockchain technology applications for Industry 4.0: A literature-based review. *Blockchain Res. Appl.* **2021**, *2*, 100027. [CrossRef]

applied sciences

MDPI

Article

Dynamic Community Discovery Method Based on Phylogenetic Planted Partition in Temporal Networks

Xiaoyang Liu [1,*], Nan Ding [1], Giacomo Fiumara [2], Pasquale De Meo [3] and Annamaria Ficara [4]

1 School of Computer Science and Engineering, Chongqing University of Technology, Chongqing 400054, China; dr.ding_1901@2019.cqut.edu.cn
2 MIFT Department, University of Messina, Viale F. S. D'Alcontres 31, 98166 Messina, Italy; giacomo.fiumara@unime.it
3 Department of Ancient and Modern Civilizations, University of Messina, Viale G. Palatucci 13, 98168 Messina, Italy; pasquale.demeo@unime.it
4 Department of Mathematics and Computer Science, University of Palermo, Via Archirafi 34, 90123 Palermo, Italy; aficara@unime.it
* Correspondence: lxy3103@163.com

Abstract: As most of the community discovery methods are researched by static thought, some community discovery algorithms cannot represent the whole dynamic network change process efficiently. This paper proposes a novel dynamic community discovery method (Phylogenetic Planted Partition Model, PPPM) for phylogenetic evolution. Firstly, the time dimension is introduced into the typical migration partition model, and all states are treated as variables, and the observation equation is constructed. Secondly, this paper takes the observation equation of the whole dynamic social network as the constraint between variables and the error function. Then, the quadratic form of the error function is minimized. Thirdly, the Levenberg–Marquardt (L–M) method is used to calculate the gradient of the error function, and the iteration is carried out. Finally, simulation experiments are carried out under the experimental environment of artificial networks and real networks. The experimental results show that: compared with FaceNet, SBM + MLE, CLBM, and PisCES, the proposed PPPM model improves accuracy by 5% and 3%, respectively. It is proven that the proposed PPPM method is robust, reasonable, and effective. This method can also be applied to the general social networking community discovery field.

Keywords: temporal networks; community discovery; phylogenetic evolution; planted of partition

Citation: Liu, X.; Ding, N.; Fiumara, G.; De Meo, P.; Ficara, A. Dynamic Community Discovery Method Based on Phylogenetic Planted Partition in Temporal Networks. *Appl. Sci.* **2022**, *12*, 3795. https://doi.org/10.3390/app12083795

Academic Editor: Agostino Forestiero

Received: 10 February 2022
Accepted: 6 April 2022
Published: 9 April 2022

1. Introduction

1.1. Background

Complex network analysis is an interdisciplinary research field which can be applied in a lot of areas such as computer science [1,2] and social, biological and physical sciences [3–5], and it is capturing the attention of many scholars. A complex network is a simple graph defined as a set of nodes connected by a set of edges. Nodes can represent individuals or organizations. Edges are relational ties between two nodes, e.g., friendship relationships between two social users. Graphs are one of the most important and powerful data structures. Complex network analysis and modeling can be used to reveal patterns of social interaction, to study recommendation systems, or protein complexes and protein functional modules. By far the most basic tasks in complex networks are node identification, link prediction, and information dissemination. These tasks have received extensive research and attention. In addition, community structure discovery is also one of the most important tasks; it is usually defined as identifying tightly connected subgraphs from a complex network. Because communities help to reveal the structure–function relationship of the network, it has been studied extensively. For example, communities within cancer networks mark key pathways associated with cancer progression [6], and the communities in the multi-layer transportation network correspond to common practices, which provides clues for airline management [7]. Therefore, a great deal of work has been carried out in the discovery of communities in the network [8–10]. A lot of work has been proposed for community discovery, existing algorithms either optimize predefined quantitative functions or acquire potential feature matrices for community detection. Typical methods include modularization-based methods [11], model-based methods [12,13] and random-walk-based methods [14–16]. S. Fortunato et al. [17,18] have conducted a comprehensive survey.

However, all these methods assume that the target network is static and ignore the timeliness of the network. In reality, many networks from society and nature are dynamic, meaning that the network structure changes over time; that is, it performs the dynamic network. More specifically, in a dynamic network, nodes may appear or disappear over time, and links between two nodes may also appear or disappear. For example, interpersonal relations often change due to individual behavior [19]. For another example, tumor cell migration leads to metastasis, which is crucial for the diagnosis and treatment of tumors [20]. Therefore, it is worthwhile to track how a community evolves in a dynamic network (also known as a dynamic or evolutionary community).

For dynamic network modeling, the most widely used method is to introduce explicit smoothing frameworks, which quantifies the similarity between snapshots in two subsequent steps by introducing the Temporal Smoothed Framework (TSF). Various TSF-based algorithms have been proposed to evolve the community by extending the static community discovery algorithm. For example, for topological connectivity, the Kim-Han algorithm [21] found dynamic communities by optimizing modularity, and the DYNMOGA method [22] is presented for the multi-objective genetic algorithm to simultaneously optimize clustering accuracy and clustering drift. Regarding matrix decomposition, the ESPC method [23] is used with matrix spectrum, the ECKF method [24] is proposed by using kernel ENMF, and the Se-NMF method [25] is used by a semi-supervised strategy to develop community testing. The Gr-NMF method [26] is adopted by graph-regularized NMF for community discovery in evolution. In the probabilistic model: FaceNet method [27] is researched by using Maximum a Posteriori (MAP) estimate, DSBM method [28] adopted the Bayesian method to obtain the evolving community by extending the random block model. According to the existing literature [29], there are six evolutionary events in the community (as shown in Figure 1), including birth, death, growth, contraction, merging, and splitting. Sometimes, a seventh event is added to these, i.e., continuing. Finally, an eight event was proposed by Cazabet and Amblard [30] and it is resurgence. A generic dynamic community discovery algorithm does not necessarily have to handle all these events, which can be differently managed in different works [31].

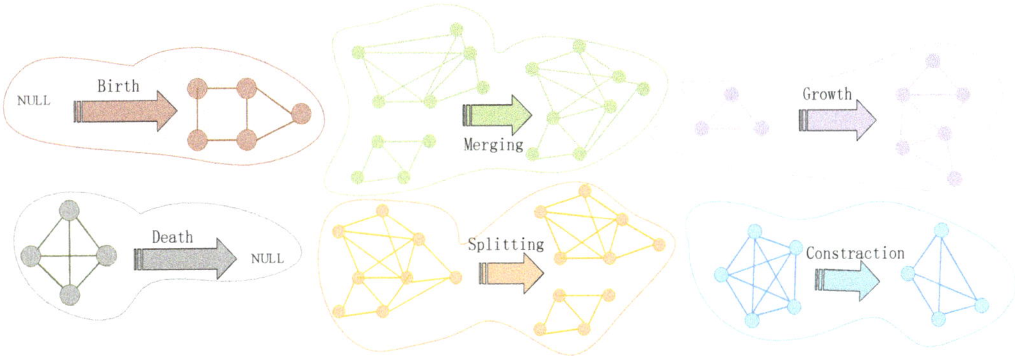

Figure 1. 6 evolutionary events in dynamic communities.

1.2. Motivation

While much work has been carried out to address the problem of dynamic community discovery, there are still some issues that need to be addressed.

Firstly, most existing dynamic network models assume that it is a hidden Markov structure, in this structure, when the current network state is given, a network snapshot at any given time is conditionally independent for all previous snapshots. This approach may not be flexible enough to replicate some of the observations in real network data.

Secondly, the dynamic system is used by filters, even for Gaussian distributions. However, after a nonlinear transformation, Gaussian terms are lost. Mean and covariance are the only measures computed by the filter. This is the result of a nonlinear transformation approximated by Gauss, and therefore this approximation may be poor.

Finally, how to combine the information of the community structure available at the previous moment with the information available at the current moment is an important question. In the traditional hidden Markov dynamic Bayesian network model, the probability of an edge appearing in a dynamic network is realized by the estimated state.

Therefore, this paper proposes a phylogenetic planted partition method, which uses the graph optimization strategy to continuously discover the evolving communities.

1.3. Our Work and Contributes

The main contributions of this paper can be summarized as follows:

(1) The time dimension is introduced into the typical stochastic block-model, and all states in the whole dynamic network system are treated as variables, and the observation equation is taken as the constraint between variables to construct an error function about the whole dynamic network system.

(2) By adopting the graph-based optimization strategy, the constraints in the entire motion trajectory can be considered once. In the linearization process, only the Jacobian matrix is calculated, and the calculation process is also relative to the entire motion trajectory. Therefore, the entire system evolution process is transformed into the nonlinear system optimization process.

(3) In natural ecosystems, inspired by the evolutionary thinking of species populations and combined with the typical probability model of stochastic block-model in community discovery, a phylogenetic planted partition method (PPPM) for dynamic community discovery is proposed.

(4) The proposed PPPM method in the two scenarios of artificial network and the real network is verified by experiments, which proves that the performance of the novel method is better than the four state-of-the-art methods (FaceNet, SBM + MLE, CLBM, and PisCES).

1.4. Organization

The remainder of this paper is structured as follows: In Section 2, related work is discussed; Section 3 introduces the proposed model in detail, describes the proposed PPPM method, and gives the derivation process. In Section 4, the experimental results of the novel PPPM method in the artificial dynamic network and real dynamic network are presented, and then compared with other existing models. Finally, in Section 5, some conclusions are given and future directions are discussed.

2. Related Work

According to the research of Aynaud et al. [32], the dynamic community discovery algorithms can be divided into four categories: coupling network, two-stage algorithm, evolutionary clustering, and probability model. However, Hartmann et al. [33] believed that all existing dynamic community discovery methods can be identified as online or offline methods. Rossetti and Cazabet [31] proposed a new survey on community detection in dynamic networks, which proposed the unique functions and challenges of dynamic community discovery algorithms.

The first kind of coupled network-based algorithm firstly builds the network by fusing edges at different times. Then, the classical static community detection algorithm is used to find the communities in the coupled network. For example, Agarwal et al. [34] discovered the ongoing events in the microblog message flow by adding edges between vertex instances at different times to build the coupling network, in which the dynamic community corresponds to the community in the built network. Because coupled networks cannot fully describe the dynamic characteristics of networks, these algorithms have been shown to accurately discover only short-cycle communities. To overcome this problem, the second kind of two-stage algorithm separated the community detection from the community dynamic, avoiding the coupling of the dynamic network.

Specifically, these algorithms used static community detection algorithms to find the community each time and then connected the community the next two times to extract the evolving community. Typical algorithms included GraphScope [35] and TRMMC [36] coupled networks and two-stage algorithms that detected dynamic communities in a dynamic network by simply extending static community detection methods and detecting dynamic communities in each operation dynamic network or static community. In general, these algorithms can achieve better performance in the case of weak network dynamics. In this case, the dynamic update method can accurately identify the dynamic community without running the community detection algorithm each time, and only need to update the previously discovered community. However, the accuracy of these algorithms is low.

The third type of dynamic community discovery method is related to clustering evolutionary, which is proposed by Chakrabarti et al. [37]. They extracted the implicit community structure in each snapshot, which is one of the most widely used methods for dynamic community discovery. The evolutionary clustering algorithm adopted the assumption of time smoothness. The community structure will not change much over a continuous-time slice. This time smoothing method can be used to overcome the randomness. Compared with other algorithms, the evolutionary community discovery algorithm aims to discover a smooth sequence of communities in a series of network snapshots (as can be seen in Figure 2). The overall objective function of the evolutionary algorithm can be decomposed into two parts: Snapshot Cost (CS) and Temporal Cost (CT) [38].

$$Cost = \alpha \cdot CS + (1 - \alpha) \cdot CT \tag{1}$$

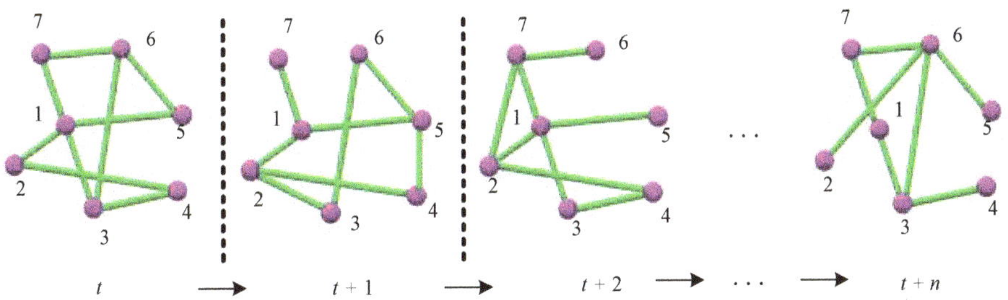

Figure 2. The series of network snapshots.

Among them, CS measures the adaptability of the community structure and network at the time t, while CT measures how similar the two community structures (the community structure is acquired at the time t to the structure is obtained at previous time $t - 1$). The parameter α is for balancing the importance between CS and CT. By introducing different object functions based on modularity, normalization of mutual information and spectrum clustering et al., this framework has been used in much of the literature [39,40] to discover communities in dynamic networks.

Folino and Pizzuti [39] formalized the dynamic community discovery algorithm as a multi-objective optimization problem, which maximized the clustering accuracy of the current time step and minimized the clustering drift from one time step to one successive time step. Ma and Dong [40] proposed two evolutionary non-negative matrix decomposition (ENMF) frameworks and proved the equivalence relation between evolutionary module density and evolutionary spectrum clustering. In addition, they introduced a semi-supervised approach, which is called sE-NMF, that incorporated prior information into the ENMF.

Chi et al. [23] extended this idea with two frameworks of evolutionary spectral clustering, which are defined as Preserving Cluster Quality (PCQ) and Preserving Cluster Membership (PCM). Both frameworks have proposed the optimization and correction cost functions, but they differ in how to define the CT. In the PCQ framework, the CT is the cost of the clustering results at the time t applied to the similarity matrix at the time $t-1$. In PCM, the CT is defined as a measure of the distance between the clustering results at the time t and $t-1$. In the PCQ:

$$\max_{Z\in\mathbb{R}^{n\times k}} \alpha tr(Z^T W^{t-1} Z) + (1-\alpha) tr(Z^T W^t Z), Z^T Z = I. \tag{2}$$

where W^{t-1} and W^t represent the adjacency matrix at the time $t-1$ and t, respectively.

$$\alpha W^{t-1} + (1-\alpha) W^t \tag{3}$$

Finally, the membership of community members can be obtained by calculating the eigenvector of Formula (3).

After the above work, an evolutionary community discovery algorithm is proposed to try to optimize the modified cost function in the definition. Since the user definition of snapshot and *CT* of community discovery results varies with community discovery algorithms, the aim of the above work is to solve the problem of how to select the parameter α, which can determine how much weight to assign to previous data or community discovery results.

Xu et al. [41] proposed an adaptive evolutionary clustering algorithm, using the following smooth approximation matrix $\hat{\Psi}^t$ to better estimate the network state.

$$\hat{\Psi}^t = \alpha^t \hat{\Psi}^{t-1} + (1-\alpha^t) W^t \tag{4}$$

where the parameter α^t controls the rate of forgetting past information, so it is also defined as a forgetting factor.

Ma et al. [26] proposed a non-negative matrix decomposition for co-regularization evolution to identify dynamic communities under a time-smoothing framework.

$$O^t = L^t + \beta Q^t + \gamma R^t \tag{5}$$

where β and γ are regularized parameters.

In recent years, researchers have proposed some excellent techniques to improve the performance of dynamic community discovery algorithms. In the probability model, the researchers have put forward an innovative model, and this paper puts forward a new dynamic model, which is suitable for dynamic Bayesian networks, namely the system evolution partition transplantation model. In this method, the model parameters are tracked by using the graph of optimization strategy. Table 1 compares some recent representative dynamic social network discovery algorithms based on a probability model.

Table 1. Comparison of our work with previous model literatures.

Ref.	Year	Approach	Theory	Dataset
[42]	2015	Maximum Likelihood Estimation	Consistency	MIT Reality Mining
[43]	2015	Kalman Filtering + Local Search	/	Facebook wall posts
[44]	2016	Expectation Maximization	/	Internet AS graphs, Friendship networks
[45]	2016	Expectation Maximization	Detectability thresholds	Synthetic
[46]	2017	Expectation Maximization	Detectability thresholds	Synthetic
[47]	2017	Time-lag corrected	Convergence rates	Synthetic
[48]	2018	Aggregating SBM subroutines + MLE	Correctness/Stage-wise convergence rates	Enron emails, Facebook friendships
[49]	2020	Exhaustive grid search	Convergence rates	Synthetic
Our work	2022	Graph-based optimization	Convergence rates	MIT Reality Mining, Enron emails

3. Meterials

3.1. Formal Definition

In this paper, a novel phylogenetic planted partition model is defined for temporal social networks by the following definitions:

Definition 1. *A social network can be represented by a graph, G, on a set of nodes, V, and a set of edges, E. Nodes and edges are represented by an adjacency matrix A, where $\rho_{ij} = 1$ represents the existence of an edge from node $i \in V$ to node $j \in V - \{i\}$, and $\rho_{ij} = 0$ represents the absence of an edge. This paper assumes the network is a directed graph, which is generally $\rho_{ij} \neq \rho_{ji}$ and has no self-loop, namely $\rho_{ii} = 0$.*

Definition 2. *The positive integer n represents the number of nodes, and $\varepsilon = (\varepsilon_1, \ldots, \varepsilon_k)$ is the probability vector on $[k] = \{1, \ldots, k\}$ (k is the number of network communities). W is a symmetric matrix whose element is $k \times k$ between [0,1]. (X, G) is defined under the Stochastic Block-Mode (SBM) $SBM(n, \theta, W)$. X is a n dimensional random vector; it is independent and identically distributed under ε. Let the community set $C_r = C_r(X) = \{C_1, \ldots, C_k\}, r \in [k]$ denotes the division of V* into *k communities. In this paper, we use the symbols p and q to indicate two generic communities, c_i and c_j, where i and j are the two nodes of a simple graph G* of *n vertices, which respectively belong to c_i (i.e., $i \in p$) and c_j (i.e., $j \in q$) at time t. Nodes i and j are connected according to a probability $\varepsilon_{c_i^t c_j^t}$, independently from the other node pairs. Therefore, the probability distribution of (X, G) [50] is as follows, where $G = (V, E)$.*

$$\Pr\{X = x\} = \prod_{i=1}^{n} \varepsilon_{c_i} = \prod_{r=1}^{k} \varepsilon_r{}^{|C_r(x)|} \tag{6}$$

$$\Pr\{E = y | X = x\} = \prod_{1 \le i < j \le n} \varepsilon_{c_i, c_j}^{y_{ij}} (1 - \varepsilon_{c_i, c_j})^{1 - y_{ij}} \tag{7}$$

$$= \prod_{1 \le p < q \le k} \varepsilon_{p,q}^{N_{pq}(x,y)} (1 - \varepsilon_{p,q})^{N_{pq}^c(x,y)} \tag{8}$$

where $N_{pq}(x, y)$ represents the observed value of the number of edges in the partition, which can be expressed as $N_{pq}(x, y) = \sum_{c_i = p} \sum_{c_j = q} y_{ij}^t$; $N_{pq}^c(x, y)$ represents the probability value of the number of edges in the partition, which can be expressed as follows.

$$N_{pq}^c(x, y) = \begin{cases} |p|(|p|-1)/2 - N_{pp}(x, y) & p = q \\ |p||q| - N_{pq}(x, y) & p \neq q \end{cases} \tag{9}$$

Equation (8) can be rewritten by:

$$\Pr\{E = y|X = x\} = \exp\{\sum_{p=1}^{k}\sum_{q=1}^{k}[N_{pq}(x,y)\log \varepsilon_{p,q} + N_{pq}^{c}(x,y)\log(1-\varepsilon_{p,q})]\} \tag{10}$$

Definition 3. *Define an evolutionary sequence of discrete time steps for social network (dynamic Bayesian network); the nodes and edges may appear or disappear with time. The temporal social networks can be expressed as $\{G^1, G^2, \ldots, G^T\}$, where $G^t = (V^t, E^t)$, superscript t represents the time step, V^t and E^t, respectively, in the time step are t collection of nodes and edges. Let $A^{(T)}$ represent the sequence of adjacency matrix on the node-set sequence $V^{(T)} = \cup_{t=1}^{T} V^t$, and let $c^{(T)}$ represent the sequence of community member membership vectors of the node. For this dynamic social network, the probability distribution of edges can be defined as:*

$$\Pr = (1-\varepsilon_{pq}^{t-1})\varepsilon_{pq}^{t}\cdots\varepsilon_{pq}^{t+d-1}(1-\varepsilon_{pq}^{t+d}) \tag{11}$$

For any pair of nodes $i \in p$ and $j \in q$ at $t-1$ and t, s.t. $\rho_{ij}^{t-1} = 1$. Namely, there is an edge from node i to the node j at the time $t-1$ and ρ_{ij}^{t} is Independent Identically Distributed (IID). The same is true for $\rho_{ij}^{t-1} = 0$. The mapping process of random sampling and probability allocation is shown in Figure 3.

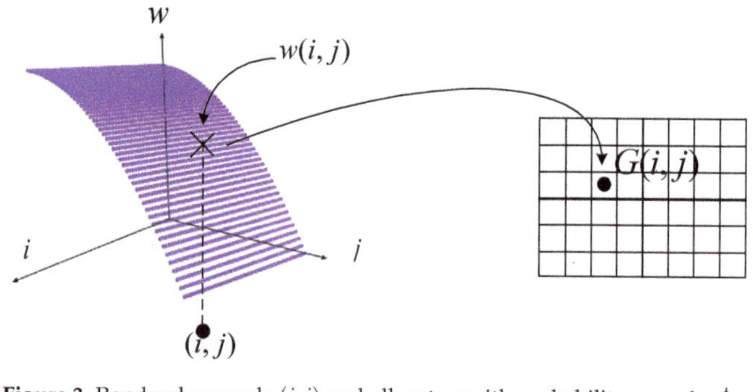

Figure 3. Randomly sample (i,j) and allocate ε with probability $\rho_{ij} = 1$, $\rho_{ij}^{t} \sim Bernoulli(\varepsilon_{c_i^t c_j^t}^{t})$.

Figure 3 shows the hypothesis of this paper. There are only two possibilities for random events: existence or non-existence of edges. In this paper, we randomly sample (i,j) and allocate ε, with probability $\rho_{ij} = 1$, $\rho_{ij}^{t} \sim Bernoulli(\varepsilon_{c_i^t c_j^t}^{t})$. Each term of the adjacency matrix A^t is independent. Therefore, Equation (10) can be rewritten as the likelihood form with parameter W^t.

$$f(A^t; W^t) = \exp\{\sum_{p=1}^{k}\sum_{q=1}^{k}[\alpha_{pq}^{t}\log\varepsilon_{pq}^{t} + \beta_{pq}^{t}\log(1-\varepsilon_{pq}^{t})]\} \tag{12}$$

where α_{pq}^{t} and β_{pq}^{t} are, respectively, denoted as follows:

$$\begin{cases} \alpha_{pq}^{t} = \sum\limits_{c_i \in p}\sum\limits_{c_j \in q}\rho_{ij}^{t} \\ \beta_{pq}^{t} = \begin{cases} |p|(|p|-1)/2 & p = q \\ |p||q| & p \neq q \end{cases} \end{cases} \tag{13}$$

3.2. A Migration Partitioning Model for Phylogenetic Evolution

The proposed dynamic community discovery method can track the status of the target over time to discover community results. Therefore, this paper constructs an observation model, which can be described by

$$S^t = W^t + Z^t, t = 0,1,2,\ldots \tag{14}$$

where Z^t is an independent Gaussian noise matrix with zero mean and variance $(\sigma^t_{pq})^2 = \varepsilon^t_{pq}(1-\varepsilon^t_{pq})/\beta^t_{pq}$. This matrix reflects the transient variations caused by noise. In this paper, we assume that $Z^t, Z^{t-1}, \ldots$ are independent of each other.

In the dynamic system model, $S^{(t)}$ expresses the set of observed values, and $W^{(t)}$ represents the state of the sequence of observed values that generate noise in the dynamic system. This paper refines the final model by modeling the evolution of specified states over time. Because ε^t is a probability between $0 \sim 1$, and this paper deals with ε^t in logarithmic form, that is, $y^t = \log(\varepsilon^t_{pq}/(1-\varepsilon^t_{pq}))$, then a time-series dynamic observation model of system evolution can be constructed as follows:

$$y^t = H^t y^{t-1} + z^t \tag{15}$$

where H^t denotes the state transition model, y^t represents the vector metric representation of matrix W^t, z^t implies the process noise, z^t is a random vector with zero mean and Θ^t is the covariance matrix. According to the vectorization expression of S^t and observation noise Z^t, the observation model (15) can be rewritten as:

$$s^t = g(y^t) + z^t \tag{16}$$

The logical activation function $g(\cdot)$ is handled by the Sigmod function, which is:

$$g(x) = \frac{1}{1+e^{-x}} \tag{17}$$

This paper assumes that the initial state of the dynamic system obeys the Gaussian distribution, namely, $y^0 \sim N(\mu^0, \Theta^0)$. The nonlinear optimization problem in the time-series dynamic observation model of system evolution is constructed in this paper, which is the problem of calculating the Maximum a Posteriori (MAP) of c^t, while for Gaussian distributions, the maximization problem can be translated into the negative logarithm problem of minimizing the target probability c^t. Therefore, Equation (12) is converted into the following logarithmic likelihood:

$$\begin{aligned} \hat{c}^t &= \log f(A^t|y^t) \\ &= \sum_{p=1}^{k}\sum_{q=1}^{k}\{\alpha^t_{pq}\log h(y^t_{pq}) + \beta^t_{pq}\log[h(1-y^t_{pq})]\} \end{aligned} \tag{18}$$

The following error function will be constructed:

$$e^t(y) = y^t - f(y^{t-1}, v^t, 0) \tag{19}$$

Then, minimize the quadratic form of the error function:

$$\min J(y) = \sum_{t=1}^{T}(\frac{1}{2}e^t(y)^T(W^t)^{-1}e^t(y)) \tag{20}$$

Finally, make the first-order expansion of $f(x)$:

$$f(x+\Delta x) \approx f(x) + J(x)\Delta x \tag{21}$$

where $J(x)$ is the derivative of $f(x)$ with respect to x, which is actually a matrix of $m \times n$, which is also Jacobian. The derivative problem can be turned into a recursive approximation problem; therefore, L–M method is adopted in this paper to determine the step size Δx, the L–M method avoids the non-singular and morbid state properties of the coefficient matrix of linear equations and can provide a more stable and accurate increment Δx. In the previous methods, as the approximate second-order Taylor expansion adopted in the GaUSs-Newton method could only have a good approximation effect near the expansion point, a trust-region is added to Δx. It should be noted that the trust-region should not be so large that the approximation is inaccurate. The approximate value in the trust region is considered to be valid; when it is outside of this region, the approximation might go wrong. The scope of the trust region is determined by the difference between the approximate model and the actual function. Determine rules: if the differences are small, let the scope be as large as possible; if the difference is large, narrow the approximation. Therefore, Equation (22) is used to judge whether the Taylor approximation is good enough or not.

$$\rho = \frac{f(x+\Delta x) - f(x)}{J(x)\Delta x} \tag{22}$$

where the numerator ρ is the decreasing value of the actual function, and the denominator is the decreasing value of the approximate model. If ρ is close to 1, then the approximation is good. If ρ is too small, meaning that the actual reduced value is far less than the approximate reduced value, then the approximate result is considered to be poor and the approximate range needs to be narrowed. On the contrary, when ρ is large, it means that the actual decline is larger than expected, and the approximate range can be enlarged.

Because the temporal dynamic observation model of system evolution constructed is nonlinear and df/dx is not easy to obtain, this paper intends to adopt an iterative method (if there is an extreme value, then convergence to approximation) to converge the approximation. The steps are shown in Algorithm 1.

Algorithm 1: Main procedure of the iterative method

1. Given an initial value x_0, radius r and parameter k
2. **for** the k-th iteration, solving:
$\min\limits_{\Delta x_k} \frac{1}{2}||f(x_k) + J(x_k)\Delta x_k||^2$, s. t. $||D\Delta x_k||_2 \leq r$
3. Compute ρ
4. **if** $\rho > 3/4$
5. $r = 2r$
6. **else if** $\rho < 1/4$
7. $r = 0.5r$
8. $x_{k+1} = x_k + \Delta x_k$
9. **if** convergence
10. **break**
11. **end**

where the limiting condition r is the radius of the trust region. In Equation (21), the incremental range is limited to a sphere of radius r, which is seen as an ellipsoid after multiplying by D. D is taken as a non-negative diagonal matrix, usually with the square root of the diagonal element $J^T J$, and it is equivalent to directly constraining Δx in the ball.

$$\min_{\Delta x_k} \frac{1}{2}||f(x_k) + J(x_k)\Delta x_k||^2 + \frac{\lambda}{2}||D\Delta x||^2 \tag{23}$$

where λ is the Lagrange multiplier. Finally, this paper needs to obtain the gradient by solving the objective function (23). Since it is an optimization problem with inequality constraints, the Lagrange multiplier is used in this paper to transform the objective function into an unconstrained optimization problem. Additionally, then the target function is transformed.

Let us expand out the square of the target function of (23).

$$\begin{aligned}
&\tfrac{1}{2}\left|\left|f(x_k)+J(x_k)\Delta x_k\right|\right|^2+\tfrac{\lambda}{2}\left|\left|D\Delta x_k\right|\right|^2\\
&=\tfrac{1}{2}(f(x_k)+J(x_k)\Delta x_k)^T(f(x_k)+J(x_k)\Delta x_k)\\
&+\tfrac{\lambda}{2}(D\Delta x_k)^T(D\Delta x_k)\\
&=\tfrac{1}{2}(\left|\left|f(x_k)\right|\right|_2^2+2f(x_k)^TJ(x_k)\Delta x_k+\Delta {x_k}^TJ(x_k)^TJ(x_k)\Delta x_k)\\
&+\tfrac{\lambda}{2}(D^T\Delta {x_k}^TD\Delta x_k)
\end{aligned} \tag{24}$$

Then, solve the derivative of Δx_k in Equation (24) and set it to zero:

$$2J(x_k)^Tf(x_k)+2J(x_k)^TJ(x_k)\Delta x_k+2\lambda D^TD\Delta x=0 \tag{25}$$

The following equations are obtained:

$$J(x_k)^TJ(x_k)\Delta x_k+\lambda D^TD\Delta x_k=-J(x_k)^Tf(x_k) \tag{26}$$

Let $J(x_k)^TJ(x_k)=H$, the right-hand side of the equation be defined as g, and the equation can be simplified as follows:

$$(H+\lambda D^TD)\Delta x_k=g \tag{27}$$

In the initial time step of the algorithm, the proposed PPPM method is initialized with the spectral clustering algorithm; that is, the initial estimation of community is generated at the time $t=1$. The advantage of using the spectral clustering algorithm as the initialization algorithm here is that it can prevent the local search from falling into c^t poor local maximum in the initial time step. The main procedure of the proposed PPPM method can be shown in Algorithm 2.

Algorithm 2: The main procedure of PPPM

Input: $G=\{G^1,G^2,\ldots,G^T\}$, k//dynamic networks and the number of communities
Output: c^t//the community

```
at t = 0
    Initialize c^0 by using spectral clustering applied on W^0
  at t > 0
    if iteration ≤ max iteration//hill-climbing algorithm
      ĉ_0^t ← −∞//negative Log of the best adjacent case till to a constant
      ċ^t ← c^t//currently being traversing case
    for i = 1 to |V^t| do//traverse all adjacent solutions
      for j = 1 to k; s.t. c_i^t ≠ j do
          ċ_i^t ← j//change community of a node
          compute y^t using Equations (15)–(17)
              compute Log ĉ_1^t using (18)
          if ĉ_1^t > ĉ_0^t then//current case is the best case
              (ĉ_0^t, c̄^t) ← (ĉ_1^t, ċ^t)
          ċ_i^t ← c_i^t//refresh community of current node
          if ĉ_0^t > ĉ^t then//the best adjacent case is better than the current best case
              (ĉ^t, c^t) ← (ĉ_0^t, c̄^t)
          else//achieve a minimum
            break
      end
end
return c^t
```

4. Results

In order to prove the rationality of the novel proposed method, four algorithms are compared, namely FaceNet [27], SBM + MLE [48], CLBM [49], and PisCES [51]. Firstly,

FaceNet was chosen because it was the first proposed dynamic web community discovery algorithm that could be compared as a baseline; secondly, SBM + MLE and CLBM were used because they are the latest proposed probabilistic model-based algorithms; finally, PisCES is also a recently proposed non-probabilistic model algorithm. In this paper, the indicators of the following two evaluation models are adopted.

(1) Adjust Rand Index (*ARI*), $ARI \in [-1.1]$, if the value of *ARI* is closer to 1, it means better results.

$$ARI = \frac{RI - \mathrm{E}(RI)}{\max(RI) - \mathrm{E}(RI)} \tag{28}$$

where $\mathrm{E}(RI)$ represents the expected value of RI and $\max(RI)$ denotes the maximum value of RI.

(2) Mean-squared errors (MSE), the smaller the value, the smaller the error, that is, the better the result.

$$MSE = \frac{1}{n}\sum_{i=1}^{m} (y_i - \hat{y}_i)^2 \tag{29}$$

where y_i is the real data, $\hat{y}_i$ expresses the fitting data, and n implies the number of samples.

Figure 4 can simulate the evolution process of an artificial dynamic network over time. The network consists of 156 nodes and 614 edges, and a total of eight time steps are set.

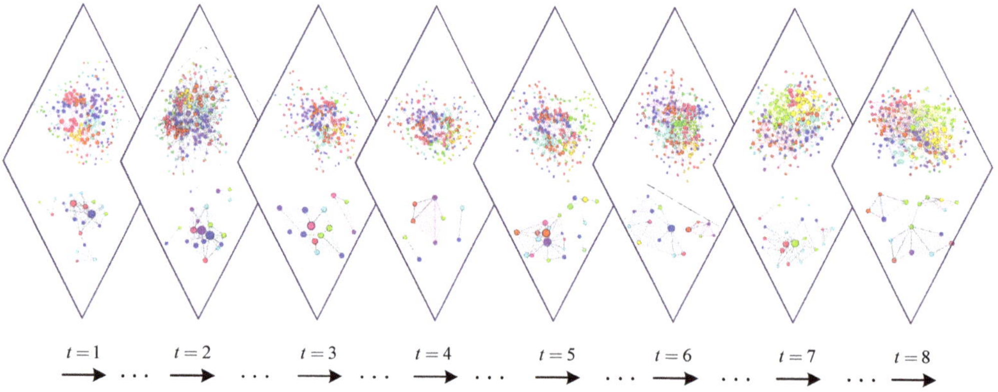

Figure 4. The simulation of the evolution process of an artificial dynamic network.

In Figure 4, there are eight rhombic blocks, and the whole dynamic social network can be represented by the evolution of these eight rhombic blocks over time. The upper part represents the dynamic network of a time step, the lower part denotes the community where the current time step may exist, and the lower part is composed of the nodes with the highest degree of nodes of each color in the network absorbing nearby nodes to form larger nodes. (Absorption rule: connected with the node with the greatest degree and with the same color). For example, in the lower half $t = 1$, there are three submodules, each of which represents a possible community. More specifically, each submodule can be composed of nodes of different colors and sizes, and each color can represent nodes with the same characteristics in the dynamic social network. it can be seen that the community structure of dynamic social networks is phylogenetic over time.

4.1. Synthetic Networks

The artificial network is generated in this paper, which consists of 128 nodes, initially divided into four communities, where each community has 32 nodes. At the initial time step, the edge probability of the system evolution migration partition is set as $\varepsilon^1_{pp} = 0.3109$ and $\varepsilon^1_{pq} = 0.0765$ ($p, q = 1, 2, 3, 4$ and $p \neq q$). The initial covariance Θ^1 is set to the identity matrix $0.04I$. The state vector G evolves according to the Gaussian random walk model, namely $H^t = I$ in Equation (15). This paper generates 25 and 50 time steps. At each time step, nodes are randomly selected to leave their communities and randomly assigned to one of the other three communities. Table 2 statistically compares the proposed PPPM method with the average *ARI* experimental results of multiple parameters of four representative models in an artificial network environment.

Table 2. The results of the proposed PPPM and representative model on the Mean ARI (synthetic data).

Time Step	Random Rate	Proposed PPPM	PisCES	CLBM	SBM + MLE	FaceNet
25	10%	**0.65702**	0.50409	0.48838	0.56099	0.46667
50		**0.56630**	0.38265	0.34544	0.49414	0.34324
25	20%	**0.72236**	0.57972	0.54694	0.66961	0.50949
50		**0.94401**	0.90616	0.89504	0.92753	0.88592
Mean		**0.72242**	0.670	0.56895	0.663068	0.55133

In Table 2, bold font indicates that the result is the best. It can be clearly seen that the proposed PPPM method has the best performance under all parameters. It can be calculated that the average performance of the novel method is improved by 0.05 compared with the other four best models.

Figure 5 shows the comparison of average ARI results between the proposed method and four representative models in an artificial network environment.

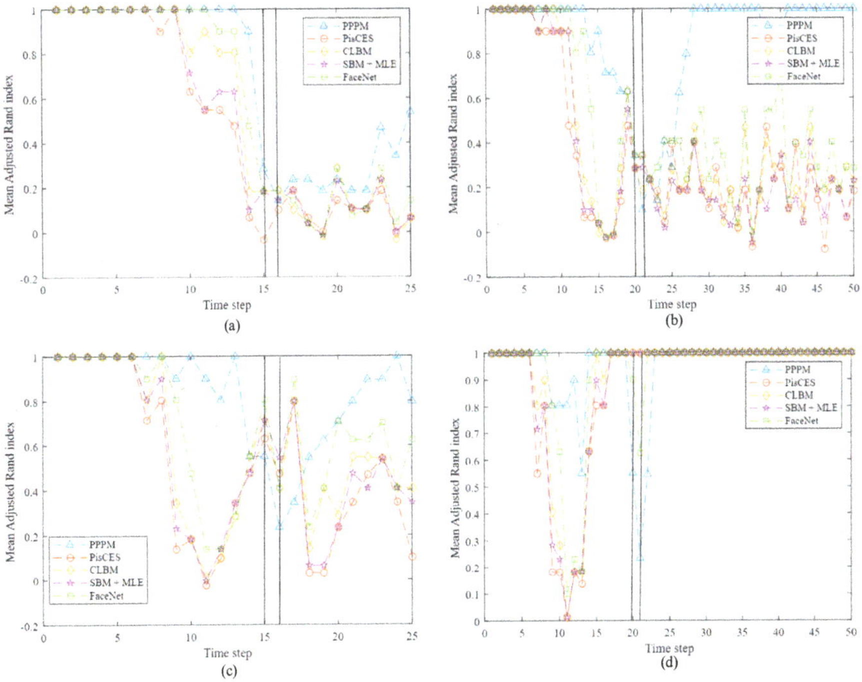

Figure 5. Comparison of the proposed model with 4 different models on Mean ARI (synthetic network). (**a**) indicates that the time step is 25 and the randomly selected parameter is 10%, (**b**) indicates that the time step is 50 and the randomly selected parameter is 10%, (**c**) indicates that the time step is 25 and the randomly selected parameter is 20%, (**d**) indicates that the time step is 50 and the randomly selected parameter is 20%.

As shown in Figure 5a, 25 time steps are generated by the artificial network. On each time step, the randomly selected parameter is set to 10%. In this experiment, the parameters of the noise item are changed at the 15th time step (the left line) and set back to the original state at the 16th time step (the right line). It is evident that in the 15th time step, the only two models with SBM and PPPM + MLE line charts show the correct change in trend, i.e., a downward trend, and the proposed PPPM method declines faster, and the increasing trend of PisPCES, CLBM, and FaceNet are unaffected and keep the previous state, after the 16th time step, which also can obviously show that compared with the other four kinds of models, the novel method callback trend is more obvious. This indicates that the proposed novel method has a more consistent response to noise terms.

In Figure 5b, the artificial network generates 50 time steps, randomly selects parameters and sets them to 10%, changes the parameters of the noise item at the 20th time step (the left line), and sets them back to the original state at the 21st time step (the right line). It is evident that in the 20th time step, the only two models with PPPM and CLBM line charts show the correct change trend, i.e., a downward trend, and PPPM declines faster, and PisPCES and FaceNet are on the rise, with SBM CLBM + MLE remaining unaffected and they to keep the previous state, after the 16th time step, which also can obviously show that compared with the other four kinds of models, PPPM callback trend is more apparent in terms of reverting to the previous state, which shows that this paper proposed the model of response that is more consistent in noise.

In Figure 5c, the artificial network generates 25 time steps, randomly selects parameters and sets them to 20%, changes the parameters of the noise item at the 15th time step (the left line), and sets them back to the original state at the 16th time step (the right line). It is obvious that at the 15th time step, the line graph of all models shows the correct trend of change, namely the downward trend. It is worth noting that the downward trend of PPPM is the most obvious. After the 16th time step, it is also obvious that compared with the other four models, the proposed novel method has a more obvious callback trend, which also indicates that the novel method has a more consistent response to the noise term.

Figure 5d shows that the artificial network generates 50 time steps, randomly selects the parameters and sets them to 20%, changes the parameters of the noise item at the 20th time step (the left line), and sets them back to the original state at the 21st time step (the right line). It is evident that in the 20th time step, only two models with PPPM and FaceNet line charts show the correct change trend downward trend, and PPPM decline faster, while the remaining three kinds of model, CLBM, SBM + MLE, PisPCES, are not affected, and keep the previous state; after the 16th time step, PPPM and FaceNet all can to go back to the previous state, which demonstrates that the proposed method has a more consistent response in noise.

In conclusion, in the artificial network, this paper proposed a dynamic community-found PPPM method compared with the other four kinds of a typical model. The model is tested in the perturbation parameter test (the noise is changed in a particular time step). The prediction accuracy of the model index (ARI) increased by 5% on average, and the experimental results show that the proposed model is robust.

4.2. *Real-World Networks*

4.2.1. MIT Reality Mining

This experiment is conducted on the MIT dataset [52]. The dataset is collected by recording the mobile phone activity of 94 students and employees over a year. The dataset built a dynamic network based on physical distance, which is measured by scanning nearby Bluetooth devices every 5 min. Data collected near the beginning and end of experiments with low participation rates are excluded in this experiment. Each time step corresponds to one week, so there are 37 time steps between August 2004 and May 2005. Figure 6 shows the mean-variance error results of the proposed novel method and four representative models under the artificial network.

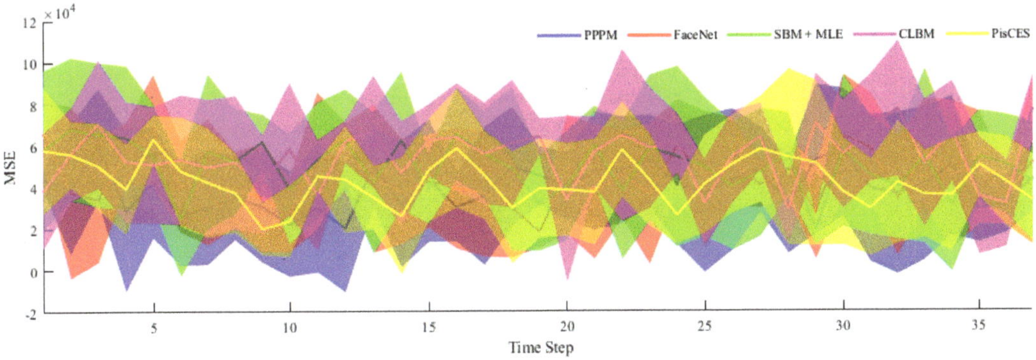

Figure 6. Comparison of the proposed method with 4 different models on MSE (synthetic network).

Figure 6 shows that, under the MSE evaluation index, the smaller the error, the better the result; that is, the closer the model image is to the x-axis. Obviously, compared with other colors (the other four models), the image with blue color (the proposed method in this paper) is closer to the x-axis; that is, the proposed PPPM method has a lower MSE value and a smaller error. Table 3 compares the average ARI results of the proposed method with those of the four representative models in the real network (MIT reality mining) environment.

Table 3. The results of the proposed PPPM method and representative model on the Mean ARI (real data).

	FaceNet	PPPM	SBM + MLE	CLBM	PisCES
Max	0.8991	**0.9555**	0.8678	0.8876	0.9412
75%	0.7125	**0.8005**	0.7856	0.6258	0.7811
Median	0.4902	**0.6523**	0.6215	0.4981	0.6536
25%	0.3154	**0.5111**	0.4992	0.2314	0.4902
Min	0.1002	**0.40**	0.2671	0.1487	0.3243

In Table 3, the bold font shows that the result is the best, you can clearly see that the proposed PPPM in all parameters (the maximum value; the first 75% of the value; the median; the first 25% of the value; minimum value) cases are the best and clear, the average performance of PPPM performance (median) than the best model is increased by 3% in the other four. Figure 7 shows the comparison of MARI values on the MIT dataset between the proposed method and four state-of-the-art models.

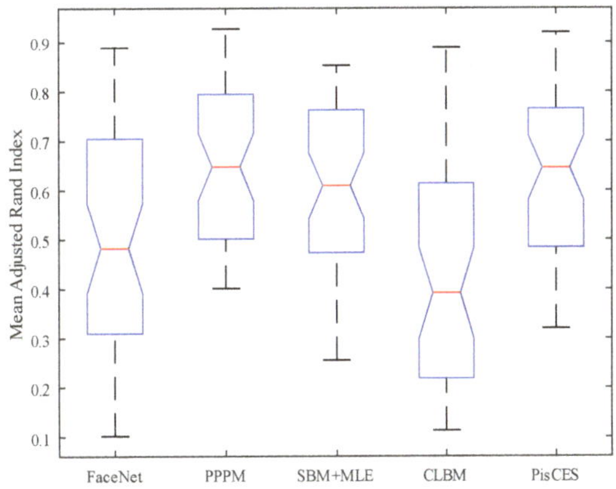

Figure 7. Comparison of the proposed model with 4 different models on MARI (reality network).

In Figure 7, the upper and lower edges of each box in the boxplot represent 25% and 75% values, respectively, and the middle red line denotes the median. It is obvious that PPPM, SEM + MLE, and the three boxes perform better than the other two model boxes. Among the three models with better performance, the PPPM box position is slightly higher than that of SEM + MLE and PisCES boxes, and the median value is also slightly higher than that of SEM + MLE and PisCES models. In conclusion, compared with the other four representative models in the real network, the proposed dynamic community PPPM method performs better under the two evaluation indexes of prediction accuracy and error.

4.2.2. Enron Email Data

The experiment is conducted on the dynamic social network, which is built by Enron [53], and it consisted of about 500,000 emails between 184 Enron employees from 1998 to 2002. The directional edge between the employee and the time point occurs if at least one email is sent within the first week. Each time step corresponds to an interval of 1 week. This dataset does not distinguish between emails sent to "recipients," "CC" or "BCC." In addition to email dataset, most employee roles (such as CEO, president, manager, employee) exist within the company and they are used as known communities. The first 56 weeks and the last 13 weeks are filtered because only a few emails are sent. Figure 8 compares the estimated community probability between a normal week and an event week. The higher the probability, the higher the community activity. Both the x-axis and the y-axis denote the estimated communities, and the color blocks on the diagonals express the activity within each community, and the color blocks of the diagonals imply the activity between each community.

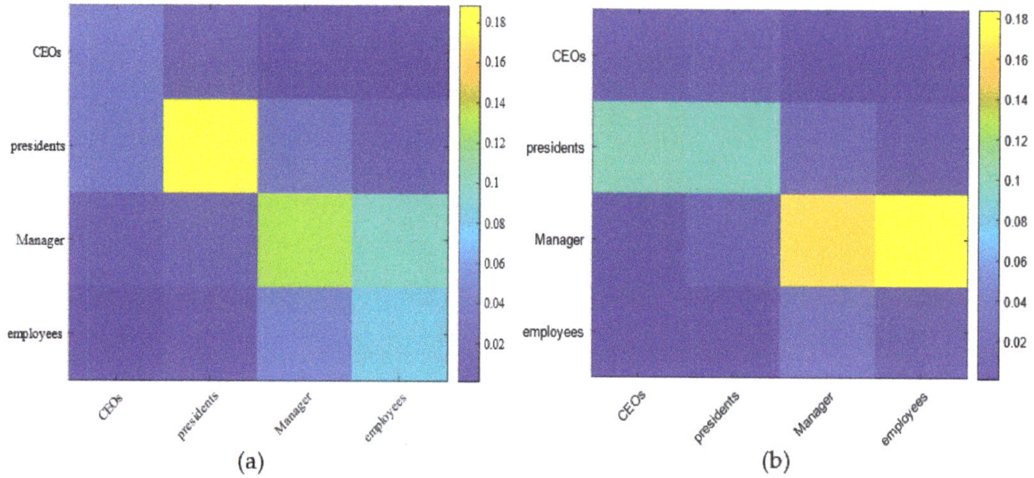

Figure 8. The comparison of community probability in normal week and event week. (**a**) indicates the normal week(week 59), (**b**) indicates the event week (the 89 weeks when CEO Jeffrey resigns).

As shown in Figure 8a, in a normal week (week 59), the president community is the most active, followed by managers and employees, and the CEO community is the least active. It is also worth noting that from the color block distribution of managers and employees, the two communities may merge into one large community. This phenomenon can be reflected in the fact that communication between department managers and employees is usually close, and managers and employees are more likely to get along with each other. As shown in Figure 8b, in the event week (the 89 weeks when CEO Jeffrey resigns), the most active community is that of the managers, followed by president community, and the brightest color block is the managers to the employee community. This is reflected in the fact that in real life when CEOs resign, the discussion is most intense among managers because it is directly related to their personal interests. Discussions between managers and employees also proliferate for the same simple reason that it is indirectly related to the employees' personal interests. Figure 9 reveals the estimated edge connections between communities in Enron's email network under the proposed dynamic community's discovery approach PPPM and shows a 95% confidence interval (note: the lines on the left and right of the figure are for weeks 59 and 89, respectively).

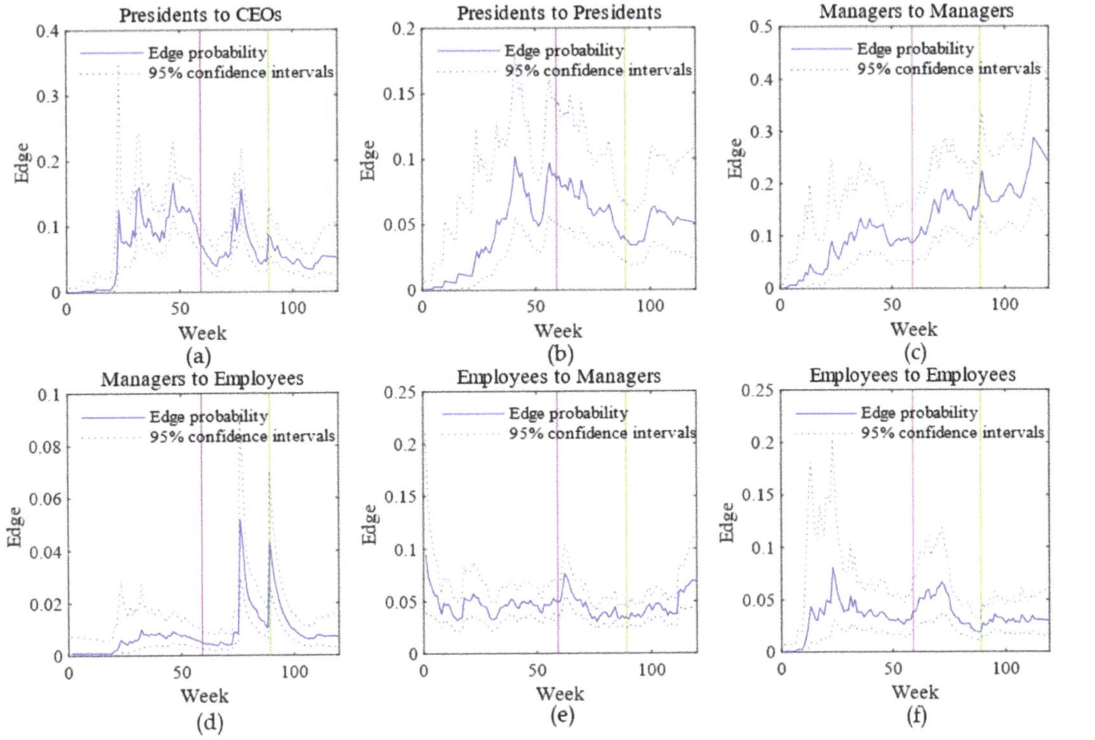

Figure 9. Probability of edges between communities on the Enron mail. (**a–f**) are the edge probabilities between different roles.

As shown in Figure 9a, it is the edge probability of presidents to CEOs; it can be seen that the presidents to CEOs edge probability increased slightly at week 59 (normal) and 89 (Jeffrey CEO resigned), which corresponds to presidents to CEOs activity (increased) from Figure 8a,b. Figure 9b shows the side probability within the president community. In the 59th and 89th weeks, the side probability inside the community shows a downward trend. This also corresponds to the active state (decreased activity) within the president community from Figure 8a,b. Figure 9c,d show the side probabilities between managers and the manager community and between the managers and the employee community, respectively. It can be seen that in the 59th and 89th weeks, the changing trend of the side probabilities of these two communities is consistent with that in Figure 9a.

Similarly, this change also corresponds to the changes in active state between managers and manager community and the employee community (increased activity) from Figure 8a,b. Figure 9e shows the edge probability between the employees and the manager community. It is not difficult to see that there is no obvious trend of change in week 59 and 89. Similarly, this situation also corresponds to the consistent change in the active state between the employees and the manager community from Figure 8a,b (there is no significant change in the activity). Finally, Figure 9f shows the edge probability between employees and the employee community. In week 59 and 89, similarly, the changing trend of edge probability of these two communities is consistent with the change in Figure 9b; namely, it displays the downward trend. At the same time, it also corresponds to the consistent change in active state between employees and the employee community from Figure 8a,b (decreased activity).

To sum up, the proposed PPPM method can well reflect some phenomena existing in the real network, and the probability estimated by the novel method can make relatively consistent predictions with the advance of time and the occurrence of specific events.

5. Conclusions

The proposed model has practical theoretical and practical significance to mine and it also simulates deeper hidden information that is present in dynamic social networks. At present, the dynamic social network community discovery method cannot effectively represent the entire dynamic network evolution process. Therefore, inspired by the evolution theory of natural biosensors, this paper proposes a community discovery method based on phylogenetic planted partition. Firstly, the time dimension is added to the transplant partition model, all states in the whole dynamic network system are treated as variables, the observation equation is used as a constraint between variables, and an error function about the whole dynamic network system is constructed. Then, the quadratic form of the error function is minimized, which can abstract the observation results of the network more realistically. Secondly, a graph optimization strategy is used to consider the constraints in the whole motion trajectory at one time, and the Jacobian matrix is calculated during the linearization process. Because the calculation process is relative to the whole motion trajectory, the whole system evolution process is transformed into a nonlinear system optimization process. The gradient of the error function is obtained by using the L–M method, and then the iteration is carried out according to the direction of the gradient; finally, the proposed method is compared with four state-of-the-art representative models under two scenarios of artificial network and real network. The experimental results show that the PPPM method has better performance than the other four representative models in building a dynamic network model and mining dynamic network hidden information.

Next, this paper will consider how to integrate the multi-layer model mechanism into the proposed model and will study dynamic network hiding information with multi-layer information in future research.

Author Contributions: Conceptualization, X.L. and N.D.; methodology, N.D.; software, N.D.; formal analysis, N.D.; writing—original draft preparation, N.D.; writing—review and editing, G.F., P.D.M. and A.F.; visualization, N.D.; supervision, X.L.; project administration, X.L.; funding acquisition, X.L. All authors have read and agreed to the published version of the manuscript.

Funding: This work was supported in part by National Social Science Fund of China (17XXW004), Science and Technology Research Project of Chongqing Municipal Education Commission (KJZD-K202001101), Humanities and Social Sciences Research Project of Chongqing Municipal Education Commission (20SKGH166),Postgraduate Innovation Fund of Chongqing University of Technology (ycx20192060), Chongqing Postgraduate Research Innovation Project CYS20343, Chongqing Ba-nan District Science and Technology Bureau Science and Technology Talents Special Project (2020.58), General Project of Chongqing Natural Science Foundation (cstc2021jcyj-msxmX0162), 2021 National Education Examination Research Project (GJK2021028).

Institutional Review Board Statement: Not applicable.

Informed Consent Statement: Not applicable.

Data Availability Statement: The calculated data presented in this work are available from the corresponding authors upon reasonable request. Figures 7–9 contain data available in Refs. [52,53].

Conflicts of Interest: The authors declare no conflict of interest.

References

1. Newman, M.E. The structure and function of complex networks. *SIAM Rev.* **2003**, *45*, 167–256. [CrossRef]
2. Dakiche, N.; Benbouzid, F.T.; Slimani, Y.; Benatchba, K. Tracking community evolution in social networks: A survey. *Inf. Process. Manag.* **2018**, *56*, 1084–1102. [CrossRef]
3. Girvan, M.; Newman, M.E. Community structure in social and biological networks. *Proc. Natl. Acad. Sci. USA* **2002**, *99*, 7821–7826. [CrossRef] [PubMed]
4. Guruharsha, K.; Rual, J.F.; Zhai, B.; Mintseris, J.; Vaidya, P.; Vaidya, N.; Beekman, C.; Wong, C.; Rhee, D.Y.; Cenaj, O. A protein complex network of Drosophila melanogaster. *Cell* **2011**, *147*, 690–703. [CrossRef]
5. Pagani, G.A.; Aiello, M. The power grid as a complex network: A survey. *Physica A* **2013**, *392*, 2688–2700. [CrossRef]
6. Sanchez, F.; Mina, M. Oncogenic signaling pathways in the cancer genome atlas. *Cell* **2018**, *173*, 321–337. [CrossRef]
7. Baccaletti, S.; Bianconi, G.; Criado, R. The structure and dynamics of multilayer networks. *Phys. Rep.* **2010**, *544*, 1–122. [CrossRef]
8. Ma, X.; Sun, P.; Zhang, Z. An integrative framework for protein interaction and methylation data to discover epigenetic modules, IEEE/ACM Trans. *Comput. Biot. Bioinf.* **2019**, *16*, 1855–1866.
9. Ma, X.; Dong, D.; Wang, Q. Community detection in multi-layer networks using joint nonnegative matrix factorization. *IEEE Trans. Knowl. Data Eng.* **2019**, *31*, 273–286. [CrossRef]
10. Huang, Z.; Rege, X. Detecting community in attributed networks by dynamically exploring node attributes and topological structure. *Knowl. Based Syst.* **2020**, *196*, 105760. [CrossRef]
11. Džamić, D.; Aloise, D.; Mladenović, N. Ascent–descent variable neighborhood decomposition search for community detection by modularity maximization. *Ann. Oper. Res.* **2019**, *272*, 273–287. [CrossRef]
12. Karrer, B.; Newman, M.E. Stochastic blockmodels and community structure in networks. *Phys. Rev. E* **2011**, *83*, 016107. [CrossRef] [PubMed]
13. Wen, Y.M.; Huang, L.; Wang, C.D.; Lin, K.Y. Direction recovery in undirected social networks based on community structure and popularity. *Inform. Sci.* **2019**, *473*, 31–43. [CrossRef]
14. He, D.; Feng, Z.; Jin, D.; Wang, X.; Zhang, W. Joint identification of network communities and semantics via integrative modeling of network topologies and node contents. In Proceedings of the Thirty-First AAAI Conference on Artificial Intelligence, San Francisco, CA, USA, 4–9 February 2017; pp. 116–124.
15. Airoldi, E.M.; Blei, D.M.; Fienberg, S.E.; Xing, E.P. Mixed membership stochastic blockmodels. *J. Mach. Learn. Res.* **2008**, *9*, 1981–2014.
16. Qiao, M.; Yu, J.; Bian, W.; Li, Q.; Tao, D. *Improving Stochastic Block Models by Incorporating Power-Law Degree Characteristic*; IJCAI: Melbourne, Australia, 2017; pp. 2620–2626.
17. Fortunato, S. Community detection in graphs. *Phys. Rep.* **2010**, *486*, 75–174. [CrossRef]
18. Fortunato, S.; Hric, D. Community detection in networks: A user guide. *Phys. Rep.* **2016**, *659*, 1–44. [CrossRef]
19. Rand, D.; Christakis, N. Dynamic social networks promote cooperation in experiments with humans. *Proc. Natl. Acad. Sci. USA* **2011**, *108*, 19193–19198. [CrossRef]
20. Chiang, A.; Massagie, J. Molecular basis of metastasis. *N. Engl. J. Med.* **2008**, *359*, 927–932. [CrossRef]
21. Kim, M.; Han, J. A particle-and-density based evolutionary clustering method for dynamic networks. *Proc. VLDB Endow.* **2009**, *2*, 622–633. [CrossRef]
22. Folino, F.; Pizzuti, C. An evolutionary multi-objective approach for community discovery in dynamic networks. *IEEE Trans. Knowl. Data Eng.* **2014**, *26*, 1838–1852. [CrossRef]
23. Chi, Y.; Song, X.; Zhou, D.; Hino, K.; Tseng, B.L. On evolutionary spectral clustering. *ACM Trans. Knowl. Data Discov.* **2009**, *3*, 1–30. [CrossRef]
24. Wang, L.; Rege, M. Low-rank kernel matrix factorization for large-scale evolutionary clustering. *IEEE Trans. Knowl. Data Eng.* **2012**, *24*, 1036–1050. [CrossRef]
25. Ma, X.; Dong, D. Evolutionary nonnegative matrix factorization algorithms for community detection in dynamic networks. *IEEE Trans. Knowl. Data Eng.* **2017**, *29*, 1045–1058. [CrossRef]
26. Ma, X.; Zhang, B.; Ma, C.; Ma, Z. Co-regularized Nonnegative Matrix Factorization for Evolving Community Detection in Dynamic Networks. *Inf. Sci.* **2020**, *528*, 265–279. [CrossRef]
27. Lin, Y.; Zhu, S. Analyzing communities and their evolutions in dynamic social networks. *ACM Trans. Knowl. Discov. Data* **2009**, *3*, 1–31. [CrossRef]
28. Yang, T.; Chi, Y. Detecting communities and their evolutions in dynamic social networks-a bayesian approach. *Mach. Learn.* **2011**, *82*, 157–189. [CrossRef]
29. Palla, G.; Barabási, A.L.; Vicsek, T. Quantifying social group evolution. *Nature* **2007**, *446*, 664–667. [CrossRef]
30. Cazabet, R.; Amblard, F. Dynamic Community Detection. In *Encyclopedia of Social Network Analysis and Mining*; Springer: New York, NY, USA, 2014; pp. 404–414.
31. Rossetti, G.; Cazabet, R. Community Discovery in Dynamic Networks: A Survey. *ACM Comput. Surv.* **2017**, *51*, 1–37. [CrossRef]
32. Aynaud, T.; Fleury, E.; Guillarme. Communities in evolving networks: Definitions, detection, and analysis techniques. In *Dynamics on and of Complex Networks*; Springer: New York, NY, USA, 2013; Volume 2, pp. 159–200.
33. Hartmann, T.; Kappes, A.; Wagner, D. Clustering Evolving Networks. In *Algorithm Engineering*; Springer: New York, NY, USA, 2016; Volume 9220, pp. 280–329.

34. Agarwal, M.; Ramamritham, K.; Bhide, M. Real time discovery of dense clusters in highly dynamic graphs: Identifying real world events in highly. dynamic environments. *Proc. VLDB Endow.* **2012**, *5*, 980–991. [CrossRef]
35. Tang, L.; Liu, H. Identifying evolving groups in dynamic multimode networks. *IEEE Trans. Knowl. Data Eng.* **2012**, *24*, 72–85. [CrossRef]
36. Sun, J.; Faloutsos, C. Graphscope: Parameter-free of large time evolving-graph. In Proceedings of the 13th Conference on Knowledge Discovery Data Mining, New York, NY, USA, 12–15 August 2007; pp. 687–696.
37. Chakrabarti, D.; Kumar, R.; Tomkins, A. Evolutionary clustering. In Proceedings of the 12th ACM SIGKDD International Conference on Knowledge Discovery and Data Mining, New York, NY, USA, 20–23 August 2006; pp. 554–560.
38. Chi, Y.; Song, X.D.; Zhou, D.Y.; Koji, H.; Belle, L.T. Evolutionary spectral clustering by incorporating temporal smoothness. In Proceedings of the 13th ACM SIGKDD International Conference on Knowledge Discovery and Data Mining, New York, NY, USA, 12–15 August 2007; pp. 153–162.
39. Folino, F.; Pizzuti, C. Multiobjective evolutionary community detection for dynamic networks. In Proceedings of the Conference on Genetic and Evolutionary Computation, Oregon, Portland, 7–11 July 2010; pp. 535–536.
40. Gong, M.G.; Zhang, L.J.; Ma, J.J. Community detection in dynamic social networks based on multi-objective immune algorithm. *J. Comput. Sci. Technol.* **2012**, *27*, 455–467. [CrossRef]
41. Xu, K.S.; Kliger, M.; Hero, A.O., III. Adaptive evolutionary clustering. *Data Min. Knowl. Discov.* **2014**, *28*, 304–336. [CrossRef]
42. Han, Q.; Kevin, X.; Edoardo, A. Consistent estimation of dynamic and multilayer block models. In Proceedings of the 32th International Conference on Machine Learning, Lille, France, 6–11 July 2015; pp. 1511–1520.
43. Kevin, X. Stochastic block transition models for dynamic networks. In Proceedings of the 18th International Conference on Artificial Intelligence and Statistics, San Diego, California, USA, 9–12 May 2015; pp. 1079–1087.
44. Zhang, X.; Moore, C.; Newman, M.E.J. Random graph models for dynamic networks. *Eur. Phys. J. B* **2017**, *90*, 1–14. [CrossRef]
45. Amir, G.; Pan, Z.; Aaron, C.; Cristopher, M.; Leto, P. Detectability Thresholds and Optimal Algorithms for Community Structure in Dynamic Networks. *Phys. Rev. X* **2016**, *6*, 031005.
46. Sharmodeep, B.; Shirshendu, C. Spectral clustering for multiple dissociative sparse networks. *arXiv* **2017**, arXiv:1805.10594.
47. Paolo, B.; Fabrizio, L.; Piero, M.; Daniele, T. Detectability thresholds in networks with dynamic link and community structure. *arXiv* **2017**, arXiv:1701.05804.
48. Mehrnaz, A.; Theja, T. Block-Structure Based Time-Series Models for Graph Sequences. *arXiv* **2018**, arXiv:1804.08796.
49. Étienne, G.; Anthony, C.; Mustapha, L.; Hanane, A.; Loïc, G. Conditional Latent Block Model: A Multivariate Time Series Clustering Approach for Autonomous Driving Validation. *arXiv* **2020**, arXiv:2008.00946.
50. Emmanuel, A. Community Detection and Stochastic Block Models: Recent Developments. *J. Mach. Learn. Res.* **2017**, *18*, 1–86.
51. Liu, F.; Choi, D. Global spectral clustering in dynamic networks. *Proc. Natl. Acad. Sci. USA* **2018**, *115*, 927–932. [CrossRef] [PubMed]
52. Eagle, N.; Pentland, A.S.; Lazer, D. Inferring friendship network structure by using mobile phone data. *Proc. Natl. Acad. Sci. USA* **2009**, *106*, 15274–15278. [CrossRef] [PubMed]
53. Klimt, B.; Yang, Y. The enron corpus: A new dataset for email classification research. In Proceedings of the European Conference on Machine Learning, Pisa, Italy, 20–24 September 2004; Springer: Berlin, Germany, 2004; pp. 217–226.

applied sciences

MDPI

Article

A Graph-Cut-Based Approach to Community Detection in Networks

Hyungsik Shin [1], Jeryang Park [2] and Dongwoo Kang [1,*]

1 School of Electronic and Electrical Engineering, Hongik University, Seoul 04066, Korea; hyungsik.shin@hongik.ac.kr
2 Department of Civil and Environmental Engineering, Hongik University, Seoul 04066, Korea; jeryang@hongik.ac.kr
* Correspondence: dkang@hongik.ac.kr

Abstract: Networks can be used to model various aspects of our lives as well as relations among many real-world entities and objects. To detect a community structure in a network can enhance our understanding of the characteristics, properties, and inner workings of the network. Therefore, there has been significant research on detecting and evaluating community structures in networks. Many fields, including social sciences, biology, engineering, computer science, and applied mathematics, have developed various methods for analyzing and detecting community structures in networks. In this paper, a new community detection algorithm, which repeats the process of dividing a community into two smaller communities by finding a minimum cut, is proposed. The proposed algorithm is applied to some example network data and shows fairly good community detection results with comparable modularity Q values.

Keywords: community detection; graph cut; betweenness centrality; modularity

Citation: Shin, H.; Park, J.; Kang, D. A Graph-Cut-Based Approach to Community Detection in Networks. *Appl. Sci.* **2022**, *12*, 6218. https://doi.org/10.3390/app12126218

Academic Editors: Giacomo Fiumara, Pasquale De Meo, Xiaoyang Liu and Annamaria Ficara

Received: 6 May 2022
Accepted: 16 June 2022
Published: 18 June 2022

1. Introduction

Graphs consisting of vertices and edges can be used to model various aspects of our lives and real-world environments. For example, social media services can model each service subscriber as a vertex and a friend relationship between two individuals as an edge connecting the two corresponding vertices. Another example would be a water supply network; a pipe network connecting many water sources and consumers can be modeled as a graph. The Internet and the world wide web can be modeled as networks as well.

As networks can represent many abstract contexts of our lives, it is a natural desire to try to discover a core structure inherent in them. Partitioning a network via grouping vertices of similar affiliations can help us to grasp the main structure of a given network with a holistic viewpoint. In other words, a community structure of a network can give us a better understanding of its characteristics, properties, and inner workings. An example of community detection of a network is shown in Figure 1.

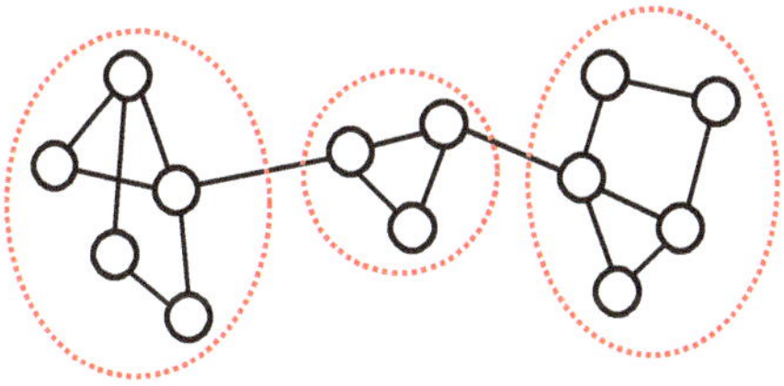

Figure 1. An example of community detection of a network.

There has been significant research on detecting and evaluating community structures in networks. Many fields, including social sciences, biology, engineering, computer science, and applied mathematics, have identified their own needs for developing methods to analyze and detect community structures in networks. Graph partitioning problems in graph theory are also closely related to the community detection problem. Many algorithms for community detection and clustering have been proposed so far, and there are a few extensive review papers on them [1,2].

The main contribution of this paper is to propose a new community detection algorithm, which repeats the process of dividing a community into two smaller communities by finding a minimum cut of the community. In order to compute the minimum cut of a community, the betweenness centrality values of vertices are used to choose the source and sink nodes. The modularity criterion is used in a greedy way to determine the local optimal partition among many candidate communities to be cut.

The rest of this paper is organized as follows. Section 2 briefly reviews prior work on community detection methods. Section 3 presents the proposed algorithm, and Section 4 illustrates the results of the algorithm applied to a few example networks. Finally, Section 5 discusses the results and concludes the paper.

2. Prior Works

Uncovering community structures inherent in networks can shed light on many aspects of our world because networks can model various relations and inner workings among entities and objects. Therefore, community detection in networks has received a lot of attention from many fields [2–7], and it is rapidly evolving. Scientists and researchers from various fields have accordingly produced a large number of research articles and study results .

The community detection problem has been studied mainly for static networks, but the problem has also been studied for dynamic networks as well as overlapping communities recently [8–14]. At their initial development phase, community detection algorithms concerned disjoint communities, where each vertex of a network belongs to only one community. However, many real-world networks such as social networks can have overlapping communities, so that a vertex is permitted to belong to several communities at the same time. Even though the problems for dynamic networks or overlapping communities are interesting and worth tackling, this paper limits its scope to methods for disjoint community detection of static networks.

Since it is unrealistic to review all the previously proposed algorithms for community detection, only a few representative algorithms are mentioned in this section. More detailed reviews and studies regarding community detection can be accessed through several extensive review articles and books [2–5,15–17].

Even though many algorithms for community detection have been proposed so far, it seems that there is no single algorithm that can detect communities universally well for many kinds of various networks [18,19]. Therefore, it is believed to be better to have multiple different algorithms and the apply suitable one to each particular network. Most algorithms proposed for disjoint community detection of static networks may be classified into three categories: traditional, modularity-based, and dynamic algorithms [2].

Traditional algorithms include hierarchical clustering [1,20–24], Girvan–Newman (GN) algorithms [25–28] and their variants [29–31], spectral clustering [32–39], and graph-partitioning-based algorithms [39,40]. Most of these algorithms are based on clustering, and they have provided basic concepts of community detection for later developments.

Modularity-based algorithms include the Guimera and Amaral algorithm [41], the fast GN algorithm [27], the Clauset algorithm [42], optimization-based algorithms [25,41,43–46], and genetic algorithms [47–51]. The concept of modularity was introduced by Newman and Girvan [25], and modularity-based algorithms try to optimize the modularity value by finding a good community structure in a network via some heuristics. The concept of modularity is briefly reviewed in Section 3.

Dynamic algorithms include those that are based on spin models, random walk, and synchronization. A random walk can be used for community detection by letting an entity make random moves along the edges of a network [52–55]. More details on dynamic algorithms can be found in review papers and books [2–5].

Since many algorithms and methods for community detection have been proposed, the task of evaluating them becomes important as well. To measure the performance of the algorithms and methods, a few standard benchmarks of complex networks with known community structures have been devised. Among these benchmarks, the GN and LFR (Lancichinetti, Fortunato, and Radicchi) benchmarks are frequently used [56,57]. These benchmark networks are synthesized with predetermined community structures, and an algorithm is evaluated by comparing the result of community detection to the known answer, i.e., the predetermined community structure. Besides these synthetic benchmarks, there are some real-world network data that are frequently used to evaluate community detection algorithms. In Section 4, some of these real-world network data are used to illustrate the performance of the algorithm proposed in this paper.

3. Methods

The community detection method proposed in this paper is based on the idea that connections between vertices of the same community tend to be denser than connections between vertices of different communities. Since a minimum cut in graph theory finds as few connections to a cut as possible, applying a minimum cut to a network may give us boundaries between different communities. In other words, a minimum cut might result in a cut-set that contains more inter-connections between communities than intra-connections within communities.

The proposed method of this paper iteratively applies the minimum cut solution to a network to generate a community structure. The method first finds two temporary source and sink nodes that are expected to be of different communities, and computes a minimum cut between the two nodes. As a result of the cut, the network is divided into two groups. Then, the same procedure is applied to each group, and quality measures of the resulting community structures of the entire network are compared to select the better result between the two. The same procedure is repeated until no further refinement of the quality of community structure is obtained. A precise description of the proposed algorithm is presented later in this section.

3.1. Network and Community Structure

To describe the problem and the proposed algorithm precisely, some basic notations are introduced.

Definition 1. *A* network *is a graph $G = (V, E)$ defined by V and E, where V and E are the set of vertices (nodes) and the set of edges, respectively. An edge is a pair of two vertices. Given a network $G = (V, E)$ and a subset $C \subset V$ of vertices, $E(C) \subset E$ is a subset of E consisting of all the edges whose endpoints belong to C.*

This paper concerns undirected graphs without loops or multiple edges. even though the proposed method may be easily extended to the case of directed graphs. Therefore, an edge $e = \{u, v\} \in E$ is a pair of two vertices $u, v \in V$, and the order of the two is not relevant unless it is noted otherwise. We also assume that the graph is *connected*; every pair of vertices (u, v) of G has a path from u to v. In this paper, networks are used interchangeably with graphs.

Definition 2. *A* community structure *$C = \{C_1, C_2, \ldots, C_k\}$ of a network $G = (V, E)$ is a* partition *of V, the set of vertices of G. In other words, $\cup_{i=1}^{k} C_i = V$ and $C_i \cap C_j = \emptyset$ for $i, j = 1, \ldots, k$ with $i \neq j$. Here, k refers to the number of communities of the community structure C.*

With these definitions, the objective of the community detection problem can be stated as finding a good community structure C when a network G is given.

3.2. Minimum Cut

Since the minimum cut in graph theory is employed in the proposed method, a brief review is provided in this section.

Definition 3. *A* cut $C = (S, T)$ *is a partition of the set of vertices V of a graph* $G = (V, E)$ *into two subsets S and T. The* cut-set *of a cut* $C = (S, T)$ *is the set* $\{\{u, v\} \in E \mid u \in S, v \in T\}$ *of edges that have one endpoint in S and the other endpoint in T. If s and t are specified vertices of the graph G, then an* s-t cut *is a cut* $C = (S, T)$ *such that* $s \in S$ *and* $t \in T$.

In this paper, we consider networks whose edges have no weight values assigned, i.e., unweighted and undirected graphs. However, graphs may have edge weights in general, and the *weight* of a cut in an undirected graph is defined as follows.

Definition 4. *In an unweighted and undirected graph, the* size *or* weight *of a cut is defined to be the number of edges that belong to the cut-set of the cut. In a weighted graph, the* value *or* weight *of a cut is defined to be the sum of the weights of the edges in the cut-set.*

Given the definitions of a *cut* and the *weight* of a cut, a minimum cut of a network is defined as follows.

Definition 5. *A* minimum cut *of an undirected graph is a cut whose weight is not larger than the weight of any other cut.*

It is well known in network science that the maximum flow of a flow network from a source vertex s to a sink vertex t is equal to the capacity of a minimum cut that separates the two vertices s and t. This result is called the max-flow min-cut theorem and was introduced by Ford and Fulkerson [58]. The Ford–Fulkerson method can be used to find a minimum cut and a maximum flow, and the Edmonds–Karp algorithm is an implementation of the method that runs in polynomial time [59,60].

Even though the max-flow min-cut theorem deals with a directed graph with capacity values assigned on edges, it is straightforward to extend the theorem to undirected and weighted graphs. Besides the Ford–Fulkerson method, there are other algorithms for finding a minimum cut of undirected edge-weighted graphs such as the Stoer–Wagner algorithm [61,62]. The Stoer–Wagner algorithm has the time complexity of $O(|V||E| + |V|^2 \log |V|)$ [62].

Since the max-flow min-cut theorem is well known and there are many algorithms for finding a minimum cut of undirected graphs, we denote such an algorithm as MINIMUM-CUT without a detailed description of the procedure.

Definition 6. *Given a connected graph* $G = (V, E)$*, a capacity function c, and source and sink vertices s and t, the procedure* MINIMUM-CUT(G, c, s, t) *is an implementation of a method that finds a minimum cut of the graph G with c, s, and t.*

The capacity function c in Definition 6 gives the weight value of each edge of a weighted graph, i.e., a capacity is used interchangeably with a weight in this paper. The MINIMUM-CUT procedure is employed in our proposed algorithm, and more details are described later.

3.3. The Betweenness Centrality

To apply MINIMUM-CUT to a graph, source and sink vertices s and t should be provided as well as the capacity c for each edge. For community detection, the source and sink vertices should be chosen so that the two vertices do not belong to the same community, because the two vertices are cut by MINIMUM-CUT. The *betweenness centrality* in graph theory is used to select two vertices in our proposed method. In particular, the two vertices of highest betweenness centrality values are chosen as the source and sink vertices in applying MINIMUM-CUT.

In graph theory, betweenness centrality is a measure of the centrality of a vertex in a graph [63]. Roughly speaking, the betweenness centrality of a vertex represents the degree of how many pairs of nodes are connected via the shortest paths passing through the vertex. The betweenness centrality $c_B(v)$ of a vertex $v \in V$ of a graph $G = (V, E)$ based on shortest paths is defined as the following

$$c_B(v) = \sum_{s,t \in V} \frac{\sigma(s,t|v)}{\sigma(s,t)}, \quad (1)$$

where $\sigma(s,t)$ is the number of shortest paths from $s \in V$ to $t \in V$, and $\sigma(s,t|v)$ is the number of those paths passing through a vertex v other than s and t [64,65]. If $s = t$, we let $\sigma(s,t) = 1$, and if $v \in \{s,t\}$, we let $\sigma(s,t|v) = 0$. On unweighted graphs, computing betweenness centrality takes $O(|V||E|)$ times using Brandes' algorithm [64].

As briefly mentioned before, the betweenness centrality values for all the vertices are computed to select the source and sink vertices s and t, which are provided as inputs to MINIMUM-CUT. Those two vertices of the highest values are chosen because it is expected that many shortest paths connecting vertices of different communities pass through these two nodes.

3.4. *The DIVIDE-INTO-TWO Algorithm*

To find a community structure of a given network, the proposed algorithm of this paper iterates by dividing a group of vertices into two smaller groups until a stopping criterion is satisfied. Before the algorithm is described precisely, we present the procedure of dividing a group of vertices into two smaller groups in Algorithm 1, called DIVIDE-INTO-TWO.

Algorithm 1 The DIVIDE-INTO-TWO algorithm

Input: A connected graph $G = (V, E)$
Output: $C = \{C_1, C_2\}$, a partition of G

procedure DIVIDE-INTO-TWO(G)
    if $|V| \leq 1$ **then**
        $C \leftarrow \{V\}$
    else
        Compute betweenness centrality $c_B(v)$ for all $v \in V$
        $s \leftarrow \arg\max_{v \in V} c_B(v)$
        $t \leftarrow \arg\max_{v \in V \setminus \{s\}} c_B(v)$
        $S \leftarrow \{v \in V \mid \{s,v\} \in E\}$
        $T \leftarrow \{v \in V \mid \{t,v\} \in E\}$
        for all $\{u,v\} \in E$ **do**
            $c_{uv} \leftarrow 1$
        end for
        for all $v \in S$ **do**
            if $v \notin T \cup \{t\}$ **then**
                $c_{sv} \leftarrow \infty$
            end if
        end for
        for all $v \in T$ **do**
            if $v \notin S \cup \{s\}$ **then**
                $c_{tv} \leftarrow \infty$
            end if
        end for
        $C \leftarrow$ MINIMUM-CUT(G, c, s, t)
    end if
end procedure

After the source s and sink t are selected by the betweenness centrality computation, DIVIDE-INTO-TWO initializes the capacity values of all the edges of the network to be one. Then, it assigns infinity as the capacity to the edges that are connected to the source or sink vertices. This assignment is to prevent the edges connected to s or t from being cut by MINIMUM-CUT. The procedure forms a group of vertices S and T, where vertices of S and T are neighbors of the source and sink nodes, respectively. The capacity value of infinity makes the vertices of groups S or T strongly connected to themselves, which prevents the edges from being cut by MINIMUM-CUT. All the other edges remain with a capacity of one, as initialized.

When assigning infinity as a capacity value to edges, it should be avoided that the source s and sink t are connected by a path whose every edge has infinity as its capacity value. If there is any path connecting the source and sink vertices where all the edges of the path have infinity as capacity, then MINIMUM-CUT cannot find a cut set of finite minimum cut value. To carefully assign infinity as capacity, DIVIDE-INTO-TWO checks whether the other vertex belongs to S or T, as described precisely in Algorithm 1.

Computing the betweenness centralities via Brandes' algorithm takes $O(|V||E|)$ time, while the MINIMUM-CUT procedure takes $O(|V||E| + |V|^2 \log |V|)$ time when using the Stoer–Wagner algorithm. The Stoer–Wagner algorithm is one of the fast algorithms for finding a minimum cut from a graph. Considering these time complexities, the DIVIDE-INTO-TWO algorithm has the same time complexity as MINIMUM-CUT. Note that all the other executions of DIVIDE-INTO-TWO except for the two main procedures can be performed in linear time. For example, selecting the source and sink vertices and forming the two sets S and T can be done in linear time. Assigning capacity values to all the edges also takes linear time.

3.5. Modularity: A Quality Measure

In most practical cases, community detection algorithms are applied to a network whose community structure is not known ahead of time. Therefore, it is necessary to have a quality measure to evaluate a community structure derived by a community detection algorithm. In other words, we need a measure to answer the question of how well the found community structure represents the underlying connection characteristics of the network.

Even though a quality measure of community structure is needed, there exists one difficulty in defining such a measure. There is no pre-defined precise answer for the community structure of a given network; two people may have different opinions about the inherent community structure of the same network. It is hard to say that a community structure of a network is strictly better than another community structure of the same network. Examples that illustrate this difficulty are presented in Section 4.

Although it is difficult to find a well-defined and unified quality measure to evaluate a particular division of a given network, some measures have been proposed so far. One of the widely used measures is *modularity* [25]. The modularity measure is defined as follows.

Definition 7. *Given a network $G = (V, E)$ with a community structure $C = \{C_1, C_2, \ldots, C_k\}$, a $k \times k$ symmetric matrix M is defined, where each element M_{ij} for $i, j = 1, \ldots, k$ is the fraction of all edges that connect vertices in community i to vertices in community j. Then, the* modularity *measure Q of the community structure C of the network G is defined by*

$$Q(C; G) = \mathrm{Tr}\, M - \left\| M^2 \right\|, \quad (2)$$

where $\|X\|$ is the sum of the elements of the matrix X.

Note that the row sum $m_i = \sum_j M_{ij}$ of the matrix M represents the fraction of edges that have at least one vertex in community i at its endpoints. If all the pre-existing edges are removed and new edges are randomly placed between vertices while preserving the row sums, the expected fraction of all the new edges that connect vertices in community i and vertices in community j equals $m_i \cdot m_j$. Therefore, the difference between the pre-existing and the expected fraction, $M_{ii} - m_i^2$, may be used as an indication of how strongly the vertices in the community i are connected to each other. In other words, it can measure how densely the vertices in the community i are linked. By adding these differences for all the k communities, the modularity measure can be defined as

$$Q = \sum_{i=1}^{k} \left(M_{ii} - m_i^2 \right), \tag{3}$$

which is shown to be equal to the right side of Equation (2).

The maximum value that Q can have is $Q = 1$, and this value indicates a network with a very strong community structure [25,66]. The modularity Q values for networks with strong community structure usually range from about 0.3 to 0.7.

Our proposed algorithm computes the modularity value after each division of the given network via graph cuts, and it determines the next step based on the calculated Q values. In particular, the Q value is used as a stopping criterion to stop iterations of community division. Therefore, the Q values are critical for the algorithm.

3.6. The Proposed Algorithm

The proposed algorithm is a divisive one, so that a given network is divided into groups of vertices in a stepwise manner. The main idea of the algorithm is that a connected network may be partitioned into two subgroups by finding a minimum cut separating the two, which is expected to reveal a good community structure of the given network. After a division of each connected component of a network, the Q value for the resulting community structure is calculated. By comparing the resulting Q value for every connected component, the algorithm selects the best partition and creates two communities out of the selected component. The algorithm iterates this process until no further improvement in the modularity value is observed. The algorithm is named MCCD (Minimum Cut-based Community Detection) and is presented precisely in Algorithm 2.

Algorithm 2 The MCCD algorithm

Input: A connected graph $G = (V, E)$
Output: $C = \{C_1, \ldots, C_k\}$, a community structure of G

procedure MCCD(G)
    $C \leftarrow \{V\}$
    $Q \leftarrow 0$
    loop
        if $|C| = |V|$ **then**
            break
        end if
        for $i \leftarrow 1$ to $|C|$ **do**
            $\{C_{i1}, C_{i2}\} \leftarrow$ DIVIDE-INTO-TWO(C_i, $E(C_i)$)
            $Q_i \leftarrow Q\left(\{C_1, \ldots, C_{i-1}, C_{i1}, C_{i2}, C_{i+1}, \ldots, C_{|C|}\}; G\right)$
        end for
        $m \leftarrow \arg\max_i Q_i$
        if $Q > Q_m$ **then**
            break
        end if
        $C \leftarrow \{C_1, \ldots, C_{m-1}, C_{m1}, C_{m2}, C_{m+1}, \ldots, C_{|C|}\}$
        $Q \leftarrow Q_m$
    end loop
end procedure

Note that the locally optimal partition $\{C_1, \ldots, C_{m-1}, C_{m1}, C_{m2}, C_{m+1}, \ldots, C_{|C|}\}$, where $m = \arg\max_i Q_i$, is selected at each iteration of the MCCD algorithm. The iteration repeats until a decrease in the modularity Q value is observed or each vertex has its own community, i.e., the entire network is partitioned into one-vertex communities. As illustrated in Section 4, however, the Q value starts decreasing after only a few iterations for many networks. In other words, many networks are expected to have a community structure consisting of not so many communities.

Even though the MCCD algorithm stops its iteration of dividing each community into two smaller communities when it detects decreasing Q values, it is also possible to continue the iteration until a desired number of communities are obtained, via a slight modification of the algorithm. This modified algorithm is precisely presented in Algorithm 3 and is called k-MCCD. The desired number of communities k, which is given as an input, is assumed to satisfy $1 \leq k \leq |V|$. As presented in Algorithm 3, k-MCCD stops its iterations when the number of communities reaches the desired number k.

Algorithm 3 The k-MCCD algorithm

Input: A connected graph $G = (V, E)$ and a desired number of communities k
Output: $C = \{C_1, \ldots, C_k\}$, a community structure of G

procedure k-MCCD(G, k)
  $C \leftarrow \{V\}$
  $Q \leftarrow 0$
  loop
    if $|C| = k$ **then**
      break
    end if
    for $i \leftarrow 1$ to $|C|$ **do**
      $\{C_{i1}, C_{i2}\} \leftarrow$ DIVIDE-INTO-TWO(C_i, $E(C_i)$)
      $Q_i \leftarrow Q\Big(\{C_1, \ldots, C_{i-1}, C_{i1}, C_{i2}, C_{i+1}, \ldots, C_{|C|}\}; G\Big)$
    end for
    $m \leftarrow \arg\max_i Q_i$
    $C \leftarrow \{C_1, \ldots, C_{m-1}, C_{m1}, C_{m2}, C_{m+1}, \ldots, C_{|C|}\}$
    $Q \leftarrow Q_m$
  end loop
end procedure

As mentioned previously, the procedure DIVIDE-INTO-TWO has the time complexity of $O(|V||E| + |V|^2 \log |V|)$. Therefore, when MCCD calls DIVIDE-INTO-TWO(C_i, $E(C_i)$) for a community C_i, it takes $O(|C_i||E(C_i)| + |C_i|^2 \log |C_i|)$ time. MCCD calls DIVIDE-INTO-TWO for each community C_i during one iteration, so the total computation time summed for every community can be given by

$$\sum_{i=1}^{|C|} O(|C_i||E(C_i)| + |C_i|^2 \log |C_i|). \tag{4}$$

The computation time of Equation (4) satisfies

$$\begin{aligned} \sum_{i=1}^{|C|} O(|C_i||E(C_i)| + |C_i|^2 \log |C_i|) &\leq \sum_{i=1}^{|C|} O(|C_i||E| + |C_i|^2 \log |C_i|) \\ &\leq O(|V||E| + |V|^2 \log |V|), \end{aligned}$$

where the convexity of a function $f(x) = yx + x^2 \log x$ for sufficiently large x is used for the second inequality.

Besides DIVIDE-INTO-TWO, another procedure should be accounted for when considering the time complexity of MCCD, which is the computation of the modularity Q_i for each community C_i. Assuming that the change in Q is tracked during the iterations, it is known that the time complexity of modularity computation is bounded by $O(|V| + |E|)$ [27]. Therefore, the modularity computation does not affect the time complexity of MCCD, because DIVIDE-INTO-TWO takes a longer time. Since MCCD iterates dividing a community into two smaller communities until the stopping criterion is satisfied, in the worst case where the iteration number is equal to $|V|$, the time complexity of MCCD is given by $O(|V|^2|E| + |V|^3 \log |V|)$, which is in polynomial time. However, many practical networks are believed to have a small number of communities, which is equal to the total number of iterations of MCCD. Therefore, the time complexity of MCCD may be bounded by $O(|V||E| + |V|^2 \log |V|)$ in those cases.

4. Results

This section describes the results of application of the proposed MCCD algorithm to four different example networks. The first example deals with a very simple network whose community structure is obvious. This example checks the validity of the proposed algorithm; if the algorithm cannot detect the obvious community structure of the simple network, it is useless to consider it any further. After the simple validity check, three more examples are used to show the performance of the MCCD algorithm. These examples are frequently used in the literature on the study of networks, so they can provide an evaluation of the performance of the proposed algorithm.

4.1. Example 1: A Simple Network

A very simple network, whose community structure is obvious, is generated as shown in Figure 2a. The network is a slightly generalized version of a graph, usually called an *n-barbell* graph, and it has two complete subgraphs that are connected by a path [67–69]. It is obvious that the network has two communities: $C_1 = \{0, 1, 2, 3, 4, 5\}$ and $C_2 = \{6, 7, 8, 9, 10, 11\}$.

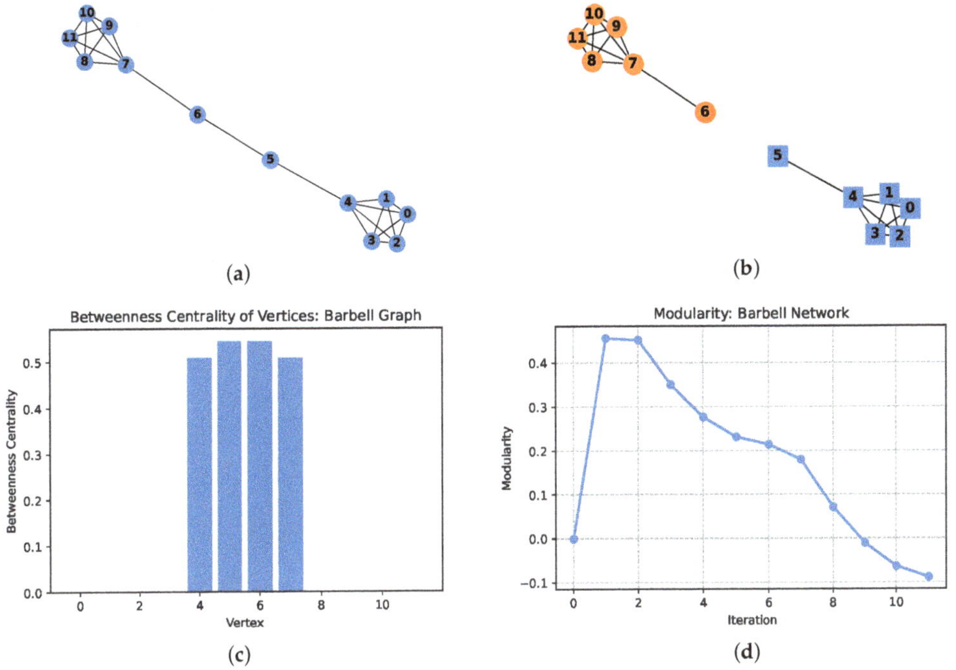

Figure 2. A barbell graph and the application of MCCD to the graph. (**a**) The barbell graph; (**b**) the two detected communities by MCCD; (**c**) the betweenness centrality values $c_B(v)$; (**d**) the modularity values after each iteration of DIVIDE-INTO-TWO.

The result of the MCCD algorithm applied to the network is shown in Figure 2b. As expected, the MCCD algorithm detects the two obvious communities, and they are represented as blue rectangles (C_1) and orange circles (C_2) in the figure. Therefore, the MCCD algorithm works well for the simple barbell graph.

The betweenness centrality value $c_B(v)$ for all vertices v of the barbell network is shown in Figure 2c. Vertices 5 and 6, which are boundary nodes of each community, have the highest betweenness centrality value because the shortest paths connecting nodes of the network pass through them more than any other nodes. Therefore, the MCCD algorithm selects the two vertices 5 and 6 as the source and sink nodes at the first iteration of the process. The algorithm then succeeds in finding the minimum cut between the two vertices and in detecting the obvious communities C_1 and C_2.

Note that the MCCD algorithm stops its iteration after the second pass because the modularity score of the second pass, which is about 0.4527, is lower than that of the first pass, which is about 0.4565. The modularity value of every iteration is shown in Figure 2d, where the MCCD algorithm is slightly modified to perform its iterations even though the modularity value Q_m of each iteration starts decreasing; the lines 13–15 of MCCD Algorithm 2 are temporarily ignored to repeat the DIVIDE-INTO-TWO procedure until every edge is cut, so that each vertex forms its own community. As can be observed in Figure 2d, the modularity value quickly decreases as MCCD repeats its iteration passes.

4.2. Example 2: Zachary's Karate Club Network

One of the frequently used real-world network data is Zachary's karate club network [7]. The karate club network represents social interactions between 34 individuals, who were the members of a karate club at a university. Wayne Zachary obtained the network data by observing social interactions between the members for a period of three years from 1970 to 1972.

Before the study began, the club had employed an instructor (Mr. Hi) for karate lessons. At the beginning of the study, there was a conflict between the instructor and the club president (John A.) over the price of karate lessons. The instructor wished to raise prices while the president wanted to stabilize them. As time passed, there arose a series of factional confrontations, and then the president fired the instructor for attempting to raise lesson prices unilaterally. The supporters of the instructor resigned and formed a new organization headed by the instructor, thus completing the fission of the club.

Figure 3a shows the karate club network, where each vertex represents a member of the club. Vertices 0 and 33 represent the instructor and the president, respectively; the members who belong to the group headed by the instructor after the fission are represented by blue square vertices and the members who belong to the other group are represented by orange circle vertices. An edge of the network is drawn if two individuals were observed to interact outside the normal activities of the club, i.e., if they could be said to be friends outside the club activities.

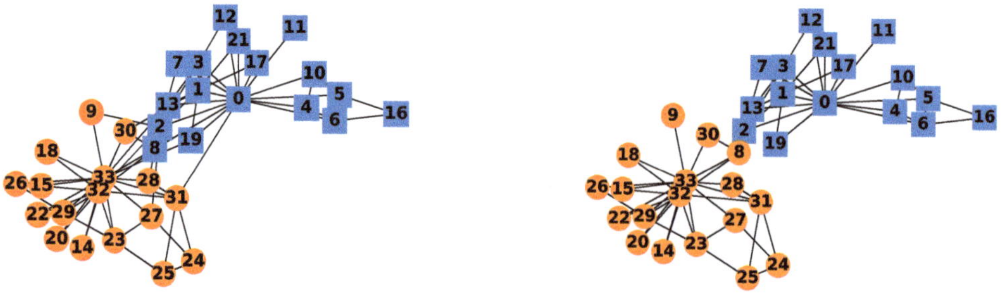

(a) (b)

Figure 3. *Cont.*

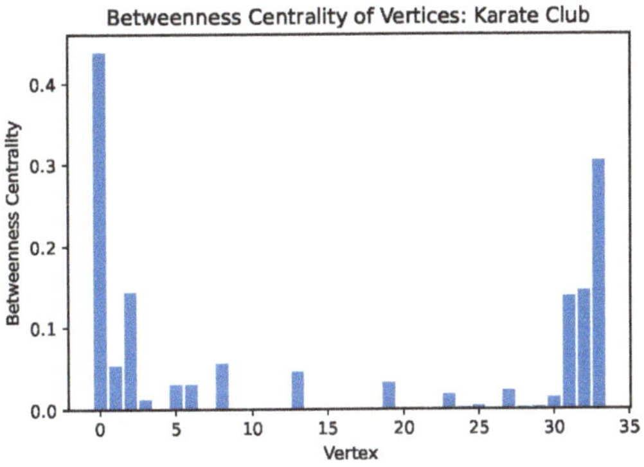

(c)

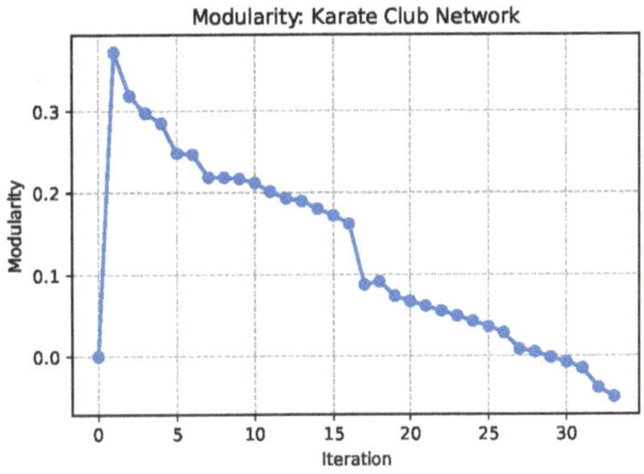

(d)

Figure 3. Zachary's karate club network and the application of MCCD to the network. (**a**) The karate club network; (**b**) the two detected communities by MCCD; (**c**) the betweenness centrality values $c_B(v)$; (**d**) the modularity values after each iteration of DIVIDE-INTO-TWO.

The result of the proposed MCCD algorithm is shown in Figure 3b. Comparing the result with the data shown in Figure 3a, only one vertex, which is labeled 8, is misclassified by MCCD; vertex 8 joined the organization headed by the instructor after the fission, but the MCCD algorithm predicts that it belongs to the other group. Therefore, it can be said that MCCD works well for the karate club network.

Figure 3c shows the betweenness centrality values $c_B(v)$ for the vertices of the network. As expected, the instructor and the president nodes have the highest values, so MCCD selects the two vertices as the *source* and *sink* nodes at its first iteration. The modularity values after each iteration of MCCD are shown in Figure 3d. The modularity value achieves its maximum value of 0.3715 after the first iteration, and the values quickly decrease as MCCD iterates its DIVIDE-INTO-TWO procedures.

To model information flow in the karate club, Zachary defined a capacitated network for the club, which includes a capacity matrix C whose entries are interpreted as representing a capacity or value of maximum possible information flow between two members of the club [7]. In other words, each edge of the network has a quantified value representing the strength/weakness of friend relationship between the two end vertices. These capacity values were assigned by analyzing data of social interactions between the club members. Using this capacitated network model, Zachary employed the max-flow min-cut theory and applied the labeling procedure of Ford and Fulkerson [58,70,71]. To apply the algorithm, the vertices 0 and 33 of the network were manually designated as the *source* and *sink* nodes.

Zachary obtained the same result as MCCD, which is expected, and explained why vertex 8 is misclassified [7]. One point to note is that, contrary to Zachary's method, MCCD automatically selects the source and sink nodes and iterates the DIVIDE-INTO-TWO operation to find the best result for community detection.

Many community detection algorithms that have been proposed previously used the karate club network data to verify their performances, and the modularity results of a few well-known algorithms are shown in Table 1. Most algorithms result in higher modularity values than MCCD, but the ground truth of the karate club network has the modularity value of 0.371 [6]. As described above, MCCD misclassifies only one vertex, so that the modularity value of MCCD is very close to the ground truth. This result illustrates that modularity is not a perfect measure even though it is a very good one.

Table 1. Modularity values for the karate club network of a few well-known community detection algorithms and MCCD: Girvan and Newman (GN) [25]; Newman (FN) [27]; Clauset, Newman, and Moore (CNM) [42]; Duch and Arenas (DA) [72]. Data from [25,27,46,72].

Algorithm	GN	FN	CNM	DA	MCCD
Modularity	0.401	0.381	0.381	0.419	0.372

4.3. Example 3: The Social Network of Bottlenose Dolphins

The third example network is the social network of bottlenose dolphins that was studied by Lusseau [73]. Lusseau conducted a study about a social network of bottlenose dolphins residing in the Doubtful Sound, Fiordland, New Zealand, from November 1994 to November 2001. Based on the observations and study, 62 dolphins were formed into a social network, and it is shown in Figure 4a [74].

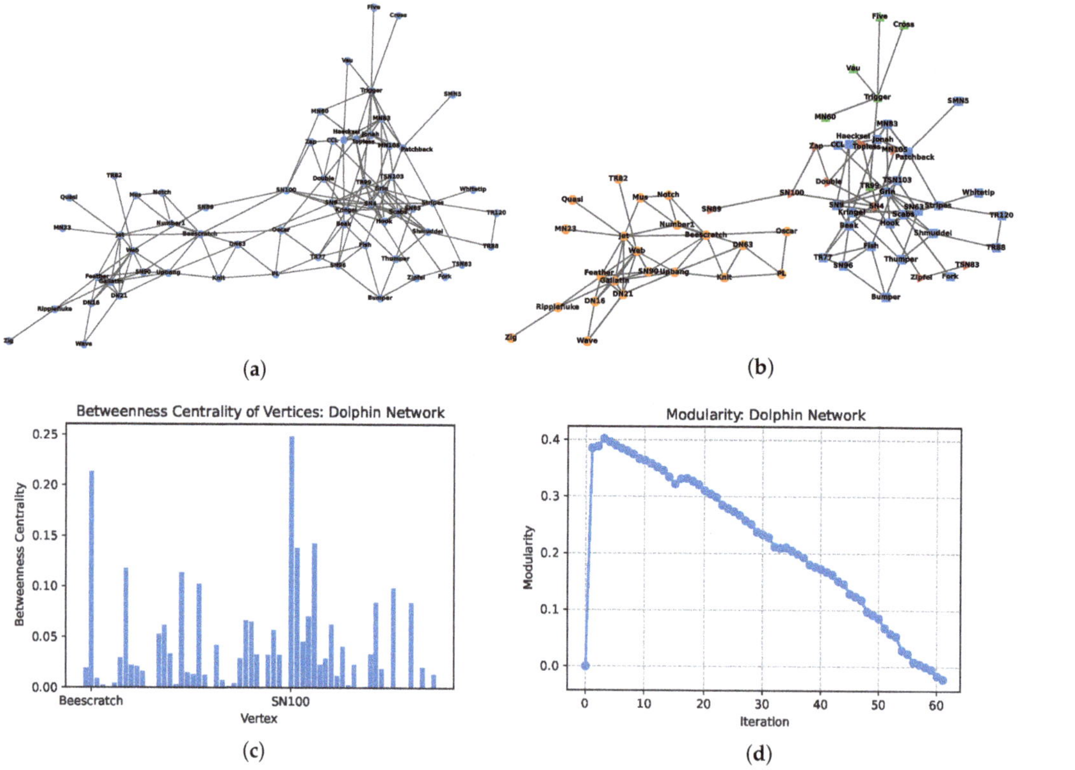

Figure 4. The social network of bottlenose dolphins in Doubtful Sound, New Zealand, and the application of MCCD to the network. (**a**) The social network of bottlenose dolphins; (**b**) the communities detected by MCCD; (**c**) the betweenness centrality values $c_B(v)$; (**d**) the modularity values after each iteration of DIVIDE-INTO-TWO.

As can be seen from Figure 4a, the community structure of the dolphin network is not random, but it is not obvious, either. In other words, it seems that there is a weak community structure in the network, and different people may have different opinions on the structure as well as the number of communities. As mentioned before, it is hard to find a universally agreed answer for the community structure of a network, and this example illustrates this difficulty.

Figure 4b shows the result of MCCD application to the network. The modularity value achieved its maximum of 0.4021 after three iterations; hence, MCCD detects four communities shown in the figure. The betweenness centrality values of the vertices and the modularity values after each iteration are shown in Figure 4c,d, respectively. It turns out that the modularity values of the first few iterations are similar, and this implies that it is hard to tell which structure is best for community detection. Even though MCCD generates the four community structure, structures of slightly different numbers of communities are also plausible. Lusseau presented a sociogram with three groups in his original paper [74]. In fact, the whole network visually seems to have two large groups, where the network is divided into two groups mostly by the gender of dolphins: males and females.

4.4. Example 4: The Characters Network of *Les Misérables*

Figure 5a shows the network of major characters in *Les Misérables*, a famous novel written by Victor Hugo. The network data was constructed by considering coappearances of characters in the same chapter of the book [75]. The 77 vertices of the network are major characters, and an edge between two vertices represents coappearance of the corresponding two characters in the same chapter. Even though Knuth assigned a value to each edge by counting the number of coappearances of the two characters, our method does not use the edge attribute values; MCCD uses the network structure only.

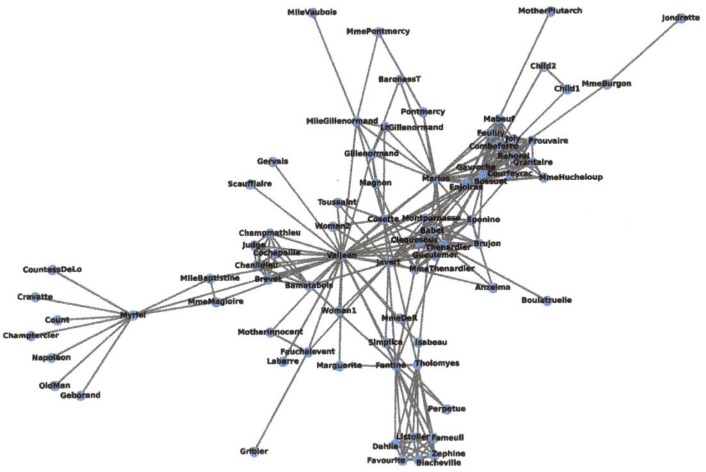

(a)

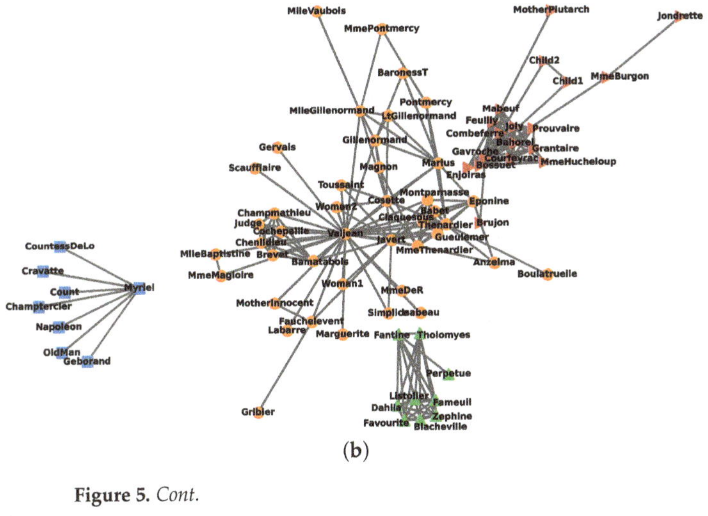

(b)

Figure 5. *Cont.*

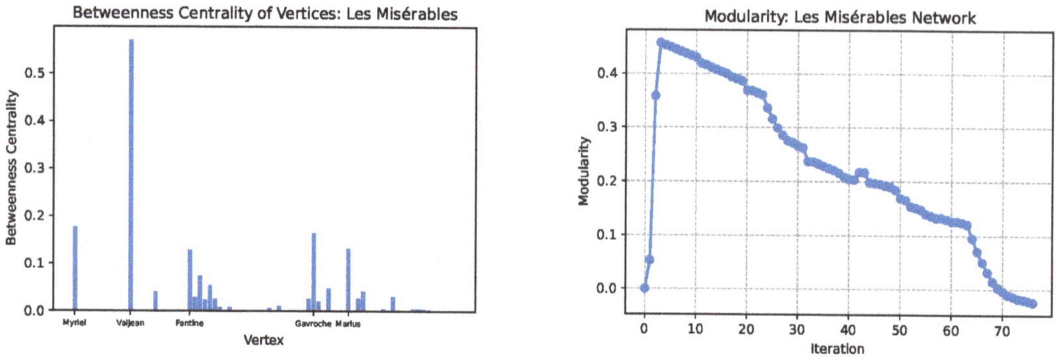

(c) (d)

Figure 5. The character network of *Les Misérables* and the application of MCCD to the network. (**a**) The network; (**b**) the communities detected by MCCD; (**c**) the betweenness centrality values $c_B(v)$; (**d**) the modularity values after each iteration of DIVIDE-INTO-TWO.

The result of MCCD application to the network is shown in Figure 5b. As shown in Figure 5c, the modularity value achieves its maximum of 0.4570 when the network is partitioned into four communities. As expected, the protagonist Jean Valjean and the police officer Javert form the central positions of the main community. Another community that consists mainly of the members of the Friend of the ABC, the fictional association of revolutionary French republican students, is also detected. Enjolras, the charismatic leader of the association, belongs to this community. Furthermore, two more communities are also detected, and it can be observed that these two are centered by the main characters Myriel and Fantine, respectively.

5. Discussion

In this paper, a new algorithm for community detection was proposed and some experimental results were provided. As previously mentioned, the problem of community detection has its inherent difficulties; there is no exact solution for the problem. People may have different opinions about the community structure of a given network. As seen in Section 4, the networks of bottlenose dolphins and of characters in the novel *Les Misérables* are complicated, so that there is no universally agreed-upon community structure. Even though the modularity measure Q can provide a criterion to estimate the quality of community structures, it is only an indirect measure, and some may find a lower-scored community structure to be better than a higher-scored one.

Considering that the problem of community detection has no exact solution, the proposed algorithm MCCD shows fairly good results with comparable Q values when applied to the example networks of Section 4. In particular, the algorithm shows satisfactory results for the barbell graph and Zachary's karate club network. For the more complicated networks in Section 4, it also detects community structures well, which can be visually observed from Figures 4 and 5.

Even though the proposed algorithm showed its feasibility via the example networks of Section 4, further evaluation of the performance of the algorithm by using various benchmarks and larger network data is needed and remains as future work. Frequently used benchmarks such as GN and LFR may be used for the evaluation. For the example networks of Section 4, MCCD takes a few seconds with typical current computers, but the time performance should be carefully investigated with larger networks and benchmarks.

Optimization of the MCCD algorithm implementation is also left as future work. Many procedures called by MCCD may be carefully modified and implemented to reduce computation time. Modifying the algorithm so that it can be applied to detect overlapping community structures may be left as another element of future work. This may not be easy, however, because graph cut methods in general divide vertices without overlapping.

Author Contributions: Conceptualization, H.S. and J.P.; methodology, H.S.; software, H.S.; validation, H.S., J.P. and D.K.; formal analysis, H.S.; investigation, J.P. and D.K.; resources, H.S. and D.K.; data curation, H.S.; writing—original draft preparation, H.S.; writing—review and editing, J.P. and D.K.; visualization, H.S.; supervision, H.S., J.P. and D.K.; project administration, D.K.; funding acquisition, H.S. All authors have read and agreed to the published version of the manuscript.

Funding: This research was supported partly by the Basic Science Research Program through the National Research Foundation of Korea (NRF) funded by the Ministry of Science and ICT (NRF-2019R1C1C1008017) and partly by the Hongik University new faculty research support fund.

Institutional Review Board Statement: Not applicable.

Informed Consent Statement: Not applicable.

Data Availability Statement: The data presented in this study are available in this article and there are no other data to share.

Conflicts of Interest: The authors declare no conflict of interest. The funders had no role in the design of the study; in the collection, analyses, or interpretation of data; in the writing of the manuscript, or in the decision to publish the results.

Abbreviations

The following abbreviations are used in this manuscript:

GN	Girvan and Newman
LFR	Lancichinetti, Fortunato, and Radicchi
MCCD	Minimum Cut-based Community Detection
CNM	Clauset, Newman, and Moore
DA	Duch and Arenas

References

1. Fortunato, S. Community detection in graphs. *Phys. Rep.* **2010**, *486*, 75–174. [CrossRef]
2. Javed, M.A.; Younis, M.S.; Latif, S.; Qadir, J.; Baig, A. Community detection in networks: A multidisciplinary review. *J. Netw. Comput. Appl.* **2018**, *108*, 87–111. [CrossRef]
3. Danon, L.; Díaz-Guilera, A.; Duch, J.; Arenas, A. Comparing community structure identification. *J. Stat. Mech. Theory Exp.* **2005**, *2005*, P09008. [CrossRef]
4. Lancichinetti, A.; Fortunato, S. Community detection algorithms: A comparative analysis. *Phys. Rev. E* **2009**, *80*, 056117. [CrossRef]
5. Yang, B.; Liu, D.; Liu, J. Discovering Communities from Social Networks: Methodologies and Applications. In *Handbook of Social Network Technologies and Applications*; Furht, B., Ed.; Springer: Boston, MA, USA, 2010; pp. 331–346. [CrossRef]
6. Cheng, J.; Leng, M.; Li, L.; Zhou, H.; Chen, X. Active Semi-Supervised Community Detection Based on Must-Link and Cannot-Link Constraints. *PLoS ONE* **2014**, *9*, e110088. [CrossRef]
7. Zachary, W.W. An Information Flow Model for Conflict and Fission in Small Groups. *J. Anthropol. Res.* **1977**, *33*, 452–473. [CrossRef]
8. Asur, S.; Parthasarathy, S.; Ucar, D. An Event-Based Framework for Characterizing the Evolutionary Behavior of Interaction Graphs. *ACM Trans. Knowl. Discov. Data* **2009**, *3*, 1–36. [CrossRef]
9. Backstrom, L.; Huttenlocher, D.; Kleinberg, J.; Lan, X. Group Formation in Large Social Networks: Membership, Growth, and Evolution. In Proceedings of the 12th ACM SIGKDD International Conference on Knowledge Discovery and Data Mining, Philadelphia, PA, USA, 20–23 August 2006; pp. 44–54. [CrossRef]
10. Dunlavy, D.M.; Kolda, T.G.; Acar, E. Temporal Link Prediction Using Matrix and Tensor Factorizations. *ACM Trans. Knowl. Discov. Data* **2011**, *5*, 1–27. [CrossRef]
11. Fu, W.; Song, L.; Xing, E.P. Dynamic Mixed Membership Blockmodel for Evolving Networks. In Proceedings of the 26th Annual International Conference on Machine Learning, Montreal, QC, Canada, 14–18 June 2009; pp. 329–336. [CrossRef]
12. Chakraborty, T.; Chakraborty, A. OverCite: Finding Overlapping Communities in Citation Network. In Proceedings of the 2013 IEEE/ACM International Conference on Advances in Social Networks Analysis and Mining, Niagara, ON, Canada, 25–28 August 2013; pp. 1124–1131. [CrossRef]

13. Yang, J.; Leskovec, J. Overlapping Community Detection at Scale: A Nonnegative Matrix Factorization Approach. In Proceedings of the Sixth ACM International Conference on Web Search and Data Mining, Rome, Italy, 4–8 Februaty 2013; pp. 587–596. [CrossRef]
14. Maity, S.; Rath, S.K. Extended Clique percolation method to detect overlapping community structure. In Proceedings of the 2014 International Conference on Advances in Computing, Communications and Informatics (ICACCI), Delhi, India, 24–27 September 2014; pp. 31–37. [CrossRef]
15. Gulbahce, N.; Lehmann, S. The art of community detection. *BioEssays* **2008**, *30*, 934–938. [CrossRef]
16. Papadopoulos, S.; Kompatsiaris, Y.; Vakali, A.; Spyridonos, P. Community detection in Social Media: Performance and application considerations. *Data Min. Knowl. Discov.* **2012**, *24*, 515–554. [CrossRef]
17. Chintalapudi, S.R.; Prasad, M.H.M.K. A survey on community detection algorithms in large scale real world networks. In Proceedings of the 2015 2nd International Conference on Computing for Sustainable Global Development (INDIACom), New Delhi, India, 11–13 March 2015; pp. 1323–1327.
18. Plantié, M.; Crampes, M. Survey on Social Community Detection. In *Social Media Retrieval*; Ramzan, N., van Zwol, R., Lee, J.S., Clüver, K., Hua, X.S., Eds.; Springer: London, UK, 2013; pp. 65–85. [CrossRef]
19. Yang, Z.; Algesheimer, R.; Tessone, C. A Comparative Analysis of Community Detection Algorithms on Artificial Networks. *Sci. Rep.* **2016**, *6*, 30750. [CrossRef] [PubMed]
20. Wu, Z.; Leahy, R. An optimal graph theoretic approach to data clustering: Theory and its application to image segmentation. *IEEE Trans. Pattern Anal. Mach. Intell.* **1993**, *15*, 1101–1113. [CrossRef]
21. Shen, H.; Cheng, X.; Cai, K.; Hu, M.B. Detect overlapping and hierarchical community structure in networks. *Phys. A Stat. Mech. Its Appl.* **2009**, *388*, 1706–1712. [CrossRef]
22. Ahn, Y.Y.; Bagrow, J.P.; Lehmann, S. Link communities reveal multiscale complexity in networks. *Nature* **2010**, *466*, 761–764. [CrossRef]
23. Barabási, A.L.; Albert, R. Emergence of Scaling in Random Networks. *Science* **1999**, *286*, 509–512. [CrossRef]
24. Hastie, T.; Friedman, J.; Tibshirani, R. *The Elements of Statistical Learning*; Springer: New York, NY, USA, 2009. [CrossRef]
25. Newman, M.E.J.; Girvan, M. Finding and evaluating community structure in networks. *Phys. Rev. E* **2004**, *69*, 026113. [CrossRef]
26. Newman, M.E.J. Properties of highly clustered networks. *Phys. Rev. E* **2003**, *68*, 026121. [CrossRef]
27. Newman, M.E.J. Fast algorithm for detecting community structure in networks. *Phys. Rev. E* **2004**, *69*, 066133. [CrossRef]
28. Newman, M.E.J. Analysis of weighted networks. *Phys. Rev. E* **2004**, *70*, 056131. [CrossRef]
29. Tyler, J.R.; Wilkinson, D.M.; Huberman, B.A. E-Mail as Spectroscopy: Automated Discovery of Community Structure within Organizations. *Inf. Soc.* **2005**, *21*, 143–153. [CrossRef]
30. Wilkinson, D.M.; Huberman, B.A. A method for finding communities of related genes. *Proc. Natl. Acad. Sci. USA* **2004**, *101*, 5241–5248. [CrossRef] [PubMed]
31. Rattigan, M.J.; Maier, M.; Jensen, D. Graph Clustering with Network Structure Indices. In Proceedings of the 24th International Conference on Machine Learning, Corvalis, OR, USA, 20–24 June 2007; pp. 783–790. [CrossRef]
32. Shi, J.; Malik, J. Normalized cuts and image segmentation. *IEEE Trans. Pattern Anal. Mach. Intell.* **2000**, *22*, 888–905.
33. von Luxburg, U. A tutorial on spectral clustering. *Stat. Comput.* **2007**, *17*, 395–416. [CrossRef]
34. Donath, W.E.; Hoffman, A.J. Lower Bounds for the Partitioning of Graphs. *IBM J. Res. Dev.* **1973**, *17*, 420–425. [CrossRef]
35. Fiedler, M. Algebraic connectivity of graphs. *Czechoslov. Math. J.* **1973**, *23*, 298–305. [CrossRef]
36. Barnard, S.T.; Pothen, A.; Simon, H. A spectral algorithm for envelope reduction of sparse matrices. *Numer. Linear Algebra Appl.* **1995**, *2*, 317–334. [CrossRef]
37. Meilă, M.; Shi, J. A random walks view of spectral segmentation. In Proceedings of the 8th International Workshop on Artificial Intelligence and Statistics, Key West, FL, USA, 4–7 January 2001; pp. 203–208.
38. Ng, A.; Jordan, M.; Weiss, Y. On Spectral Clustering: Analysis and an algorithm. In Proceedings of the Advances in Neural Information Processing Systems, Vancouver, BC, Canada, 3–8 December 2001; Volume 14.
39. Pothen, A. Graph Partitioning Algorithms with Applications to Scientific Computing. In *Parallel Numerical Algorithms*; Keyes, D.E., Sameh, A., Venkatakrishnan, V., Eds.; Springer: Dordrecht, The Netherlands, 1997; pp. 323–368. [CrossRef]
40. Kernighan, B.W.; Lin, S. An Efficient Heuristic Procedure for Partitioning Graphs. *Bell Syst. Tech. J.* **1970**, *49*, 291–307. [CrossRef]
41. Guimerà, R.; Nunes Amaral, L.A. Functional cartography of complex metabolic networks. *Nature* **2005**, *433*, 895–900. [CrossRef]
42. Clauset, A.; Newman, M.E.J.; Moore, C. Finding community structure in very large networks. *Phys. Rev. E* **2004**, *70*, 066111. [CrossRef]
43. Newman, M.E.J. Detecting community structure in networks. *Eur. Phys. J. B Condens. Matter* **2004**, *38*, 321–330. [CrossRef]
44. Boettcher, S.; Percus, A.G. Extremal optimization for graph partitioning. *Phys. Rev. E* **2001**, *64*, 026114. [CrossRef] [PubMed]
45. Liu, J.; Liu, T. Detecting community structure in complex networks using simulated annealing with k-means algorithms. *Phys. A Stat. Mech. Its Appl.* **2010**, *389*, 2300–2309. [CrossRef]
46. Newman, M.E.J. Modularity and community structure in networks. *Proc. Natl. Acad. Sci. USA* **2006**, *103*, 8577–8582. [CrossRef]
47. Liu, C.; Liu, J.; Jiang, Z. A Multiobjective Evolutionary Algorithm Based on Similarity for Community Detection From Signed Social Networks. *IEEE Trans. Cybern.* **2014**, *44*, 2274–2287. [CrossRef]
48. Pizzuti, C. GA-Net: A Genetic Algorithm for Community Detection in Social Networks. In Proceedings of the Parallel Problem Solving from Nature—PPSN X, Dortmund, Germany, 13–17 September 2008; pp. 1081–1090.

49. Gong, M.; Fu, B.; Jiao, L.; Du, H. Memetic algorithm for community detection in networks. *Phys. Rev. E* **2011**, *84*, 056101. [CrossRef]
50. Gong, M.; Ma, L.; Zhang, Q.; Jiao, L. Community detection in networks by using multiobjective evolutionary algorithm with decomposition. *Phys. A Stat. Mech. Appl.* **2012**, *391*, 4050–4060. [CrossRef]
51. Zeng, Y.; Liu, J. Community Detection from Signed Social Networks Using a Multi-objective Evolutionary Algorithm. In Proceedings of the 18th Asia Pacific Symposium on Intelligent and Evolutionary Systems, Singapore, 10–12 November 2014; Volume 1; pp. 259–270.
52. Hughes, B.D. *Random Walks and Random Environments: Random Walks*; Oxford University Press: Oxford, UK, 1995; Volume 1.
53. Zhou, H. Distance, dissimilarity index, and network community structure. *Phys. Rev. E* **2003**, *67*, 061901. [CrossRef]
54. Zhou, H.; Lipowsky, R. Network Brownian Motion: A New Method to Measure Vertex-Vertex Proximity and to Identify Communities and Subcommunities. In Proceedings of the Computational Science—ICCS 2004, Kraków, Poland, 6–9 June 2004; pp. 1062–1069.
55. Pons, P.; Latapy, M. Computing Communities in Large Networks Using Random Walks. In Proceedings of the Computer and Information Sciences—ISCIS 2005, Istanbul, Turkey, 26–28 October 2005; pp. 284–293.
56. Girvan, M.; Newman, M.E.J. Community structure in social and biological networks. *Proc. Natl. Acad. Sci. USA* **2002**, *99*, 7821–7826. [CrossRef]
57. Lancichinetti, A.; Fortunato, S.; Radicchi, F. Benchmark graphs for testing community detection algorithms. *Phys. Rev. E* **2008**, *78*, 046110. [CrossRef]
58. Fulkerson, D.R.; Ford, L.R. *Flows in Networks*; Princeton University Press: Princeton, NJ, USA, 1962.
59. Dinic, E.A. Algorithm for solution of a problem of maximum flow in networks with power estimation. *Soviet Math. Doklady* **1970**, *11*, 1277–1280.
60. Edmonds, J.; Karp, R.M. Theoretical improvements in algorithmic efficiency for network flow problems. *J. ACM* **1972**, *19*, 248–264. [CrossRef]
61. Nagamochi, H.; Ibaraki, T. Computing edge-connectivity in multigraphs and capacitated graphs. *SIAM J. Discret. Math.* **1992**, *5*, 54–66. [CrossRef]
62. Stoer, M.; Wagner, F. A Simple Min-Cut Algorithm. *J. ACM* **1997**, *44*, 585–591. [CrossRef]
63. Freeman, L.C. A Set of Measures of Centrality Based on Betweenness. *Sociometry* **1977**, *40*, 35. [CrossRef]
64. Brandes, U. A faster algorithm for betweenness centrality. *J. Math. Sociol.* **2001**, *25*, 163–177. [CrossRef]
65. Brandes, U. On variants of shortest-path betweenness centrality and their generic computation. *Soc. Netw.* **2008**, *30*, 136–145. [CrossRef]
66. Newman, M.E.J. Mixing patterns in networks. *Phys. Rev. E* **2003**, *67*, 026126. [CrossRef]
67. Wilf, H.S. The Editor's Corner: The White Screen Problem. *Am. Math. Mon.* **1989**, *96*, 704. [CrossRef]
68. Ghosh, A.; Boyd, S.; Saberi, A. Minimizing Effective Resistance of a Graph. *SIAM Rev.* **2008**, *50*, 37–66. [CrossRef]
69. Herbster, M.; Pontil, M. Prediction on a Graph with a Perceptron. In Proceedings of the Advances in Neural Information Processing Systems 19 (NIPS 2006), Vancouver, BC, Canada, 4–7 December 2006.
70. Ford, L.R.; Fulkerson, D.R. Maximal Flow Through a Network. *Can. J. Math.* **1956**, *8*, 399–404. [CrossRef]
71. Ford, L.R.; Fulkerson, D.R. A Simple Algorithm for Finding Maximal Network Flows and an Application to the Hitchcock Problem. *Can. J. Math.* **1957**, *9*, 210–218. [CrossRef]
72. Duch, J.; Arenas, A. Community detection in complex networks using extremal optimization. *Phys. Rev. E* **2005**, *72*, 027104. [CrossRef] [PubMed]
73. Lusseau, D. The emergent properties of a dolphin social network. *Proc. R. Soc. Lond. Ser. B: Biol. Sci.* **2003**, *270*, S186–S188. [CrossRef] [PubMed]
74. Lusseau, D.; Schneider, K.; Boisseau, O.J.; Haase, P.; Slooten, E.; Dawson, S.M. The bottlenose dolphin community of Doubtful Sound features a large proportion of long-lasting associations. *Behav. Ecol. Sociobiol.* **2003**, *54*, 396–405. [CrossRef]
75. Knuth, D.E. *The Stanford GraphBase: A Platform for Combinatorial Computing*; ACM Press: New York, NY, USA, 1993; Volume 1.

applied sciences

MDPI

Article

Unsupervised Community Detection Algorithm with Stochastic Competitive Learning Incorporating Local Node Similarity

Jian Huang and Yijun Gu *

College of Information and Cyber Security, People's Public Security University of China, Beijing 100038, China; 2021211442@stu.ppsuc.edu.cn
* Correspondence: guyijun@ppsuc.edu.cn

Abstract: Community detection is an important task in the analysis of complex networks, which is significant for mining and analyzing the organization and function of networks. As an unsupervised learning algorithm based on the particle competition mechanism, stochastic competitive learning has been applied in the field of community detection in complex networks, but still has several limitations. In order to improve the stability and accuracy of stochastic competitive learning and solve the problem of community detection, we propose an unsupervised community detection algorithm LNSSCL (Local Node Similarity-Integrated Stochastic Competitive Learning). The algorithm calculates node degree as well as Salton similarity metrics to determine the starting position of particle walk; local node similarity is incorporated into the particle preferential walk rule; the particle is dynamically adjusted to control capability increments according to the control range; particles select the node with the strongest control capability within the node to be resurrected; and the LNSSCL algorithm introduces a node affiliation selection step to adjust the node community labels. Experimental comparisons with 12 representative community detection algorithms on real network datasets and synthetic networks show that the LNSSCL algorithm is overall better than other compared algorithms in terms of standardized mutual information (NMI) and modularity (Q). The improvement effect for the stochastic competition learning algorithm is evident, and it can effectively accomplish the community detection task in complex networks.

Keywords: unsupervised learning; community detection; local node similarity; particle competition; stochastic competitive learning; complex networks

Citation: Huang, J.; Gu, Y. Unsupervised Community Detection Algorithm with Stochastic Competitive Learning Incorporating Local Node Similarity. *Appl. Sci.* **2023**, *13*, 10496. https://doi.org/10.3390/app131810496

Academic Editors: Pasquale De Meo, Giacomo Fiumara, Xiaoyang Liu and Annamaria Ficara

Received: 8 August 2023
Revised: 14 September 2023
Accepted: 15 September 2023
Published: 20 September 2023

1. Introduction

With the advancement of information technology, many complex systems in real life can often be described and represented in the form of complex networks, such as social networks, citation networks, scientist collaboration networks, and protein interaction networks. The majority of these real-life networks typically exhibit distinct community structures, where a network consists of multiple communities, and the connections between nodes within a community are highly dense, while connections between nodes of different communities are relatively sparse [1]. Community detection is a fundamental task in the analysis of complex networks, aiming to partition the entire network into several communities. This process holds significant importance for studying and analyzing the organizational structure and functionality of networks, as well as uncovering latent patterns within them.

There has been a great deal of research in detecting and evaluating community structures in complex networks. In pursuit of this fundamental task of community detection, researchers have proposed numerous community detection algorithms based on various methods such as graph partitioning, statistical inference, clustering, modularity optimization, dynamics, and deep learning. More detailed reviews of community detection can be available through several more extensive review articles [2–6]. Among many methods, dynamics-based methods constitute a significant branch of community detection algorithms, which reveal the community structure by modeling the interactions between nodes in a network. Currently, some of the mainstream dynamic-based community detection algorithms include label propagation, random walk, Markov clustering, dynamic distance, and particle competition [7].

To further improve the accuracy and stability of community detection algorithms, competitive learning is applied to the field of community detection. Competition is a natural process observed in the natural world and many social systems with limited resources. Competitive learning, as a crucial machine learning strategy, has been widely applied in artificial neural networks for achieving unsupervised learning. Early developments include Adaptive Resonance Theory networks [8], Self-Organizing Feature Map networks [9], Learning Vector Quantization neural networks [10], Dual Propagation neural networks [11], and Differential Competitive Learning [12]. The competition among particles in complex networks can generate intricate patterns formed by predefined interactions among individuals within a system. The simple interactions between particles in a network can construct the complex behaviors of the entire network, enabling functions like community detection and node classification. Particle competition-based community detection methods were first proposed by Quiles and Zhao [13]. In this approach, a set of particles is initially placed randomly in the network. These particles engage in random walks and compete with each other based on predefined rules to occupy nodes. When each community is dominated by only one particle, the process reaches dynamic equilibrium, thus accomplishing the community detection task. Silva et al. [14] introduced the Stochastic Competitive Learning (SCL) algorithm for unsupervised learning, refining the walk rules of the particle competition model. Particles move in the network based on a convex combination of random walk and preferential walk rules. After entering a silent state due to energy depletion, particles execute a jump resurrection step. The final attribution of nodes to communities is determined based on the relative control abilities of particles, achieving the community detection task. The SCL algorithm represents a nonlinear stochastic dynamical system, characterized by adaptability, local motion reflecting the whole, and has contributed to the advancement of complex network dynamics. Subsequently, stochastic competitive learning has found extensive application in various fields such as label noise detection [15], overlapping community detection [16], prediction of the number of sentiment evolution communities [17], graph anomaly detection [18], image segmentation [19], and assisting the visually impaired [20]. These applications have demonstrated the feasibility, rationality, and effectiveness of applying stochastic competitive learning to the domain of complex network community detection.

Despite the commendable effectiveness of the stochastic competitive learning algorithm in various fields, it has been found that the algorithm still has some shortcomings in the community detection task. Firstly, the random selection of initial positions for particles in stochastic competitive learning leads to unstable community detection outcomes. This randomness might result in an overly concentrated placement of different particles, affecting convergence speed and subsequently diminishing the quality of community detection. Secondly, the stochastic nature of the preferential walk process of particles and the uncertainty in selecting resurrection positions upon energy depletion contribute to suboptimal final results. Lastly, the constant increment in particle control ability leads to potential misjudgments in affiliating boundary nodes between communities of varying scales. The above issues make stochastic competitive learning underperform on community discovery tasks.

In order to address the aforementioned issues, this paper proposes the Local Node Similarity-Integrated Stochastic Competitive Learning algorithm LNSSCL for unsupervised community detection. The algorithm first integrates the node degree and Salton metrics to determine the starting point of particle walk in the network. During the particle walk process, the wandering direction is guided by the walk rule that incorporates the local similarity of nodes. At the same time, the control ability of the particle is dynamically adjusted according to its current control range. When a particle runs out of energy, it is assigned a unique resurrection position. After the wandering is completed, the community label is adjusted through the node affiliation selection step to obtain the final community discovery result. The main objectives and contributions of this study are as follows:

(1) Determining the initial positions of particles based on node degrees and the Salton similarity index, ensuring fixed and dispersed particle placements to mitigate intense early-stage competition and subsequently accelerate convergence speed;
(2) Incorporating the proposed node similarity measure to enhance the deterministic and directional aspects of particle preferential walk rules; refining the rules for selecting particle resurrection positions; introducing a node affiliation selection step to refine the final community detection results and enhance algorithm stability;
(3) Dynamically adapting the increment of particle control ability according to the particle's current control range, thereby improving the effectiveness of detecting communities of varying sizes within the network;
(4) The LNSSCL algorithm is experimentally compared with 12 representative algorithms on real network datasets and synthetic networks. The results demonstrate that the proposed algorithm enhances the community detection performance of stochastic competitive learning and, overall, outperforms other algorithms.

The remainder of this paper is organized as follows. Section 2 presents related work. Section 3 introduces some related preliminary knowledge. Section 4 describes the details and main framework of the proposed LNSSCL algorithm. The experiments are shown and discussed in Section 5. Section 6 gives the conclusions of this paper.

2. Related Work

Complex networks can model various relationships and internal operating mechanisms between entities and objects in the real world. Detecting the community structure present in the network can help us further reveal aspects of the real world. Therefore, community detection in complex networks has received attention from many fields and is rapidly evolving. Among the many methods proposed, Dynamics-based methods utilize the dynamic properties of complex networks. Most of these algorithms have linear time complexity and can be better scaled to large-scale networks. Since it is impractical to review all previously proposed algorithms for community detection, this section mentions only some representative algorithms from recent years.

Roghani et al. [21] introduced a community detection algorithm based on local balance label diffusion. They assigned importance scores to each node using a novel local similarity measure, selected initial core nodes, and expanded communities by balancing the diffusion of labels from core to boundary nodes, achieving rapid convergence in large-scale networks with stable and accurate results. Toth et al. [22] proposed the Synwalk algorithm, which incorporates the concept of random blocks into random walk-based community detection algorithms, combining the strengths of representative algorithms like Walktrap [23] and Infomap [24], yielding promising results. Yang et al. [25] introduced a method of enhancing Markov similarity, which utilizes the steady-state Markov transition of the initial network to derive an enhanced Markov similarity matrix. By partitioning the network into initial community structures based on the Markov similarity index and subsequently merging small communities, tightly connected communities are obtained. Jokar et al. [26] proposed a community discovery algorithm based on the synergy of label propagation and simulated annealing, which achieved good results. You et al. [27] proposed a three-stage community discovery algorithm TS, which obtained good results through central node identification, label propagation, and community combination. Fahimeh et al. [28] proposed a community detection algorithm that utilizes both local and global network information. The algorithm consists of four components: preprocessing, master community composition, community merging, and optimal community structure selection. Zhang et al. [29] propose a graph layout-based label propagation algorithm to reveal communities in a network, using multiple graph layout information to detect accurate communities and improve stability.

Chin et al. [30] proposed the semi synchronization constrained label propagation algorithm SSCLPA, which implements various constraints to improve the stability of LPA. Fei et al. [31] proposed a novel network core structure extraction algorithm for community detection (CSEA) using variational autoencoders to discover community structures more accurately. Li et al. [32] developed a new community detection method and proposed a new relaxation formulation with a low-rank double stochastic matrix factorization and a corresponding multiplicative optimization-minimization algorithm for efficient optimization.

3. Background

In this section, we introduce some related preliminary knowledge, including basic definition, local node similarity, and the theory of stochastic competitive learning.

3.1. Basic Definition

Given a complex network $G = (V, E)$, where $V = \{v_i | 1 \leq i \leq n\}$ is the set of nodes and $E = \{(v_i, v_j) | 1 \leq i \neq j \leq m\}$ is the set of edges. The number of nodes is n and the number of edges is m. Unless otherwise specified, this paper solely focuses on the analysis of undirected simple graphs. The neighborhood of node v_i is defined as $N(v_i) = \{v_j \in V | (v_i, v_j) \in E\}$, and the degree of node v_i is defined as $d(v_i) = |N(v_i)|$. Let $\mathbf{A}$ be the adjacency matrix of network G, an n-order matrix, defined as follows:

$$a_{ij} = \begin{cases} 1, & (v_i, v_j) \in E \\ 0, & (v_i, v_j) \notin E \end{cases} \tag{1}$$

3.2. Local Node Similarity

In the analysis of complex networks, node similarity metrics are commonly employed to assess the degree of similarity between nodes. Generalizing from the classical triadic closure principle in social network analysis, it is understood that in a given complex network, the greater the number of common neighbors between two nodes, the more similar these nodes are. The specific definition of the common neighbor of nodes v_i and v_j is as follows:

$$CN(v_i, v_j) = N(v_i) \cap N(v_j) \tag{2}$$

Based on the local structure, node similarity metrics are derived from the concept of common neighbors and encompass various indices such as the Salton index, Jaccard index, Sorenson index, Hub Promoted Index, Hub Depressed Index, Leicht-Holme-Newman Index, Preferential Attachment Index, Adamic-Adar Index, and Resource Allocation Index [33]. The higher the local similarity between nodes, the higher the probability that they belong to the same community, and vice versa. Node similarity metrics based on local structure also offer the advantage of lower computational complexity and have been introduced into the task of complex network community detection.

3.3. Stochastic Competitive Learning

Stochastic Competitive Learning, as a classical particle competition model, constitutes a competitive dynamical system composed of multiple particles, achieving community detection through unsupervised learning [14].

In Stochastic Competitive Learning, multiple particles are randomly placed within the nodes of the network. Each particle serves as a community indicator, while the nodes in the network are treated as territories to be contended. The primary objective of particles is to expand their territories by continually traversing the network and gaining control over new nodes, while simultaneously strengthening their control over already dominated nodes. Due to the finite number of nodes in the network, natural competition arises among particles. When a particle visits any node, it enhances its control over the current node, consequently weakening the control of other competing particles over that node. Ultimately, each particle's control range tends to stabilize, leading to convergence. By analyzing the control ranges of particles after convergence, the underlying community structure of complex networks is unveiled [13].

As a stochastic nonlinear dynamical system, Stochastic Competitive Learning describes the state of the entire dynamic system through vectors $\mathbf{p(t)}$, $\mathbf{E(t)}$, $\mathbf{S(t)}$, and matrix $\mathbf{Nu(t)}$. Among these, vector $\mathbf{p(t)}$ represents the current positions of each particle within the network; vector $\mathbf{E(t)}$ signifies the energy possessed by each particle. When a particle visits a node under its control, its energy increases by Δ, whereas visiting a node controlled by a competing particle reduces its energy by Δ. This mechanism limits each particle's roaming range, thus minimizing remote and redundant network access. Vector $\mathbf{S(t)}$ denotes the dynamic state of each particle: particles with energy are in an active state, continuously traversing the network; when energy depletes, particles enter a dormant state and randomly jump to one of the nodes under their control for revival. Matrix $\mathbf{Nu(t)}$ records the visitation counts of each particle for all nodes in the network. The more a particle visits a particular node, the greater its control over that node. The particle with the highest visitation count for a node attains control over it.

In Stochastic Competitive Learning, particles in an active state navigate through the network following a convex combination of random and preferential walk rules [34]. The particle walking rule, denoted as $P^{(k)}_{transition}(i,j,t)$, is defined as follows:

$$P^{(k)}_{transition}(i,j,t) \triangleq \lambda P^{(k)}_{pref}(i,j,t) + (1-\lambda)P^{(k)}_{rand}(i,j) \tag{3}$$

where i denotes node v_i. j denotes node v_j. t represents the moment. k indicates the particle. $P^{(k)}_{pref}(i,j,t)$ denotes the particle preferential walk rule. $P^{(k)}_{rand}(i,j)$ represents the particle random rule. $\lambda \in [0,1]$ represents the probability of a particle performing preferential walk, regulating the balance between random and preferential walking. When $\lambda = 1$, the particle exclusively follows preferential walking; when $\lambda \in (0,1)$, the particle performs a combination of both random and preferential walking; and when $\lambda = 0$, the particle solely engages in random walking. The random walking mode guides the particle's exploratory behavior, where the particle randomly visits neighboring nodes without considering their control capacity. This mode reflects the particle's randomness, and the equation for random walking is defined as follows:

$$P^{(k)}_{rand}(i,j) \triangleq \frac{a_{ij}}{\sum\limits_{u \in N(v_i)} a_{iu}} \tag{4}$$

The preferential walking mode guides the particle's defensive behavior, where the particle prioritizes visiting nodes it already controls, rather than nodes that are not yet under its control. This mode reflects the particle's determinism, and the equation for preferential walking is defined as follows:

$$P^{(k)}_{pref}(i,j,t) \triangleq \frac{a_{ij}\overline{Nu}^{(k)}_j(t)}{\sum\limits_{u \in N(v_i)} a_{iu}\overline{Nu}^{(k)}_u(t)} \tag{5}$$

where $\overline{Nu}^{(k)}_j(t)$ represents the current control capacity of particle k over node v_j, determined by the proportion of visits that the particle makes to the node. The equation is defined as follows:

$$\overline{Nu}^{(k)}_j(t) \triangleq \frac{Nu^{(k)}_j(t)}{\sum\limits_{u \in K} Nu^{(u)}_j(t)} \tag{6}$$

When the entire system reaches the convergence criterion, particles cease their wandering. The convergence criterion for the system is defined as follows:

$$\|\overline{\mathbf{Nu}}(\mathbf{t}) - \overline{\mathbf{Nu}}(\mathbf{t\text{-}1})\|_{\infty} < \varepsilon \tag{7}$$

where the convergence factor ε typically takes a value of 0.05. Finally, the community structure is revealed based on the control ranges of individual particles after convergence.

To address the issue of determining the number of particles, stochastic competitive learning employs the particle average maximum control capability metric to establish a reasonable particle placement count within the network, thereby determining a suitable community quantity [35]. The definition of the particle average maximum control capability metric is as follows:

$$\langle R(t)\rangle = \frac{1}{|V|}\sum_{u=1}^{V} \max_{s\in K}\left(\overline{Nu}_u^{(s)}(t)\right) \tag{8}$$

where $\max_{s\in K}\left(\overline{Nu}_u^{(s)}(t)\right)$ represents the maximum control capability exerted by particle s on node u. For a network with a community count of K, if the number of particles placed is exactly K, each particle will dominate a community without excessively interfering with the control regions of other particles. Therefore, $\langle R(t)\rangle$ will take its maximum value. When the particle count is less than the actual number of communities, each particle competes with others to control larger communities, resulting in the attenuation of their control over nodes and causing a decrease in $\langle R(t)\rangle$. When the particle count exceeds the actual community count, particles unavoidably fiercely compete for control over the same group of nodes, leading to a decrease in $\langle R(t)\rangle$. In conclusion, when $\langle R(t)\rangle$ is maximized, the corresponding optimal number of particles placed is the best quantity. The specific method is as follows: gradually increase the placed particle count from 2 to $K + 1$ and record the R value when the system converges under different particle counts. The optimal number of particles placed corresponds to the particle count that yields the maximum $\langle R(t)\rangle$ value.

4. LNSSCL Algorithm

To enhance the stability and accuracy of community detection results, improvements have been made in various aspects such as particle initialization positions, particle preferential walking rules, particle control ability increments, particle resurrection position selection, and the introduction of node affiliation selection. In light of these enhancements, we propose the Unsupervised Community Detection Algorithm with Stochastic Competitive Learning Incorporating Local Node Similarity, which integrates local node similarity into the stochastic competitive learning framework.

4.1. Determining Particle Initialization Positions

The Stochastic Competitive Learning algorithm stipulates that each particle randomly selects a different node in the network as its starting position for walking. The random uncertainty in particle initialization can lead to unstable community detection outcomes. Additionally, this initialization approach might result in particles' starting positions clustering within a single community, intensifying the competitive relationships among particles during their walks. This situation requires a considerable amount of time for convergence. Addressing these concerns, the random placement for initialization is abandoned. Instead, each particle's initial position is determined based on the node's degree and the Salton similarity index between nodes. This approach aims to distribute particles across different communities as much as possible, accelerating the convergence rate of particle walks and enhancing the stability of community detection outcomes.

Node degree is commonly used to measure the importance of a node within the entire network, while the Salton similarity index is often employed to gauge the similarity between a node and its neighboring nodes. It is defined as follows:

$$Salton(i,j) = \frac{|N(v_i) \cap N(v_j)|}{\sqrt{d(v_i) \times d(v_j)}} \tag{9}$$

Combining the two aforementioned metrics, the rules for determining particle initialization positions are as follows. Firstly, arrange all nodes in the network in descending order based on their degree values. Select the node with the highest degree value as the starting position for the first particle's walk. Next, calculate the average Salton similarity index between the node where the already determined starting-position particle is located and all other nodes in the network. Choose the node with the smallest average Salton similarity index as the starting position for the next particle. This process is then repeated iteratively to progressively determine the starting positions for the remaining particles. Finally, when the starting position for each particle is determined, the particle initialization process is completed. Figure 1 depicts the particle initialization position under the condition of three particles. As can be seen from the figure, particles 1, 2, and 3 are dispersed and placed in the network after the position initialization step. $p^{(k)}(0)$ denotes the starting position of particle k. The rules for determining particle initialization positions can be expressed as follows:

$$p^{(k)}(0) = \begin{cases} \operatorname{argmax}(d(v_i)), & k = 1 \\ \operatorname{argmin}\left(\frac{\sum_{u=1}^{k-1} Salton(j, p^{(u)}(0))}{k}\right), & 2 \le k \end{cases} \tag{10}$$

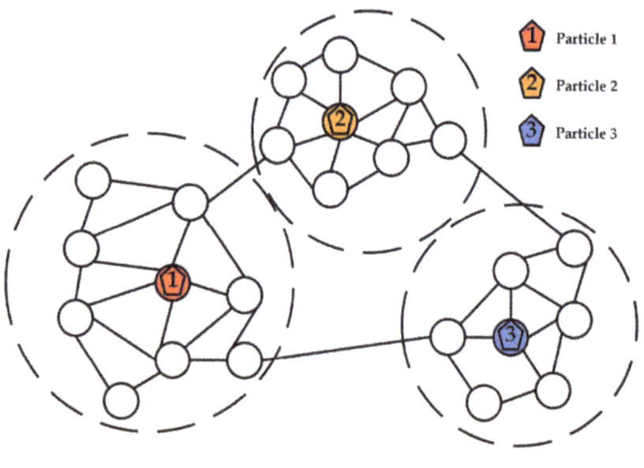

Figure 1. Schematic of particle initialization position.

4.2. Incorporating Node Local Similarity into Particle Preferential Movement Rule

The stochastic competitive learning algorithm stipulates that particles navigate through the network based on a convex combination of random walk and preferential walk rules. The preferential walk rule ensures that particles preferentially access nodes under their control, reflecting the deterministic nature of particle traversal, and numerically equivalent to the particle's control ability ratio. However, this rule solely focuses on the particle's control ability over nodes, without considering the influence of node local similarity indicators. This may lead to a relatively high degree of randomness and weak inclination in the initial direction of preferential walk, thereby affecting the stability and accuracy of community detection results. To address these issues, an enhancement to the particle preferential walk rule is introduced by incorporating node similarity, enabling nodes with greater similarity to be more likely visited by particles. This modification enhances the directionality and determinism of particle traversal during the walk.

The improved node similarity for the enhanced particle preferential walk rule not only considers cases where nodes share common neighbors, but also accounts for situations where nodes lack common neighbors. When there are shared neighbors between nodes, a similarity index considering both common neighbors and degree difference is employed to measure the degree of similarity between two nodes, as defined below:

$$Sim(i,j) = \frac{|N(v_i) \cap N(v_j)|}{|N(v_i) \cap N(v_j)| + |N(v_i) \cup N(v_j)|} \times \frac{|N(v_i) \cap N(v_j)|}{1 + |N_s - N_h|} \tag{11}$$

For nodes v_i and v_j, if $|N(v_i)| \geq |N(v_j)|$, then N_s is set to $N(v_j)$, and N_h is set to $N(v_i)$; if $|N(v_i)| < |N(v_j)|$, then N_s is set to $N(v_i)$, and N_h is set to $N(v_j)$. When there are no shared neighbors between nodes, further consideration is needed for nodes with a degree value of 1. For nodes without shared neighbors and with a degree value of 1, since their behavior is solely related to their unique first-order neighbor, their similarity value is set to 1. For nodes without shared neighbors and with a degree value greater than 1, their similarity is associated with the node's degree value. Based on the negative correlation between node degree and the unfavorable Hub Depressed Index, the degree value is inversely related to its similarity. The equation for calculating the similarity between nodes without shared neighbors is defined as follows:

$$Sim(i,j) = \begin{cases} 1, & (d(v_i) = 1) \vee (d(v_j) = 1) \\ \frac{a_{ij}}{\max\{d(v_i), d(v_j)\}}, & d(v_i) \neq 1, d(v_j) \neq 1 \end{cases} \tag{12}$$

Building upon this, the equation for the particle's preferential walk rule incorporating node similarity is provided:

$$P^{(k)}_{pref}(i,j,t) \triangleq \frac{a_{ij}\overline{Nu_j^{(k)}}(t)(1 + Sim(i,j))}{\sum\limits_{u \in N(v_i)} a_{iu}\overline{Nu_u^{(k)}}(t)(1 + Sim(i,u))} \tag{13}$$

where $Sim(i,j)$ represents the similarity index between nodes v_i and v_j, and $\overline{Nu_j^{(k)}}(t)$ represents the control capacity of particle k over node v_j. The improved preferential walk rule takes into account both the particle's control capacity over nodes and the similarity index between nodes as equally significant factors. This approach avoids the issue of randomness in the preferential walk direction that arises after particle initialization, thereby enhancing the inclination and certainty of particle movement throughout the entire preferential walk process.

4.3. Dynamically Adjusting Particle Control Capacity Increment

In the Stochastic Competitive Learning algorithm, the control capacity of particles is quantified as the proportion of node visits, thereby the number of times a particle visits a node determines its control capacity over that node. When particle k visits node v_i, the equation for the change in the particle's visit count to that node is given by:

$$Nu_i^{(k)}(t+1) = Nu_i^{(k)}(t) + 1 \tag{14}$$

According to Equation (14), it can be observed that the increment of particle control capacity remains constant at 1. This would lead to particles having the same level of competitive increment during the walking process. This uniform competitive increment among different particles could potentially result in similar community sizes controlled by different particles. Consequently, this might lead to instances where representative particles of smaller communities erroneously compete for nodes at the boundaries of larger communities. However, many real-life complex systems, represented as complex networks, often encompass communities of varying sizes. The constant increment in particle control capacity could potentially yield suboptimal results in the final community detection outcome. Figure 2 depicts the possible encroachment of a small community into a large community node when the particle control capacity increment is constant. The size of community 1 in the figure is actually larger than that of community 2. However, because of the constant particle control capacity increment, it makes it possible for the range of communities controlled by each particle to converge to the same size. This then causes nodes that should belong to community 2 to be misclassified to community 2.

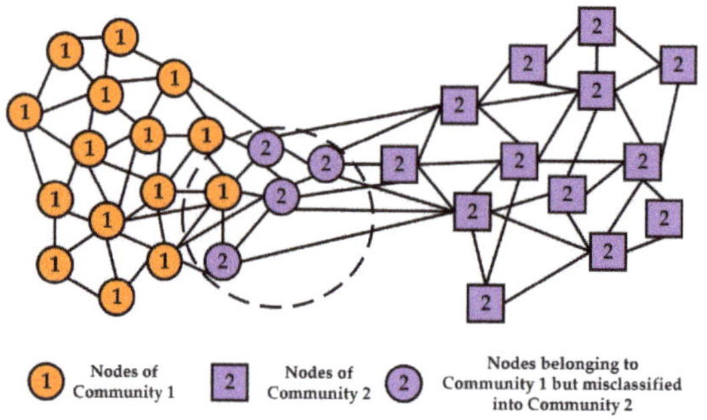

Figure 2. Schematic of error community detection when particle control capacity increment is constant.

Addressing the aforementioned issues, an improvement is made to the particle control capacity increment in order to enhance the effectiveness of discovering communities of varying sizes within the network. The enhanced particle control capacity increment is dynamically adjusted based on the current control range of the particle, and the specific equation is provided below:

$$Nu_i^{(k)}(t+1) = Nu_i^{(k)}(t) + 1 + \frac{\left|C^{(k)}(t)\right|}{|V|} \tag{15}$$

where $\left|C^{(k)}(t)\right|$ represents the current control range of particle k, which indicates the number of nodes currently under the control of particle k. From Equation (15), it can be observed that the particle's control capacity increment is positively correlated with its current control range. This relationship can unveil community structures of different sizes within the network and prevent particles from erroneously encroaching upon nodes located at the boundaries of communities.

4.4. Determining Particle Resurrection Locations and Node Affiliation Selection

In stochastic competitive learning, when a particle visits a node under its control, its energy increases. On the other hand, when it visits a node controlled by a competing particle, its energy decreases. This energy manipulation serves to constrain the particle's walking range, thus reducing long-range and redundant accesses in the network. If a particle frequently visits nodes controlled by competing particles, its energy will continuously decrease until it is exhausted and enters a dormant state. Subsequently, the particle will randomly jump to a node within its control range to revive and recharge.

Clearly, the choice of the particle's revival location has a high degree of randomness, which can lead to unstable community detection results. To address this issue, based on the particle's control over nodes, we select the node with the highest control capability as the unique revival location among the nodes it already controls, eliminating the uncertainty in location selection. If a particle currently doesn't control any nodes, it will randomly choose any node in the network for revival. The improved particle revival location selection is shown in Figure 3, where the energy of particle 1 is depleted due to its traversal into the control region of particle 3. After the improvement, particle 1 will no longer randomly jump to any controlled node within the dashed box, but will instead jump to the node indicated by the dashed arrow (assuming that node has the highest control capability value for particle 1).

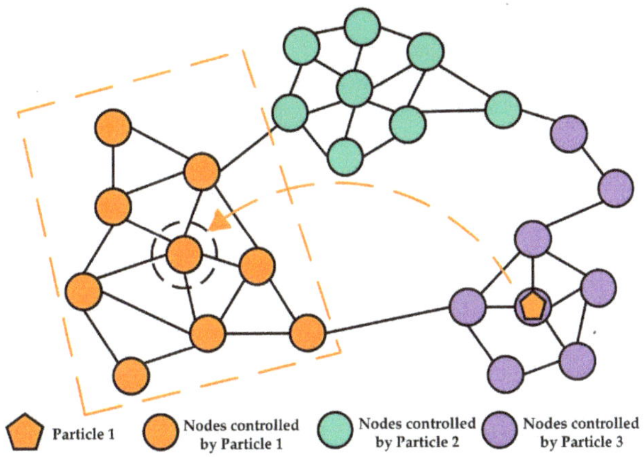

Figure 3. Schematic of particle resurrection location selection.

Once the algorithm reaches the convergence criterion, particles will cease their walks. Based on the control capabilities of each particle over nodes, all nodes in the network are assigned to the corresponding communities represented by particles. However, due to the potential randomness and potential misclassifications in the steps executed by particles before stopping their walks, a node membership selection step is introduced. By considering the frequency of community labels among neighboring nodes, this step ensures that each node is correctly assigned to its appropriate community, further optimizing the community detection outcomes. Specifically, for each node, the occurrence frequency of community labels among its neighboring nodes is observed. If the most frequent neighboring community label is unique, it is selected as the community label for that node. If the most frequent neighboring community label is not unique, an influence score is computed for each community, and the community label with the highest influence score is selected. The influence score $Effect_{C_k}$ for a community is calculated as shown in the equation below:

$$Effect_{C_k} = \sum_{j \in N(v_i)} Sim(i,j), j \in C_k \tag{16}$$

where C_k is one of the most frequent communities, and $N(v_i)$ represents the set of neighboring nodes of node v_i.

4.5. Algorithm Description

Algorithm 1 describes the method of the LNSSCL algorithm; the pseudocode is shown below.

Algorithm 1 LNSSCL algorithm

Input: Graph $G = (V, E)$
The probability of preferential walk for particles λ,
Particle energy increment Δ, Convergence factor ε
Output: The number of communities K, The set of communities C
1: $t = 1$
2: $K = 2$
3: **repeat**
1: **for** each particle k **do**:
5: calculate the initial positions of particles $p^{(k)}(0)$ using Equation (10)
6: **end for**
7: **repeat**
8: **for** $k = 1$ to K **do**:
9: calculate the particle's random walk probability $P^{(k)}_{rand}(i, j)$ using Equation (4)
10: calculate the particle's preferential walk probability $P^{(k)}_{pref}(i, j, t)$ using Equation (11), (12), and (13)
11: calculate the particle's walk probability $P^{(k)}_{transition}(i, j, t)$ using Equation (3)
12: particles walk based on the walk probability and dynamically adjust the particle's control increment using Equation (15)
13: **if** $E^{(k)}(0) \leq 0$:
14: particle performs the revival step by jumping to the within their control range that possesses the control capability for revival and re-energization.
15: **end if**
16: **end for**
17: update $\mathbf{Nu(t)}$, $\overline{\mathbf{Nu(t)}}$, $\mathbf{E(t)}$, $\mathbf{S(t)}$
18: $t = t + 1$
19: **until** Equation (7) is satisfied
20: calculating and record the average maximum control capability indicator for particles $\langle R(t) \rangle$ using Equation (8).
21: $K = K + 1$
22: **until** $\langle R(t) \rangle$ reaches its maximum value
23: assign the number of particles corresponding to step 22 to K
24: **for** each node $v_i \in V$**do**:
25: Assign the corresponding community label to node v_i based on the magnitude relationship of $\overline{Nu_i}(t)$
26: **end for**
27: **for** each node $v_i \in V$**do**:
28: get the set of neighboring nodes $N(v_i)$ for node v_i and count the frequency of appearance of community labels for each neighboring node
29: **if** the most frequently occurring neighboring community label is unique:
30: the label of node v_i is updated to the most frequently occurring community label.
31: **else**:
32: calculate the power score of each most frequent community $Effect_{C_k}$ using Equation (16)
33: the label of node v_i is updated to the community label with the highest community effectiveness score
34: **end if**
35: **end for**
36: **return** $K - 1, C$

4.6. Time Complexity Analysis

For a complex network $G = (V, E)$, assuming the average degree of nodes is $\overline{d}$, the number of nodes is n, the number of edges is m, and the common neighbors between two nodes is c. The determination of particle starting positions involves calculating node degrees and the Salton similarity index between nodes, with a time complexity of $O(m)$.

Active particles wandering in the network require calculating the probability for each particle to move from the current node to neighboring nodes. The random walk probability for each particle only requires computing node degrees, while the preferential walk probability needs to calculate the similarity index between the current node and its neighbors, with a time complexity of $O\left(Kc\overline{d}\right)$. When a particle's energy is exhausted, the revival step involves maintaining a hash table to store each particle's control nodes and their corresponding control capability values. Finding the node with the maximum control capability for jumping has a time complexity of $O(1)$. Updating the particle control matrix has a time complexity of $O(K^2)$. Since each node in the network is visited at least once by a particle, the total time complexity of particle wandering is $O\left(Kc\overline{d}n + K^2n\right)$. To determine the optimal number of particles, the algorithm needs to gradually change the particle count from 2 to K', where K' is a constant slightly larger than the actual number of communities in the network. Therefore, the entire particle wandering process of the LNSSCL algorithm has a time complexity of $O\left(K^2c\overline{d}n + K^3n\right)$. After the particle wandering process concludes, assigning community labels to all nodes based on their control capability values requires a time complexity of $O(n)$. The node membership selection step involves each node selecting its community label based on the frequency of community labels among its neighboring nodes, with a time complexity of $O\left(\left(\overline{d} + c\overline{d}^2\right)n\right)$. Since complex networks are usually sparse networks, $\overline{d} \ll n$. In summary, the time complexity of the LNSSCL algorithm is $O\left(m + K^2c\overline{d}n + K^3n + \left(\overline{d} + c\overline{d}^2\right)n\right) \approx O(M(m+n))$, where M is a constant. The time complexity of the LNSSCL algorithm is linearly related to the sum of the number of nodes and edges in the network, making the algorithm highly scalable on large-scale networks.

5. Experiments and Discussions

To test the effectiveness of the LNSSCL algorithm, experiments were conducted on real network datasets and synthetic networks, comparing the proposed algorithm with 12 representative community detection algorithms. The selected benchmark algorithms for experimentation include community detection algorithms based on random walk such as Walktrap [23] and Infomap [24]; modularity-based algorithms CNM [36], Louvain [37], and Leiden [38]; label propagation-based algorithms LPA [39], TS [27], GLLPA [29], and SSCLPA [30]; hierarchical clustering algorithm Paris [40], Markov chain-based community detection algorithm MSC [25], and the stochastic competitive learning algorithm based on the particle competition mechanism SCL [14].

5.1. Experimental Environment and Initial Parameters

The algorithm was implemented using NetworkX and scikit-learn. The specific experimental environment is shown in Table 1.

Table 1. Experimental environment parameters.

Hardware/Software	Configuration
OS	Windows 11 Home
CPU	Intel(R) Core(TM) i5-11400H @ 2.70GHz
RAM	16 GB
GPU	NVIDIA GeForce RTX 3050 Ti Laptop
Anaconda	3.9.12
Python	3.7.12
NetworkX	2.6.3
scikit-learn	1.0.2

For the setting of the initial parameters, we refer to the value range of the literature [35]. The value range of the particle preferential wandering probability parameter λ is $[0.2, 0.8]$, and we set its initial value to 0.6; the value range of the particle energy update value

Δ is $[0.1, 0.4]$, and we set its initial value to 0.3; the convergence factor ε is set to 0.05; and the minimum value of the particle energy $E_{\min}$ is set to 0, and the maximum value of the particle energy $E_{\max}$ is set to 1.

5.2. Datasets

The experiment utilized 11 real network datasets, including four labeled real network datasets and seven unlabeled real network datasets. The labeled real network datasets consist of the Karate network [41], Dolphins network [42], Polbooks network [43], and Football network [1]. The unlabeled real network datasets include the Lesmis network [44], Jazz network [45], Email network [46], Netscience network [47], Power Grid network [48], Facebook network [49], and PGP network [50]. The basic information is presented in Table 2.

Table 2. Basic information of real network datasets.

Dataset	Nodes	Edges	Community
Karate	34	78	2
Dolphins	62	159	2
Polbooks	105	441	3
Football	115	613	12
Lesmis	77	254	-
Jazz	198	5484	-
Email	1133	5451	-
Netscience	1589	2742	-
Power Grid	4941	6594	-
Facebook	4039	88,234	-
PGP	10,680	24,316	-

Likewise, to expand experiments, the LFR benchmark is used as synthetic networks [51]. We generate different scale networks on the LFR test network model for experiments. The specific parameter settings of the LFR network are shown in Table 3, where $\overline{d}$ is average degree, $d_{\max}$ denotes maximum degree, min$c$ represents minimum community size, max$c$ is maximum community size, and $tau1$ and $tau2$ are the parameters for power law distribution. μ is the mixed parameter. The larger the mixed parameter, the more difficult the community division.

Table 3. The parameters for LFR network construction.

Network	n	$\overline{d}$	$d_{\max}$	min$c$	max$c$	$tau1$	$tau2$	μ
LFR1	1000	15	50	20	100	2	1.1	0.1–0.8
LFR2	4000	15	50	20	100	2	1.1	0.1–0.8

5.3. Evaluation Index

In this paper, we use two widely adopted evaluation metrics for community detection algorithms to assess the quality of the algorithm's community detection results. For labeled real network datasets, we utilize Normalized Mutual Information (NMI) [52] and modularity [36] to evaluate the community detection results of each algorithm. For unlabeled real-world network datasets, since the ground-truth community structures of these network datasets are still unknown, we assess the quality of the detected results in terms of the modularity only. For LFR networks, we use NMI to evaluate the community detection results of each algorithm.

The NMI score is defined as shown in Equation (17):

$$NMI(X,Y) = \frac{-2\sum_{i=1}^{C_X}\sum_{j=1}^{C_Y} C_{ij}\mathrm{lb}\left(\frac{C_{ij}N}{C_iC_j}\right)}{\sum_{i=1}^{C_X} C_i\mathrm{lb}\left(\frac{C_i}{N}\right) + \sum_{j=1}^{C_Y} C_j\mathrm{lb}\left(\frac{C_j}{N}\right)} \quad (17)$$

where X represents the ground truth community structure and Y represents the community detection results of the algorithm. C_X denotes the number of true communities in the ground truth, and C_Y represents the number of communities detected by the algorithm. **C** represents the confusion matrix, where rows represent the ground truth community structure and columns represent the algorithm's community detection results. C_{ij} represents the number of common nodes between the true community i in X and the community j detected in Y. C_i represents the sum of row i in matrix **C**, and C_j represents the sum of row j in matrix **C**. N stands for the total number of nodes in the network. $NMI \in [0, 1]$. A higher NMI value indicates a better agreement between the algorithm's community detection results and the true community structure, thus implying a better performance of community detection.

The modularity(Q) is defined as shown in Equation (18):

$$Q = \frac{1}{2m} \sum_{i,j} \left(a_{ij} - \frac{d(v_i) \times d(v_j)}{2m} \right) \times \delta(c_i, c_j) \tag{18}$$

where m denotes the number of edges in the network, a_{ij} denotes the connection status between node v_i and node v_j; c_i denotes the community label of node v_i, c_j denotes the community label of node v_j; $\delta(c_i, c_j)$ denotes the Kronecker function, which takes the value 1 if c_j and c_j are the same; otherwise, it takes the value 0. Generally, a larger modularity value implies a more distinct community structure.

5.4. Experimental Results and Analysis

5.4.1. Experimental Results and Analysis on Labeled Real Network Datasets

On the four labeled real network datasets, namely Karate, Dolphins, Polbooks, and Football, we conducted comparative experiments between the proposed LNSSCL algorithm and 12 other representative community detection algorithms. We evaluated the community detection results of each algorithm using NMI score and modularity Q. The experimental results are shown in Tables 4 and 5.

Table 4. Comparison of the NMI of each algorithm on labeled real network datasets. The largest NMI are in bold.

Algorithm	Karate	Dolphins	Polbooks	Football
Walktrap	0.504	0.582	0.543	0.887
Infomap	0.699	0.417	0.529	0.911
CNM	0.692	0.557	0.531	0.698
Louvain	0.587	0.484	0.569	0.885
Leiden	0.687	0.581	0.574	0.890
LPA	0.445	0.595	0.534	0.870
TS	0.710	0.888	0.550	0.900
GLLPA	0.753	0.790	0.580	0.909
SSCLPA	0.826	0.616	0.493	0.919
Paris	0.835	0.780	0.565	0.831
MSC	0.836	0.777	0.539	0.921
SCL	0.821	0.816	0.552	0.861
LNSSCL	**0.848**	**0.899**	**0.610**	**0.937**

Table 5. Comparison of modularity Q of each algorithm on labeled real network datasets. The largest Q are in bold.

Algorithm	Karate	Dolphins	Polbooks	Football
Walktrap	0.353	0.489	0.507	0.603
Infomap	0.401	**0.532**	0.521	0.600
CNM	0.381	0.495	0.502	0.550
Louvain	0.415	0.516	0.526	0.602
Leiden	0.420	0.527	**0.527**	0.602
LPA	0.375	0.499	0.481	0.583
TS	0.420	0.381	0.520	0.600
GLLPA	**0.438**	0.496	0.507	0.608
SSCLPA	0.415	0.525	0.518	0.601
Paris	0.372	0.380	0.426	0.510
MSC	0.418	0.495	0.519	0.600
SCL	0.371	0.463	0.508	0.581
LNSSCL	0.424	**0.532**	0.524	**0.613**

As can be seen in Tables 4 and 5, the LNSSCL algorithm achieves the highest NMI values on all labeled real network datasets and has some degree of improvement over the other algorithms. Whereas, for the comparison of modularity Q, the LNSSCL algorithm does not take the optimal value on all the datasets. Though the proposed algorithm took the highest value only on Dolphins and Football datasets, the modularity scores on Karate and Polbooks datasets were close to the highest level.

The NMI measures the similarity between the algorithm's output community segmentation results and the real online community structure. The larger the NMI, the higher the similarity between the algorithm's output community segmentation results and the real online community structure. It can be seen that the community detection results of the LNSSCL algorithm on the four labeled real network datasets of Karate, Dolphins, Polbooks, and Football are closest to the real community structure of the above networks. In addition, the LNSSCL algorithm improves its NMI values by 3.3%, 10.2%, 10.5%, and 8.8% on the Karate, Dolphins, Polbooks, and Football datasets, respectively, compared to the SCL algorithm. This demonstrates the effectiveness of a series of improvements to the SCL algorithm by the LNSSCL algorithm, which improves the stability and accuracy of the SCL algorithm.

For other algorithms, CNM, Louvain, and Leiden algorithms seek to maximize the modularity of the whole network, which obtains good modularity scores but fails to discover the real community structure well. LPA, TS, GLLPA, and SSCLPA algorithms tend to have a certain degree of randomness in the node order of label updating and the label propagation process, making the final community. The computational complexity of Walktrap and Infomap, two randomized wandering algorithms, is relatively high and sensitive to the parameters set by themselves. Due to the small size of the dataset used in the experiments and the existence of small communities, although they have higher modularity scores, they failed to achieve higher NMI scores, and the consistency with the real community structure is insufficient.

Overall, the LNSSCL algorithm works best for community detection on the Karate, Dolphins, Polbooks, and Football datasets.

5.4.2. Experimental Results and Analysis on Unlabeled Real Network Datasets

On the seven unlabeled real network datasets, namely Lesmis, Jazz, Email, Netscience, Power Grid, Facebook, and PGP, the proposed LNSSCL algorithm was compared against 12 representative learning algorithms through experimental evaluation. The assessment of various algorithm's community detection outcomes was conducted using modularity Q, and the experimental results are shown in Table 6.

Table 6. Comparison of modularity Q of each algorithm on unlabeled real network datasets. The largest Q are in bold.

Algorithm	Lesmis	Jazz	Email	Netscience	Power Grid	Facebook	PGP
Walktrap	0.521	0.438	0.531	0.956	0.831	0.812	0.832
Infomap	0.546	0.442	0.538	0.930	0.759	0.706	0.821
CNM	0.501	0.439	0.503	0.955	0.935	0.777	0.853
Louvain	0.527	0.282	0.463	0.907	0.627	0.835	0.885
Leiden	0.558	0.443	0.563	0.959	0.935	0.836	0.887
LPA	0.560	0.445	0.574	0.960	0.939	0.737	0.745
TS	0.535	0.440	0.550	0.924	0.930	0.796	0.767
GLLPA	0.556	0.440	0.292	0.920	0.784	0.785	0.786
SSCLPA	0.557	0.406	0.513	0.906	0.838	0.779	0.801
Paris	0.504	0.276	0.451	0.876	0.405	0.586	0.679
MSC	0.544	0.447	0.565	0.948	0.890	0.812	0.872
SCL	0.539	0.411	0.550	0.932	0.863	0.804	0.815
LNSSCL	**0.575**	**0.463**	**0.585**	**0.971**	**0.952**	**0.847**	**0.896**

Since the unlabeled network dataset is without real community structure, we can accomplish community delineation by identifying structural features with strong internal connections and sparse external connections. Modularity measures the strength of community structure in the network and evaluates the results of algorithmic community delineation when the dataset has no real community structure. The larger the modularity degree is, the better the quality of community detection and the stronger the connection within the community.

As can be seen in Table 6, the LNSSCL algorithm achieves the maximum modularity on all seven unlabeled network datasets used in the experiments, outperforming other comparison algorithms. In addition, these network datasets are of different types, sizes, and sparsities, and the high modularity performance of the LNSCCL algorithm reflects the algorithm's good generalization and universality. In particular, the performance on two larger datasets, Facebook and PGP, shows that the algorithm has some scalability. For other algorithms, CNM, Louvain, and other algorithms oriented to maximize the modularity achieved high modularity on the Netscience, Power Grid, Facebook, and PGP datasets, with Leiden in particular being the most prominent, but did not show the best performance on the smaller datasets. LPA, TS, GLLPA, and SSCLPA algorithms are not as effective as the stable LNSSCL algorithm in the experiments, because some randomness in the node order of the label update and label propagation process cannot show stable and accurate performance. The MSC algorithm constructs a steady-state Markov similarity augmented matrix, which is capable of stable and efficient community delineation, and achieves high modularity in the experiments. Further, the NSSCL algorithm improves its modularity values by 6.7%, 12.7%, 6.4%, 4.2%, 10.3%, 5.3%, and 9.9% on Lesmis, Jazz, Email, Netscience, Power Grid, Facebook, and PGP datasets, respectively, as compared to the SCL algorithm. This shows that the LNSSCL algorithm improves the stability and accuracy of the original SCL algorithm for community detection to some extent.

5.4.3. Experimental Results and Analysis on synthetic networks

In order to better measure the performance of the algorithm, we generate networks of different sizes for experiments on the LFR test network model. In this case, the number of nodes in the LFR1 network is 1000 and the number of nodes in the LFR2 network is 4000. Since the real community structure of the LFR network is known, the performance of the algorithm is measured using NMI. Among the important parameters used to create the LFR network, the mixing parameter μ is used to represent the complexity of the community structure and determines the clarity of the community structure. As the mixing parameter μ increases, the community structure becomes more complex and the difficulty of recognizing the community increases. The experimental results of the LFR network with different mixing parameter μ are shown in Figure 4, where the horizontal coordinates represent the individual values of the mixing parameter μ and the vertical coordinates represent the NMI.

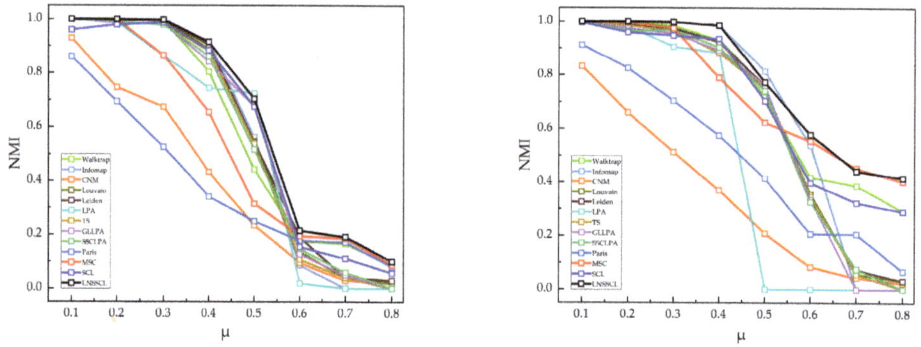

(**a**)LFR 1 (**b**)LFR 2

Figure 4. Experimental results of NMI under synthetic networks: (**a**) experimental results of LFR1 network where n = 1000; (**b**) experimental results of LFR2 network where n = 4000.

As can be seen in Figure 4, the performance of each algorithm decreases to varying degrees in both sets of networks as the mixing parameter μ increases, but the community detection performance of the LNSSCL algorithm still outperforms most of the compared algorithms.

When the mixing parameter μ is less than 0.4, the algorithms have higher NMI values, except for the CNM and Paris algorithms. When μ is greater than 0.4, the NMIs all show a decrease. Among them, label propagation-based algorithms such as LPA have very obvious changes in decline, especially the LPA algorithm, which has an NMI value of 0 in both networks when μ is not less than 0.6. The reason for the analysis is that label propagation-based algorithms have a certain degree of randomness, which is prone to cause low performance and instability in community detection when the community structure is not clear enough. Comparing the synthetic networks LFR1 and LFR2, the NMI of most of the algorithms increased with the increase in the number of nodes. However, the performance of the CNM algorithm shows a decrease, which is attributed to the fact that the algorithm may have the problem of resolution limitation when dealing with networks with larger communities, and is unable to decompose large communities into smaller sub-communities. From Figure 4, it can be found that the NMI value of LNSSCL tends to 1 when μ = 0.1, and the NMI value of LNSSCL is maximum when μ = 0.8. It indicates that the performance of LNSSCL decays the slowest, i.e., the performance of LNSSCL is more stable.

In summary, the community detection results of the LNSSCL algorithm are better on synthetic network datasets generated with different parameters.

5.4.4. Parameter Sensitivity Analysis

In this section, we focus on the effects of the parameters λ and Δ on the LNSSCL algorithm. λ represents the probability of a particle performing preferential walk, regulating the balance between random and preferential walking. When the parameter λ takes a relatively low value, the particles are more inclined to wander randomly, visit new nodes, and expand the community range, but are prone to the problem of too much randomness. When the parameter λ takes a relatively high value, the particle is more inclined to preferentially wander, frequently visit the nodes that have been controlled, and consolidate the community scope, but it is easy to fall into the localized area and it cannot visit new nodes. From Figure 5, it can be seen that the parameter λ can achieve the best community detection performance by appropriately increasing the value of the parameter when balancing random wandering and preferential wandering ($\lambda = 0.5$). For the Polbooks, Football, and Jazz datasets, the best performance is achieved when the value of λ is 0.6, while the Email dataset takes the value of λ as 0.5.

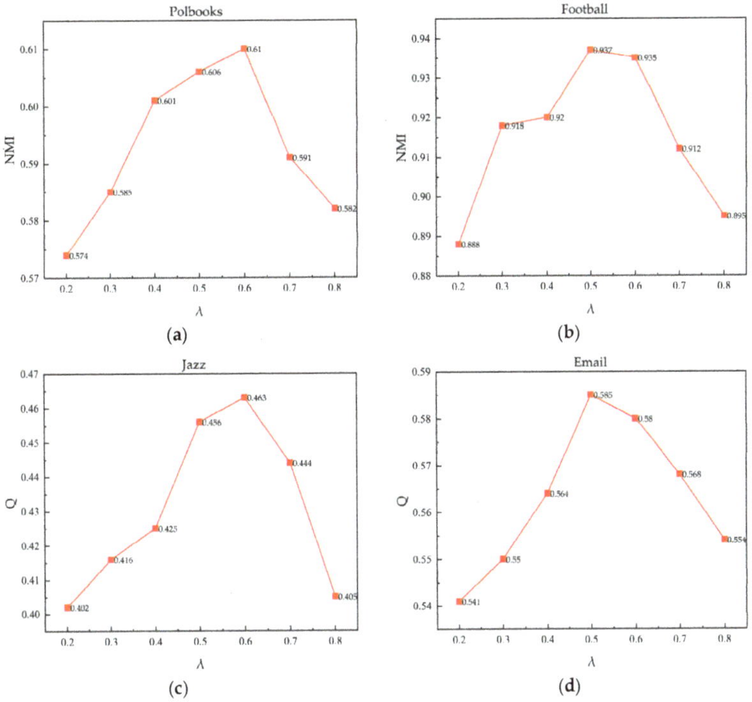

Figure 5. Parameter sensitivity analysis for λ: (**a**) experimental results of Polbooks; (**b**) experimental results of Football; (**c**) experimental results of Jazz; (**d**) experimental results of Email.

The parameter Δ is responsible for updating the particle's energy value. When Δ is very small, the particle is not penalized enough by the energy it receives for visiting nodes that are not under its control, so the particle's energy will not be depleted during its wanderings. The particle will frequently visit nodes that should belong to the nodes to which the competing particles belong and enter the core of other communities. As a result, all nodes in the network will be in constant competition, unable to establish and consolidate community boundaries, and the final community detection will be less effective. When Δ is very large, the particle will simply run out of energy once it visits a node controlled by a competing particle. The particle will frequently enter the resurrection phase and is not expected to move away from its initial position to take control of other nodes. As can be seen in Figure 6, a parameter Δ of 0.3 achieves the best performance on all four datasets of the experiment. The different values of the parameter Δ have roughly the same trend in affecting the performance on different datasets in the experiment.

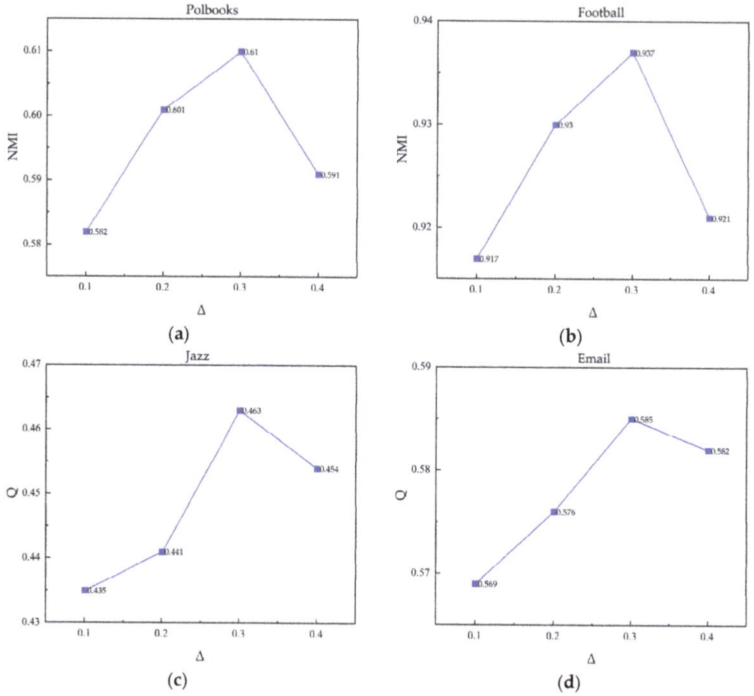

Figure 6. Parameter sensitivity analysis for Δ: (**a**) experimental results of Polbooks; (**b**) experimental results of Football; (**c**) experimental results of Jazz; (**d**) experimental results of Email.

According to Figures 5 and 6, the algorithm is somewhat sensitive to the values of λ and Δ. Numerically, there is little fluctuation within a small interval of relative range. Based on the results of the parameter sensitivity analysis experiments, we set λ to 0.6 and Δ to 0.3 for experiments on other datasets.

6. Conclusions and Future Work

This paper introduces a novel unsupervised community detection algorithm named LNSSCL, which incorporates node local similarity into the process of stochastic competitive learning. Firstly, the algorithm determines the starting position of particles' walks by calculating the degree value of nodes as well as the Salton similarity index. At the same time, the fusion of node similarity optimizes the particle preferential walk rule. During the particle wandering process, the particle control capacity increment is dynamically adjusted according to the control range of each particle. When a particle runs out of energy, the particle selects the node with the largest control power within its control range for resurrection. After the particle stops wandering, the nodes in the network are selected for affiliation based on the frequency of occurrence of community labels of neighboring nodes and the effectiveness score of neighboring communities, and the community detection results are finally obtained. Comparative experiments on real network datasets and synthetic networks show that the LNSSCL algorithm is effective in improving the SCL algorithm. Compared with other representative algorithms, the LNSSCL algorithm has better quality of community detection and is able to reveal a more reasonable community structure.

Nevertheless, the LNSSCL algorithm also has some defects. Compared with the SCL algorithm, the algorithm performs multiple node local similarity calculations during the community detection process, which requires more computational cost and complexity, and may have a larger time overhead on ultra-large networks. In the selection of some hyperparameters of the algorithm, no special parameter tuning method is used; the parameters are tuned manually. Further, the networks studied in this paper are simple undirected networks and not complex networks that are closer to the real world, such as directed, weighted, or attribute. In the next step of this study, we can consider optimizing the similarity calculation to further reduce the time overhead, adopting a dynamic adaptive hyper-parameter tuning strategy instead of traditional parameter tuning. GNN is introduced to fuse attribute information and structural features to obtain node representations and calculate node representation similarity to study the community detection strategy for attribute networks.

Author Contributions: Conceptualization, J.H. and Y.G.; methodology, J.H.; formal analysis, J.H.; investigation, J.H. and Y.G.; data curation, J.H.; writing—original draft preparation, J.H.; writing—review and editing, J.H. and Y.G.; supervision, Y.G. All authors have read and agreed to the published version of the manuscript.

Funding: This research was funded by the Double-class Special Project on Cyberspace Security and Law Enforcement Technology of the Chinese People's Public Security University (2023SYL07).

Institutional Review Board Statement: Not applicable.

Informed Consent Statement: Not applicable.

Data Availability Statement: Data openly available in http://www-personal.umich.edu/~mejn/netdata/ (accessed on 1 March 2023) and https://deim.urv.cat/~alexandre.arenas/data/welcome.htm (accessed on 1 March 2023).

Acknowledgments: The authors thank the anonymous reviewers for their insightful comments that helped improve the quality of this study.

Conflicts of Interest: The authors declare no conflict of interest.

References

1. Girvan, M.; Newman, M.E. Community Structure in Social and Biological Networks. *Proc. Natl. Acad. Sci. USA* **2002**, *99*, 7821–7826. [CrossRef]
2. Fortunato, S. Community Detection in Graphs. *Phys. Rep.* **2010**, *486*, 75–174. [CrossRef]
3. Fortunato, S.; Hric, D. Community Detection in Networks: A User Guide. *Phys. Rep.* **2016**, *659*, 1–44. [CrossRef]
4. Javed, M.A.; Younis, M.S.; Latif, S.; Qadir, J.; Baig, A. Community Detection in Networks: A Multidisciplinary Review. *J. Netw. Comput. Appl.* **2018**, *108*, 87–111. [CrossRef]
5. Jin, D.; Yu, Z.; Jiao, P.; Pan, S.; He, D.; Wu, J.; Yu, P.; Zhang, W. A Survey of Community Detection Approaches: From Statistical Modeling to Deep Learning. *IEEE Trans. Knowl. Data Eng.* **2021**, *35*, 1149–1170. [CrossRef]
6. Xing, S.; Shan, X.; Fanzhen, L.; Jia, W.; Jian, Y.; Chuan, Z.; Wenbin, H.; Cecile, P.; Surya, N.; Di, J. A Comprehensive Survey on Community Detection with Deep Learning. *IEEE Trans. Neural Netw. Learn. Syst* **2022**, 1–21. [CrossRef]
7. Wang, B.; Gu, Y.; Zheng, D. Community Detection in Error-Prone Environments Based on Particle Cooperation and Competition with Distance Dynamics. *Phys. A Stat. Mech. Its Appl.* **2022**, *607*, 128178. [CrossRef]
8. Grossberg, S. Competitive Learning: From Interactive Activation to Adaptive Resonance. *Cogn. Sci.* **1987**, *11*, 23–63. [CrossRef]
9. Kohonen, T. The Self-Organizing Map. *Proc. IEEE* **1990**, *78*, 1464–1480. [CrossRef]
10. Kohonen, T.; Kohonen, T. Learning Vector Quantization. *Self-Organ. Maps* **1995**, *30*, 175–189.
11. Hecht-Nielsen, R. Counterpropagation Networks. *Appl. Opt.* **1987**, *26*, 4979–4984. [CrossRef] [PubMed]
12. Kosko, B. Stochastic Competitive Learning. In Proceedings of the 1990 IJCNN International Joint Conference on Neural Networks, San Diego, CA, USA, 17–21 June 1990; IEEE: New York, NY, USA, 1990; pp. 215–226.
13. Quiles, M.G.; Zhao, L.; Alonso, R.L.; Romero, R.A. Particle Competition for Complex Network Community Detection. *Chaos Interdiscip. J. Nonlinear Sci.* **2008**, *18*, 033107. [CrossRef] [PubMed]
14. Silva, T.C.; Zhao, L. Stochastic Competitive Learning in Complex Networks. *IEEE Trans. Neural Netw. Learn. Syst.* **2012**, *23*, 385–398. [CrossRef] [PubMed]
15. Silva, T.C.; Zhao, L. Detecting and Preventing Error Propagation via Competitive Learning. *Neural Netw.* **2013**, *41*, 70–84. [CrossRef] [PubMed]
16. Silva, T.C.; Zhao, L. Uncovering Overlapping Cluster Structures via Stochastic Competitive Learning. *Inf. Sci.* **2013**, *247*, 40–61. [CrossRef]
17. Li, W.; Gu, Y.; Yin, D.; Xia, T.; Wang, J. Research on the Community Number Evolution Model of Public Opinion Based on Stochastic Competitive Learning. *IEEE Access* **2020**, *8*, 46267–46277. [CrossRef]

18. Zamoner, F.W.; Zhao, L. A Network-Based Semi-Supervised Outlier Detection Technique Using Particle Competition and Cooperation. In Proceedings of the 2013 Brazilian Conference on Intelligent Systems, Fortaleza, Brazil, 19–24 October 2013; IEEE: New York, NY, USA, 2013; pp. 225–230.
19. Breve, F.; Quiles, M.G.; Zhao, L. Interactive Image Segmentation Using Particle Competition and Cooperation. In Proceedings of the 2015 international joint conference on neural networks (IJCNN), Killarney, Ireland, 12–17 July 2015; IEEE: New York, NY, USA, 2015; pp. 1–8.
20. Breve, F.; Fischer, C.N. Visually Impaired Aid Using Convolutional Neural Networks, Transfer Learning, and Particle Competition and Cooperation. In Proceedings of the 2020 International Joint Conference on Neural Networks (IJCNN), Glasgow, UK, 19–24 July 2020; IEEE: New York, NY, USA, 2020; pp. 1–8.
21. Roghani, H.; Bouyer, A. A Fast Local Balanced Label Diffusion Algorithm for Community Detection in Social Networks. *IEEE Trans. Knowl. Data Eng.* **2022**, *35*, 5472–5484. [CrossRef]
22. Toth, C.; Helic, D.; Geiger, B.C. Synwalk: Community Detection via Random Walk Modelling. *Data Min. Knowl. Discov.* **2022**, *36*, 739–780. [CrossRef]
23. Pons, P.; Latapy, M. Computing Communities in Large Networks Using Random Walks. In Proceedings of the Computer and Information Sciences-ISCIS 2005: 20th International Symposium, Istanbul, Turkey, 26–28 October 2005; Proceedings 20; Springer: Berlin/Heidelberg, Germany, 2005; pp. 284–293.
24. Rosvall, M.; Bergstrom, C.T. Maps of Random Walks on Complex Networks Reveal Community Structure. *Proc. Natl. Acad. Sci. USA* **2008**, *105*, 1118–1123. [CrossRef]
25. Yang, X.-H.; Ma, G.-F.; Zeng, X.-Y.; Pang, Y.; Zhou, Y.; Zhang, Y.-D.; Ye, L. Community Detection Based on Markov Similarity Enhancement. *IEEE Trans. Circuits Syst. II Express Briefs* **2023**, *70*, 3664–3668. [CrossRef]
26. Jokar, E.; Mosleh, M.; Kheyrandish, M. Discovering Community Structure in Social Networks Based on the Synergy of Label Propagation and Simulated Annealing. *Multimed. Tools Appl.* **2022**, *81*, 21449–21470. [CrossRef]
27. You, X.; Ma, Y.; Liu, Z. A Three-Stage Algorithm on Community Detection in Social Networks. *Knowl.-Based Syst.* **2020**, *187*, 104822. [CrossRef]
28. Dabaghi-Zarandi, F.; KamaliPour, P. Community Detection in Complex Network Based on an Improved Random Algorithm Using Local and Global Network Information. *J. Netw. Comput. Appl.* **2022**, *206*, 103492. [CrossRef]
29. Zhang, Y.; Liu, Y.; Jin, R.; Tao, J.; Chen, L.; Wu, X. Gllpa: A Graph Layout Based Label Propagation Algorithm for Community Detection. *Knowl.-Based Syst.* **2020**, *206*, 106363. [CrossRef]
30. Chin, J.H.; Ratnavelu, K. Community Detection Using Constrained Label Propagation Algorithm with Nodes Exemption. *Computing* **2022**, *104*, 339–358. [CrossRef]
31. Fei, R.; Wan, Y.; Hu, B.; Li, A.; Li, Q. A Novel Network Core Structure Extraction Algorithm Utilized Variational Autoencoder for Community Detection. *Expert Syst. Appl.* **2023**, *222*, 119775. [CrossRef]
32. Li, H.-J.; Wang, L.; Zhang, Y.; Perc, M. Optimization of Identifiability for Efficient Community Detection. *New J. Phys.* **2020**, *22*, 063035. [CrossRef]
33. Lü, L.; Zhou, T. Link Prediction in Complex Networks: A Survey. *Phys. A Stat. Mech. Its Appl.* **2011**, *390*, 1150–1170. [CrossRef]
34. Silva, T.C.; Amancio, D.R. Network-Based Stochastic Competitive Learning Approach to Disambiguation in Collaborative Networks. *Chaos: Interdiscip. J. Nonlinear Sci.* **2013**, *23*, 013139. [CrossRef]
35. Silva, T.C.; Zhao, L. Case Study of Network-Based Unsupervised Learning: Stochastic Competitive Learning in Networks. In *Machine Learning in Complex Networks*; Springer International Publishing: Cham, Switzerland, 2016; pp. 241–290. ISBN 978-3-319-17289-7.
36. Clauset, A.; Newman, M.E.; Moore, C. Finding Community Structure in Very Large Networks. *Phys. Rev. E* **2004**, *70*, 066111. [CrossRef]
37. Blondel, V.D.; Guillaume, J.-L.; Lambiotte, R.; Lefebvre, E. Fast Unfolding of Communities in Large Networks. *J. Stat. Mech. Theory Exp.* **2008**, *2008*, P10008. [CrossRef]
38. Traag, V.A.; Waltman, L.; Van Eck, N.J. From Louvain to Leiden: Guaranteeing Well-Connected Communities. *Sci. Rep.* **2019**, *9*, 5233. [CrossRef]
39. Raghavan, U.N.; Albert, R.; Kumara, S. Near Linear Time Algorithm to Detect Community Structures in Large-Scale Networks. *Phys. Rev. E* **2007**, *76*, 036106. [CrossRef] [PubMed]
40. Bonald, T.; Charpentier, B.; Galland, A.; Hollocou, A. Hierarchical Graph Clustering Using Node Pair Sampling. *arXiv* **2018**, arXiv:1806.01664.
41. Zachary, W.W. An Information Flow Model for Conflict and Fission in Small Groups. *J. Anthropol. Res.* **1977**, *33*, 452–473. [CrossRef]
42. Lusseau, D.; Schneider, K.; Boisseau, O.J.; Haase, P.; Slooten, E.; Dawson, S.M. The Bottlenose Dolphin Community of Doubtful Sound Features a Large Proportion of Long-Lasting Associations: Can Geographic Isolation Explain This Unique Trait? *Behav. Ecol. Sociobiol.* **2003**, *54*, 396–405. [CrossRef]
43. Newman, M.E. Modularity and Community Structure in Networks. *Proc. Natl. Acad. Sci. USA* **2006**, *103*, 8577–8582. [CrossRef]
44. Knuth, D.E. *The Stanford GraphBase: A Platform for Combinatorial Computing*; AcM Press: New York, NY, USA, 1993; Volume 1.
45. Gleiser, P.M.; Danon, L. Community Structure in Jazz. *Adv. Complex Syst.* **2003**, *6*, 565–573. [CrossRef]

46. Guimera, R.; Danon, L.; Diaz-Guilera, A.; Giralt, F.; Arenas, A. Self-Similar Community Structure in a Network of Human Interactions. *Phys. Rev. E* **2003**, *68*, 065103. [CrossRef]
47. Newman, M.E. Finding Community Structure in Networks Using the Eigenvectors of Matrices. *Phys. Rev. E* **2006**, *74*, 036104. [CrossRef]
48. Watts, D.J.; Strogatz, S.H. Collective Dynamics of 'Small-World'Networks. *Nature* **1998**, *393*, 440–442. [CrossRef] [PubMed]
49. Leskovec, J.; Mcauley, J. Learning to Discover Social Circles in Ego Networks. *Adv. Neural Inf. Process. Syst.* **2012**, *25*, 539–547.
50. Boguná, M.; Pastor-Satorras, R.; Díaz-Guilera, A.; Arenas, A. Models of Social Networks Based on Social Distance Attachment. *Phys. Rev. E* **2004**, *70*, 056122. [CrossRef] [PubMed]
51. Lancichinetti, A.; Fortunato, S.; Radicchi, F. Benchmark Graphs for Testing Community Detection Algorithms. *Phys. Rev. E* **2008**, *78*, 046110. [CrossRef]
52. Danon, L.; Diaz-Guilera, A.; Duch, J.; Arenas, A. Comparing Community Structure Identification. *J. Stat. Mech. Theory Exp.* **2005**, *2005*, P09008. [CrossRef]

applied sciences

MDPI

Article

A Graph Convolution Collaborative Filtering Integrating Social Relations Recommendation Method

Min Ma, Qiong Cao * and Xiaoyang Liu

School of Computer Science and Engineering, Chongqing University of Technology, Chongqing 400054, China
* Correspondence: caoqiong@cqut.edu.cn

Abstract: Traditional collaborative filtering recommendation algorithms only consider the interaction between users and items leading to low recommendation accuracy. Aiming to solve this problem, a graph convolution collaborative filtering recommendation method integrating social relations is proposed. Firstly, a social recommendation model based on graph convolution representation learning and general collaborative filtering (SRGCF) is constructed; then, based on this model, a social relationship recommendation algorithm (SRRA) is proposed; secondly, the algorithm learns the representations of users and items by linear propagation on the user–item bipartite graph; then the user representations are updated by learning the representations with social information through the neighbor aggregation operation in the social network to form the final user representations. Finally, the prediction scores are calculated, and the recommendation list is generated. The comparative experimental results on four real-world datasets show that: the proposed SRRA algorithm performs the best over existing baselines on Recall@10 and NDCG@10; specifically, SRRA improved by an average of 4.40% and 9.62% compared to DICER and GraphRec, respectively, which validates that the proposed SRGCF model and SRRA algorithm are reasonable and effective.

Keywords: social relations; collaborative filtering; graph convolutional network; recommendation system

Citation: Ma, M.; Cao, Q.; Liu, X. A Graph Convolution Collaborative Filtering Integrating Social Relations Recommendation Method. *Appl. Sci.* **2022**, *12*, 11653. https://doi.org/10.3390/app122211653

Academic Editor: Gianluca Lax

Received: 30 October 2022
Accepted: 15 November 2022
Published: 16 November 2022

1. Introduction

1.1. Background

During the age of information explosion, recommender systems have become widely used and effective method to identify the most valuable one in a massive amount of data. A recommender system (in short, RS) aims at estimating the likelihood of interactions between target users and candidates based on interactive history [1,2]. RSs first learn the users' and items' representations (also called embeddings), and then use these representations to predict how a target user will like a specific item.

The first successful algorithm to generate recommendations is Collaborative Filtering (CF) which only on user provided ratings. Traditional CF methods suffer from data sparsity. Matrix Factorization (in short MF) techniques are a viable method to alleviate data-sparsity: specifically, MF methods decompose the (high-dimensional) user–rating matrix into the product of two low-dimensional user–factor and item-factor matrices such that the inner product of the vectors associated with a user and items explains observed ratings. MF methods are also effective to cope with the cold-start problem (i.e., how to generate predictions for new members of a recommender system for which historical data are poor).

Graph Convolution Networks (GCN) have been successfully applied to improve the accuracy of an RS. However, recent studies [3] prove that two common operations in the design of GCNs (namely the task of transforming features and nonlinear activation) provide a little contribution to the performance of an RS. He et al. proposed a new GCN model for supporting CF tasks called LightGCN; the LightGCN architecture includes only the neighborhood aggregation step and, thus, it learns user and item embeddings by linear propagation on the user–item graph. It finally computes the weighted sum of embeddings at all layers to produce final embeddings.

Although LightGCN achieves high levels of accuracy, it only takes data about historical interactions between users and items; many RS can manage a wealth of data describing user-to-user relationships (such as friendship or trust relationships) which are, de facto, ignored by LightGCN. As a consequence, the semantics of information mined by LightGCN is relatively simple and we ask if social data can be profitably integrated in the recommendation task to increase the level of accuracy and, potentially, to make the recommendation process more scalable. Motivated by theories from social science about influence and homophily, many researchers suggested to integrate user–item ratings with information describing social relationships (such as friendship or trust relationships); in fact, users' preferences as well as the decisions they take are often influenced by the behaviors from their peers and, thus, the predictive accuracy of an RS is magnified if social relationships are somewhat taken into account.

An RS which considers social information is often called social recommender system. Graph theory provide powerful tools to represent user interactions among users as well as interactions between users and items. On one hand, in fact, we could define a user–user interaction network (often called social graph), which maps user onto nodes and in which edges cab represent friendship or trust relationships. On the other hand, we could use a bipartite graph (often called user–item interaction graph) to model users and items as nodes while edges are relevant to encode a variety of interactions such as purchases, clicks, and comments. Several social recommender systems which jointly leverage social graph and interactive graph have been proposed so far. The first is to extend matrix factorization methods into the interactive graph and the social graph, and the second is to apply Graph Neural Networks (in short, GNNs) to obtain meaningful representations.

1.2. Motivations

Despite it is useful to integrate social data into RS, we believe that there is still room to enhance the accuracy of an RS. We introduce a new GCN architecture which integrates social data in the recommendation process Our approach aims at solving the following problems:

(1) Heterogeneous data are difficult to use: the data used in social recommendation often contain both user interaction data and user social data. Heterogeneity of data implies that we are in charge of handling representations of different objects (items and users); as a consequence, we deal with nodes which are not in the same embedded space, and, thus, these nodes are hard to be fused.
(2) High-order semantic information is hard to extract: For instance, high-order semantics describe relationships that users are indirectly connected to in the user–user social network. It is thus crucial to capture complex long-term dependencies between nodes. The more iteration layers of graph convolution architecture, the higher order semantic information will be extracted. However, excessive iteration layers will cause excessive smoothness.
(3) Difficulties in fusing multiple semantic information: Social recommender systems manage both social network and interactive graph and it also has the task that effectively integrate the information coming from both of these graphs is still open research.

1.3. Our Contributions

The following contributions have made:

(1) We innovatively integrate social relations into the training of graph convolution-based collaborative filtering recommendation method. Specifically, we propose a graph convolution collaborative filtering recommendation model integrating social relations (called SRGCF). The SRGCF model learns node embeddings by integrating high-order semantic information about social behaviors as well as interactions.
(2) We propose a recommender algorithm (called SRRA) running on top of the SRGCF model. The SRRA algorithm models the high-order relations in interactive data and

social data, respectively, fuses these two types of high-order semantic information at the same layer and then forms the final embeddings. Generated embeddings are finally used in recommendation task.

(3) We experimentally compared the SRRA algorithm with baselines on a range of real-life and large datasets. Our experiments indicate the superiority of our model against baselines.

2. Related Work

2.1. Traditional CF Recommendation Algorithm

In e-commerce industry, collaborative filtering (in short, CF) has been widely used, and in the past two decades many CF algorithms have emerged in academia and industry. Roughly speaking, CF algorithms is classified into two categories: neighborhood-based CF algorithms [4] and model-based recommendation algorithms.

Neighborhood-based CF algorithms [5–8] can find potentially relevant items from user past behaviors without any domain knowledge, and they can be further classified into two specific types which are user-based CF and item-based CF. The key of neighborhood-based CF methods is how to calculate similarity and sum up these scores.

The primary idea behind the model-based CF algorithms [9] is to embed users as well as items into the same embedding space, and then make prediction through the inner-product of their embeddings. Using data mining and machine learning techniques, model-based approaches predict unknown scores by finding patterns in training data.

However, the accuracy of traditional CF recommendation method is limited because they make only use of interactions between users and items to predict unknown ratings.

2.2. Social Recommendation Algorithm

Users in rating platforms are often allowed to create explicit relationships between other users affiliated with the same platform. Examples of these relationships are friendship and trust. Some researchers [10,11] suggest to incorporate social relations in the recommendation process to better deal with data sparsity in the rating matrix. The resulting recommendation algorithm is often called social recommender system while the user–user social network is often called social graph.

Most traditional social recommender systems leverage CF technique. In Figure 1 we report the general structure of a social recommender systems.

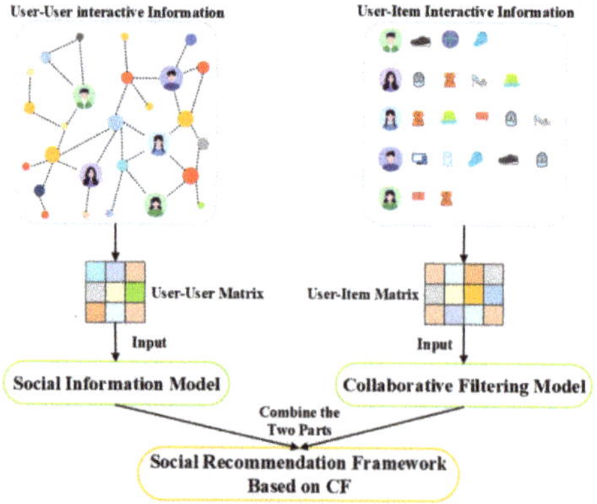

Figure 1. Social recommendation framework based on CF.

Figure 1 shows that a social recommender has two inputs, namely interactive information and social information.

According to different fusion mechanisms of these two types of data, social recommender systems can be classified into two categories: regularization-based and feature-sharing based methods. The regularization-based social recommendation algorithm has the hypothesis that users trust friends in their social circle more than strangers and tend to conform to their preferences. The regularization-based recommendation algorithm projects social data and ratings into the same space and restricts each other so that users can consider their social influence before making decisions. Two important examples of regularization-based algorithms are SocialMF [12] and CUNE [13]. Regularized social recommendation algorithms indirectly simulate social network, while, due to the indirect modeling of social information, there is a low degree of overlap and correlation between user–item interaction information and social information, which leads to a weak integration of social information and ratings.

The basic assumption of feature sharing recommendation algorithms is that user feature vectors in interactive space and social space can be projected into the same space. TrustSVD [14] and SoRec [15] are two examples of feature-sharing social recommendation algorithms. Feature sharing based recommendation algorithm can generate accurate prediction. However, the current mainstream algorithms only use original social information, it means they cannot make full use of social data.

2.3. Graph Embedding Based Recommendation

Network embedding, also referred to as graph embedding, is a process of mapping graph data into a dense vector that is usually low-dimensional, so that the obtained vectors can have representation and reasoning ability in vector space [16,17]. Network embedding can be used as the input of machine learning model and then be applied to the recommendation task.

Graph Embedding can retain the structure information of nodes in the graph, that is, the more similar the structure is in the graph, the closer its position in the vector space will be [18,19]. The principle of graph embedding is shown in Figure 2.

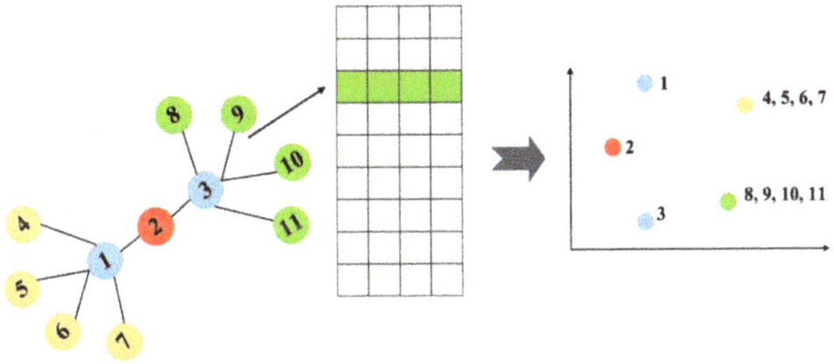

Figure 2. Illustration of graph embedding principle.

Figure 2 shows that node 1 and node 3 are similar in structure, so they maintain a symmetric position in vector space; nodes 4, 5, 6, and 7 are structurally equivalent, so they have the same position in vector space. Graph embedding based recommendation algorithms has two categories: homogenous graph embedding based and heterogeneous graph embedding based. A homogenous graph contains nodes and edges of only one type, and it only needs to aggregate neighbors of a single type to update the node representation. These algorithms are mostly based on random walk, such as Deepwalk [20], which uses truncation random walk sequence to represent the node nearest neighbor, and Node2vec [21], an improved version of Deepwalk. These algorithms only work on homogenous networks. Unfortunately, most real-world datasets can be modeled as heterogeneous graphs naturally. Thus, recommendation algorithms based on heterogeneous networks attract more attention.

In recent years, many experts and scholars [22,23] have studied the transformation of recommendation tasks into heterogeneous graph data mining tasks because real-world datasets can often be abstracted into heterogeneous graphs. Heterogeneous information networks (HIN) include various types of nodes and edges. Figure 3 is an example diagram based on HIN recommendation system.

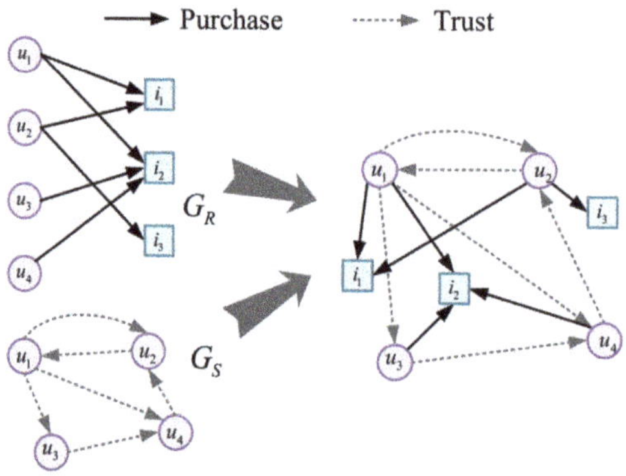

Figure 3. Illustration of a HIN recommendation system.

Figure 3 shows that a HIN contains multiple (two or more) types of entities linked by multiple (two or more) relationships.

In what follows we describe some HIN approaches which jointly consider social data and user interactions with item to generate recommendations.

In the research of Fan et al. [24], interactions and opinions are captured jointly in the interactive graph with GraphRec, which models two graphs (interactive graph and social graph) and heterogeneous edges. Wu et al. [25] proposed a DiffNet for the analysis of how the user in the social diffusion process is affected. DiffNet only modifies users' latent vectors while it does not update items' latent vectors which are independent of social influence. Yu et al. [26] present a social recommender system called MHCN, which uses hypergraphs to capture high-order social information. Each motif is encoded by a dedicated channel of a hypergraph convolutional network, and user embedding is calculated by aggregating the embeddings learned by each channel. Huang et al. [27] propose KCGN. KCGN models interdependencies between items as a triplet and it uses a coupled graph neural architecture to learn embeddings. In addition, it can automatically learn the temporal evolution of the interactive graph. Most traditional social recommenders learn embeddings in Euclidian space. Such a choice, however, is not entirely satisfactory to capture latent structural properties in graphs. In fact, both the interactive graph and the social graph display a tree-like structure that is hard to embed into a Euclidian space. To this purpose, Wang et al. [28] applied hyperbolic embeddings to represent users and items and they introduce a system, called HyperSoRec. Zhao et al. [29] describe a framework called BFHAN, which is able to improve node representations in graphs with a power-law degree distribution, and to handle various relationships of nodes associated with users. Fu et al. [30] introduce the DICER system. DICER first constructs an item-item and a user–user similarity in a weighted undirected graph way. A relation-aware graph neural network (RGNN) module is applied on the item-item graph (as well as user social network and user–user similarity graph) to obtain better users and items representations. Zhang et al. [31] describe a social recommender system called MG-HIF which constructs meta-paths and applies discrete cross-correlation to learn representations of user–item pairs; MG-HIF applies generative adversarial networks (GANs) on the social graph to learn latent friendship relationships. In addition, it uses two attentions models to fuse information from both graphs.

3. Preliminaries

3.1. Social High-Order Connectivity

Social relationships have high-order connectivity, as shown in Figure 4c.

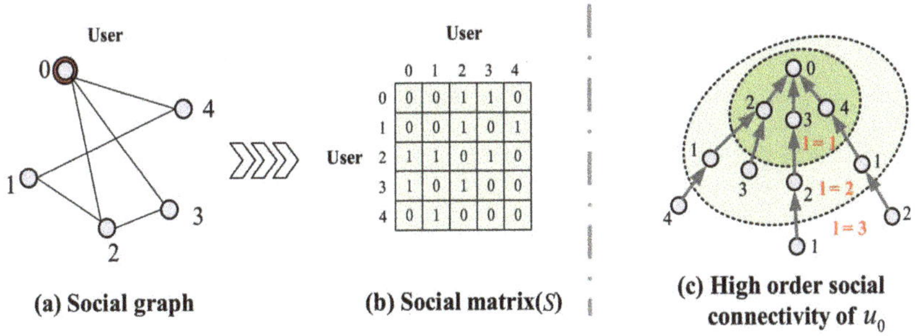

Figure 4. Social relationships.

The target node u_0 is marked with a double circle in the social graph.

Let us consider the path $u_0 \leftarrow u_2 \leftarrow u_1$, u_0 and u_1 are not directly connected, indicating that u_1 may be a potential friend of u_0. In all the pathways that can reach u_0, the closer a node is to u_0, the more paths it occupies, and the greater the influence on u_0.

3.2. *Interactive High-Order Connectivity*

Interaction relationships also have high-order connectivity, as shown in Figure 5c. Let us concentrate on a target user, say u_0, marked with a double circle in the left subgraph of the user–projected interaction diagram.

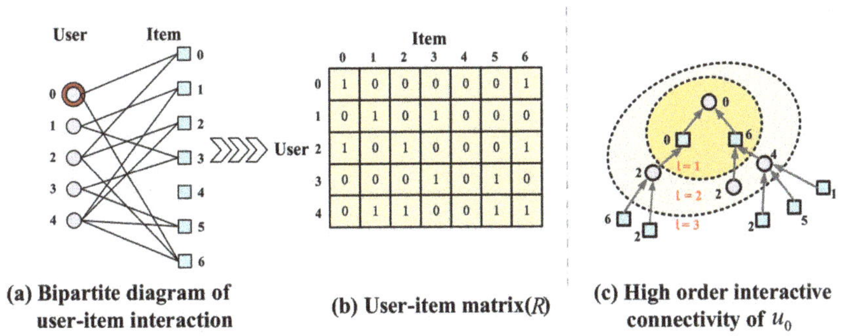

Figure 5. User–item interaction.

The subgraph on the right shows the tree structure expanded obtained by running a BFS search from u_0. High-order connectivity indicates the existence of a path to u_0 of length l greater than 1. This high-order connectivity contains rich semantic information with collaborative signals. For example, path $u_0 \leftarrow i_6 \leftarrow u_4$ represents the behavioral similarity between u_0 and u_4 because the longer path $u_0 \leftarrow i_6 \leftarrow u_4 \leftarrow i_2$ indicates that u_0 is likely to adopt i_2 because its similar user u_4 has previously interacted with i_2. Moreover, from the path of $l = 3$, u_0 is more likely to be interested in i_2 than i_5, because $< i_2, u_0 >$ has two paths connected, while $< i_5, u_0 >$ has only one.

4. Proposed Recommendation Method

4.1. *Recommendation Model Design*

For the purpose of extracting the higher-order relationships in interactive data and social data and fully integrate them to learn high-quality representations, we propose the SRGCF (Social Recommendation Graph Collaborative Filtering) model. Figure 6 shows the overall architecture of SRGCF.

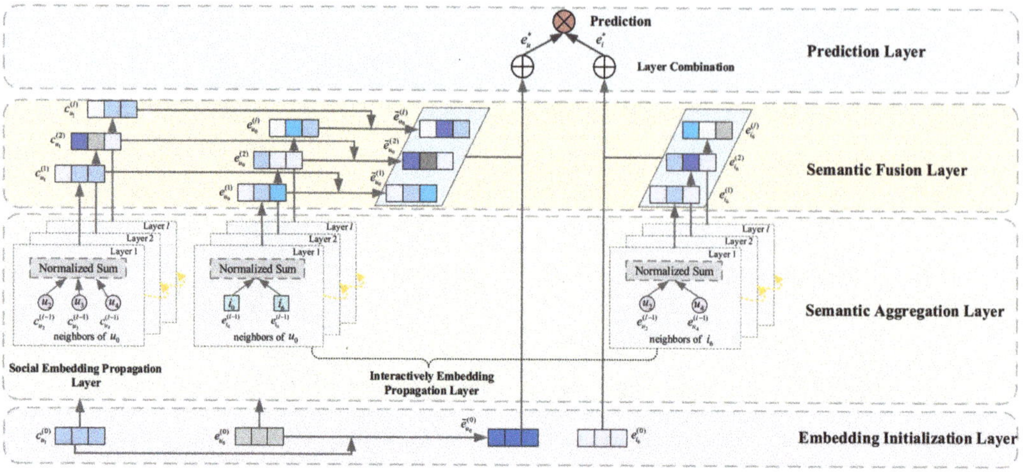

Figure 6. The overall frame structure of the proposed SRGCF model.

The SRGCF model first initializes node embeddings by using the initialization embedding layer. Second, semantic aggregation operations are carried out on the social embedding propagation layer and the interactive embedding propagation layer in the semantic aggregation layer to refine the embedding of users and items, derived from both graphs are fused in the semantic fusion layer. Then weighted sum the user and item embeddings of every layer, respectively, to form the final embeddings. Finally, the prediction layer is applied for producing recommendations.

4.1.1. Embedding Initialization Layer

Embedding matrix of the nodes are randomly initialized and the initial embeddings $e_u^{(0)} \in \mathbb{R}^d$ and $e_i^{(0)} \in \mathbb{R}^d$ of user u and item i and they can be queried, where d is the dimension of nodes' embeddings.

4.1.2. Semantic Aggregation Layer

We propose a semantic aggregation layer in order to aggregate and update the nodes' embeddings, as a result, it is a good way to keep high-level semantic information. We first introduce the concept of first-order semantic aggregation in semantic aggregation layer, and then extend it to high-level semantic aggregation to realize high-level semantic aggregation.

(1) First-order Semantic Aggregation

By iteratively aggregating the neighbor's features, GCN generates new representation of the target node. In SRGCF model, the interactive embedding propagation layer aggregates the embeddings of interacted items to refine the embedding of users. First-order semantic aggregation is reported in (1) and (2).

$$e_u = \underset{i \in H_u}{AGG}(e_i) \tag{1}$$

$$e_i = \underset{u \in H_i}{AGG}(e_u) \tag{2}$$

where $AGG(\cdot)$ denotes aggregation function; H_i represents the set of first-order neighbors of item i, that is, the set of users that have interacted with item i. Similarly, H_u represents the set of first-order neighbors of item u. Equations (1) and (2) indicate that in the interaction, e_u derived from an embedded set of its immediate neighbors, as is e_i.

The social embedding propagation layer refines user embedding by aggregating information from users' friends. The first-order semantic aggregation process is shown in (3).

$$e_u = \underset{v \in F_u}{AGG}(e_v) \tag{3}$$

where F_u represents the friends' collection of user u. It indicates that in social interaction, the embedded e_u of user u is generated through the aggregation of embedded e_v of first-order neighbor (social interaction).

(2) High-order Semantic Aggregation

The semantic aggregation layer achieves the aggregation of higher-order semantics by stacking multiple first-order semantic aggregation layers. It includes semantic aggregation for social embedding propagation layer (SEPL) and interactive embeddings propagation layer (IEPL).

(a) Semantic Aggregation for SEPL

According to social high-order connectivity, stacking l layers can fuse the information from each of l-order neighbors. Semantic aggregation for social embedding propagation layer captures higher-order friend signals by stacking multiple social embedding propagation layers to enhance user embeddings. The mathematical expression of this process is shown in (4) and (5).

$$c_u^{(l+1)} = \sum_{v \in F_u} \frac{1}{\sqrt{|F_u|}\sqrt{|F_v|}} c_v^{(l)} \tag{4}$$

$$c_v^{(l+1)} = \sum_{u \in F_v} \frac{1}{\sqrt{|F_v|}\sqrt{|F_u|}} c_u^{(l)} \tag{5}$$

where $c_u^{(l)}$ denotes the embedding of u at the l-th layer from G_S, and F_u denotes the set of friends of u.

(b) Semantic Aggregation for IEPL

It can be seen from interaction high-order connectivity that stacking even layers (i.e., from the user, the length of path is even) can capture the similarity information of user behavior, while stacking odd layers can capture the potential interaction information of users to items. Semantic aggregation for interaction embedding propagation layer captures collaborative signals of high-order connectivity in interaction data by stacking each interaction embedding propagation layer, thus enhancing embeddings. Expression of this process is shown in (6) and (7).

$$e_i^{(l+1)} = \sum_{u \in N_i} \frac{1}{\sqrt{|H_i|}\sqrt{|H_u|}} e_u^{(l)} \tag{6}$$

$$e_u^{(l+1)} = \sum_{i \in N_u} \frac{1}{\sqrt{|H_u|}\sqrt{|H_i|}} e_i^{(l)} \tag{7}$$

where $e_u^{(l)}$ and $e_i^{(l)}$ represent u's and i's embedding at l-th layer from G_R, respectively.

4.1.3. Semantic Fusion Layer

User embeddings can be enhanced by integrating social embedding propagation layer and interactive embedding propagation layer with certain social information.

After obtaining the social semantic aggregation embedding and interactive semantic aggregation embedding, respectively, the user embeddings of each layer are fused, and the fusion process is shown in (8).

$$\widetilde{e}_u^{(l)} = g(e_u^{(l)}, c_u^{(l)}) \tag{8}$$

where $\widetilde{e}_u^{(l)}$ denotes the l-th layer embedding of u from G_R and G_S, and we let $e_u^{(0)} = c_u^{(0)}$. $g(\cdot)$ is a fusion function, which can be implemented in many ways, and we adopt (9) to fuse:

$$\widetilde{e}_u^{(l)} = norm(sum(e_u^{(l)}, c_u^{(l)})) \tag{9}$$

where $sum(\cdot)$ denotes element-wise summation. Intuitively, it is an operation to enhance the signal representation, and also can keep the feature space unchanged; $norm(\cdot)$ denotes row-regularization operation that normalizes user vectors.

Then, the final user embedding e_u^* and item embedding e_i^* is obtained by fusing embeddings of all layers:

$$e_u^* = \sum_{l=0}^{L} \alpha_l \widetilde{e}_u^{(l)}; e_i^* = \sum_{l=0}^{L} \beta_l e_i^{(l)} \tag{10}$$

where e_u^* denotes u's final embedding, e_i^* denotes the i's final embedding, and L denotes total number of layers. In line with LightGCN, set α and β as $1/(L+1)$. The settings of these two parameters are flexible, and attention mechanism can be applied to learn them.

4.1.4. Prediction Layer

The last part of the model recommends products to users according to the embedding of items. We use the inner-product form for prediction:

$$\hat{y}_{ui} = e_u^{*T} e_i^* \tag{11}$$

Then BPR loss [32] was calculated and model parameters were optimized as shown in Equation (12).

$$J = \sum_{(u,i,j)\in O} -\ln \sigma(\hat{y}_{ui} - \hat{y}_{uj}) + \lambda \|\Theta\|_2^2 \tag{12}$$

where $O = \{(u,i,j)|(u,i) \in R^+, (u,j) \in R^-\}$ represents pair-wised training data, R^+ expresses interactions that exist in history, and R^- denotes interaction that does not. Θ are model parameters, where the model parameters only include the initial embedding vectors $e_u^{(0)}$ and $e_i^{(0)}$. λ is used to prevent overfitting.

4.2. The Proposed SRRA Recommendation Algorithm

In order to facilitate implementation, SRRA algorithm is proposed under the framework of SRGCF model, which is implemented in the form of matrix (see Algorithm 1 for details).

The interactive matrix is denoted as $R \in \mathbb{R}^{M\times N}$, M are the numbers of user and N are the numbers of item, R_{ui} equals to 1 if u have interaction with i, if not R_{ui} equals to 0. Then the adjacency matrix A of G_R is:

$$A = \begin{pmatrix} 0 & R \\ R^T & 0 \end{pmatrix} \tag{13}$$

Let the embedding matrix of layer 0 be $E^{(0)} \in \mathbb{R}^{(M+N)\times d}$, where d is the dimension of embedding vector, and the $(l+1)$-th layer matrix can be computed as:

$$E^{(l+1)} = (D^{-\frac{1}{2}} A D^{-\frac{1}{2}}) E^{(l)} \tag{14}$$

where D is the degree matrix of A, which is a diagonal matrix and its dimension is $(M+N) \times (M+N)$. Each element D_{ii} represents the number of non-zero values of the i-th row vector in A.

Algorithm 1: Social Relationship Recommendation Algorithm (SRRA).

Step1: calculate the embeddings of users and items
Input: R, S, M, N, d, l
Initialize: $E^{(0)} = C^{(0)}, K, \alpha = \beta = 1/(l+1)$
calculate A and B by R, S, respectively
calculate D and P by A, B, respectively
$E_U{}^{(0)}, E_I{}^{(0)} \leftarrow E^{(0)}, \cdots C_U{}^{(0)} \leftarrow C^{(0)}$
$\widetilde{E}_U \leftarrow E_U{}^{(0)}, C_U{}^{(0)},$
set $E_I{}^* = E_I{}^{(0)}$
For $l \in L$:
calculate $E^{(l+1)}$ and $C^{(l+1)}$ by $E^{(l)}$, $C^{(l)}$, respectively
get $E_U{}^{(l+1)}, E_I{}^{(l+1)}, C_U{}^{(l+1)}$ by split $E^{(l+1)}$, $C^{(l+1)}$, respectively
$\widetilde{E}_U{}^{(l+1)} \leftarrow E_U{}^{(l+1)}, C_U{}^{(l+1)}$
$E_U{}^* += \widetilde{E}_U{}^{(l+1)}, E_U{}^* += E_I{}^{(l+1)}$
End For
$E_U{}^* = \alpha E_U{}^*, E_I{}^* = \beta E_U{}^*$
Step2: calculate the loss of SRRA
set $L_{BPR} = 0$
For $u \in U$:
$e_u{}^{(0)} = lookup(E_U{}^{(0)}, u)$ // find the initial vector of user u from $E_U{}^{(0)}$
$L_{BPR} += \left\| e_u{}^{(0)} \right\|^2$ // add the regularization item into loss
For $i \in R_u{}^+$: // iterate over the positive example item set for user u
$e_i{}^* = lookup(E_I{}^*, i)$ // find the vector of item i from $E_I{}^*$
$\hat{y}_{ui} = e_u{}^{*T} e_i{}^*$ // calculate the score of positive samples
For $j \in R_u^-$: // iterate over the negative example item set for user u
$e_j{}^* = lookup(E_I{}^*, j)$
$\hat{y}_{uj} = e_u{}^{*T} e_j{}^*$
$L_{BPR} += (-\ln \sigma(\hat{y}_{ui} - \hat{y}_{uj}))$// calculate the BPR loss
End For
End For
Step3: generate recommendations
Train the algorithm until it converges
According to the predicted score, select Top 10 items for recommendation
Return Recall, NDCG

Similarly, the social matrix is denoted as $S \in \mathbb{R}^{M \times M}$, where S_{uv} is 0 if u and v are friends, otherwise S_{uv} is 1. The adjacency matrix B of G_S is:

$$B = \begin{pmatrix} 0 & S \\ S^T & 0 \end{pmatrix} \tag{15}$$

Let the embedding matrix of layer 0 be $C^{(0)} \in \mathbb{R}^{(M+M) \times d}$, and the user matrix of layer $l+1$ can be obtained as shown in (16).

$$C^{(l+1)} = (P^{-\frac{1}{2}} B P^{-\frac{1}{2}}) C^{(l)} \tag{16}$$

where P is the degree matrix of matrix B.

Then, due to $E^{(l)} = stack(E_U{}^{(l)}, E_I{}^{(l)})$, $E^{(l)}$ can be divided into user's and item's matrices, denoted as $E_U{}^{(l)}$ and $E_I{}^{(l)}$, respectively. Similarly, due to $C^{(l)} = stack(C_U{}^{(l)}, C_U{}^{(l)})$, $C^{(l)}$ can be divided into two parts, both of which are user embedding matrix, where $C_U{}^{(l)}, E_U{}^{(l)} \in \mathbb{R}^{M \times d}$ and $E_I{}^{(k)} \in \mathbb{R}^{N \times d}$.

Finally, the l-th layer user representation is calculated as:

$$\widetilde{E}_U{}^{(l)} = norm(sum(E_U{}^{(l)}, C_U{}^{(l)})) \tag{17}$$

The final representations can be obtained through integrating the representation of each layer:

$$E_U^* = \sum_{l=0}^{L} \alpha_l \widetilde{E}_U^{(l)}, E_I^* = \sum_{l=0}^{L} \beta_l E_I^{(l)} \tag{18}$$

We use inner-product to compute score:

$$\hat{Y} = {E_U^*}^T E_I^* \tag{19}$$

4.3. Model Training

The loss function calculated by BPR is shown in (20).

$$L_{BPR} = -\sum_{u=1}^{M} \sum_{i \in H_u} \sum_{j \notin H_u} \ln \sigma(\hat{y}_{ui} - \hat{y}_{uj}) + \lambda \left\| E^{(0)} \right\|^2 \tag{20}$$

Adam [33] is used as the optimizer of loss function. Its primary characteristic is that it can self-adapt learning rates.

5. Experiment

5.1. Experiment Setup

5.1.1. Datasets

We used four datasets in this paper. The following is an introduction to these datasets, and their statistical details are summarized in Table 1.

- Brightkite. A position sharing platform with social networking platform where users share their locations through check-ins. It includes check-in data as well as social data.
- Gowalla. A position sharing platform similar to Brightkite. This dataset includes check-in data and user social data.
- Epinions. A consumer review website which allows users to clicked items and add trust users. This dataset contains users' rating data and trust network data.
- LastFM. A social music platform for music sharing. This dataset includes data about users' listening to music and users' relationships.

Table 1. Statistical details of four datasets.

Dataset	Brightkite	Gowalla	Epinions	LastFM
#User	6310	14,923	12,392	1860
#Item	317,448	756,595	112,267	17,583
#Interaction	1,392,069	2,825,857	742,682	92,601
#Connection	27,754	82,112	198,264	24,800
R-Density	6.9495×10^{-4}	2.5028×10^{-4}	5.3384×10^{-4}	2.8315×10^{-4}
S-Density	6.9705×10^{-4}	3.6872×10^{-4}	1.2911×10^{-3}	7.1685×10^{-3}

5.1.2. Baselines

For the purpose of illustrating how effective our model is, we compared SRRA with three types of approaches: one social recommendation model based on deep learning (DL), three social recommendation models based on DL and GNN, and one recommendation model based on GCN, which is shown in Table 2.

- LightGCN [3]: It is effective to extract the collaborative signal explicitly in the embedding process by modeling high-order connectivity in interactive graphs.
- DSCF [34]: It utilizes information provided by distant neighbors and explicitly captures the neighbor's different opinions towards items.
- DiffNet [25]: It is a GNN model which analyzes how users make their decisions based on recursive social diffusion.

- GraphRec [24]: It captures interactions and opinions in G_R and it also models two graphs (e.g., G_R and G_S) and the strength of heterogeneity in a coherent way.
- DICER [30]: It models user and item by introducing high-order neighbor information, and draws the most relevant interactive information based on deep context.

Table 2. Comparison of the model's characteristics.

Methods	Social Recommendation	DL	Graph-Based	
			GNN	GCN
LightGCN				√
DSCF	√	√		
DiffNet	√	√	√	
GraphRec	√	√	√	
DICER	√	√	√	
SRRA	√			√

5.1.3. Evaluation Metrics

We utilize two widely adopted metrics Recall@K and NDCG@K in comparisons, since we try to recommend the Top-K list items for each user. Specifically, Recall measures the percentage of the test data that users actually like from the Top-K list. In addition, NDCG is a position-aware ranking metric that measures how the hit items are placed and gives a higher score if they are at the top of the list.

5.1.4. Experiments Details

We use 80% of Brightkite, Gowalla, Epinions and LastFM for training, 10% for tuning hyper-parameters, and 10% for testing final performance. Parameters for all methods are randomly initialized with standard normal distribution. In addition, initialization and tuning of parameters for the baseline algorithms followed the procedures described in the corresponding papers. With batch size 1024, we tested each value in {8, 16, 32, 64, 128, 256} for embedding size d, and we also find the proper value for learning rate and L2 regularization factor in {0.0005, 0.001, 0.005, 0.01, 0.05, 0.1} and $\{1 \times 10^{-6}, 1 \times 10^{-5}, \ldots, 1 \times 10^{-2}\}$, respectively. The aggregation factors α_l and β_l of each layer were set as $1/(L+1)$, where L represents the total number of layers.

5.2. Overall Comparison

We compare all methods in this subsection. In Table 3, we show performance comparison between SRRA and baselines. The following conclusions can be drawn:

First, methods that incorporate social relations outperform that does not. In Table 3, for example, DSCF, DiffNet, GraphRec, DICER, and SRRA outperform LightGCN. This demonstrates that social information is effective and helpful by being incorporated into recommender systems.

Second, our method SRRA achieves the best performance on these four datasets. Specifically, in comparison to DICER, a GNN and DL-based social recommendation model, SRRA scores better by an average improvement of 2.27%, 2.85%, 5.58%, and 6.90%; and to GraphRec, a very expressive GNN-based social recommendation model, SRRA scores better by an average improvement of 5.55%, 6.44%, 13.42% and 13.08% on the four datasets, respectively. We guess a possible reason is that, for Brightkite and Gowalla, as they are social networks related to location, the activities and consumption preferences for users in this type of social platform is not easy to be affected; and for Epinions and LastFM, people strongly rely on social relations to acquire correct review of goods and lists of music they will listen to. It is possible to attribute our improved model over the baseline to two factors: (1) our model use GCN architecture to extract the social ties and interactive ties in a high-order way, which can leverage the relative information from multi-hop neighbors and high-order collaborative information propagated over the user interaction graph;

(2) we fuse all-order social and collaborative information when modeling user and item representation which generates improved user and item representation.

Table 3. Performance comparison between SRRA and baselines.

			Recall@10			
Datasets \ Methods	**LightGCN**	**DSCF**	**DiffNet**	**GraphRec**	**DICER**	**SRRA**
Brightkite	0.1642	0.1895	0.1962	0.2172	0.2235	**0.2293**
Gowalla	0.2083	0.2253	0.2399	0.2779	0.2886	**0.3011**
Epinions	0.2269	0.2613	0.2874	0.2845	0.3155	**0.3341**
LastFM	0.2519	0.2742	0.2932	0.2876	0.3059	**0.3272**
			NDCG@10			
Datasets \ Methods	**LightGCN**	**DSCF**	**DiffNet**	**GraphRec**	**DICER**	**SRRA**
Brightkite	0.1321	0.1393	0.1539	0.1612	0.1672	**0.1701**
Gowalla	0.1355	0.1482	0.1667	0.1724	0.1744	**0.1782**
Epinions	0.1425	0.1598	0.1642	0.1709	0.1737	**0.1824**
LastFM	0.1431	0.1563	0.1628	0.1862	0.1953	**0.2086**

5.3. Parameter Analysis

For the proposed model, there are two crucial parameters: the number of layers l and the embedding size d. In this section, we first change only one parameter and fix the others, and then observe how the performance changes.

Effect of the number of layers l. Take Gowalla and Epinions for example, we set the l from 1 to 5 to measure the impact of different layers, and then we can obtain the performance with the different number of layers that showed in Figure 7. We observed that performance increases and then decreases as the number of layers grows. When the number of layers grows from 1 to 4, performance of SRRA is improved. However, performance starts to become worse when the number of layers is 5. It demonstrates that too many layers may cause over smoothness that is a common problem existing in graph convolution methods. Thus, in order to prevent this, we need to use the proper number of layers.

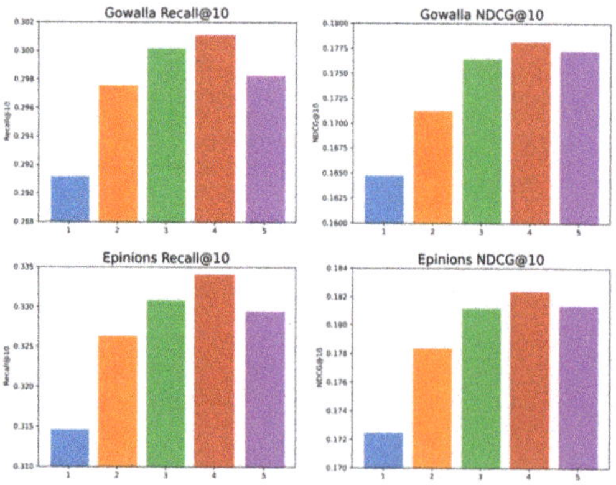

Figure 7. Effect of #layers l on Gowalla and Epinions.

Effect of embedding size d. In this subsection, take Gowalla and Epinions for example, we analyze how the embedding size of e_u and e_i affect proposed model. On these two datasets, Figure 8 compares the performance of our proposed model when varying its embedding size *d*.

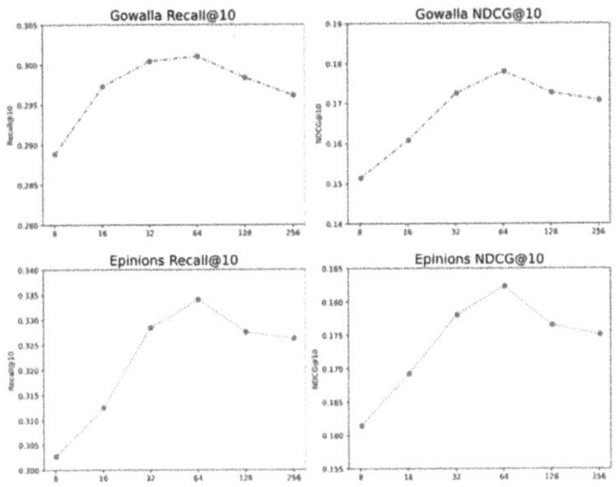

Figure 8. Effect of embedding size *d* on Gowalla and Epinions.

Accordingly, as embedding size increases, performance first become better, then worse. If the size grows from 8 to 64, the performance improves obviously. However, the performance of SRRA starts to deteriorate when the embedding size is 128. It demonstrates that a large embedding size is likely to produce powerful representations. Nevertheless, if the length of embeddings is too long, our model will become more complex. Thus, we must find an appropriate embedding size to make a trade-off, and we find that 64 is the optimal value.

6. Conclusions

In this work, we proposed a new social recommendation method which leverages graph convolution technique and integrates social relations. Firstly, we construct the architecture of a general collaborative filtering social recommendation model based on graph convolution (SRGCF), which consists of four parts, which are initialization embedding layer, semantic aggregation layer, semantic fusion layer and prediction layer, respectively. The semantic aggregation layer and semantic fusion layer are the core of SRGCF, which play the role of extracting high-order semantic information and integrating various semantic information, respectively. Then, we propose a feasible SRRA algorithm on top of the architecture, which can model interactions as well as social relations. It can use richer social information to mine the potential relationship, so as to improve the performance of recommendations. Comparative experiments on four datasets have proven the effectiveness of the proposed model.

Different from previous work, we try to explore how to use graph neural network method and introduce social auxiliary information to construct recommendation model in order to learn better representation. The graph-based model is superior to the traditional recommendation model because it can learn not only the representation of entities but also the relationships between them. However, limited by the shortcomings of graph neural network itself, such as excessive smoothing after several iterations, entity representation may not be fully learned, which requires some optimization in model design. In the future, we plan to optimize the model architecture by increasing the coupling between social modeling and interactive modeling, so that the representation learning is more adequate. We will also try to explore the advantages of other graphical representation learning techniques to improve the learning ability of the model.

Author Contributions: Conceptualization, M.M.; methodology, X.L.; software, Q.C.; formal analysis, Q.C.; writing—original draft preparation, Q.C.; writing—review and editing, Q.C. All authors have read and agreed to the published version of the manuscript.

Funding: This research was funded by Science and Technology Research Project of Chongqing Municipal Education Commission, grant number KJZDK202001101; Chongqing Postgraduate Research Innovation Project, grant number gzlcx20223205; General Project of Chongqing Natural Science Foundation, grant number cstc2021jcyj-msxmX0162; 2021 National Education Examination Research Project, grant number GJK2021028.

Institutional Review Board Statement: Not applicable.

Informed Consent Statement: Not applicable.

Data Availability Statement: The calculated data presented in this work are available from the corresponding authors upon reasonable request.

Acknowledgments: The author would like to thank the anonymous reviewers for their valuable comments on our paper.

Conflicts of Interest: The authors declare no conflict of interest.

Nomenclature

U	the set of users
I	the set of items
G_R	user–item interaction graph
Gs	user–user social graph
d	the dimension of node embedding
H_i	neighbors of item i on G_R
H_u	neighbors of user u on Gs
l	#layer
L	total number of layers
F_u	the friends of user u
$e_u^{(l)}$	the embedding of user u at the l-th layer from G_R
$c_u^{(l)}$	the embedding of user u at the l-th layer from Gs
$\widetilde{e}_u^{(l)}$	the l-th layer embedding of user u from G_R and Gs
e_u^*	final embedding of user u
e_i^*	final embedding of item i
M	the numbers of users
N	the numbers of items
R	user–item interaction matrix
S	social matrix
A	adjacency matrix of G_R
B	adjacency matrix of Gs
D	degree matrix of matrix A
P	degree matrix of matrix B
R^+	observable interactions
R^-	unobserved interactions
Θ	model parameters
$E^{(l)}$	the l-th layer matrix of GCN on G_R
$C^{(l)}$	the l-th layer matrix of GCN on Gs
$E_U{}^*$	final embedding matrix of users
$E_I{}^*$	final embedding matrix of items

References

1. Liu, X.; Li, X.; Fiumara, G.; De Meo, P. Link prediction approach combined graph neural network with capsule network. *Expert Syst. Appl.* **2023**, *212*, 118737. [CrossRef]
2. Liu, X.; Zhao, Z.; Zhang, Y.; Liu, C.; Yang, F. Social Network Rumor Detection Method Combining Dual-Attention Mechanism with Graph Convolutional Network. *IEEE Trans. Comput. Soc. Syst.* **2022**. [CrossRef]
3. He, X.; Deng, K.; Wang, X.; Li, Y.; Zhang, Y.; Wang, M. LightGCN: Simplifying and Powering Graph Convolution Network for Recommendation. In Proceedings of the 43rd International ACM SIGIR Conference on Research and Development in Information Retrieval (SIGIR'20), Virtual Event, 25–30 July 2020; pp. 639–648.

4. Liu, T.; Deng, X.; He, Z.; Long, Y. TCD-CF: Triple cross-domain collaborative filtering recommendation. *Pattern Recognit. Lett.* **2021**, *149*, 185–192. [CrossRef]
5. Nassar, N.; Jafar, A.; Rahhal, Y. A novel deep multi-criteria collaborative filtering model for recommendation system. *Knowl. Based Syst.* **2020**, *187*, 32–39. [CrossRef]
6. He, X.; Liao, L.; Zhang, H.; Nie, L.; Hu, X.; Chua, T. Neural Collaborative Filtering. In Proceedings of the 26th International Conference on World Wide Web (WWW'17), Perth, Australia, 3–7 May 2017; pp. 173–182.
7. Wang, H.; Wang, N.; Yeung, D. Collaborative Deep Learning for Recommender Systems. In Proceedings of the 21th ACM SIGKDD International Conference on Knowledge Discovery and Data Mining (KDD'15), Sydney, Australia, 10–13 August 2015; pp. 1235–1244.
8. Xue, F.; He, X.; Wang, X.; Xu, J.; Liu, K.; Hong, R. Deep Item-based Collaborative Filtering for Top-N Recommendation. *ACM Trans. Inf. Syst.* **2019**, *37*, 1–25. [CrossRef]
9. Khojamli, H.; Razmara, J. Survey of similarity functions on neighborhood-based collaborative filtering. *Expert Syst. Appl.* **2021**, *185*, 142–155. [CrossRef]
10. Sanz-Cruzado, J.; Castells, P.; Macdonald, C.; Ounis, I. Effective contact recommendation in social networks by adaptation of information retrieval models. *Inf. Process. Manag.* **2020**, *57*, 1633–1647. [CrossRef]
11. Sun, Z.; Guo, Q.; Yang, J.; Fang, H.; Guo, G.; Zhang, J.; Burke, R. Research commentary on recommendations with side information: A survey and research directions. *Electron. Commer. Res. Appl.* **2019**, *37*, 43–50. [CrossRef]
12. Jamali, M.; Ester, M. A Matrix Factorization Technique with Trust Propagation for Recommendation in Social Networks. *Assoc. Comput. Mach.* **2010**, *1*, 135–142.
13. Nitesh, C.; Wei, W. Collaborative User Network Embedding for Social Recommender Systems. In Proceedings of the 2017 SIAM International Conference on Data Mining, Sandy, UT, USA, 27–29 April 2017; pp. 381–389.
14. Guibing, G.; Jie, Z.; Yorke, S.N. TrustSVD: Collaborative Filtering with Both the Explicit and Implicit Influence of User Trust and of Item Ratings. In Proceedings of the 29th AAAI Conference on Artificial Intelligence, New York, NY, USA; 2015; pp. 123–129.
15. Ma, H.; Yang, H.; Lyu, M.R.; King, I. SoRec: Social Recommendation using Probabilistic Matrix Factorization. *Neurocomputing* **2019**, *341*, 931–940.
16. Cui, P.; Wang, X.; Pei, J.; Zhu, W. A Survey on Network Embedding. *IEEE Trans. Knowl. Data Eng.* **2019**, *31*, 833–852. [CrossRef]
17. Palash, G.; Emilio, F. Graph Embedding Techniques, Applications, and Performance: A Survey. *Knowledge- Based Syst.* **2017**, *151*, 78–84.
18. Wang, D.; Cui, P.; Zhu, W. Structural Deep Network Embedding. In Proceedings of the 22nd ACM SIGKDD International Conference on Knowledge Discovery and Data Mining, San Francisco, CA, USA, 13–17 August 2016; pp. 1225–1234.
19. Tang, J.; Qu, M.; Wang, M.; Zhang, M.; Yan, J.; Mei, Q. LINE: Large-scale Information Network Embedding. In Proceedings of the International Conference on World Wide Web, Florence, Italy, 18–22 May 2015; Volume 3, pp. 1067–1077.
20. Perozzi, B.; Al-Rfou, R.; Skiena, S. Deepwalk: Online learning of social representations. In Proceedings of the 20th ACM SIGKDD International Conference on Knowledge Discovery and Data Mining, New York, NY, USA, 24–27 August 2014; pp. 701–710.
21. Grover, A.; Leskovec, J. node2vec: Scalable Feature Learning for Networks. In Proceedings of the 22nd ACM SIGKDD International Conference on Knowledge Discovery and Data Mining (KDD'16), San Francisco, CA, USA, 13–17 August 2016; pp. 855–864.
22. De Paola, A.; Gaglio, S.; Giammanco, A.; Lo Re, G.; Morana, M. A multi-agent system for itinerary suggestion in smart environments. *CAAI Trans. Intell. Technol.* **2021**, *6*, 377–393. [CrossRef]
23. Weng, L.; Zhang, Q.; Lin, Z.; Wu, L. Harnessing heterogeneous social networks for better recommendations: A grey relational analysis approach. *Expert Syst. Appl.* **2021**, *174*, 142–154. [CrossRef]
24. Fan, W.; Ma, Y.; Li, Q.; He, Y.; Zhao, Y.E.; Tang, J.; Yin, D. Graph Neural Networks for Social Recommendation. In Proceedings of the World Wide Web Conference (WWW 2019), San Francisco, CA, USA, 13–17 May 2019; pp. 417–426.
25. Wu, L.; Sun, P.; Fu, Y.; Hong, R.; Wang, X.; Wang, M. A Neural Influence Diffusion Model for Social Recommendation. In Proceedings of the 42nd International ACM SIGIR Conference on Research and Development in Information Retrieval, SIGIR 2019, Paris, France, 21–25 July 2019; pp. 235–244.
26. Yu, J.; Yin, H.; Li, J.; Wang, Q.; Hung, N.Q.V.; Zhang, X. Self-supervised multi-channel hypergraph convolutional network for social recommendation. In Proceedings of the Web Conference 2021, Ljubljana, Slovenia, 19–23 April 2021; pp. 413–424.
27. Huang, C.; Xu, H.; Xu, Y.; Dai, P.; Xia, L.; Lu, M.; Bo, L.; Xing, H.; Lai, X.; Ye, Y. Knowledge-aware coupled graph neural network for social recommendation. In Proceedings of the 35th AAAI Conference on Artificial Intelligence (AAAI), Virtual, 2–9 February 2021.
28. Wang, H.; Lian, D.; Tong, H.; Liu, Q.; Huang, Z.; Chen, E. HyperSoRec: Exploiting Hyperbolic User and Item Representations with Multiple Aspects for Social-aware Recommendation. *ACM Trans. Inf. Syst.* **2022**, *40*, 1–28. [CrossRef]
29. Zhao, M.; Deng, Q.; Wang, K.; Wu, R.; Tao, J.; Fan, C.; Chen, L.; Cui, P. Bilateral Filtering Graph Convolutional Network for Multi-relational Social Recommendation in the Power-law Networks. *ACM Trans. Inf. Syst.* **2022**, *40*, 1–24. [CrossRef]
30. Fu, B.; Zhang, W.; Hu, G.; Dai, X.; Huang, S.; Chen, J. Dual Side Deep Context-aware Modulation for Social Recommendation. In Proceedings of the Web Conference 2021, Ljubljana, Slovenia, 19–23 April 2021; pp. 2524–2534.
31. Zhang, C.; Wang, Y.; Zhu, L.; Song, J.; Yin, H. Multi-graph heterogeneous interaction fusion for social recommendation. *ACM Trans. Inf. Syst. TOIS* **2021**, *40*, 1–26. [CrossRef]

32. Rendle, S.; Freudenthaler, C.; Gantner, Z.; Schmidt-Thieme, L. BPR: Bayesian Personalized Ranking from Implicit Feedback. In Proceedings of the 25th Conference on Uncertainty in Artificial Intelligence, Montreal, QC, Canada, 18–21 June 2009; pp. 452–461.
33. Kingma, D.P.; Ba, J.L. Adam: A Method for Stochastic Optimization. In Proceedings of the International Conference on Learning Representations, San Diego, CA, USA, 7–9 May 2015.
34. Fan, W.; Ma, Y.; Yin, D.; Wang, J.; Tang, J.; Li, Q. Deep Social Collaborative Filtering. In Proceedings of the 13th ACM Conference on Recommender Systems (RecSys'19), Copenhagen, Denmark, 16–20 September 2019; pp. 305–313.

applied sciences

MDPI

Article

Directed Network Disassembly Method Based on Non-Backtracking Matrix

Jinlong Ma [1,2], Peng Wang [1,2] and Huijia Li [3,*]

1 School of Information Science and Engineering, Hebei University of Science and Technology, Shijiazhuang 050018, China
2 Hebei Technology Innovation Center of Intelligent Internet of Things, Shijiazhuang 050018, China
3 School of Science, Beijing University of Posts and Telecommunication, Beijing 100876, China
* Correspondence: lihuijia0808@gmail.com

Abstract: Network disassembly refers to the removal of the minimum set of nodes to split the network into disconnected sub-part to achieve effective control of the network. However, most of the existing work only focuses on the disassembly of undirected networks, and there are few studies on directed networks, because when the edges in the network are directed, the application of the existing methods will lead to a higher cost of disassembly. Aiming at fixing the problem, an effective edge module disassembly method based on a non-backtracking matrix is proposed. This method combines the edge module spectrum partition and directed network disassembly problem to find the minimum set of key points connecting different edge modules for removal. This method is applied to large-scale artificial and real networks to verify its effectiveness. Multiple experimental results show that the proposed method has great advantages in disassembly accuracy and computational efficiency.

Keywords: directed network dismantling; non-backtracking matrix; spectral partition; minimal dismantling set

Citation: Ma, J.; Wang, P.; Li, H. Directed Network Disassembly Method Based on Non-Backtracking Matrix. *Appl. Sci.* **2022**, *12*, 12047. https://doi.org/10.3390/app122312047

Academic Editor: Keun Ho Ryu

Received: 12 September 2022
Accepted: 17 November 2022
Published: 25 November 2022

1. Introduction

In complexity science, a network (denoted by $G = (V, E)$ in graph theory) is composed of a node set V consisting of n nodes and an edge set E consisting of m edges between the nodes. Many real-world networks such as the Internet, WWW, large-scale power networks, transportation networks and interpersonal networks can be modeled in this concise way [1]. Using this method, these networks can be regarded as a collection of nodes with independent characteristics interconnected with other individuals. Each individual is regarded as a node in the network, and the connection between nodes is regarded as the edge of the network. This abstract method can intuitively show the topology of the real network, and also provides an effective research method for understanding the state and the function of the real network [2].

However, with the continuous development of technology and society, epidemic viruses [3], computer viruses [4], misinformation [5], or corruption [6] have more serious negative effects in the human world. However, removing or deactivating a part of the key nodes through the network dismantling method in the network to decompose the network into several isolated sub-parts can effectively protect the robustness of the network, control the dynamic behavior of the network, and curb the negative effects in the network mentioned above. Previous studies proved that this method to remove or deactivate the key nodes can effectively curb the spread of epidemics in the population [7], prevent the spread of misinformation through social networks [8] and prevent the spread of viruses in computer networks [9]. Some studies on complex networks choose a set of node subsets S in the network with an optimal method, and explore the influence of removing S on the network characteristics. For example, exploring how the maximum connected (strong) subset of the network will change after removing S, in the example of epidemics or network viruses transmission, if S is isolated or infected first, the impact on the speed of virus transmission can be determined [10]?

In the actual situation, it will produce a certain cost consumption C when selecting and removing the node subset S in the network. In the epidemic propagation model, the vaccination of the node requires a certain socioeconomic cost. Removing different node sets in the computer virus or public opinion propagation network consumes different resources in the actual situation. Therefore, a combinatorial optimization problem is generated, whereby under the influence of the constraint removal node subset S in the network, the removal cost C is minimized. Additionally, removing nodes will destroy the network structure and affect the function of the network, so it is necessary to remove as few nodes as possible and find a set of nodes with the lowest removal cost C, that is, the minimum disassembly set (we consider the most common situations where the disassembly cost is the unit cost; the minimum disassembly set is the set with the least number of nodes) and remove it. After the network is decomposed into multiple sub-parts that are not connected, network disassembly is achieved [11]. Finding the minimum disassembly set is an NP-hard problem [12]. For this kind of problem, only effective approximation algorithms or heuristic algorithms can be found at present. For the disassembly problem, there have been some recursive algorithms based on the degree or centrality of the nodes. For example, in 2015, a heuristic algorithm called 'collective' influence (CI) [13] was proposed, which determines the ownership of the nodes in an undirected random network according to the degree of nodes and the degree of local neighbor nodes; in 2016, Alfredo Braunstein proposed a three-segment minimum sum (hereafter referred to as the Min-Sum) method for dismantling large random undirected networks [11]; for large undirected random networks, this method is a more effective dismantling algorithm. In 2019, Ren proposed a general network disassembly (hereafter referred to as the GND) method for undirected weight networks [14]. In addition, some relatively new disassembly methods and analyses of disassembly [15,16] are provided, such as the disassembly algorithm based on the message passing model (2020) [17] and the sensitive disassembly method of neighbor connection (2020) [18].

Most of the existing disassembly algorithms are carried out in undirected networks, while there are few disassembly algorithms for directed networks. However, in many real-world networks (such as WWW networks, acquaintance relationships, network email networks, text association networks and article citation networks, etc.), the edges between nodes are unidirectionally connected, and there is no mutual relationship in undirected networks [19]. The existing disassembly methods are sometimes not suitable for the disassembly of these directed networks. Compared with disassembly in the undirected network, disassembly in the directed network needs to consider the direction of the edges between the nodes in the network. For example, in a public opinion network, when an incoming node (Innode) connected by a directed edge e (also known as an arc in graph theory) is a communicator, the outgoing node of this edge (Outnode) is not likely to be propagated by the node. When applying the undirected network-based disassembly method to disassemble the network, this one-way relationship between nodes is sometimes ignored, resulting in the removal of this edge e, and causing unnecessary disassembly to affect the disassembly effect.

The influence of the internal mechanism of this network on the disassembly is ignored in the traditional disassembly method. To solve this problem, the non-backtracking matrix representing the edge adjacency relationship is applied as the operator of spectral division, whereby it retains the directionality of the node relationship, and combines the disassembly of the directed network with the spectral division method. An improved spectral disassembly method for directed networks (hereafter referred to as the DIR method) is proposed; the edges in the directed network are directly applied as the disassembly unit, and the strongest connected subgraph of the directed network is used as the disassembly subject. The spectral characteristics of the edge adjacency matrix (non-backtracking matrix) are applied to the bipartition of the edge modules in the maximum strongly connected subset.

The overlapping node set of the node sets connecting the two edge modules is then found as the minimum disassembly set to disassemble the network until the disassembly scale reaches the specified disassembly scale of the network (the maximum number of nodes in the strongly connected subgraph). The excellent characteristics of the non-backtracking matrix can be made full use of by using the DIR method, and the DIR method can greatly protect the topology of the network during the disassembly process. Furthermore, it is verified by experiments that the DIR method is suitable for the disassembly of directed networks. Finally, the influence of different disassembly methods on the network structure is analyzed by analyzing the changes of network indexes such as the clustering coefficient, assortativity coefficient and modularity function in the disassembly process, and it is verified that the application of edge module partition to disassemble the network can greatly retain the structural information in the networks.

2. Related Works

As described in the previous section, many network dismantling methods have been proposed in recent years. Next, I will introduce two methods that are compared with this article, namely the GND algorithm and Min-Sum algorithm.

GND method. This method considers the case where the node removal cost is equal to the node weight and is not a unit cost. First, perform spectral division of the Laplacian matrix for which the operators are node weights of the network. After the division is successful, the node weight coverage algorithm is applied to the edges connecting different divisions to find the minimum weight point set that can divide the network, so as to find the minimum cost disassembly set. Compared with the previous algorithm, GND is more general and applicable. It considers the influence of node weights in undirected networks on the network disassembly problem. However, the operator in the GND method is a node adjacency matrix.

Min-Sum method. Braunstein et al. proposed a three-stage Min-Sum algorithm to dismantling networks. They first decycle a network with a variant of the Min-Sum message passing algorithm. After all cycles are broken, they break the remaining trees into small components until the largest component is smaller than the desired threshold. Finally, they refine the node set of network dismantling by moving some of them back to the original network. However, this kind of method tends to delete irrelevant nodes during the loop removal step and then moves them back to the original network in the following node re-inserting step, which reduces the disassembly efficiency.

However, when analyzing directed networks, most spectral methods using node adjacency matrices will use symmetric adjacency matrices to make the network undirected [20]. This processing method will inevitably lose some information in the network [21], resulting in the search process for the minimum set of nodes to be disassembled in the directed network which will add some unnecessary nodes and cause unnecessary disassembly.

3. Preliminaries

In this section, we will provide a simple disassembly flow chart and introduce the knowledge of non-backtracking matrices so that readers can understand the proposed method more easily.

3.1. Model

As shown in Figure 1, by applying the DIR method to disassemble the directed network in the figure, according to the spectral characteristics of the non-backtracking matrix, the edge of the directed network is divided into two different red and blue modules; overlapping nodes 5 and 6 that connect these two different edge modules were found. By removing nodes 5 and 6 and disconnecting the edges connected to them, the directed network can be divided into two disconnected sub-parts. Compared with the previous disassembly method, this disassembly method for removing overlapping nodes of edge modules requires fewer disassembly steps, does not need to find the minimum node coverage set, and the corresponding disassembly cost is lower (the nodes found by the traditional decomposition method are 5, 6, 7, 12), which is more suitable for directed networks.

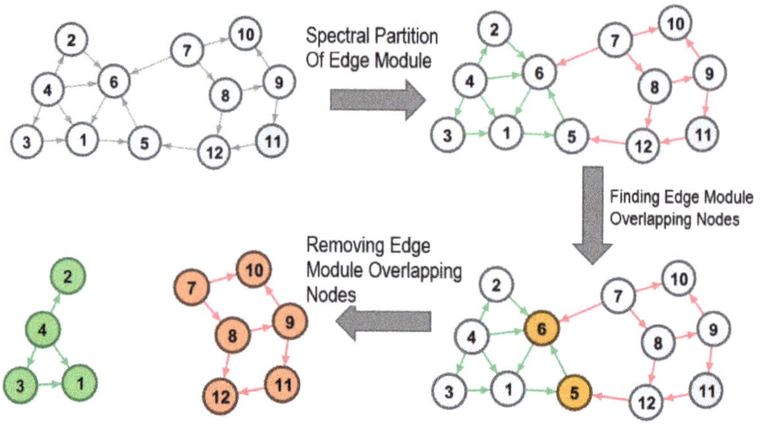

Figure 1. Directed network disassembly flow chart.

3.2. Non-Backtracking Matrix

In a directed network $\boldsymbol{G}$, i, j, k and l are all nodes in $\boldsymbol{V}$, according to the definition of non-backtracking random walk, but only when $j = k$ and $i \neq l$, directed edge $i \to j$ is connected to another directed edge $k \to l$. In a directed network, $\boldsymbol{B}$ is a $m * m$ non-backtracking matrix. This non-backtracking matrix is used to represent the adjacency relationship of edges in a directed network, defined as

$$\boldsymbol{B}_{i \to j, k \to i} = \begin{cases} 1, & \text{if } j = k \text{ and } i \neq l \\ 0, & \text{other cases} \end{cases} \tag{1}$$

The non-backtracking matrix $\boldsymbol{B}$ is different from the adjacency matrix $\boldsymbol{A}$, where $\boldsymbol{B}$ takes each directed edge as an element, and represents the adjacency relationship between the edges in the matrix; therefore, it is also called the edge adjacency matrix. The excellent properties of the non-backtracking operator have been shown above [22], and the spectral characteristics of the non-backtracking matrix have better performance in the network than the node adjacency matrix $\boldsymbol{A}$ or other matrices, especially in terms of the strong separation of its second eigenvector for the network structure division. At the same time, directed networks in the real world tend to have relatively sparse structures and large scales. The non-backtracking matrix $\boldsymbol{B}$ also performs well in sparse networks compared to the node adjacency matrix $\boldsymbol{A}$. The adjacency matrix of the edge, $\boldsymbol{B}$, stores the relationship between the edges in the network, and is not sensitive to the information of the nodes in the network so as to avoid the tendency to remove the nodes with a relatively large degree during dismantling and cause damage to the connected subset of the network [23], thus retaining the structural information in the directed network to the greatest extent. It is also proved by experiments that applying the non-backtracking matrix to disassemble the one-way connection relationship of the edges in the directed network can reduce the disturbance of the node's topology information to the selected disassembly node set, and effectively avoid the problem of network information loss when directly using the directed network adjacency matrix as the spectral algorithm operator.

4. Method

In this section, we propose a method that combines edge module partition with network disassembly to construct a network disassembly algorithm in the directed networks. The non-backtracking matrix is used to store the adjacent information of the edges, and the non-backtracking matrix is used as the operator to construct the minimum number of edges in the disassembly function. The edge module is divided by solving the approximate second eigenvector of the function matrix; after the division, the minimum number of edge sets connecting the different edge modules and the node set where the modules overlap in the edge sets are determined. By removing this node set, the connection between different modules is destroyed, and disassembly is finally achieved.

4.1. Disassemble the Objective Function

In this section, we consider the general case of dividing a network in two modules of equal size according to the nature of the edges, minimizing the number of edges between two different modules. The non-backtracking matrix is used to store the edge adjacency information in the directed network, because in the disassembly problem, we will eventually remove all overlapping nodes on different edge modules, and the weight of the edge does not affect the selection of the minimum node set; therefore, we set the weight of each edge as the unit weight. We divide m edges in the edge network into two equal-sized $\frac{m}{2}$ modules according to the corresponding characteristic. We define an index variable $s_{i\to j} \in \boldsymbol{R}^m$ for any directed edge $i \to j, i, j \in \mathbf{N}$ in the network, and assume that if this edge $i \to j$ belongs to partition module 1, then $s_{i\to j} = 1$; if edge $i \to j$ belongs to partition module 2, then $s_{i\to j} = -1$. So, we obtain

$$\frac{1}{2}\left(s_{i\to j}s_{j\to k}+1\right) = \begin{cases} 1, & \text{If two connected edges } i \to j, j \to k (i \neq k) \text{ belong to the same edge module} \\ 0, & \text{other cases} \end{cases} \tag{2}$$

Equation (2) is used to determine whether two connected edges belong to the same module. Combined with Equation (2), we use the non-backtracking matrix as an operator to obtain the objective function of the minimum number of disassembled edges, which is used to find a set of edges that connect two different modules with the smallest number:

$$\begin{aligned} \min: R &= \sum_{i\to j,k\to l} \boldsymbol{B}_{i\to j,k\to l} - \sum_{i\to j,k\to l} \frac{1}{2}\left(s_{i\to j}s_{k\to l}+1\right)\boldsymbol{B}_{i\to j,k\to l} \\ &= \frac{1}{2}\sum_{i\to j,k\to l}\left(1 - s_{i\to j}s_{k\to l}\right)\boldsymbol{B}_{i\to j,k\to l} \\ &= \frac{1}{2}\sum_{i\to j,k\to l}\left(d_{i\to j}\delta_{k\to l,i\to j} - \boldsymbol{B}_{i\to j,k\to l}\right)s_{i\to j}s_{k\to l} \\ &= \frac{1}{2}s^T\boldsymbol{B}'s \end{aligned} \tag{3}$$

$$\text{s.t.} \begin{cases} 1^T s = 0, \\ s_{i\to j} \in R, i \to j \in E \end{cases} \tag{4}$$

where $\boldsymbol{B}' = \boldsymbol{D_B} - \boldsymbol{B}$, $\boldsymbol{D_B}$ is a diagonal matrix, $(\boldsymbol{D_B})_{i\to j,i\to j} = \sum_{k\to l}\boldsymbol{B}_{i\to j,k\to l}$. Equation (3) represents the difference obtained by the logarithm of the minimized total connected edges minus the logarithm of the edges connected inside the edge module. When two connected edges are divided into different edge modules, $s_{i\to j}s_{j\to k} = -1, \boldsymbol{B}_{i\to j,j\to k} = 1$, the nodes connecting the two edges needs to be removed; on the contrary, when two connected edges are divided into the same edge module, the nodes connecting the two edges do not need to be removed. Finally, the set of nodes that need to be removed corresponding to the set of partitions that minimize R is the minimum disassembly set. We specify the number of nodes whose disassembly cost is the minimum disassembly set.

$1^T s = 0$ in Equation (4) ensures that the two modules are of equal size. Unfortunately, this optimization problem is an NP-hard problem. For this problem, the approximate solution can be found by relaxing constraint $s_{i\to j} \in \{-1,1\}$ to $s_{i\to j} \in R$. According to the Courant--Fisher theory[24], the solution of this relaxation constraint minimum optimization problem can be found by analyzing the eigenvector $v^{(2)}$ corresponding to the second smallest eigenvalue λ_2 of $\boldsymbol{B'}$. So, if node j connects two edges $i \to j, j \to k (i \neq k)$ corresponding to the value of the second smallest eigenvector, one of the second smallest eigenvectors are non-negative $\left(v_{i \to j}^{(2)} \geq 0\right)$, and the other's second smallest eigenvector is negative $\left(v_{i \to j}^{(2)} < 0\right)$; this node will be removed. Removing all such nodes in the network can decompose the network into two sub-parts.

4.2. Divide Vector

Because the large-scale network has many edges, its corresponding second eigenvector of $\boldsymbol{B}$ is difficult to obtain accurately [25]. The traditional power-law iterative model is applied to perform a simple and refined approximation algorithm for the second smallest eigenvalue. Matrix $\boldsymbol{B'}$ has m real non-negative eigenvalues $\lambda_1 \leq \lambda_2 \leq \ldots \leq \lambda_m$, and the corresponding eigenvectors are $v^{(1)}, v^{(2)}, \ldots, v^{(m)}$, which are orthonormal bases in $\boldsymbol{R^m}$ space. We define the maximum degree of elements in matrix $\boldsymbol{B}$ as $d_{\max}$, x, y represents the row and column of the matrix, and the upper bound of the spectrum can be obtained by calculating the 1-norm.

$$\begin{aligned}
\lambda_m &\leq \max_{\|v\|_1=1} \|(\boldsymbol{D_B} - \boldsymbol{B})v\|_1 \\
&= \max_{\|v\|_1=1} \sum_{x=1}^{m} \left| v_x \sum_{y=1}^{m} \boldsymbol{B}_{xy} - \sum_{y=1}^{m} v_y \boldsymbol{B}_{xy} \right| \\
&\leq \max_{\|v\|_1=1} \sum_{x=1}^{m} \sum_{y=1}^{m} \left| v_x \boldsymbol{B}_{xy} \right| + \sum_{x=1}^{m} \sum_{y=1}^{m} \left| v_x \boldsymbol{B}_{xy} \right| \\
&= \max_{\|v\|_1=1} \|\boldsymbol{B}v\|_1 + \|\boldsymbol{B}v\|_1 \\
&= 2d_{\max}
\end{aligned} \tag{5}$$

The upper bound of the spectrum calculated by Equation (5) is $\lambda_m \leq 2d_{\max}$. In order to calculate the approximate second eigenvector, we calculate the matrix $\boldsymbol{H} = 2d_{\max} - \boldsymbol{B'}$, which has the same eigenvector as $\boldsymbol{B'}$. Therefore, the corresponding eigenvalue is now converted into the calculation $0 \leq \xi_m = 2d_{\max} - \lambda_m \leq \ldots \leq \xi_1 = 2d_{\max}$, in which the eigenvector $v^{(2)}$ corresponding to the second largest eigenvalue ξ_2 is calculated. Then, we find the eigenvector of $\boldsymbol{H}$ corresponding to the eigenvalue λ_2 using the following power-law iterative algorithm.

Algorithm 1 can find an approximate eigenvector corresponding to λ_2; we can use our orthogonal eigenvector basis to represent any random vector $v = \sum_{i=1}^{m} \varphi_i v^{(i)}$; the second step of the algorithm can guarantee $\varphi_1 = 0$ and $\varphi_2 \neq 0$. Finally, by multiplying the vector v by the linear operator $\boldsymbol{H}^k$, we obtain

$$\boldsymbol{H}^k v = \sum_{i=2}^{m} \varphi_i v^{(i)} \propto \varphi_2 v^{(2)} + \sum_{i=3}^{m} \varphi_i \left(\frac{\xi_i}{\xi_2}\right)^k v^{(i)} \tag{6}$$

Since $\lambda_3 > \lambda_2$, there is $\left|\frac{\xi_i}{\xi_2}\right| < 1$, and we obtain $\varphi_i \left(\frac{\xi_i}{\xi_2}\right)^k v^{(i)} \to 0$. When the scale of the index k (the number of iterations) of the operator $\boldsymbol{H}$ is $O\left(\log(m)^{1+\varepsilon}\right)$, v tends to be the expected value $E\left[\left|\lambda_2 - \frac{v^T B v}{v^T v}\right|\right] \to 0$ of the eigenvalue λ_2 corresponding to $\boldsymbol{B'}$, where m is the number of edges of the real network.

Algorithm 1: Approximate feature vector algorithm

input: Non-backtracking matrix $\boldsymbol{B}$, network edge number m, $v_1 = (1, 1, \ldots, 1)^T$
output: Approximate second eigenvector v
1: Randomly select vector v on the unit sphere;
2: $v \leftarrow v - \frac{v^T v}{v^T v_1} \cdot v_1$;
3: For $i = 1$ to $\tau(m)$;
4: $v \leftarrow \frac{\boldsymbol{H}v}{\|\boldsymbol{H}v\|}$;
5: End for;
6: Return v.

4.3. Directed Network Disassembly

Algorithm 2 provides a recursive solution that can repeatedly disassemble a network to a specified scale. The number of nodes in the maximal strongly connected subset GSC is defined as the disassembly scale. In this algorithm, we intend to disassemble the directed network until the disassembly scale is smaller than the target scale C. The above algorithm is also defined according to this idea. The input of the algorithm is the node-edge topology of the directed network. The final output is the minimum node disassembly set and the required disassembly cost when the directed network is disassembled to a specified scale; in the first step of the algorithm, the maximal strongly connected subset of the network is taken as the disassembly subject of the directed network, which can filter out the nodes and edges in the network that are not related to network disassembly; this can further improve the disassembly efficiency of the directed network. The selection of the strongly connected subset as the disassembly subject can be directly compared with the connected subset of the commonly used undirected network, which can avoid undirectional networks to meet the undirected network disassembly conditions and cause redundant disassembly. The process is controlled by judging the size of the strongest connected subset of the network; the minimum disassembly set and disassembly cost are initialized to 0 in step 2; the Laplacian matrix of the non-backtracking matrix of the maximum strongly connected subgraph is generated in step 3 for the next division of the edge module; in the fourth step, the eigenvector corresponding to the second eigenvalue is obtained by calculating $\boldsymbol{H} = 2d_{\max} - \boldsymbol{B}'$ and applying the eigenvector approximation algorithm, which is used to divide the edge module; the overlapping node set between edge modules is found and removed in the entire network $\boldsymbol{G}$ in step 5 and 6; the node to be removed is added to the disassembly set and the maximum strongly connected subset and disassembly set of the network in step 7 are updated; the minimum disassembly set and disassembly cost are updated in step 8; whether the maximum strongly connected subset size of the network reaches the target disassembly size is determined in step 9. This recursive algorithm can obtain the set of nodes that disassemble the directed network into a minimum set of connections between different edge modules of a specified size.

Algorithm 2: Directed network disassembly algorithm (DIR method)

input: Network G
output: Minimum disassembly set L_s, minimum disassembly cost c
1: Select the maximum strongly connected subgraph GSC in the network and calculate its non-backtracking matrix $\boldsymbol{B}_{GSC}$ according to Equation (1);
2: Initialize L_s, c to 0;
3: Calculating $\boldsymbol{B}'$ corresponding to $\boldsymbol{B}_{GSC}$ by Equation (3);
4: Use algorithm 1 to obtain the division vector v and divide the maximum strongly connected subgraph into two edge modules;
5: Find the edge set connecting the two edge modules and create a partition subgraph;
6: Find the overlapping node set S in the partitioned subgraph;
7: Remove S from network G and update network G;
8: Merge S into L_s and update L_s, c, and $\boldsymbol{B}_{GSC}$;
9: If the size of the largest strongly connected subgraph $GSC_{size} <$ target disassembly size C, return L_s and c;
Otherwise, go back to step 3.

4.4. Algorithm Complexity

The time complexity of the approximate feature vector is equal to the number of iterations $\tau(m)$ multiplied by the product of matrix $\boldsymbol{H}$ and vector v, namely $O(\tau(m)m^2)$, where m is the number of network edges.

The complexity of performing a bisection for the entire network is $O(m^2\tau(m))$. The complexity of performing another bisection on the two modules with an approximate size of $m/2$ after the division is $2 \cdot O\left(\left(\frac{m}{2}\right)^2\tau(m)\right)$. The complexity of another bisection for the four modules with an approximate size of $m/4$ after division is $4 \cdot O\left(\left(\frac{m}{4}\right)^2\tau(m)\right)$. Until $O(GSC) = 1$, the complexity of another bisection for $m/2 = 2^{\log_2(m)-1}$ modules with an approximate scale of 2 after division is $2^{\log_2(m)-1}O\left(\left(\frac{m}{2^{\log_2(m)-1}}\right)^2\tau(m)\right)$. The total time complexity is as follows:

$$\begin{aligned}
&O\left(m^2\tau(m)\right) + 2 \cdot O\left(\left(\frac{m}{2}\right)^2\tau(m)\right) + 4 \cdot O\left(\left(\frac{m}{4}\right)^2\tau(m)\right) \\
&+ \ldots + 2^{\log_2(m)-1}O\left(\left(\frac{m}{2^{\log_2(m)-1}}\right)^2\tau(m)\right) \\
&= \sum_{i=0}^{\log_2(m)-1} 2^i O\left(\left(\frac{m}{2^i}\right)^2\tau(m)\right) = O\left(m^2\tau(m)\right)\left(\sum_{i=0}^{\log_2(m)-1} 1\right)^2 \\
&= O\left(m^2\tau(m)\log^2(m)\right)
\end{aligned} \tag{7}$$

The computational complexity of the dismantling recursive algorithm is $O\left(m^2\tau(m)\log^2(m)\right)$. For a sparse network, $\tau(m) = \log(m)^{1+\varepsilon}$ at moment $\varepsilon > 0$ and there is an upper bound $1/\log\left(\left|\frac{\xi_2}{\lambda_3}\right|\right)$ [14]; therefore, a better dismantling effect can be obtained. The computational complexity is $O(m^2\log(m)^{3+\varepsilon})$ at this moment.

In the algorithm, the space required for each non-backtracking matrix is $O(m^2)$, the recursive depth is $O(\log(m))$, and the required space complexity is $O(m^2\log(m))$, where m is the number of network edges.

5. Experimental Results

In order to verify the applicability of the DIR method in directed networks, it is used in artificial directed ER networks, BA networks and real networks, and the disassembly results are compared with two commonly used methods (GND algorithm and Min-Sum method). The dataset of Table 1 is selected in the experiment, and the experimental comparison is carried out in different artificial directed networks and large-scale real networks (for the convenience of comparison, the disassembly scale and disassembly cost in this paper are both proportional).

Table 1. Network dataset.

Network Name	Number of Nodes n	Number of Edges m	Node Connection Probability p	Average Degree
ER random network	1000	approximately equal to 10,000	0.01	10
BA random network	1000	approximately equal to 10,000		10
Email-EU-core network	1005	24,929		24.80
Weki-vote network	8297	103,689		12.50

Some scholars, e.g., Ren [14] have proved that the GND method has a higher disassembly efficiency than other algorithms such as Min-Sum and information transfer in undirected networks when the network disassembly cost is the unit cost (number of nodes) and non-unit cost (based on node degree). When the disassembly scale is the same, the

GND method has a lower disassembly cost than other algorithms, and the GND method can destroy the network structure with a smaller disassembly cost. This method has better performance than other algorithms in the disassembly of undirected networks. The DIR method is compared with the GND method and Min-Sum method, considering that the disassembly cost is the unit cost (i.e., the number of nodes).

Figures 2 and 3 are the disassembly results of different methods in different directed networks, where the corresponding curves of disassembly scale and disassembly cost are provided. The ordinate disassembly scale is the proportion of the number of nodes in the largest (strongly) connected subgraph, and the abscissa disassembly cost is the proportion of the number of nodes in the smallest disassembly set when disassembly is at the scale shown in the ordinate. As shown in Figure 2, in a dense ER random network, when the required disassembly size is less than 0.25, the disassembly cost of the DIR method is smaller than that of the GND and Min-Sum methods; in the artificial BA directed network with an average degree of 10, the disassembly cost of the DIR method is significantly lower than that of the other two methods. Additionally, in the relatively sparse real directed network (Figure 2), when the network is disassembled to the same specified scale, the DIR method has a lower cost than other methods. The reason for the difference is that methods such as GND and Min-Sum take the largest connected subgraph in the network as the disassembly subject. The DIR method takes the largest strongly connected subset of the network as the disassembly subject. When the network is dense enough, the size of the largest strongly connected subgraph and the connected subgraph in the directed network is not hugely different; however, in the relatively sparse network (such as the Weki network), the cost of applying the maximum strongly connected subgraph of the directed network for disassembly is significantly lower than that of the GND and Min-Sum methods. The experiments show that the DIR method has the advantage of lower cost in directed network disassembly, which shows the efficiency of the DIR method in directed network disassembly.

Next, we explore the impact of different disassembly methods on the network structure. By applying different disassembly methods to different networks and comparing the clustering coefficient [26] and assortativity coefficient [27] of the disassembly process network, the superiority of the DIR method to retain the network structure information to a great extent is proved.

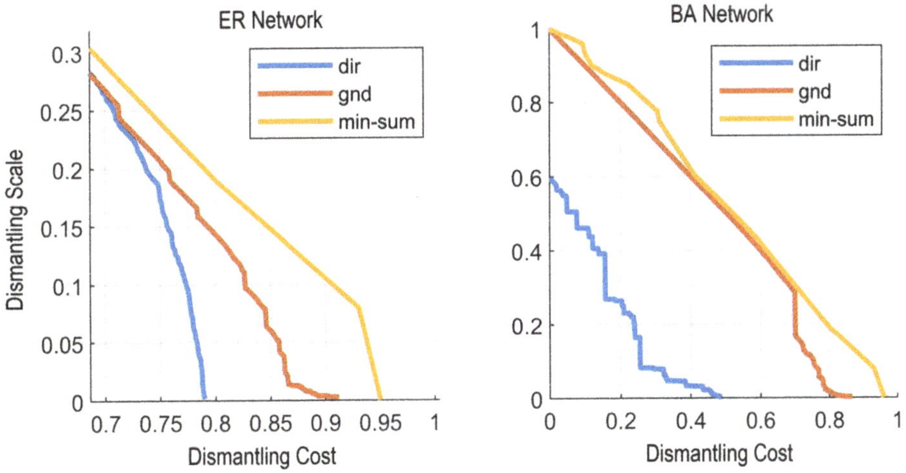

Figure 2. Curve graph of the disassembly cost and disassembly scale of directed ER random network and directed BA random network.

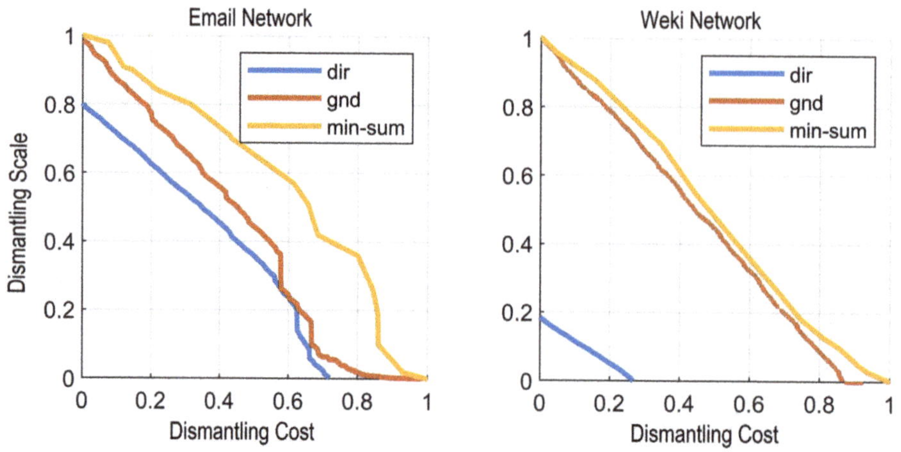

Figure 3. Curve graph of directed real network disassembly cost and disassembly scale.

The clustering coefficient in graph theory is used to measure the degree of node aggregation. There is evidence that in most real-world networks, especially in social networks, nodes tend to create relatively tightly connected groups; this possibility is often greater than the average probability of randomly establishing a relationship between two nodes. A network such as $\boldsymbol{G} = (\boldsymbol{V}, \boldsymbol{E})$ is formally composed of a set of nodes and edges between nodes, with edges connecting nodes. The neighborhood N_i of a node v_i is defined as its adjacent node, $N_i = \{v_j : e_{ij} \in E \vee e_{ji} \in E\}$. The local clustering coefficient C_i of a node in a directed network is

$$C_i = \frac{\left|\left\{e_{jk} : v_j, v_k \in N_i, e_{jk} \in E\right\}\right|}{k_i(k_i - 1)} \quad (8)$$

As an alternative to the global clustering coefficient, Watt and Strogatz [19] use the average of local clustering coefficients of all vertices as the overall clustering level of the network.

$$C = \frac{1}{n}\sum_i C_i \quad (9)$$

Here, we compare the influence of the DIR and the other two disassembly methods on the degrees of node connection in the network by observing the change of the global clustering coefficient of the network during the disassembly process, and then explore the impact on the network structure.

The experiment first disassembles the artificial ER random network and the BA network; the relationship between the disassembly cost and the clustering coefficient is shown in Figure 4. When the disassembly cost is less than 0.7 in the artificial ER random network, the curve of the average clustering coefficient corresponding to the DIR method is more stable than the curve of the GND and Min-Sum method, and it also reaches a stable state first in the BA network. The DIR method has less disturbances for the clustering coefficient of the whole network compared to the GND and Min-Sum method, which reflects the superiority of removing overlapping nodes between modules by dividing the edge modules. The influence of the DIR method on the network structure is smaller than that of directly deleting nodes in the network; in the ER random network, the three methods will increase the network clustering coefficient with the disassembly in a certain period of time. This is because the disassembly has caused an increasing number of nodes in the network to appear in clusters. The result of the experiment in the real network is shown in Figure 5. It can be seen that in the real-world directed network, with the increase in the disassembly cost, the three disassembly methods will reduce the clustering coefficient in the network, which is related to the sparsity of the real network. The dismantling of the nodes of the real network will reduce the agglomeration between nodes and the connection between groups will become sparse; however, it can still be seen from the graph that the curve corresponding to the DIR method is more gentle than the GND and Min-Sum methods. The influence of this aggregation phenomenon on the real network during the disassembly process is smaller than that of the two methods; and relatively speaking, the Min-Sum method will have a more obvious impact on the aggregation phenomenon of the network, because the Min-Sum method tends to remove nodes with large degrees in the network and is less able to protect the structural information of the network.

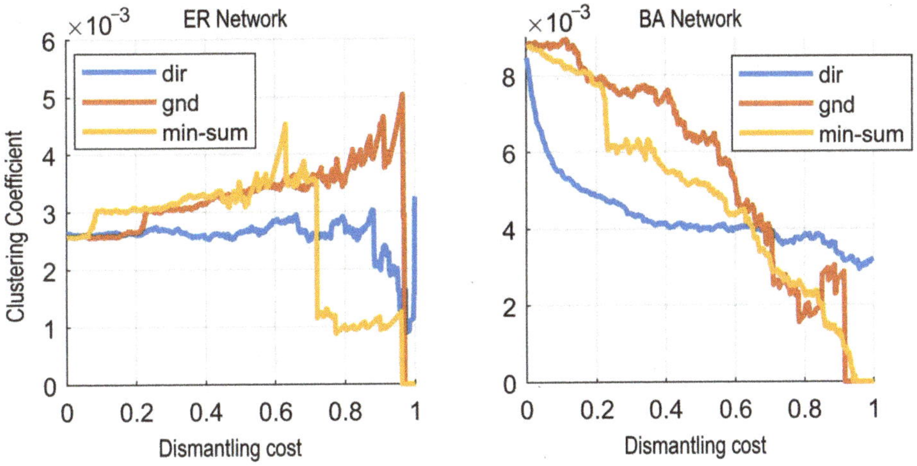

Figure 4. Curve graph of disassembly cost and clustering coefficient of directed ER random network and directed BA random network.

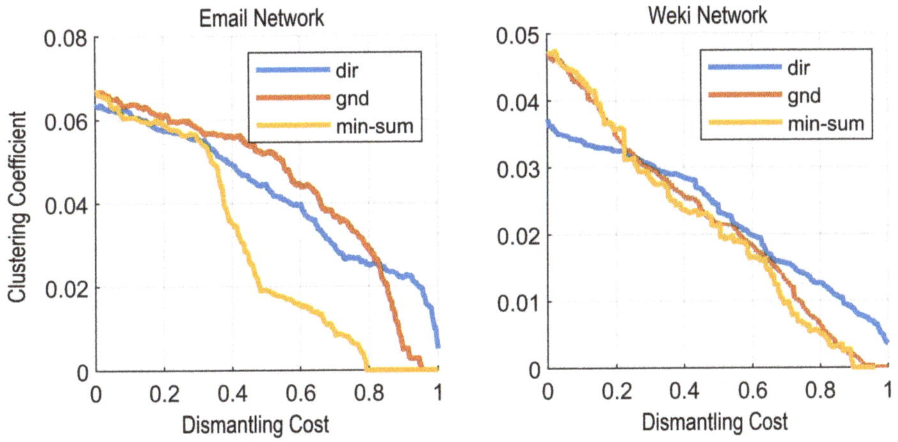

Figure 5. Curve graph of real network disassembly cost and clustering coefficient.

The coefficient of assortativity is used to measure whether the network is assortative or disassortative. It is used to investigate whether the nodes with similar values of degree in the network tend to be connected to its approximate nodes. It can be characterized by the Pearson coefficient r (degree-degree correlation). $r > 0$ indicates that the entire network presents an assortative structure, and the nodes with large degrees tend to be connected to the nodes with large degrees. $r < 0$ indicates that the entire network presents disassortativity, and $r = 0$ indicates that there is no correlation in the network structure. In the experiment, the change of the network structure by the dismantling of the DIR method is analyzed by observing the influence of the dismantling process on the network assimilation index.

As shown in Figures 6 and 7, the changes of the assortativity in the network of the DIR method are less in number than in the other two methods. Whether in the ER random network or in the real network, when the disassembly cost is less than 0.7, the blue curve is more gentle, the change of the global assortativity of the network is smaller, and the influence of removing the overlapping nodes between the edge modules on the assortativity of the network is smaller than that of the GND and Min-Sum methods.

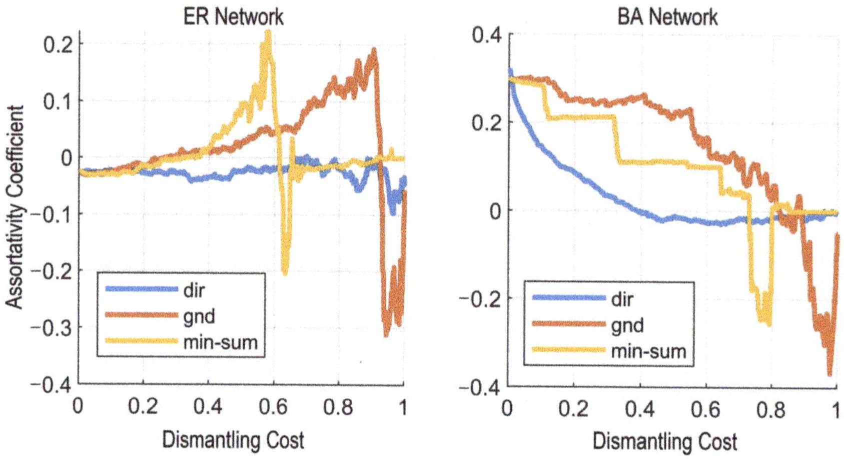

Figure 6. Curve graph of disassembly cost and assortative coefficient of directed ER random network and directed BA random network.

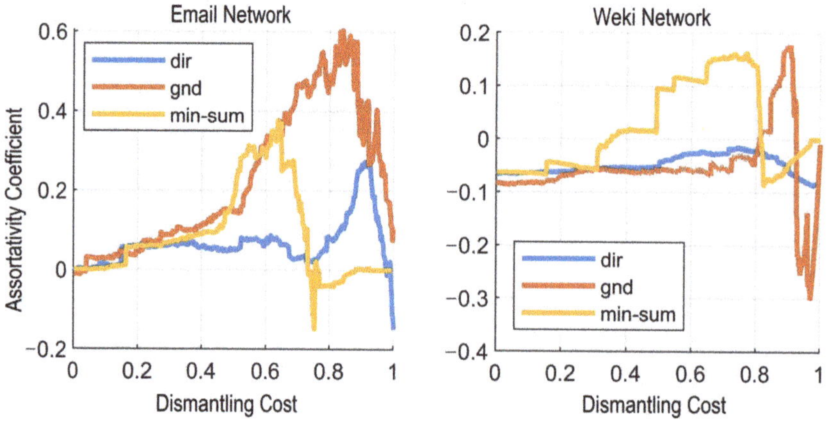

Figure 7. Curve graph of real network disassembly cost and assortativity coefficient.

The module degree [28] is used as a performance index to measure the community division. It is used to see the impact on the structure of the network community when we disassemble the network. The module degree function is $Q = \frac{1}{2}\sum_{i,j} a_{ij}\delta(c_i, c_j)$, where a_{ij} is an element in the point adjacency matrix A, c_i, c_j is the community to which node i and j belong to, and $\delta(c_i, c_j)$ is the membership function. If i and j are in the same community, it is 1, otherwise it is 0.

The calculation of the module degree Q in the disassembly process of the DIR method is shown in Figure 8. It can be seen that when the disassembly cost is less than 0.8 in the picture, the module degree Q increases with the increase in the disassembly cost. It shows that the removal node in the disassembly process also deletes the inter-group edges between different communities, which plays a certain role in promoting effective community division. When the cost is greater than 0.88, the disassembly will destroy the inter-group edges within the community and cause the modularity Q to decrease sharply.

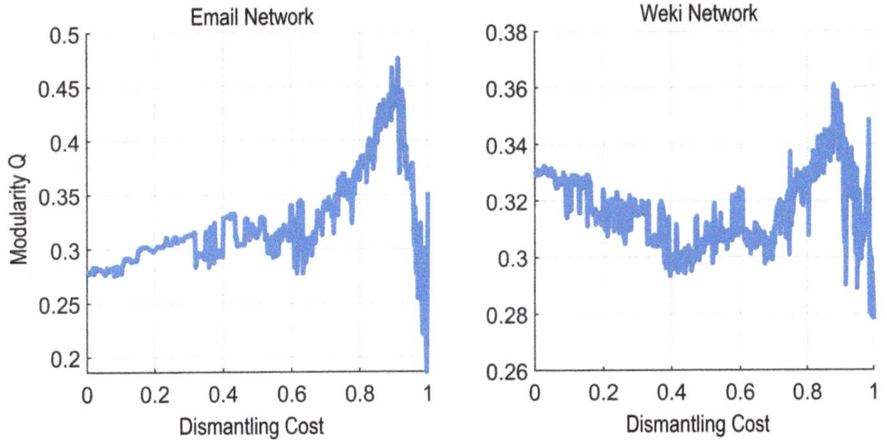

Figure 8. Curve graph of disassembly cost and module degree Q of real network.

In summary, the directed network disassembly (DIR) method we proposed in all the experiments has a higher disassembly efficiency than the GND and Min-Sum methods in both artificial directed networks and real directed networks. When the network is disassembled to the same scale, the DIR method incurs the lowest cost; at the same time, by comparing the clustering coefficient and the assortative coefficient in the disassembly process, it is also proved that the DIR method can reduce the influence of disassembly on the network clustering coefficient and the assortative coefficient in the disassembly process, and can also effectively retain the information in the network; the influence of the DIR method on network modularity is also explored through experiments. When the disassembly scale is less than a given threshold, the DIR method has a certain promoting effect on network community division.

6. Conclusions

An effective disassembly method is proposed for the disassembly of directed networks; the method combines edge module division with network disassembly, using a non-backtracking matrix to construct the function of the minimum number of edges for edge disassembly, and finds the overlapping nodes between edge modules to obtain the approximate solution of the eigenvector of the cost function. Different from the traditional undirected network disassembly method, the DIR method considers the unidirectional relationship between nodes in the directed network, makes full use of the excellent spectral characteristics of the non-backtracking matrix to divide the directed network, ensures the efficiency of disassembly and reduces the impact of disassembly on the overall structure of the network by removing the overlapping nodes between the edge modules during the network disassembly process. By comparing the DIR method with other methods in different artificial directed networks and real directed networks, it is proved that the DIR method is efficient in the network disassembly of directed networks. At the same time, it is also verified that the partition of the edge module applied to the application of non-backtracking operators in the network disassembly leads to a low disturbance of the network structure. The experimental results show that using this method to disassemble the network can achieve lower costs and protect the structure information in the directed network to a great extent.

Author Contributions: Conceptualization, H.L. and J.M.; methodology, H.L.; software, H.L.; validation, J.M. and P.W.; supervision, H.L. and J.M.; formal analysis, J.M. and P.W.; writing—original draft preparation, H.L. and P.W.; funding acquisition, H.L. and J.M. All authors have read and agreed to the published version of the manuscript.

Funding: This research was funded by National Natural Science Foundation of China under Grant 71871233, Science and Technology Project of Hebei Education Department under Grant ZD2022031, Research on the Development of Social Science of Hebei Province under Grant 20220202181 and Fundamental Research Funds for the Hebei Universities under Grant 2021YWF08.

Institutional Review Board Statement: Not appliable.

Informed Consent Statement: Not appliable.

Data Availability Statement: Not appliable.

Conflicts of Interest: The authors declare no conflict of interest. The funders had no role in the design of the study; in the collection, analyses, or interpretation of data; in the writing of the manuscript; or in the decision to publish the results.

References

1. Barrat, A.; Barthelemy, M.; Vespignani, A. *Dynamical Processes on Complex Networks*; Cambridge University Press: Cambridge, UK, 2008.
2. Ma, J.; Li, M.; Li, H.J. Traffic dynamics on multilayer networks with different speeds. *IEEE Trans. Circuits Syst. II Express Briefs* **2022**, *69*, 1697–1701. [CrossRef]
3. Brockmann, D.; Helbing, D. The hidden geometry of complex, network-driven contagion phenomena. *Science* **2013**, *342*, 1337–1342. [CrossRef] [PubMed]
4. Helbing, D. Globally networked risks and how to respond. *Nature* **2013**, *497*, 51–59. [CrossRef] [PubMed]
5. Del Vicario, M.; Bessi, A.; Zollo, F.; Petroni, F.; Scala, A.; Caldarelli, G.; Eugene Stanley, H.; Quattrociocchi, W. The spreading of misinformation online. *Proc. Natl. Acad. Sci. USA* **2016**, *113*, 554–559. [CrossRef]
6. Gallotti, R.; Valle, F.; Castaldo, N.; Sacco, P.; De Domenico, M. Assessing the risks of 'infodemics' in response to COVID-19 epidemics. *Nat. Hum. Behav.* **2020**, *4*, 1285–1293. [CrossRef] [PubMed]
7. Pastor-Satorras, R.; Vespignani, A. Epidemic spreading in scale-free networks. *Phys. Rev. Lett.* **2001**, *86*, 3200. [CrossRef]
8. Kempe, D.; Kleinberg, J.; Tardos, É. Maximizing the spread of influence through a social network. In Proceedings of the ninth ACM SIGKDD International Conference on Knowledge Discovery and Data Mining, Washington, DC, USA, 24–27 August 2003; pp. 137–146.
9. Cohen, R.; Erez, K.; Ben-Avraham, D.; Havlin, S. Breakdown of the internet under intentional attack. *Phys. Rev. Lett.* **2001**, *86*, 3682. [CrossRef] [PubMed]
10. Shunjiang, N.I.; Wenguo, W.; Hui, Z. Simulation of strategies for large-scale spread containment of infectious diseases. *J. Tsinghua Univ. Sci. Technol.* **2016**, *56*, 97–101.
11. Braunstein, A.; Dall'Asta, L.; Semerjian, G.; Zdeborová, L. Network dismantling. *Proc. Natl. Acad. Sci. USA* **2016**, *113*, 12368–12373. [CrossRef] [PubMed]
12. Knuth, D.E. Postscript about NP-hard problems. *ACM SIGACT News* **1974**, *6*, 15–16. [CrossRef]
13. Morone, F.; Makse, H.A. Influence maximization in complex networks through optimal percolation. *Nature* **2015**, *524*, 65–68. [CrossRef]
14. Ren, X.L.; Gleinig, N.; Helbing, D.; Antulov-Fantulin, N. Generalized network dismantling. *Proc. Natl. Acad. Sci. USA* **2019**, *116*, 6554–6559. [CrossRef]
15. Zdeborová, L.; Zhang, P.; Zhou, H.J. Fast and simple decycling and dismantling of networks. *Sci. Rep.* **2016**, *6*, 37954. [CrossRef] [PubMed]
16. Wandelt, S.; Sun, X.; Feng, D.; Zanin, M.; Havlin, S. A comparative analysis of approaches to network-dismantling. *Sci. Rep.* **2018**, *8*, 13513.
17. Zhao, D.; Yang, S.; Han, X.; Zhang, S.; Wang, Z. Dismantling and vertex cover of network through message passing. *IEEE Trans. Circuits Syst. II Express Briefs* **2020**, *67*, 2732–2736. [CrossRef]
18. Wang, Z.; Sun, C.; Yuan, G.; Rui, X.; Yang, X. A neighborhood link sensitive dismantling method for social networks. *J. Comput. Sci.* **2020**, *43*, 101129. [CrossRef]
19. Dorogovtsev, S.N.; Mendes, J.F.F. Evolution of networks. *Adv. Phys.* **2002**, *51*, 1079–1187. [CrossRef]
20. Shi, J.; Malik, J. Normalized cuts and image segmentation. *IEEE Trans. Pattern Anal. Mach. Intell.* **2000**, *22*, 888-905.
21. Zheng, Q.; Skillicorn, D.B. Spectral embedding of directed networks. *Soc. Netw. Anal. Min.* **2016**, *6*, 76. [CrossRef]

22. Krzakala, F.; Moore, C.; Mossel, E.; Neeman, J.; Sly, A.; Zdeborová, L.; Zhang, P. Spectral redemption in clustering sparse networks. *Proc. Natl. Acad. Sci. USA* **2013**, *110*, 20935–20940. [CrossRef]
23. Bollobás, B.; Janson, S.; Riordan, O. The phase transition in inhomogeneous random graphs. *Random Struct. Algorithms* **2007**, *31*, 3–122. [CrossRef]
24. Ikebe, Y.; Inagaki, T.; Miyamoto, S. The monotonicity theorem, Cauchy's interlace theorem, and the Courant-Fischer theorem. *Am. Math. Mon.* **1987**, *94*, 352–354.
25. Pothen, A.; Simon, H.D.; Liou, K.P. Partitioning sparse matrices with eigenvectors of graphs. *SIAM J. Matrix Anal. Appl.* **1990**, *11*, 430–452. [CrossRef]
26. Watts, D.J.; Strogatz, S.H. Collective dynamics of 'small-world'networks. *Nature* **1998**, *393*, 440–442. [CrossRef]
27. Noldus, R.; Van Mieghem, P. Assortativity in complex networks. *J. Complex Netw.* **2015**, *3*, 507–542. [CrossRef]
28. Newman, M.E.J. Modularity and community structure in networks. *Proc. Natl. Acad. Sci. USA* **2006**, *103*, 8577–8582. [CrossRef]

applied sciences

MDPI

Article

TRAL: A Tag-Aware Recommendation Algorithm Based on Attention Learning

Yi Zuo, Shengzong Liu *, Yun Zhou and Huanhua Liu

School of Information Technology and Management, Hunan University of Finance and Economics, Changsha 410205, China
* Correspondence: lsz@hufe.edu.cn

Abstract: A social tagging system improves recommendation performance by introducing tags as auxiliary information. These tags are text descriptions of target items provided by individual users, which can be arbitrary words or phrases, so they can provide more abundant information about user interests and item characteristics. However, there are many problems to be solved in tag information, such as data sparsity, ambiguity, and redundancy. In addition, it is difficult to capture multi-aspect user interests and item characteristics from these tags, which is essential to the recommendation performance. In the view of these situations, we propose a tag-aware recommendation model based on attention learning, which can capture diverse tag-based potential features for users and items. The proposed model adopts the embedding method to produce dense tag-based feature vectors for each user and each item. To compress these vectors into a fixed-length feature vector, we construct an attention pooling layer that can automatically allocate different weights to different features according to their importance. We concatenate the feature vectors of users and items as the input of a multi-layer fully connected network to learn non-linear high-level interaction features. In addition, a generalized linear model is also conducted to extract low-level interaction features. By integrating these features of different types, the proposed model can provide more accurate recommendations. We establish extensive experiments on two real-world datasets to validate the effect of the proposed model. Comparable results show that our model perform better than several state-of-the-art tag-aware recommendation methods in terms of HR and NDCG metrics. Further ablation studies also demonstrate the effectiveness of attention learning.

Keywords: attention learning; tag information; tag-aware recommendation

Citation: Zuo, Y.; Liu, S.; Zhou, Y.; Liu, H. TRAL: A Tag-Aware Recommendation Algorithm Based on Attention Learning. *Appl. Sci.* **2023**, *13*, 814. https://doi.org/10.3390/app13020814

Academic Editors: Giacomo Fiumara, Pasquale De Meo, Xiaoyang Liu and Annamaria Ficara

Received: 14 December 2022
Revised: 3 January 2023
Accepted: 4 January 2023
Published: 6 January 2023

1. Introduction

A recommendation system (RS) has been considered as an extremely effective instrument to tackle the problem of information overload, because it can provide personalized services for individual users by analyzing their interests, preferences, and needs [1]. Many algorithms have been proposed to generate personalized recommendations. To enhance algorithm performance, other superior side information has been incorporated into the recommender system in recent years. In particular, tag-aware recommender systems (TRS) allow users to mark custom tags for relevant items. In this way, TRS can build the implicit relationship between users and items through a wide range of tags. These tags are generally composed of concise words or phrases defined by users, providing good supplementary information for describing user preferences and item characteristics [2]. Thus far, TRS have successfully found applications in many online business services, such as books, movies, music, videos, and social media.

Although the introduction of tags can advance the recommendation performance, some new problems will inevitably arise. For example, most users may only mark a few tags to a few items, resulting in sparse data. In addition, since users can take any word or phrase as a tag, it is easy to cause redundancy and ambiguity in the tag latent space [3–5]. Since user-defined tags are the key factor in expressing user interests and item features, whether the tag information can be effectively processed is crucial to ensure the recommendation performance. Accordingly, the clustering techniques are introduced to extend the traditional collaborative filtering (CF) into TRS [3,4]. The goal of clustering is to aggregate redundant tags. However, it is hard to calculate the similarity of tags, especially when the tag space is extremely sparse. The tool WordNet [6], known as an online English lexical database, is also absorbed to compute the similarity for improved tag-aware recommendation [7]. Nevertheless, manually defining a valid dictionary is time-consuming and words in dictionaries are usually limited. More importantly, these methods cannot generate high-quality recommendation results. The reason may be that the used learning methods are shallow structures that are insufficient to mine the potential meaning of tags.

To obtain more abstract latent features, researchers begin to leverage the deep network model, which has been proved as its most powerful feature expression ability in many fields [8]. For instance, ACF [5] adopts the deep autoencoders to extract low-dimensional dense user features based on tags. These tag-based features are then utilized by user-based CF to generate recommendations. Experimental results show that ACF is obviously better than the clustering-based CF. DSPR [9] uses deep neural networks to obtain the abstract tag-based features of users and items and maximizes similarities between users and their associated items based on those features. TRSDL [10] employs deep neural networks and recurrent neural networks to learn the non-linear latent features of items and users, respectively. Then, the rating prediction is conducted based on these latent features.

In TRS, users may mark various tags to different items, indicating their diverse interests. Similarly, items are assigned multifarious tags that describe their various characteristics. However, in most deep network-based recommendation algorithms, user-defined tags are first transformed into multi-hot feature vectors, and then compressed into a fixed-length representation vector for a given user or a specific item by sum or average pooling, and finally concatenated together to feed into a multi-layer perceptron (MLP) to learn the non-linear relations. In other words, multi-aspect user preferences or item characteristics are compressed into a certain fixed-length feature vector. In order to represent diverse characteristics, the dimension of the feature representation should be large enough to have sufficient expression ability. However, this will significantly increase the scale of learning parameters, causing computing and storage burden.

In addition, as is known to us, the user's preference on a target item comes from the fact that certain characteristics of the item exactly match some specific interests of the user. Therefore, it is not suitable to compress all the diverse interests of a user into the same representation vector when estimating a candidate item, as not all features are equally useful. The useless features may even produce unnecessary noises and deteriorate the recommendation performance. In short, ingenious approaches that can differentiate the importance of different features are required to extract tag-based latent features. Furthermore, although deep networks can automatically learn more expressive feature representations, it is not easy to extract appropriate low-dimensional dense representations for users and items when the potential tag space is very sparse. As discussed in [11–13], both low-level and high-level feature interactions should play important roles for recommendation performance, since such interactions of features behind user preferences and item features are highly sophisticated.

To process the above-mentioned issues, we develop a tag-aware recommendation algorithm based on attention learning (TRAL), which adopts the attention mechanism to discriminate the importance of different features from tag space. Firstly, we utilize the tag embedding technique to extract low-dimensional dense features from the user-tag matrix and item-tag matrix. Secondly, to acquire more abstract and effective representation vectors for each user or item, the attention-based pooling layer is employed to compress these features to a single representation. In this way, different features are assigned different weights according to their importance. Therefore, tag-based features can make different contributions to the final prediction. More importantly, the use of the attention mechanism means that the importance of different features can be automatically learned without any human domain knowledge. Thirdly, the extracted representation vectors for users and items are concatenated together to feed into a general MLP. The high-level feature interactions are, hence, further learned for improving the recommendation performance. In addition, to make full use of low-level feature interactions, the generalized linear model is also introduced. Finally, we combine the representation vectors obtained from the linear model and the depth model and input them into a common logic loss function for joint training.

To sum up, the main contributions of our work are listed as follows:

- We point out the limitation of using simple compression methods to obtain the fixed-length tag-based vector that represents multi-aspect user preferences or item features. To this end, we develop a new tag-aware recommendation algorithm which introduces the attention network to adaptively learn the importance of different features.
- We combine a generalized linear model and a deep neural network so as to take advantages of both low-level and high-level feature interactions.
- We perform extensive experiments on two real-world datasets, demonstrating the rationality and effectiveness of the proposed TRAL.

The rest of this paper is organized as follows. Section 2 briefly summarizes the related work. Section 3 introduces some preliminaries. We elaborate on the proposed TRAL in Section 4 and conduct experiments in Section 5. Conclusions are given in Section 6.

2. Related Work

Naturally, many traditional recommendation algorithms are extended to TRS. For example, Nakamoto et al. [14] proposed a tag-based contextual CF model which modifies the user similarity computation and the item score prediction according to the tagging information. Marinho et al. [15] projected the ternary relation of the user-item-tag to a lower-dimensional space where CF can be applied to provide recommendations. Zhen et al. [16] incorporated tagging information seamlessly into the model-based CF method by regularizing the matrix factorization procedure. Chen et al. [17] developed a tag-based CF model that adopts topic modeling to capture the semantic information of tags for users and items, respectively. Wang et al. [18] devised a robust and efficient probabilistic model based on Bayesian principle for tag-aware recommendation.

To tackle the problem of ambiguity and redundancy in tag information, other kinds of methods have been widely investigated. Shepitsen et al. [3] designed a personalized tag-aware recommendation algorithm based on hierarchical clustering. Through the clustering method, redundant tags can be aggregated, and the user's preferences can be better understood by measuring the importance of associated tag clusters. Symeonidis et al. [19] developed a general tag-aware framework to model the three types of entities: user, items, and tags. The modeled 3-dimensional data is first represented by a 3-order tensor and the dimension reduction is then performed via a higher-order singular-value decomposition. To address the problem of high dimension and sparsity of tagging information, Li et al. [20] developed a novel tag-aware recommendation framework based on Bayesian personalized ranking (BPR) with matrix factorization, where the tag mapping scheme was designed to capture low-dimensional dense features for users and items. Different from the method based on dimension reduction, Zhang et al. [21] developed an integrated diffusion-based recommendation model directly based on user-item-tag tripartite graphs. In the recent work, Pan et al. [22] designed a social tag expansion model to alleviate the tag sparsity problem. The model can explore relations among tags and assign proper weights to the expanded tags. By updating the user profile dynamically through the assigned weights, better recommendation performance can be gained. In [23], the topic optimization was introduced into CF to further enhance both the effectiveness and the efficiency of tag-aware recommendations. In the proposed method, the tags' topic model is established and then used to find the latent preference of users and the latent affiliation of items on topics.

Due to its powerful ability for feature extraction, deep learning has been widely employed in TRS recently. Zuo et al. [5] developed a tag-aware deep model, where tag-based latent features for users are learned by the deep autoencoders. Xu et al. [24] developed a novel tag-aware recommendation model which adopts deep-semantic similarity-based neural networks to extract tag-based representations for users and items. In addition, negative sampling technique is applied so as to enhance the efficiency of the training process. Based on this model, autoencoders are integrated to further accelerate the learning process in [9]. Liang et al. [10] proposed a hybrid tag-aware model by combining deep neural networks and recurrent neural networks for rating prediction. The task of deep neural networks is to capture abstract representation of item characteristics, while the aim of recurrent neural networks is to model user dynamic preferences. Huang et al. [25] proposed a novel tag-based recommendation model that combines the attention network, the stacked autoencoder and MLP to provide recommendations. In the proposed model, a neural attention network is conducted to overcome the difficulty of assigning tag weights for personalized users. Chen et al. [26] designed an attentive intersection model which integrates the neural attention network and factorization machine. The proposed model fully utilizes the intersection between user and item tags to learn conjunct features. Recently, Ahmadian et al. [27] proposed a new tag-aware algorithm that employs deep neural networks to model the representation of trust relationships and tag information.

3. Preliminaries

Generally, users are allowed to assign certain items with personalized tags in TRS. These different tags can indicate user interests and item characteristics from several angles. By fully exploring the rich tagging information, TRS can further capture the connotation of tags, abstract features of items and predict preferences of users, thereby improving the quality of recommendations. Suppose the size of user set U, item set I, and tag set T are $|U|$,$|I|$, and $|T|$, respectively. The user tagging behavior can be formally defined as a tuple $F = (U, T, I, Y)$, in which Y indicates the internal relations between users, items and tags. More specifically, we can use the following 3-order tensor to represent Y: $Y = y_{(u,i,t)} \in \mathbb{R}^{|U|\times|I|\times|T|}$. If a given user u labels an item i with tag t, the corresponding $y_{(u,i,t)} = 1$, otherwise $y_{(u,i,t)} = 0$.

Given a user u and a tag t, we can compute the number of times that u has marked items with t. Analogously, the number of times that the item i has been labeled with tag t is also calculated. In this way, the user–item–tag tensor is decomposed into two adjacent matrices: user–tag matrix and item–tag matrix, as shown in Figure 1. Each row of the user–tag matrix represents the tag-based feature for one user, while each row of the item–tag matrix indicates the tag-based feature for one item. Note that each user often utilizes many tags, and each item is usually annotated by several tags. Consequently, tag-based multi-valued discrete vectors are obtained for users and items, respectively. The aim of the proposed model is to generate the personalized ranked item list for each user based on these tag-based features, also known as the top-n recommendation.

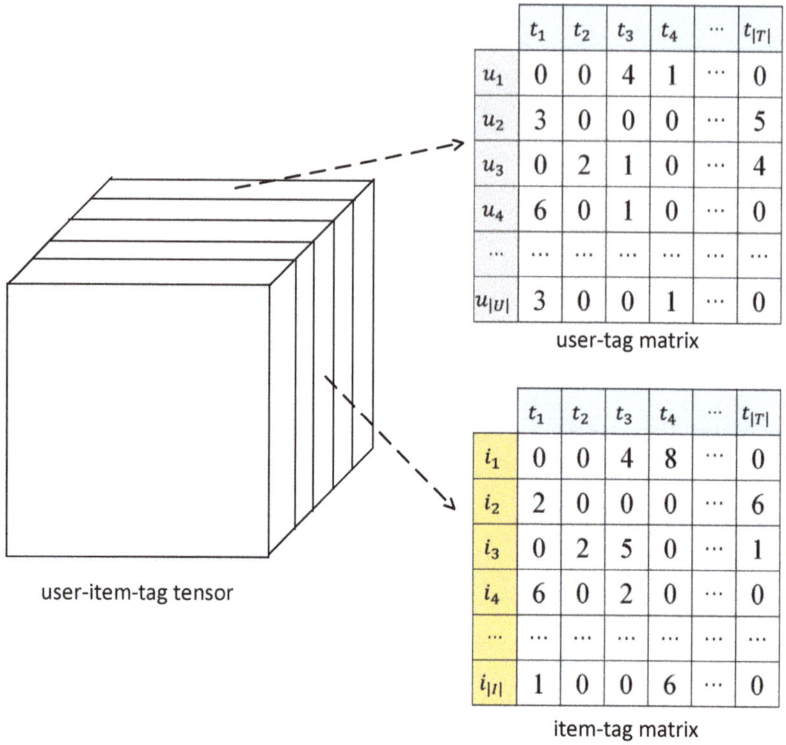

Figure 1. An example of obtaining the user–tag matrix and the item–tag matrix by decomposing the user–item–tag tensor.

4. Method

In this section, we describe the proposed TRAL in detail, the overall structure of which is presented in Figure 2. There are three main modules in the framework: (1) the deep component that integrates a neural attention network and fully connected layers to capture higher-order tag-based features for users and items; (2) the wide component that conducts the generalized linear model to learn low-order features; and (3) the predict layer that combines high-order and low-order features and leverages joint optimization for generating personalized recommendations. More specifically, we first obtain the tag-based representation vectors for users and items from the user–item–tag tensor. These vectors are then used as the input of the deep component and the wide component, respectively, to capture high-order and low-order interaction features. Finally, we integrate these features of different types in the predict layer to provide high-quality tag-aware recommendation.

4.1. The Deep Component

In TRS, user interests and item characteristics are hidden in tagging behavior data. It is remarkable to capture latent tag-based features for users and items, which is the key to advance the performance of recommendations. Consequently, we design a deep component to make the best of deep learning in representation and combination. As presented in Figure 2, the deep component is composed of three main layers: embedding layer, attention layer, and interaction layer.

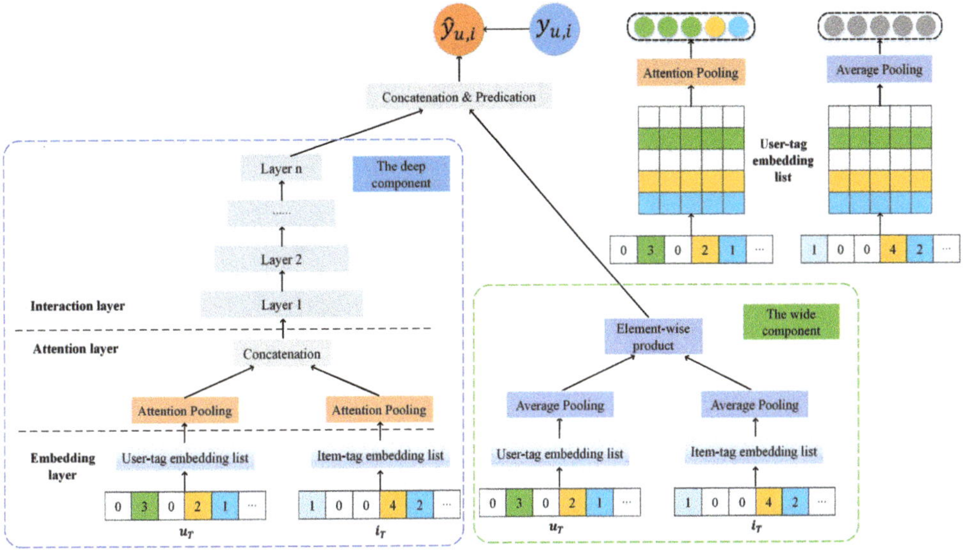

Figure 2. The overall structure of the proposed model. The upper right corner illustrates the difference between average pooling and attention pooling. The weights of different feature vectors are set to the same value in average pooling, while these weights will be automatically learned according to their importance in attention pooling.

4.1.1. Embedding Layer

In order to transform the high-dimensional sparse binary features into low-dimensional dense vectors, we introduce the widely used embedding method, which is inspired by representations for words and phrases [28]. Concretely, we first construct a tag-based embedding dictionary for users:

$$E^u = [e_1^u, \cdots, e_j^u, \cdots, e_{|T|}^u] \in \mathbb{R}^{d \times |T|} \tag{1}$$

where $e_j^u \in \mathbb{R}^d$ represents the embedding vector for tag j that is relative to users and d is the embedding size. Then, for a given user, we extract the corresponding row of the user-tag matrix to produce its user-tag vector u_T, which is apparently a multi-valued discrete vector. Assuming that $u_T[x] >= 1$ for $x \in \{j_1, j_2, \cdots, j_k\}$, we can acquire the tag-based embedding representation according to the table lookup mechanism, which is a list of embedding vectors: $h_t^U = \{e_{j_1}^U, e_{j_2}^U, \cdots, e_{j_k}^U\}$. In particular, in order to capture more accurate potential features based on tags, we also embed tag frequencies, since they can reveal the degree of user preference and item properties [29]. Finally, the resulting embedding list is as follows:

$$h_t^U = \{e_{j_1}^U || e_{j_1}^F, e_{j_2}^U || e_{j_2}^F, \cdots, e_{j_k}^U || e_{j_k}^F\} \tag{2}$$

where $||$ denotes the concatenation operation and $e_{j_1}^F$ represents the embedding of tag frequency which is divided into discrete buckets in the pre-processing. For a target item, a similar method can be conducted to generate its embedding representation h_t^I. Although tags describing users and items belong to the same set, the latent feature of users and items are obviously different. For this reason, we use two different embedding dictionaries for users and items, respectively.

4.1.2. Attention Layer

In a real-life scenario, there are obvious differences in users' behavior habits and cultural backgrounds. Therefore, the number and the content of tags for the same item marked by different users will be significantly different. Note that the input dimensions of full connected neural networks are required to be consistent. Additional operators should be taken to convert the variable list of embedding vectors to a fixed-length vector. A common approach is to construct a pooling layer, where the list of embedding vectors is compressed by a weighted sum operator. Suppose that the list of embedding vectors for a given user is $h_t^U = \{e_1, e_2, \cdots, e_L\}$, the compressed feature vector is obtained by the following weighted sum operation:

$$v_t^U = \sum_{j=1}^{L} a_j e_j \tag{3}$$

where a_j is the weight indicating the importance of the corresponding tag-based embedding vector e_j. Two widely adopted strategies are sum pooling and average pooling, which treat each embedding vector equally and set all weights to the same value.

Clearly, user tagging behaviors play critical role in modeling user preferences in TRS. For a certain user, the compressed fix-length feature vector by sum or average pooling remains constant no matter what the predicted item is. It requires that the tag-based feature vector is capable to express multiple interests of users. To this end, the embedding size should be expanded to large enough, resulting in the increase in computation burden. Moreover, we argue that not all user tagging information are equally important and effective when predicting a target item. Those tag-based features that are less useful should naturally be given a lower weight, which is the main limitation of sum or average pooling.

Based on the above considerations, we resort to the attention mechanism, which allows the weight to be calculated automatically from data. The rationale is that the contributions of different embedding vectors should be taken into consideration when compressing them into a single representation vector. In this work, we construct a two-layer neural network to realize the attention mechanism. Specifically, the weight of each tagging feature vector is first calculated by:

$$\hat{a}_j = q^T ReLU(W_a e_j + b) \tag{4}$$

where $W_a \in \mathbb{R}^{m \times d}$, $b \in \mathbb{R}^m$, $q \in \mathbb{R}^m$ are learnable parameters, and m represents the number of hidden layer neurons in the attention network, which is called attention factor. The final attention weights are then normalized by a softmax function:

$$a_j = \frac{\exp(\hat{a}_j)}{\sum_{j=1}^{L} \exp(\hat{a}_j)} \tag{5}$$

In this way, a fixed-length user representation v_t^U can be adaptively derived by discriminating the importance of different embedding feature vectors. An attention network with similar structures is created so as to generate a fixed-length item representation v_t^I.

4.1.3. Interaction Layer

To capture the high-order interaction features, the obtained tag-based user and item representations from the attention layer are concatenated and further fed into a fully connected neural network with multiple layers. In this way, we can enhance the flexibility and non-linearity of our model to learn the interactions between user v_t^U and item v_t^I, compared with the simple element-wise product operations. Formally, given the concatenated feature vector $Z_0 = [v_t^U || v_t^I]$, the update rule of the neural network in the k-th layer can be defined as:

$$Z_k = \sigma_k(W_k Z_{k-1} + b_k) \tag{6}$$

where W_k, b_k and σ_k represent the weight matrix, the bias vector, and the activation function for the k-th layer, respectively. For the activation function, we apply ReLU (Rectifier), which is proved to be non-saturated and well-suited for sparse data [30]. It is worth mentioning that the network is established by using a classic tower structure, in which the number of neurons in each layer gradually decreases from bottom to top. Consequently, after performing the computation layer by layer, we can obtain more abstract representation from the tag-based user-item interactions.

4.2. The Wide Component

Low-order feature interactions are beneficial to the recommendation results and can be used as an effective supplement to high-order features. Therefore, the wide component adopts a generalized matrix factorization method [11]. Suppose the latent vectors for user u and item i are v_u and v_i, respectively. The representation vector for low order feature interactions is calculated by:

$$Z_w = v_u \odot v_i \tag{7}$$

where $\odot$ means the element-wise product of two vectors. To obtain the tag-based latent vectors for users and items, we adopt a similar approach as used in the embedded layer of the deep component. Specifically, two embedding dictionaries for users and items are first established, respectively. Then, a list of embedding vectors representing a given user or item is obtained by extracting the corresponding rows from the embedding dictionary. To obtain a fixed-length latent vector, we finally perform a simple average pooling. It should be noted that we do not use the attention network in the wide component. The reason for this lies in two aspects. On the one hand, the wide component focuses on learning low order features by making full use of the generalized linear model, so the attention network is unnecessary. On the other hand, it will bring more parameters, making training more difficult, and increase the possibility of over-fitting. Specifically, in order to further improve the training efficiency, we share the tag embeddings used in the deep component.

4.3. Joint Optimization

In our work, two different components are established for capturing high-order and low-order feature interactions, respectively. To further combine the advantages of both, we perform the joint optimization. More specifically, the outputs of the two components are concatenated and then input into the predict layer for joint training. Let Z_w and Z_d denote the tag-based interaction feature from the wide and deep component, respectively. The prediction of the combined model is formally defined as:

$$\hat{y_{u,i}} = \sigma(W^T[Z_w||Z_d] + b) \tag{8}$$

where $\sigma(\cdot)$ is the sigmoid function, W is the weight matrix, b is the bias term, and $\hat{y_{u,i}}$ is the predicted score that measures how much users u like item i.

The commonly used negative log-likelihood function is taken as the loss function, which can be defined as:

$$L = -\sum_{(u,i)\in S^+ \cup S^-} (y_{u,i} \log \hat{y_{u,i}} + (1 - y_{u,i}) \log(1 - \hat{y_{u,i}})) \tag{9}$$

where S^+ and S^- denote the positive and negative sample set, respectively. If the sample $(u,i) \in S^+$, $y_{u,i}$ is set to 1. Otherwise, $y_{u,i}$ is set to 0. For a given user u, the positive sample (u,i^+) can be easily obtained from the observed interactions, while the negative sample (u,i^-) is selected from the non-interacted items. If all unobserved interactions are treated as negative samples, the amount of calculation will inevitably increase dramatically. To cope with this problem, we adopt the negative sampling technique [11,24,31], which generates negative instances by randomly sampling from the unobserved interactions based on a uniform distribution. More concretely, for each positive sample, we randomly select a certain number of negative samples.

Moreover, we utilize the L_2 regularization to avoid over-fitting. The final objective function can be defined as:

$$L = -\sum_{(u,i)\in S^+ \cup S^-} (y_{u,i} \log \hat{y_{u,i}} + (1 - y_{u,i}) \log(1 - \hat{y_{u,i}})) + \lambda \|W\|_2 \tag{10}$$

where λ indicates the strength of regularization. Joint optimization is finally achieved by back-propagating the gradients from the output to the wide and deep components of the proposed model simultaneously with the help of mini-batch stochastic optimization. To further reduce the computational load, we introduce the mini-batch aware regularization [32], which only computes the L_2 norm on the parameters of non-zero sparse features in the current mini-batch.

5. Experiments

In this section, we elaborate on our experiments, including datasets, evaluation metrics, baselines, parameter settings, comparison results, and related analysis.

5.1. Dataset Description and Evaluation Metrics

To measure the proposed tag-aware model, we conduct a series of experiments on the following two public datasets: Delicious and LastFM, which are both published on the website of HetRec [33].

- Delicious is a dataset obtained from the Delicious social bookmarking system, which allows users to annotate web bookmarks with various tags. In this dataset, bookmarks are regarded as items to be recommended.
- LastFM is a dataset collected from Last.fm online music system, where users are encouraged to tag music artists they have listened. In this dataset, artists are treated as items to be recommended.

Note that user–item interactions are established by the tagging behaviors. Following the same assumption as in [5,24], we consider that an item is liked by those users which have tagged this item. In addition, those infrequent tags used less than 5 times in Last.Fm and 15 times in Delicious are eliminated to alleviate sparsity of tagging information [5]. Specific statistics of the two processed datasets are summarized in Table 1. The task of the proposed model is to provide recommendations based on user–item interactions and tagging information.

Table 1. Statistics of the two datasets.

Dataset	#Users	#Items	#Tags	#Assignments
Delicious	1843	65,877	3508	339,744
LastFM	1808	12,212	2305	175,641

To measure the results of the recommendation, we perform the common leave-one-out evaluation [11,34]. More specifically, for each user, we take the last interacted item as the positive test instance and leave the remaining interactions for training. Moreover, we sample 99 items as negative instances for each user randomly from the item set that are not interacted by this user. Adding the positive instance, a test set of 100 items is obtained. Instead of using all the non-interactive items as negative instances, the random sampling strategy can dramatically reduce the amount of calculation [35,36]. After predicting the relevant scores of each item in the test set, the recommendation model will provide a top-n ranked list for each user. The performance of the ranked list is finally estimated by Hit Ratio (HR) and Normalized Discounted Cumulative Gain (NDCG) [37]. HR considers whether the positive item appears in the top-n list, while NDCG measures the quality of ranking by computing the position of the positive item in the list. The higher score of HR and NDCG indicate better recommendation results.

5.2. Baselines and Parameter Settings

To show the effectiveness of the proposed TRAL, we compare the recommendation performance with the following baselines.

- CCF: CCF uses hierarchical clustering to obtain different tag clusters, each of which can be viewed as the representation of a certain topic area [3]. Cluster-based feature vectors for users and items are generated and the relevance relation between them can be estimated.
- ACF: ACF introduces the deep autoencoders to derive tag-based user latent features, on which user-based CF is performed to provide recommendations [5].
- NCF: It is a general framework to employ neural network architectures for CF [11]. By replacing the inner product with a MLP, NCF can learn non-linear interactions between users and items.
- DSPR: DSPR adopts MLPs with shared parameters to extract latent features for users and items based on tagging information [24]. Deep-semantic similarities between target users and their relative items can be computed to generate the ranked recommendation list.
- TRSDL: It is a tag-aware recommendation method, which introduces pre-trained word embeddings to represent tag information and learns latent features of users and items via deep structures [10].
- BPR-T: It is a ranking-based collaborative filtering model which incorporates the tag mapping scheme and the Bayesian ranking optimization [20].
- STEM: STEM establishes a new social tag expansion model to tackle the problem of tag sparsity, thereby improving the recommendation accuracy [22].

To guarantee the recommendation performance, we randomly select one interaction for each user as the validate set to determine hyper-parameters of each model. For the sake of fairness, each model is optimized by the mini-batch Adam [38]. The learning rate of each model is tuned in {0.0001, 0.0005, 0.001, 0.005, 0.01, 0.05}, while the batch size is searched from {128, 256, 512, 1024}. The number of negative samples is search from 1 to 10. The maximum number of iterations for optimization is fixed to 300 for all the models. Early stopping strategy is also applied in the light of the performance on the validation set. For the proposed TRAL, the embedding dimension of tags and the attention factor are set as 32 and 16, respectively. The embedding dimension of tag frequency is fixed to 8. In addition, we construct a tower network architecture with three layers. The c is searched from $\{1 \times 10^{-5}, 1 \times 10^{-4}, 1 \times 10^{-3}, 1 \times 10^{-2}, 1 \times 10^{-1}\}$. The specific parameters of our model are listed in Table 2. The proposed model is implemented with Pytorch and all experiments are conducted on a PC configured with an Intel Core I9-10900X @3.40GHz CPU with 32 GB memory, and an Nvidia GeForce RTX 3080 Ti GPU with 12 GB memory.

Table 2. Specific parameters of the proposed TRAL used in experiment.

MLP and Attention Learning	embedding size of tags	32
	embedding size of tag frequency	8
	size of hidden layers	[80, 40, 20]
	attention factor	16
Training Process	optimizer	Adam
	learning rate	0.001
	maximum number of iterations	300
	batch size	256
	regularization	L_2 norm
	number of negative samples	4

5.3. Performance Comparison

Experimental results of eight recommendation models on two public datasets are presented in Table 3, where we show the best results in boldface and best baseline results in underline. Additionally, we calculate the improvement (imp.) of the proposed model compared to the best baseline. It is clear that the proposed TRAL consistently surpasses other approaches in all evaluation metrics. For example, the performance of TRAL achieves 5.9% and 2.6% improvement over the best baseline in the light of HR@10 and NDCG@10 on Delicious.

Table 3. Overall performance comparison on two datasets in terms of HR and NDCG metrics. Boldface represents the highest score and underline denotes the best result of the baselines. The improvement (imp.) of the proposed model compared to the best baseline is calculated.

Dataset	Delicious				LastFM			
Metrics	**HR@10**	**NDCG@10**	**HR@20**	**NDCG@20**	**HR@10**	**NDCG@10**	**HR@20**	**NDCG@20**
CCF	0.6103	0.3851	0.6346	0.4123	0.5420	0.2548	0.5618	0.2652
ACF	0.6524	0.4216	0.6812	0.4520	0.5624	0.2651	0.5816	0.2764
NCF	0.6836	0.4435	0.6970	0.4712	0.5961	0.2872	0.6108	0.3056
DSPR	<u>0.8041</u>	<u>0.6182</u>	<u>0.8166</u>	<u>0.6315</u>	0.6854	0.3125	0.7012	0.3204
TRSDL	0.7925	0.6012	0.8104	0.6214	<u>0.7052</u>	<u>0.3356</u>	<u>0.7126</u>	<u>0.3468</u>
BPR-T	0.7123	0.5556	0.7492	0.5725	0.6532	0.3173	0.6750	0.3325
STEM	0.7458	0.5423	0.7643	0.5680	0.6726	0.3027	0.6953	0.3252
TRAL	**0.8518**	**0.6345**	**0.8618**	**0.6505**	**0.7336**	**0.3621**	**0.7582**	**0.3820**
Imp.	5.9%	2.6%	5.5%	3.0%	4.0%	7.9%	6.4%	10.2%

It is worth noting that CCF performs worst in most metrics among these baselines, which adequately reveals that the clustering method is insufficient to capture accurate abstract representation for users or items compared to deep learning strategies. In addition, the performance of DSPR, TRSDL, and TRAL are significantly better than ACF. The reason is that ACF only employs the autoencoder to capture low dimensional feature representations of users by constantly optimizing the reconstruction error. In contrast, other deep learning models directly optimize the correlation between users and items so as to extract more accurate feature representation. Furthermore, we can see that NCF behaves marginally better than ACF, but it is obviously worse than other deep learning recommendation methods with the help of tagging information. This convincingly proves the important role of tags as auxiliary information in improving the recommendation performance. The performance of BPR-T and STEM is obviously better than that of NCF, but worse than the proposed TRAL, which indicates that the deep learning model should be well designed for current problems to bring competitive results.

To summarize, the main reasons why the proposed model is superior to other baselines are as follows: (1) constructing the deep architecture to capture effective tag-based features; (2) exploiting the attention mechanism to distinguish the importance of different tag-based features adaptively; and (3) combining the deep and the wide component to learn the high-order and low-order interactions between user and item latent features.

5.4. Ablation Studies

In this section, we carry out several ablation studies on the proposed components or strategies in our model, including the effect of the attention network and the combination of the two components.

5.4.1. Effect of the Attention Network

To investigate the effect of the attention network, we replace it with average pooling to generate a variant method called TRAL-no-A. Figure 3 presents the recommendation results on two datasets in terms of HR and NDCG metrics. It can be observed that TRAL performs significantly better than TRAL-no-A on both metrics, indicating that the recommendation performance will be seriously degraded without the attention network. Benefiting from the attention mechanism, different tag-based features are automatically compressed into a fixed-length vector with different weights. More accurate representations of users and items are thus derived to facilitate subsequent recommendations.

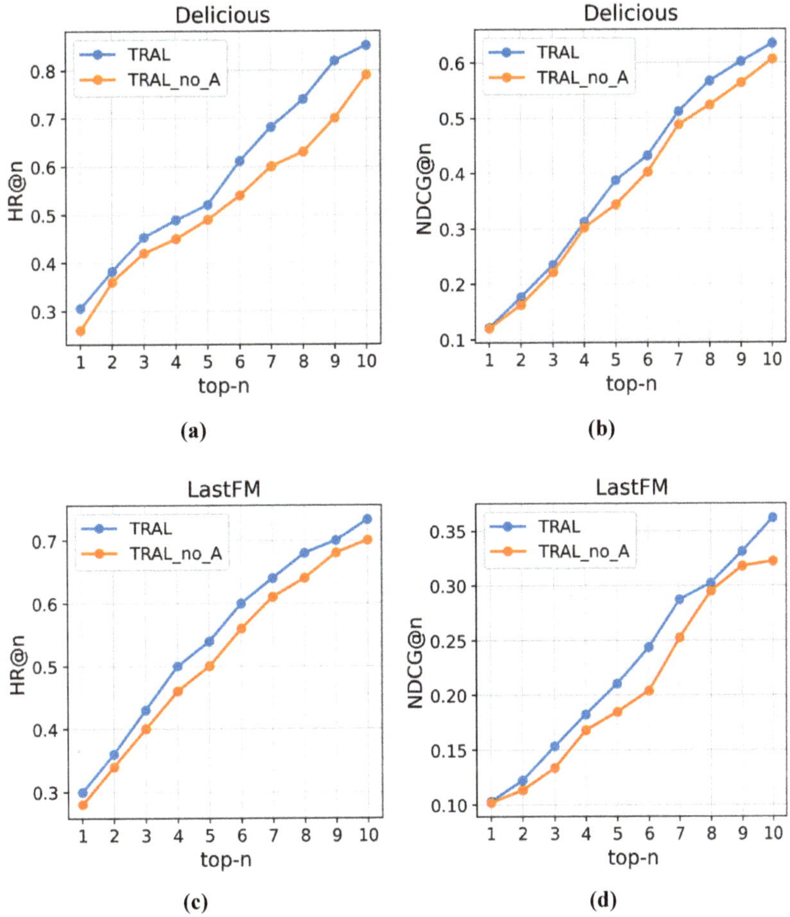

Figure 3. Performance comparison in terms of HR@n and NDCG@n on the two datasets. (**a**) HR@n on Delicious. (**b**) NDCG@n on Delicious. (**c**) HR@n on LastFM. (**d**) NDCG@n on LastFM.

As the model capability of the attention network is affected by the attention factor, we conducted several experiments to further study the impact of the attention mechanism. Figure 4 displays the results of different attention factors on the two datasets in terms of NDCG@10. Note that for Delicious and LastFM, the range of NDCG values under different attention factors falls within [0.62, 0.64] and [0.35, 0.37], respectively. The results show that the performance of TRAL is relatively stable across different attention factors on both datasets. It demonstrates that the design of attention network can make the model have strong robustness while improving the algorithm performance.

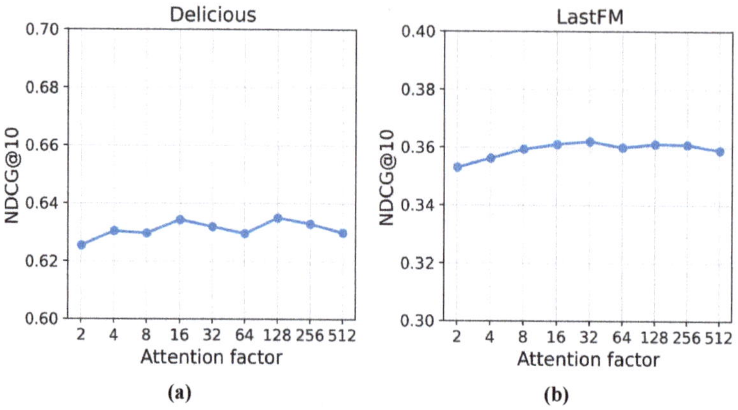

Figure 4. Recommendation results of different attention factors. (**a**) NDCG@10 on Delicious. (**b**) NDCG@10 on LastFM.

5.4.2. Effect of Combining the Two Components

To verify the effectiveness of combining the wide and the deep component, we compare the performance of the proposed model with that of its two variants, in which only the wide or the deep component is employed, named TRS-w and TRS-d, respectively. In addition, we also compare the results with the model using the wide component and the attention network, termed it as TRS-w-a. Table 4 presents the comparison results. It is clear that removing either the wide or the deep component will lead to a significant decline in algorithm performance, revealing the rationality of combining the two components. Among the three variants, TRS-d achieve the best results, which is due to the better expression ability of deep learning. Note that TRS-w performs slightly worse than TRS-w-a, indicating the positive effect of the attention mechanism.

Table 4. Ablation results on the two datasets.

Dataset	**Delicious**		**LastFM**	
Metrics	**HR@10**	**NDCG@10**	**HR@10**	**NDCG@10**
TRS-w	0.6125	0.3420	0.5671	0.2683
TRS-d	0.8093	0.5052	0.6924	0.3458
TRS-w-a	0.6340	0.3654	0.6021	0.2735
TRAL	0.8518	0.6345	0.7336	0.3621

5.5. Parameter Analysis

5.5.1. Number of Negative Samples

To examine the influence of the number of negative samples on recommendation performance, we search the number ranging from 1 to 10. Figure 5 displays the experiment results on Delicious and LastFM in terms of NDCG@10. In addition, the results of NCF and DSPR are also plotted. We can see that the performance of the proposed TRAL is significantly better than NCF and DSPR for different numbers of negative samples. It is worth noting that there is no fixed optimal value for all datasets or all models. When only one negative sample is used for each positive sample, the recommended performance is obviously not good enough, while too many samples will lead to performance degradation. A suitable number of negative samples is around 3 to 6. In our work, we set the number to 4, which is also used in the previous experiments [11].

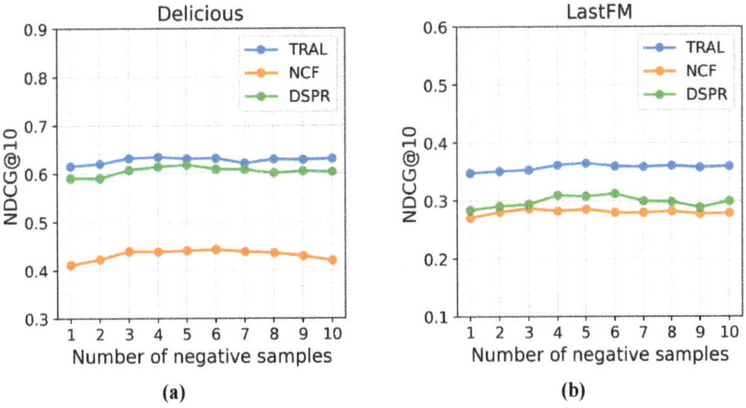

Figure 5. Recommendation results of different number of negative samples. (**a**) NDCG@10 on Delicious. (**b**) NDCG@10 on LastFM.

5.5.2. Embedding Size

To investigate the impact of the embedding size, we establish experiments by varying its value in the range of {8, 16, 32, 64, and 128}. Experimental results are summarized in Figure 6. From the results, we can given several observations. When the embedded dimension is relatively small, the expression ability is insufficient to model the interactions between users and items. With the increase in the embedding size, the performance of the proposed model is gradually improved. However, after the dimension increases to a certain value, the model cannot achieve significant improvement. In particular, we even find slight performance degradation on LastFM when using a large value of embedding size. To balance the performance and the computational cost, we set the embedding size to 32 in our work.

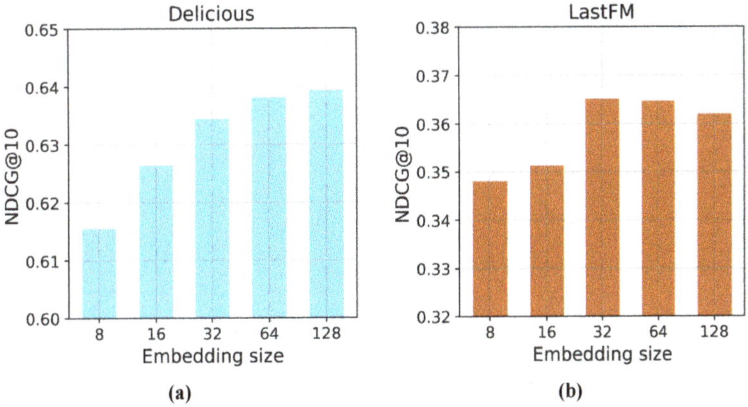

Figure 6. Recommendation results of different embedding sizes. (**a**) NDCG@10 on Delicious. (**b**) NDCG@10 on LastFM.

6. Conclusions

In social tagging systems, a great variety of tags are utilized to describe item characteristics and user preferences. In order to effectively handle tagging information, we propose a novel tag-aware recommendation model based on attention learning. The proposed model constructs a deep component to extract high-level interaction features by integrating an attention network and a multi-layer fully connected network. The aim of the attention network is to make different features contribute differently according to their importance. Furthermore, we establish a wide component to capture low-level interaction features. By combining these different features, we can obtain more accurate representation for users and items, thus improving the recommendation performance. Experimental results demonstrated that the proposed model performs significantly better than other comparison algorithms.

However, there may be some limitations. Firstly, the proposed model is capable of addressing tagging information effectively when these tags are sufficient to express the accurate user interests and item characteristics. Unfortunately, the tagging behavior data in the real recommendation scenario is rather sparse. Secondly, the proposed model cannot deal with the cold-start problem, which refers to how to provide recommendations for new users or fresh items. The obvious reason is that these new users or items have no tagging information at all. Thirdly, the proposed model ignores the sequence information of users, which can indicate the drift of user interests.

To further improve the proposed model, future research directions thus focus on the following aspects. (1) We will investigate some new techniques achieve appropriate data augmentation, such as tag expansion [22] and graph data augmentation [39]. (2) To solve the cold-start problem, we can combine other side information, including images, texts and social relations. Moreover, the introduction of additional information will boost the recommendation performance if the hybrid algorithm is well designed. (3) In order to accurately capture the changes of users' interests over time, we need to design new component which can extract sequence information.

Author Contributions: The overall study supervised by S.L.; Methodology, hardware, software, and preparing the original draft by Y.Z. (Yi Zuo); Review and editing by Y.Z. (Yun Zhou) and H.L. All authors have read and agreed to the published version of the manuscript.

Funding: This work is supported by National Natural Science Foundation of China (Grant no. 61902117 and Grant no. 72073041), the National Natural Science Foundation of Hunan Province (Grant No. 2020JJ5010), and Scientific research project of Hunan Provincial Department of Education (Grant no. 22A0667).

Institutional Review Board Statement: Not applicable.

Informed Consent Statement: Not applicable.

Data Availability Statement: The data that support the findings of this study are available from the corresponding author, upon reasonable request.

Conflicts of Interest: The authors declare no conflict of interest.

References

1. Bobadilla, J.; Ortega, F.; Hernando, A.; Gutiérrez, A. Recommender systems survey. *Knowl.-Based Syst.* **2013**, *46*, 109–132. [CrossRef]
2. Zhang, Z.K.; Zhou, T.; Zhang, Y.C. Tag-aware recommender systems: A state-of-the-art survey. *J. Comput. Sci. Technol.* **2011**, *26*, 767–777. [CrossRef]
3. Shepitsen, A.; Gemmell, J.; Mobasher, B.; Burke, R. Personalized recommendation in social tagging systems using hierarchical clustering. In Proceedings of the 2008 ACM Conference on Recommender Systems, Lausanne, Switzerland, 23–25 October 2008; pp. 259–266.
4. Tso-Sutter, K.H.; Marinho, L.B.; Schmidt-Thieme, L. Tag-aware recommender systems by fusion of collaborative filtering algorithms. In Proceedings of the 2008 ACM Symposium on Applied Computing, Fortaleza, Brazil, 16–20 March 2008; pp. 1995–1999.
5. Zuo, Y.; Zeng, J.; Gong, M.; Jiao, L. Tag-aware recommender systems based on deep neural networks. *Neurocomputing* **2016**, *204*, 51–60. [CrossRef]
6. Miller, G.A. WordNet: A lexical database for English. *Commun. ACM* **1995**, *38*, 39–41. [CrossRef]
7. Zhao, S.; Du, N.; Nauerz, A.; Zhang, X.; Yuan, Q.; Fu, R. Improved recommendation based on collaborative tagging behaviors. In Proceedings of the 13th International Conference on Intelligent User Interfaces, Gran Canaria, Spain, 13–16 January 2008; pp. 413–416.
8. Schmidhuber, J. Deep learning in neural networks: An overview. *Neural Netw.* **2015**, *61*, 85–117.

9. Xu, Z.; Lukasiewicz, T.; Chen, C.; Miao, Y.; Meng, X. Tag-aware personalized recommendation using a hybrid deep model. In Proceedings of the Twenty-Sixth International Joint Conference on Artificial Intelligence, Melbourne, Australia, 19–25 August 2017.
10. Liang, N.; Zheng, H.T.; Chen, J.Y.; Sangaiah, A.K.; Zhao, C.Z. Trsdl: Tag-aware recommender system based on deep learning–intelligent computing systems. *Appl. Sci.* **2018**, *8*, 799. [CrossRef]
11. He, X.; Liao, L.; Zhang, H.; Nie, L.; Hu, X.; Chua, T.S. Neural collaborative filtering. In Proceedings of the 26th International Conference on World Wide Web, Perth, Australia, 3–7 April 2017; pp. 173–182.
12. Cheng, H.T.; Koc, L.; Harmsen, J.; Shaked, T.; Chandra, T.; Aradhye, H.; Anderson, G.; Corrado, G.; Chai, W.; Ispir, M.; et al. Wide & deep learning for recommender systems. In Proceedings of the 1st Workshop on Deep Learning for Recommender Systems, Boston, MA, USA, 15 September 2016; pp. 7–10.
13. Guo, H.; Tang, R.; Ye, Y.; Li, Z.; He, X. DeepFM: A factorization-machine based neural network for CTR prediction. *arXiv* **2017**, arXiv:1703.04247.
14. Nakamoto, R.; Nakajima, S.; Miyazaki, J.; Uemura, S. Tag-based contextual collaborative filtering. *IAENG Int. J. Comput. Sci.* **2007**, *34*, 2.
15. Marinho, L.B.; Schmidt-Thieme, L. Collaborative tag recommendations. In *Data Analysis, Machine Learning and Applications*; Springer: Berlin/Heidelberg, Germany, 2008; pp. 533–540.
16. Zhen, Y.; Li, W.J.; Yeung, D.Y. TagiCoFi: Tag informed collaborative filtering. In Proceedings of the Third ACM Conference on Recommender Systems, New York, NY, USA, 23–25 October 2009; pp. 69–76.
17. Chen, C.; Zheng, X.; Wang, Y.; Hong, F.; Chen, D. Capturing semantic correlation for item recommendation in tagging systems. In Proceedings of the AAAI Conference on Artificial Intelligence, Phoenix, Arizona, 12–17 February 2016; Volume 30.
18. Wang, Z.; Deng, Z. Tag recommendation based on bayesian principle. In Proceedings of the International Conference on Advanced Data Mining and Applications, Chongqing, China, 19–21 November 2010; Springer: Berlin/Heidelberg, Germany, 2010; pp. 191–201.
19. Symeonidis, P.; Nanopoulos, A.; Manolopoulos, Y. Tag recommendations based on tensor dimensionality reduction. In Proceedings of the 2008 ACM Conference on Recommender Systems, Lausanne, Switzerland, 23–25 October 2008; pp. 43–50.
20. Li, H.; Diao, X.; Cao, J.; Zhang, L.; Feng, Q. Tag-aware recommendation based on Bayesian personalized ranking and feature mapping. *Intell. Data Anal.* **2019**, *23*, 641–659. [CrossRef]
21. Zhang, Z.K.; Zhou, T.; Zhang, Y.C. Personalized recommendation via integrated diffusion on user–item–tag tripartite graphs. *Phys. A Stat. Mech. Its Appl.* **2010**, *389*, 179–186. [CrossRef]
22. Pan, Y.; Huo, Y.; Tang, J.; Zeng, Y.; Chen, B. Exploiting relational tag expansion for dynamic user profile in a tag-aware ranking recommender system. *Inf. Sci.* **2021**, *545*, 448–464. [CrossRef]
23. Pan, X.; Zeng, X.; Ding, L. Topic optimization–incorporated collaborative recommendation for social tagging. *Data Technol. Appl.* **2022**, 1–20. [CrossRef]
24. Xu, Z.; Chen, C.; Lukasiewicz, T.; Miao, Y.; Meng, X. Tag-aware personalized recommendation using a deep-semantic similarity model with negative sampling. In Proceedings of the 25th ACM International on Conference on Information and Knowledge Management, Indianapolis, IN, USA, 24–28 October 2016; pp. 1921–1924.
25. Huang, R.; Wang, N.; Han, C.; Yu, F.; Cui, L. TNAM: A tag-aware neural attention model for Top-N recommendation. *Neurocomputing* **2020**, *385*, 1–12. [CrossRef]
26. Chen, B.; Ding, Y.; Xin, X.; Li, Y.; Wang, Y.; Wang, D. AIRec: Attentive intersection model for tag-aware recommendation. *Neurocomputing* **2021**, *421*, 105–114. [CrossRef]
27. Ahmadian, S.; Ahmadian, M.; Jalili, M. A deep learning based trust-and tag-aware recommender system. *Neurocomputing* **2022**, *488*, 557–571. [CrossRef]
28. Mikolov, T.; Sutskever, I.; Chen, K.; Corrado, G.S.; Dean, J. Distributed representations of words and phrases and their compositionality. *arXiv* **2013**, arXiv:1310.4546. https://doi.org/10.48550/arXiv.1310.4546.
29. Liu, H. Resource recommendation via user tagging behavior analysis. *Clust. Comput.* **2019**, *22*, 885–894. [CrossRef]
30. Glorot, X.; Bordes, A.; Bengio, Y. Deep sparse rectifier neural networks. In Proceedings of the Fourteenth International Conference on Artificial Intelligence and Statistics. JMLR Workshop and Conference Proceedings, Fort Lauderdale, FL, USA, 11–13 April 2011; pp. 315–323.
31. Xiao, J.; Ye, H.; He, X.; Zhang, H.; Wu, F.; Chua, T.S. Attentional factorization machines: Learning the weight of feature interactions via attention networks. *arXiv* **2017**, arXiv:1708.04617.
32. Zhou, G.; Zhu, X.; Song, C.; Fan, Y.; Zhu, H.; Ma, X.; Yan, Y.; Jin, J.; Li, H.; Gai, K. Deep interest network for click-through rate prediction. In Proceedings of the 24th ACM SIGKDD International Conference on Knowledge Discovery & Data Mining, London, UK, 19–23 August 2018; pp. 1059–1068.
33. Cantador, I.; Brusilovsky, P.; Kuflik, T. Second workshop on information heterogeneity and fusion in recommender systems (HetRec2011). In Proceedings of the Fifth ACM Conference on Recommender Systems, Chicago, IL, USA, 23–27 October 2011; pp. 387–388.
34. Bayer, I.; He, X.; Kanagal, B.; Rendle, S. A generic coordinate descent framework for learning from implicit feedback. In Proceedings of the 26th International Conference on World Wide Web, Perth, Australia, 3–7 April 2017; pp. 1341–1350.
35. Elkahky, A.M.; Song, Y.; He, X. A multi-view deep learning approach for cross domain user modeling in recommendation systems. In Proceedings of the 24th International Conference on World Wide Web, Florence, Italy, 18–22 May 2015; pp. 278–288.

36. Koren, Y. Factorization meets the neighborhood: A multifaceted collaborative filtering model. In Proceedings of the 14th ACM SIGKDD International Conference on Knowledge Discovery and Data Mining, Las Vegas, NV, USA, 24–27 August 2008; pp. 426–434.
37. Fayyaz, Z.; Ebrahimian, M.; Nawara, D.; Ibrahim, A.; Kashef, R. Recommendation systems: Algorithms, challenges, metrics, and business opportunities. *Appl. Sci.* **2020**, *10*, 7748. [CrossRef]
38. Kingma, D.P.; Ba, J. Adam: A method for stochastic optimization. *arXiv* **2014**, arXiv:1412.6980.
39. Han, X.; Jiang, Z.; Liu, N.; Hu, X. G-Mixup: Graph Data Augmentation for Graph Classification. *arXiv* **2022**, arXiv:2202.07179.

applied sciences

MDPI

Article

Graph-Augmentation-Free Self-Supervised Learning for Social Recommendation

Nan Xiang [1,2,*], Xiaoxia Ma [1], Huiling Liu [1], Xiao Tang [1] and Lu Wang [1,2]

1 Liangjiang International College, Chongqing University of Technology, Chongqing 401135, China
2 Chongqing Jialing Special Equipment Co., Ltd., Chongqing 400032, China
* Correspondence: xiangnan@cqut.edu.cn

Abstract: Social recommendation systems can improve recommendation quality in cases of sparse user–item interaction data, which has attracted the industry's attention. In reality, social recommendation systems mostly mine real user preferences from social networks. However, trust relationships in social networks are complex and it is difficult to extract valuable user preference information, which worsens recommendation performance. To address this problem, this paper proposes a social recommendation algorithm based on multi-graph contrastive learning. To ensure the reliability of user preferences, the algorithm builds multiple enhanced user relationship views of the user's social network and encodes multi-view high-order relationship learning node representations using graph and hypergraph convolutional networks. Considering the effect of the long-tail phenomenon, graph-augmentation-free self-supervised learning is used as an auxiliary task to contrastively enhance node representations by adding uniform noise to each layer of encoder embeddings. Three open datasets were used to evaluate the algorithm, and it was compared to well-known recommendation systems. The experimental studies demonstrated the superiority of the algorithm.

Keywords: social recommendation; self-supervised learning; graph convolutional network

Citation: Xiang, N.; Ma, X.; Liu, H.; Tang, X.; Wang, L. Graph-Augmentation-Free Self-Supervised Learning for Social Recommendation. *Appl. Sci.* **2023**, *13*, 3034. https://doi.org/10.3390/app13053034

Academic Editors: Giacomo Fiumara, Xiaoyang Liu, Annamaria Ficara and Pasquale De Meo

Received: 31 January 2023
Revised: 23 February 2023
Accepted: 24 February 2023
Published: 27 February 2023

1. Introduction

With the advancement of the Internet, social platforms, such as WeChat, Weibo, and Twitter, have become an essential part of people's daily lives. More and more users like to express their opinions and present their hobbies on these platforms, and the interactions between users result in a wide range of consumption behaviors. Moreover, social homogeneity [1] and social influence theory [2] demonstrate that connected users in social networks have similar interest preferences and continue to influence one another as information spreads. Based on these findings, social relations are frequently integrated into recommender systems as a powerful supplement to user–item interaction information to address the problem of data sparsity [3], and numerous social recommendation methods have been developed. Social recommendation algorithms based on graph neural networks have demonstrated improved performance recently and helped advance recommendation technology; however, these models still have certain drawbacks.

1. **Interaction data are sparse and noisy.** Most recommendation models utilize supervised learning techniques [4,5], which substantially rely on user–item interaction data and are unable to develop high-quality user–item representations when data are sparse. As a result, cold-start problems usually occur. In addition, GNN-based recommendation algorithms must aggregate and propagate node embeddings and their neighbors during training, which amplifies the impact of interaction noise (i.e., user mis-click behavior), resulting in confusion with regard to user preferences.

2. **The effect of the long-tail phenomenon.** Due to the skewed distribution of interactive data [4], the recommendation algorithm only emphasizes a portion of some users' mainstream interests, resulting in underfitting of the sample tail distribution and trapping of the user's interest in the "filter bubble" [6], which is known as the long-tail phenomenon.

3. Noise in social relationships. Existing recommendation models are generally based on the assumption that users in social networks have similar item preferences. However, the formation of social relationships is a complicated process that can be based on interests or pure social relationships. Users may have completely different preferences for certain items, but this is not always the case. The model becomes noisier as a result of this assumption, which makes it difficult to effectively incorporate user characteristics and recommendation targets in social networks.

Present work. In view of the above limitations and challenges, this paper proposes an improved social recommendation model—graph-augmentation-free self-supervised learning for social recommendation (GAFSRec). Here, we applied self-supervised contrastive learning to recommendation with the goal of increasing the mutual information for the same user/item view, thereby reducing the reliance on labels and resolving the first and second issues mentioned above. Most contrastive learning techniques currently in use (such as random edge/node loss) improve the consistency of nodes across views through structural perturbations [4,7–9]. However, the most recent research demonstrates [10] that data augmentation can still be accomplished without structural perturbation by adding the proper amount of random uniform noise to the original image. As a result, we employed the method of adding sufficiently small and uniform noise to the graph convolution layer of the recommendation task to achieve cross-layer comparative learning. This technique can operate in the embedding space, making it more effective and simpler to use, and it can subtly attenuate the long-tail phenomenon. The third problem was solved by using both explicit and implicit social relationships, employing hypergraph convolutional networks to mine users' high-order social relationships, and adding the role of key opinion leaders to prevent extra social noise from affecting user preferences.

To improve efficiency, we employed a multi-task training technique with recommendation as the primary task and self-supervised contrastive learning as an auxiliary task. We first created four views using a social network graph and a user–item interaction graph: user–item interaction graph, explicit friend social graph, implicit friend social graph, and user–item sharing graph with explicit friends. Graph encoders (a graph convolutional network and hypergraph convolutional network) were then built for each view to learn the users' high-order relational representation. To avoid the difficulties caused by data sparsity in modeling, we incorporated cross-layer contrastive learning without graph augmentation into GAFSRec, amplified the variance via a graph convolutional neural network, and regularized the representation of recommendations with contrast-augmented views. Finally, we combined the recommendation task and the self-supervised task within the framework of master-assisted learning. The performance of the recommendation task was significantly improved after jointly optimizing these two tasks and utilizing the interactions between all components.

The main contributions of this paper can be summarized as follows:

1. We designed a high-order heterogeneous graph based on motifs, integrated social relations and item ratings, comprehensively modeled relational information in the network, and undertook modeling through graph convolution to capture high-order relations between users;
2. We incorporated the cross-layer self-supervised contrastive learning task without graph augmentation into network training, enabling it to run more efficiently while ensuring the reliability of recommendations;
3. We conducted extensive experiments with multiple real datasets, and the comparative results showed that the proposed model was superior and that the model was effective in ablation experiments.

The rest of this paper is organized as follows. Section 2 presents related work. Section 3 describes the framework for the multi-graph contrastive social recommendation model. Section 4 presents the experimental results and analysis. Finally, Section 5 brings the paper to a close.

2. Related Work

2.1. Graph-Based Recommendation

To obtain high-order collaborative signals, graph-based recommendation algorithms employ multi-hop interaction topologies [11,12]. The success of graph convolutional neural network technology served as inspiration for Feng et al. [10] to develop a hypergraph representation learning framework that captures the relationship between high-order data by developing hyperedge convolution operations. Hypergraphs have a flexible topology and can be used to model complex and high-order dependencies. Additionally, some recently created recommender systems, such as HyperRec [13], MHCN [9], and HCCF [14], have begun to use hypergraph topologies to capture high-order interactions among nodes. HyperRec [13] applies numerous convolutional layers to hypergraphs to capture multi-level connections in order to describe short-term item correlation features. In order to mine high-order user relationship patterns, Yu et al. [9] presented a multi-channel hypergraph convolutional network (MHCN) and created numerous topic-induced hypergraphs. They also employed self-supervised learning to boost the learning of mutual information shared between channels. The application of hypergraphs to social recommendations is made possible by the retention of more high-level information. By simultaneously collecting local and global cooperation information through a hypergraph-enhanced cross-view contrastive learning architecture, HCCF [14] overcomes the over-smoothing problem. In comparison, our system makes use of hypergraph encoders to help create heterogeneous networks, maintain the collaborative relationship between users and items, and enhance the ability to identify user preferences.

2.2. Self-Supervised Contrastive Learning

Self-supervised learning is divided into two categories: generative models and contrastive models. The goal of creating a model is to reconstruct the input so that the input and output are as similar as possible using techniques such as GAN [15] and VAE [16] autoencoders. Contrastive models have recently emerged as an effective self-supervised framework for capturing feature representation consistency across different viewpoints.

Model contrasts are classified into three types [17]: structure-level contrasts, feature-level contrasts, and model-level contrasts. Local–local and global–global structural-level comparison targets exist. SGL [8] is a typical representative, changing the structure of graphs through local perturbations (node loss, edge loss, and random walk). These data enhancement and pre-training methods can extract more supervisory signals from the original graph data, allowing the graph neural network to learn node representation more effectively. Due to dataset constraints, feature-level comparisons have received less attention. Model-level comparison can be implemented end-to-end relatively easily, and SimGCL [10] directly adds noise to the embedding space for enhancement. Experiments show that, regardless of whether graph enhancement is used or not, optimization of the contrast loss can help learn representations. To model node embeddings, we used a cross-layer contrastive learning method without graph augmentation in this study.

3. Proposed Model

3.1. Preliminaries

Let $U = \{u_1, u_2, \cdots, u_m\}$ $(|U| = m)$ represent the collection of users and $I = \{i_1, i_2, \cdots, i_n\}$ $(|I| = n)$ represent the collection of items. Since we are focused on item recommendation, we define $\boldsymbol{R} \in \mathbb{R}^{m \times n}$ to represent the user–item interaction binary matrix. For each pair (u, i), $r_{ui} = 1$ indicates that user u has interacted with item i and, conversely, $r_{ui} = 0$ indicates that user u has not interacted with item i or that user u is not interested in item i. We represent social relations using directed social networks, where $\boldsymbol{S} \in \mathbb{R}^{m \times m}$ represents an asymmetric relation matrix. Additionally, $\left\{\mathbf{Z}_u^{(1)}, \mathbf{Z}_u^{(2)}, \cdots, \mathbf{Z}_u^{(l)}\right\} \in \mathbb{R}^{m \times d}$ and $\left\{\mathbf{Z}_i^{(1)}, \mathbf{Z}_i^{(2)}, \cdots, \mathbf{Z}_i^{(l)}\right\} \in \mathbb{R}^{n \times d}$ denote the embeddings of the size d-dimensional users and items learned in each layer, respectively. This article uses bold uppercase letters for matrices and bold lowercase letters for vectors.

Definition 1. *Hypergraph.*

Let $G = (V, E)$ represent a hypergraph, V a vertex set containing N vertices in the hypergraph, and E an edge set containing M hyperedges. Each hyperedge $\varepsilon \in E$ contains two or more vertices and is assigned a positive weight $W_{\varepsilon\varepsilon}$. All weights form a diagonal matrix $W \in \mathbb{R}^{M\times M}$. A hypergraph is represented by an incidence matrix $H \in \mathbb{R}^{N\times M}$, which is defined as follows:

$$H_{i,j} = \begin{cases} 1, & \text{if } v_i \in V \\ 0, & \text{if } v_i \notin V \end{cases} \tag{1}$$

If the hyperedge $\varepsilon_j \in E$ contains a vertex, then $H_{i,j} = 1$; otherwise, $H_{i,j} = 0$. The degrees of the vertices and edges of a hypergraph are represented as follows: $D_v = \sum_{\varepsilon=1}^{M} W_{\varepsilon\varepsilon} H_{i\varepsilon}$; $D_e = \sum_{i=1}^{N} H_{i\varepsilon}$.

3.2. High-Order Social Information Exploitation

In this study, we used two graphs as data sources: a user–item interaction graph G_r and a user social network graph.

We aligned the two networks into a heterogeneous network and divided it into three sets of views—an explicit friend social graph, an implicit friend social graph, and items shared by users with explicit friends' graphs—in order to establish high-order associations between users in the network. The project-sharing graph of users and explicit friends describes a user's interest in sharing items with friends, which can also serve as relationship-strengthening. The social graph of explicit friends describes a user's interest in expanding their social circle. The social graph of implicit friends describes the similar interests a user shares with similar but unfamiliar users and can alleviate the negative impact of unreliable social relations.

Taking into account the fact that there are some significant social network structures that have impacts on the authority and reputation of nodes in the representation of higher-order relationships as network motifs, we used the motif-based PageRank [18] framework. PageRank is a general algorithm for ranking users in social networks [19]. It can be utilized as a measurement standard for opinion-leader mining, impact, and credibility analyses by assessing the authority of network nodes. However, only edge-based relations are exploited, with higher-order structures in complex networks being neglected. This aspect is improved by the motif-based PageRank algorithm. As shown by Figure 1, which covers the fundamental and significant user social types, the user–item interaction graph splits the explicit friend social graph into seven motifs. M8, also defined as the user's implicit friend social network, represents strangers who share the user's interests. The relationship M9–M10, generally described as users' and explicit friends' item-sharing graph, is, at the same time, extended in accordance with friends' shared buying behaviors.

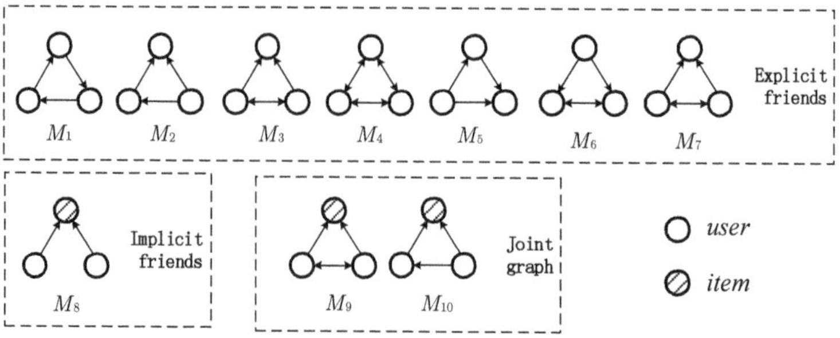

Figure 1. Directed triangle pattern theme for a social network.

When given any motif set M_k, we can calculate the adjacency matrix of the motif using Table 1.

$$(A_M)_{i,j} = \sum_{i \in U, j \in U} 1\ (i, j\ occur\ in\ M_k) \tag{2}$$

In Table 1, $B = S \odot S^T$ and $U = S - B$ are the adjacency matrices of two-way and one-way social networks respectively. Without considering self-connections, $A_e = \sum_{k=1}^{7} A_{M_K}$, $A_i = A_{M_8}$, and $A_j = A_{M_9} + A_{M_{10}}$, where, in $A_j = A_{M_9} + A_{M_{10}}$, we only keep values greater than 5 for the reliability of the implicit friend pair experiment. Furthermore, the adjacency matrix for the user–items graph is $\boldsymbol{A}_r = \boldsymbol{R}$.

Table 1. Computation of motif-based adjacency matrices.

Motif	Matrix Computation	A_{M_i}
M_1	$C = (U \cdot U) \odot U^T$	$C + C^T$
M_2	$C = (B \cdot U) \odot U^T + (U \cdot B) \odot U^T + (U \cdot U) \odot B$	$C + C^T$
M_3	$C = (B \cdot B) \odot U + (B \cdot U) \odot B + (U \cdot B) \odot B$	$C + C^T$
M_4	$C = (B \cdot B) \odot B$	C
M_5	$C = (U \cdot U) \odot U^T + (U \cdot U^T) \odot U + (U^T \cdot U) \odot U$	$C + C^T$
M_6	$C = (U \cdot B) \odot U + (B \cdot U) \odot U^T + (U^T \cdot U) \odot B$	C
M_7	$C = (U^T \cdot B) \odot U^T + (B \cdot U) \odot U + (U \cdot U^T) \odot B$	C
M_8	$C = R \cdot R^T$	C
M_9	$C = (R \cdot R^T) \odot B$	C
M_{10}	$C = (R \cdot R^T) \odot U$	$C + C^T$

3.3. Graph Collaborative Filtering BackBone

In this section, we present our model GAFSRec. In Figure 2, the schematic overview of our model is illustrated.

Due to its strong ability to capture node dependencies, we adopted LightGCN [20] to aggregate neighborhood information. Formally, a general GCN is constructed using the following hierarchical propagation:

$$\boldsymbol{Z}^{(l)} = \boldsymbol{D}^{-1}\boldsymbol{A}\boldsymbol{Z}^{(l-1)} \tag{3}$$

where $Z^{(l)}$ is the l-th layer embedding of the node, and $G_q \in \{G_r, G_e, G_i, G_j\}$ are the four views, with A being the adjacency matrix and D the degree diagonal matrix of A. We encoded the original node vectors into the embedding vectors required by each view through gating functions.

$$\boldsymbol{Z}^{(0)}_{q \in \{r,e,i,j\}} = \boldsymbol{Z}^{(0)} \odot \sigma(\boldsymbol{Z}^{(0)}\boldsymbol{W}^q_g + \boldsymbol{b}^q_g) \tag{4}$$

where σ, $\boldsymbol{W}^q_g \in \mathbb{R}^{d \times d}$, and $\boldsymbol{b}^q_g \in \mathbb{R}^d$ are the activation function, weight matrix, and bias vector. $q \in \{r, e, i, j\}$ represents four views. $\boldsymbol{Z}^{(0)}$ denotes the initial embedding vector and $\odot$ represents the dot product.

We used the generalization ability of hypergraph modeling to capture more effective high-level user information. Therefore, in the encoder, a hypergraph convolutional neural network [21] was used.

$$\boldsymbol{Z}^{(l)} = \boldsymbol{D}^{-1}\boldsymbol{A}\boldsymbol{Z}^{(l-1)} = \boldsymbol{D}_v^{-1}\boldsymbol{H}\boldsymbol{W}\boldsymbol{D}_e^{-1}\boldsymbol{H}^T\boldsymbol{Z}^{(l-1)} \tag{5}$$

where $\boldsymbol{A} = \boldsymbol{H}\boldsymbol{W}\boldsymbol{D}_e^{-1}\boldsymbol{H}^T$ is the Laplacian matrix of the hypergraph convolutional network; H is the incidence matrix of each hypergraph; $\boldsymbol{D}_v$ and $\boldsymbol{D}_e$ are the degree matrix of the nodes and the degree matrix of the hyperedges; and σ, Θ, and W are activation functions, learnable filter matrices, and the parameters of the diagonal matrix.

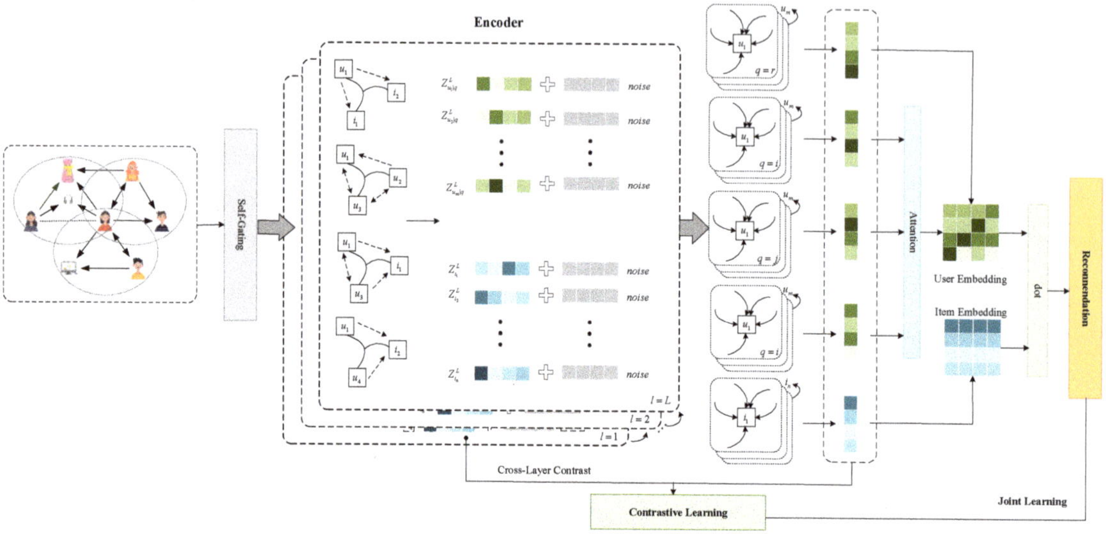

Figure 2. The overall system framework for GAFSRec.

Considering node influence and credibility, we adopted motifs [18] to construct a hypergraph. Given the complexity of the actual construction of the Laplacian matrix for hypergraph convolution (which includes a large number of graph-induced hyperedges), matrix multiplication can be considered for simplification.

Finally, the hypergraph convolutional neural network can be expressed as: $\boldsymbol{Z}_{q\in\{r,e,i,j\}}^{(l)} = \widetilde{\boldsymbol{D}}_q^{-1}\boldsymbol{A}_q\boldsymbol{Z}_q^{(l-1)}$, where $\widetilde{\boldsymbol{D}}_q \in \mathbb{R}^{m\times m}$ is the degree matrix of $\boldsymbol{A}_q$. It can be seen from this that the graph convolutional neural network is a special case of the hypergraph neural network.

After l-layer propagation, we used the weighting functions and the readout function to output the representations of all layers, obtaining the following representations:

$$\boldsymbol{Z}_{u|q} = \frac{1}{L+1}\sum_{l=0}^{L}\boldsymbol{Z}_{u|q}^{(l)},\ \boldsymbol{Z}_{i|q} = \frac{1}{L+1}\sum_{l=0}^{L}\boldsymbol{Z}_{i|q}^{(l)}$$

We applied an attention approach [22] to learn the weights α and aggregate the user embedding vectors for augmented views.

$$\alpha_q = \frac{\exp(\boldsymbol{a}^T\boldsymbol{W}_a\boldsymbol{Z}_{u|q})}{\sum\limits_{q'\in\{e,i,j\}}\exp(\boldsymbol{a}^T\boldsymbol{W}_a\boldsymbol{Z}_{u|q'})}$$

where a and W are trainable parameters. The final user embedding $\boldsymbol{Z}_u$ and item embedding $\boldsymbol{Z}_i$ look like this:

$$\boldsymbol{Z}_u = \textstyle\sum_{q\in\{e,i,j\}}\alpha_q\boldsymbol{Z}_{u|q} + \sum\limits_{l=0}^{L}\boldsymbol{Z}_{u|r}^{(l)}$$
$$\boldsymbol{Z}_i = \tfrac{1}{L+1}\sum\limits_{l=0}^{L}\boldsymbol{Z}_{i|r}^{(l)}$$

The ranking score generated by this model recommendation is defined as the inner product with user and item embeddings:

$$\widehat{y}_{u,i} = \mathbf{Z}_u^T \mathbf{Z}_i$$

To optimize the parameters in the model, we adopted the Bayesian loss function [23]. The main reason was that the Bayesian loss function considers the comparison of pairwise preferences between observed interactions and unobserved interactions. The loss function of this model is:

$$\mathcal{L}_{BPR} = -\sum_{i \in I(u), j \notin I(u)} \ln \sigma(\widehat{y}_{u,i} - \widehat{y}_{u,j}) + \beta \|\Phi\|_2^2$$

where Φ is a trainable parameter and β is a regularization coefficient that controls the strength of L2 regularization, prevents overfitting, and is a sigmoid function. Each input datum is a triplet sample $< u, i, j >$; this triplet includes the current user u, the positive item i purchased by u, and the randomly drawn negative item j. The negative item j is the user u's unliked or unknown items.

3.4. Graph-Augmentation-Free Self-Supervised Learning

Data augmentation is the premise of self-supervised contrastive learning models, and they can obtain a more uniform representation by perturbing the structure and optimizing the contrastive loss. With regard to the learning of graph representations, a study on SimGCL [10] showed that it is the uniformity of distribution rather than dropout-based graph augmentation that has the greatest impact on the performance of self-supervised graph learning. Therefore, we considered adding random noise to the embedding to create a self-supervised signal that could enhance the performance of the contrastive learning.

$$\widetilde{\mathbf{Z}}^{(l)} = \mathbf{D}^{-1}\mathbf{A}\widetilde{\mathbf{Z}}^{(l-1)} + \varepsilon^{(l)}$$

where ε is the added random uniform noise vector, and the noise direction is in the same direction as the embedding vector, $\varepsilon = x \odot sign(z),\ x \in \mathbb{R}^d \sim U(0,1)$. The embedding representation added with perturbation retains most of the information of the original representation, as well as some invariance.

By applying different scales of embedding vectors to the current node embedding, the perturbed embedding vectors can then be fed into the encoder. The embedding vector of the final perturbed node is expressed as:

$$\widetilde{\mathbf{Z}} = \frac{1}{L+1}\sum_{l=0}^{L} \widetilde{\mathbf{Z}}^{(l)}$$

We regarded augmented views from the same node as positive examples and augmented views from different nodes as negative examples. Positive auxiliary supervision promotes the consistency of predictions among different views of the same node, while negative supervision strengthens the divergence between different nodes. Formally, we adopted the contrastive loss InfoNCE [24] to maximize the consistency of positive examples and minimize the consistency of negative examples:

$$\mathcal{L}_{cl}^{U} = -\sum_{u \in \mathcal{U}} \log \frac{\exp(sim(\widetilde{z}_u^{(k)} \cdot \widetilde{z}_u)/\tau)}{\sum_{v \in \mathcal{U}} \exp(sim(\widetilde{z}_u^{(k)} \cdot \widetilde{z}_v)/\tau)}$$

$$\mathcal{L}_{cl}^{I} = -\sum_{i \in \mathcal{I}} \log \frac{\exp(sim(\widetilde{z}_i^{(k)} \cdot \widetilde{z}_i)/\tau)}{\sum_{j \in \mathcal{I}} \exp(sim(\widetilde{z}_i^{(k)} \cdot \widetilde{z}_j)/\tau)}$$

where $\widetilde{z}_u^{(k)} = \widetilde{z}_u^{(k)} / \|\widetilde{z}_u^{(k)}\|_2$ represents the k-th layer L2-regularized embedding vector compared with the final layer embedding, $sim(a, b)$ represents the dot-product cosine similarity between normalized embeddings, and b. τ is the temperature parameter. The total loss function for self-supervised learning includes the contrastive loss for each view user and the item contrastive loss, as shown in the following equation:

$$\mathcal{L}_{\text{ssl}} = \sum_{q \in \{r,e.i,j\}} \mathcal{L}_{cl}^{q,U} + \mathcal{L}_{cl}^{r,I}$$

To improve recommendation through contrastive learning, we utilized a multi-task training strategy to jointly optimize the recommendation task and the self-supervised learning task, as shown in the following equation:

$$\mathcal{L} = \mathcal{L}_{BPR} + \lambda \mathcal{L}_{ssl}$$

where λ is the hyperparameter used to control the auxiliary task.

3.5. Complexity Analysis

In this section, we analyze the theoretical complexity of GAFSRec. Since the time complexity of LightGCN-based convolutional graph encoding is $O(L \times |\boldsymbol{R}| \times d)$, the total encoder complexity of this architecture is less than $4 \times O(L \times |\boldsymbol{R}| \times d)$ because the adjacency matrix of the auxiliary encoder is sparser than that of the user–item interaction graph. The time costs of the gate function and the aggregation layer are both $O(m \times d^2)$. This architecture adopts BPR loss; each batch contains B interactions, and the time cost is $O(2 \times B \times d)$. Since the contrasting between positive/negative samples in contrastive learning increases the time cost, cross-layer self-supervised contrastive learning contributes a time complexity of $5 \times O(B \times M \times d)$, where M represents the number of nodes in a batch. Since this model does not involve graph augmentation, the complexity of GAFSRec is much lower than that of graph-augmented social recommendation models. In our experiments, with the same embedding size and using the Douban-Book dataset, MHCN took 34 s per epoch and GAFSRec only took 11 s. Detailed information on the experiments can be found in Section 4.2.1.

4. Experiments and Results

In this section, we describe the extensive experiments we conducted to validate GAFS-Rec. The experiments were conducted in order to answer the following three questions: (1) Does GAFSRec outperform state-of-the-art baselines? (2) Does each component in GAFSRec play a role? (3) What are the effects of hyperparameters on the GAFSRec model?

4.1. Experimental Settings

4.1.1. Datasets

We conducted experiments using three real-world datasets: Douban-Book [25], Yelp [26], and Ciao [5]. The statistical data for the datasets are shown in Table 2. In accordance with the summary provided by Tao et al. [27], we conducted statistical analyses of the three datasets, which were helpful for the analysis of the results of the subsequent experiments. It was found that the higher the level of social diffusion was (greater than 1), the greater the possibility of similar preferences among users was, and the higher the effective social density was, the lower the scoring density was, indicating that explicit social relationships are very important for recommendation. We performed fivefold cross-validation with the three datasets and report the averaged results.

Table 2. Data statistics for social recommendation datasets.

	Douban-Book	Yelp	Ciao
Users (U-I graph)	12,859	19,539	7375
Users (U-U graph)	12,748	30,934	7317
Items	22,294	22,228	105,114
Feedback	598,420	450,884	284,086
Valid user pairs	48,542,437	51,061,951	5,052,316
Social pairs	169,150	864,157	111,781
Valid social pairs	77,508	368,405	56,267
Candidate user pairs	165,353,881	381,772,521	54,390,625
Valid ratio	29.357%	13.375%	9.289%
Valid social ratio	45.822%	42.632%	50.337%
Social density	0.104%	0.090%	0.209%
Rating density	0.209%	0.104%	0.037%
Social diffusity level	1.561	3.187	5.419
Valid social density	0.160%	0.721%	1.114%

4.1.2. Baselines

We evaluated GAFSRec against 13 baselines spanning different recommendations:

- MF-based collaborative filtering models;
- BPR [23]: a popular recommendation model based on Bayesian personalized ranking;
- SBPR [28]: an MF-based social recommendation model that extends BPR and utilizes social relations to model the relative order of candidate items;
- GNN-based collaborative filtering frameworks;
- NGCF [7]: a complex GCN-based recommendation model that generates user/item representations by aggregating feature embeddings with high-order connection information;
- LightGCN [20]: a general recommendation model based on GCN, improved on the basis of NGCF by removing linear changes and activation functions;
- DiffNET++ [29]: a GNN-based social recommendation method that simultaneously simulates the recursive dynamic social diffusion of user space and item space;
- Recommendation with hypergraph neural networks;
- MHCN [9]: a social recommendation method based on a motif-based hypergraph. It aggregates high-level user information through hypergraph convolutional multi-channel embedding and utilizes auxiliary tasks to maximize the mutual information between nodes and graphs and generate self-supervised signals;
- HyperRec [13]: this method leverages the hypergraph structure to model the relationship between users and their interactive items by considering multi-order information in dynamic environments;
- HCCF [14]: a hypergraph-guided self-supervised learning recommendation model that jointly captures local and global collaborations through a hypergraph-enhanced cross-view contrastive learning architecture;
- Self-supervised learning for recommendation;
- SEPT [25]: a social recommendation model that utilizes multiple views to generate supervisory signals;
- SGL [8]: the most typical self-supervised comparative learning recommendation model, which uses structural perturbation to generate comparative views and maximizes the consistency between nodes. The experiment in this study used the structural perturbation method involving missing edges;
- BUIR [30]: this method adopts two encoders that learn from each other and randomly generates augmented views for supervised training;
- SimGCL [10]: the latest self-supervised contrastive learning recommendation model. It uses the no-image-enhancement method and only adds the final embedding obtained by adding uniform noise in the embedding space for comparison;
- NCL [31]: a prototypal structural contrastive learning recommendation model.

4.1.3. Metrics

To evaluate all the models, we chose two evaluation metrics: Recall@K and NDCG@K. Recall@K employs the proportions of each user's favorite items appearing in the recommended items. Normalized discounted cumulative gain (NDCG) means that the scores for the relevance of each recommendation result are accumulated and used as the score for the entire recommendation list. The greater the relevance of the recommendation result is, the greater the DCG is. The recall rate is defined as: $\text{Recall@K} = \frac{\sum_{i=1}^{K} rel_i}{\min(K,|y_u^{test}|)}$. The normalized discounted cumulative gain is defined as $\text{DCG@K} = \sum_{i=1}^{K} \frac{2^{rel_i}-1}{\log_2(i+1)}$ $\text{NDCG@K} = \frac{DCG@K}{iDCG@K}$.

4.1.4. Settings

To ensure a fair comparison, we checked the best hyperparameter settings reported in the original papers for the baselines and then used grid search to fine-tune all the hyperparameters for the baselines. For the general setup of all baselines, all embeddings were initialized with Xavier. The embedding size was 64, the L2 regularization parameter was $\beta = 0.01$, and the batch size was 2048. We optimized all models using Adam with a learning rate of 0.001 and let the temperature $\tau = 0.2$.

4.2. Recommendation Performance

Next, we verified whether GAFSRec could outperform the baselines and achieve the expected performance. The purpose of social recommendation is to alleviate data sparsity. We also conducted a cold-start experiment with the entire training set. The cold-start experiment dataset only included the data for cold-start users with fewer than ten historical purchase records. The results are shown in Tables 3 and 4. It can be observed that GAFSRec outperformed the baselines in general and cold-start cases with all datasets.

4.2.1. Comparison of Training Efficiency

In this section, we report the actual training time to verify the theoretical plausibility of the method. The reported data were collected using a workstation equipped with an Intel(R)Core™ i9-9900K CPU and a GeForce RTX 2080Ti GPU. The model depths of all methods were set to two layers.

As shown in Figure 3, compared with the MHCN and SGL models, GAFSRec demonstrated longer computation time due to the parallel processing of hierarchical self-supervised learning tasks in the MHCN model, and the running time also increased with the number of datasets. For SGL-ED, only the user–item interaction dataset was computed, and most of the runtime was spent on building the perturbation graph. In the large Douban-Book dataset especially, the speed of GAFSRec was three times faster than that of MHCN, thus demonstrating the strong advantage of graph-augmentation-free cross-layer contrastive learning in terms of operational efficiency.

4.2.2. Comparison of Ability to Promote Long-Tail Items

As mentioned in the introduction, GNN-based recommendation models are easily affected by the long-tail problem. To verify that GAFSRec could entirely alleviate the long-tail problem, we divided the test set into ten groups according to popularity. Each group contained the same numbers of interactions, and the larger the ID of a group was, the more popular the items were. Then, we set the number of layers to two for the experiments and verified the long-tail recommendation ability of the model by observing Recall@20.

Table 3. General recommendation performance comparison.

Dataset	Metric	BPR	SBPR	NGCF	LightGCN	DiffNET++	HyperRec	MHCN	HCCF	BUIR	SLRec	SGL	SimGCL	NCL	GAFSRec	Improvement
Douban-Book	Recall@20	0.0889	0.0918	0.1167	0.0936	0.0988	0.1394	0.1487	0.1666	0.0982	0.1400	0.1617	0.1629	0.1650	**0.1781**	6.90%
	NDCG@20	0.0682	0.0717	0.0943	0.0770	0.0791	0.1260	0.1329	0.1421	0.0793	0.1263	0.1412	0.1422	0.1416	**0.1637**	15.17%
Yelp	Recall@20	0.0554	0.0665	0.0891	0.0820	0.0852	0.1120	0.1174	0.1157	0.0834	0.1126	0.1146	0.1145	0.1148	**0.1292**	12.37%
	NDCG@20	0.0289	0.0403	0.0533	0.0472	0.0496	0.0690	0.0746	0.0750	0.0482	0.0696	0.0732	0.0729	0.0741	**0.0826**	10.26%
Ciao	Recall@20	0.0412	0.0312	0.0517	0.0555	0.0477	0.0591	0.0629	0.0647	0.0528	0.0602	0.0635	0.0642	0.0625	**0.0692**	7.07%
	NDCG@20	0.0310	0.0252	0.0379	0.0426	0.0352	0.0400	0.0483	0.0502	0.0419	0.0468	0.0492	0.0501	0.0494	**0.0522**	3.88%

Table 4. Cold-start recommendation performance comparison.

Dataset	Method	BPR	SBPR	NGCF	LightGCN	DiffNET++	MHCN	HCCF	SGL	SimGCL	NCL	GAFSRec	Improvement
Douban-Book	Recall@20	0.0754	0.1045	0.1135	0.1108	0.1105	0.1755	0.1877	0.1929	0.1924	0.1725	**0.1972**	2.79%
	NDCG@20	0.0388	0.0879	0.0644	0.0631	0.0590	0.1043	0.1098	0.1154	0.1164	0.1024	**0.1233**	5.94%
Yelp	Recall@20	0.0619	0.0696	0.0881	0.0881	0.0917	0.1258	0.1271	0.1322	0.1212	0.1150	**0.1330**	5.89%
	NDCG@20	0.0296	0.0427	0.0415	0.0402	0.0420	0.0610	0.0662	0.0709	0.0625	0.0571	**0.0669**	11.57%
Ciao	Recall@20	0.0416	0.0413	0.0576	0.0544	0.0519	0.0638	0.0638	0.0605	0.0627	0.0635	**0.0731**	14.51%
	NDCG@20	0.0209	0.0251	0.0330	0.0324	0.0308	0.0340	0.0331	0.0338	0.0339	0.0338	**0.0374**	9.82%

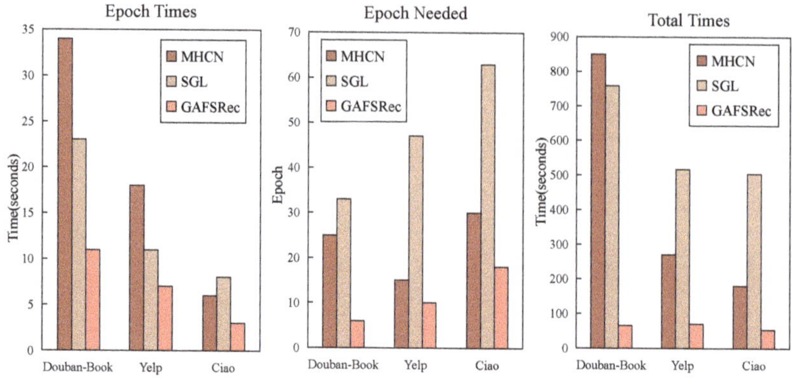

Figure 3. The training speeds of the compared methods.

As shown in Figure 4, LightGCN tended to recommend popular items and had the lowest recommendation ability for long-tail items (as illustrated in the small figure). Due to sparse interaction signals, it was difficult for LightGCN to obtain high-quality representations of long-tail items. However, with SimGCL, by optimizing the consistency of InfoNCE-loss learning representations, long-tail problems could be avoided as much as was possible, and excellent recommendation performance could be achieved. MHCN also optimized the InfoNCE loss through global and local comparisons but did not learn a more consistent representation, resulting in inferior performance to SimGCL when recommending long-tail items. In contrast, GAFSRec showed outstanding advantages when recommending long-tail products (such as GroupIDs 1, 2, and 3) and the highest recall value, but it was not as good as other models for GroupID 10. It can be seen that learning a more uniform representation by optimizing the InfoNCE loss can enable models to debias and alleviate the long-tail phenomenon, as well as increasing freshness and helping to meet user needs.

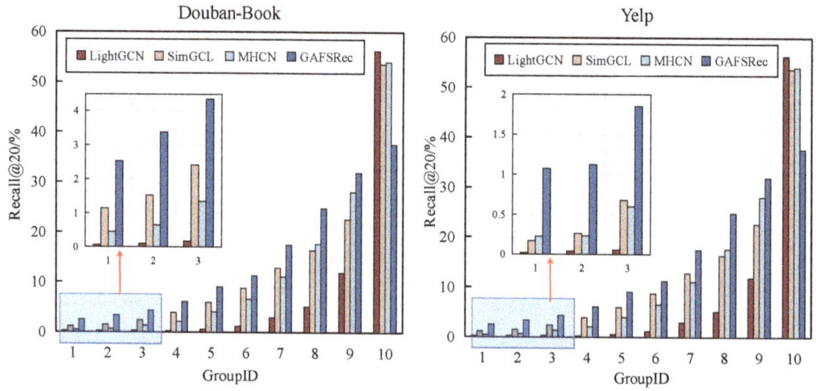

Figure 4. The ability to promote long-tail items.

4.3. *Ablation Study*

4.3.1. Investigation of Multi-Graph Setting

To investigate the influence of high-order relationships in user social networks on recommendation, we first investigated the impact of individual views on recommendations by removing any one of the three social relationships' views and leaving two remaining. As can be seen from Figure 5, removing any view resulted in performance degradation.

The bars in the figure (except the complete bar) represent the cases with the corresponding views removed, while the complete bar represents the complete model.

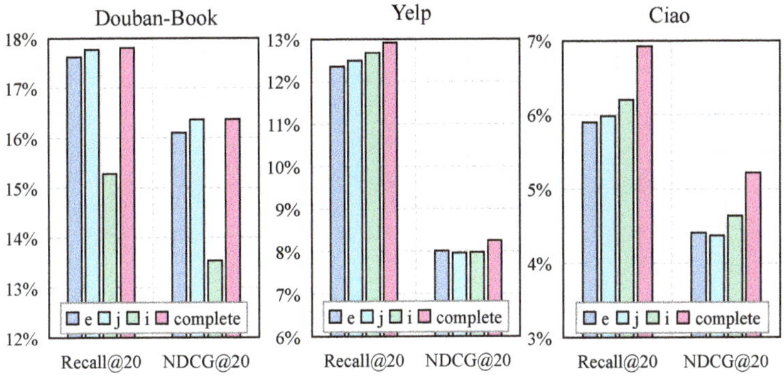

Figure 5. Contributions of each graph with different datasets.

Obviously, for the Douban-Book dataset, due to the low effective social density of the dataset, the explicit user view had little influence on the recommendation, while the recommendation performance of the view lacking the implicit user dropped sharply, indicating that the implicit user was the user. Items that met user needs were recommended, proving the key role of implicit users for recommendation. For the Yelp and Ciao datasets, due to the higher effective social density of the datasets, explicit user views had a greater impact on the recommendation. Figure 6 visualizes each graph's attention score (median attention score for training set users), revealing that the implicit user had the highest attention score, while the explicit user and the joint graph had low attention scores. According to the findings shown in Figures 5 and 6, implicit users contributed significantly to the analysis of user preferences while explicit users played a greater role when the social density was high, and the joint graph did not necessarily bring greater benefits due to social noise.

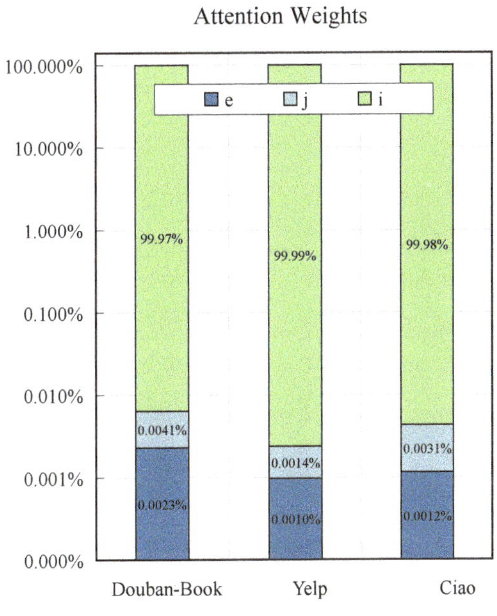

Figure 6. Attention weights for each graph with different datasets.

4.3.2. Investigation of Contrastive Learning Setting

In this study, we investigated the impact of contrastive learning on recommendation through two GCLSRec variants:

1. Removing social recommendation for contrastive learning tasks (without CL);
2. Disabling cross-layer comparison and using the final embedding of each layer embedding to add uniform noise and construct two sets of views for comparison of the learning task recommendations (CL-c).

The evaluation results are reported in Table 5. It can be observed that not adding self-supervised contrastive learning (without CL) reduced the accuracy of recommendation, and when recommending top 40 items, the recommendation performance only improved slightly, as the aggregation mechanism of GCN can obscure high-order connection information. While model recommendation by constructing contrasting views individually (CL-ours) worked well, constructing contrasting views individually (CL-c) led to high model complexity and was time-consuming. However, GAFSRec with cross-layer comparative learning could not only guarantee recommendation accuracy but also had remarkably high recommendation efficiency.

Table 5. Performance comparison of different GCLSRec variants.

Dataset	Douban-Book		Yelp		Ciao	
Metric	**Recall**	**NDCG**	**Recall**	**NDCG**	**Recall**	**NDCG**
			Top 20			
Without CL	15.318%	13.371%	9.926%	6.175%	5.704%	4.265%
CL-c	17.716%	16.716%	12.870%	8.152%	6.644%	4.942%
CL-ours	**17.806%**	**16.370%**	**12.922%**	**8.256%**	**6.922%**	**5.219%**
			Top 40			
Without CL	15.958%	11.379%	14.123%	7.147%	8.633%	5.402%
CL-c	23.038%	18.055%	18.178%	9.815%	8.987%	5.737%
CL-ours	**22.943%**	**17.724%**	**18.516%**	**10.045%**	**8.821%**	**5.628%**

4.4. Parameter Sensitivity Analysis

4.4.1. Influence of λ and ε

We observed the effect on the recommendation performance when the combination of λ and ε was changed. When λ was [0, 0.01, 0.05, 0.1, 0.2, 0.5, 1], ε was [0, 0.01, 0.05, 0.1, 0.2, 0.5, 1], and the number of model layers was set to two.

As shown in Figure 7, when fixing ε at 0.1, all parameters showed similar trend changes, and the best performance was achieved at specific λ values (Douban-Book dataset $\lambda = 0.1$, Yelp dataset $\lambda = 0.05$, Ciao dataset $\lambda = 0.01$). However, when λ was too large ($\lambda = 0.5, 1$), a significant performance drop could be seen.

We fixed λ at the best values for the three datasets, as reported in Figure 8, and then adjusted ε to observe the performance change. When $\varepsilon = 0.01$, the best recommendation performance (Douban-Book dataset and Yelp dataset) was achieved. However, when ε was too large or too small, the recommendation was especially vulnerable to changes in ε, so the performance declined faster. What can be seen from this is that GAFSRec was more sensitive to changes in λ, and with $\varepsilon = 0.01$, the model could maintain stable performance.

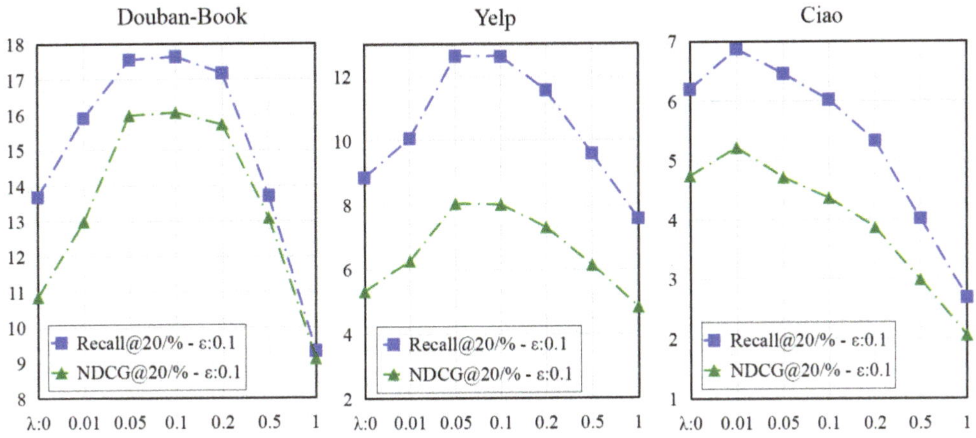

Figure 7. The influence of λ.

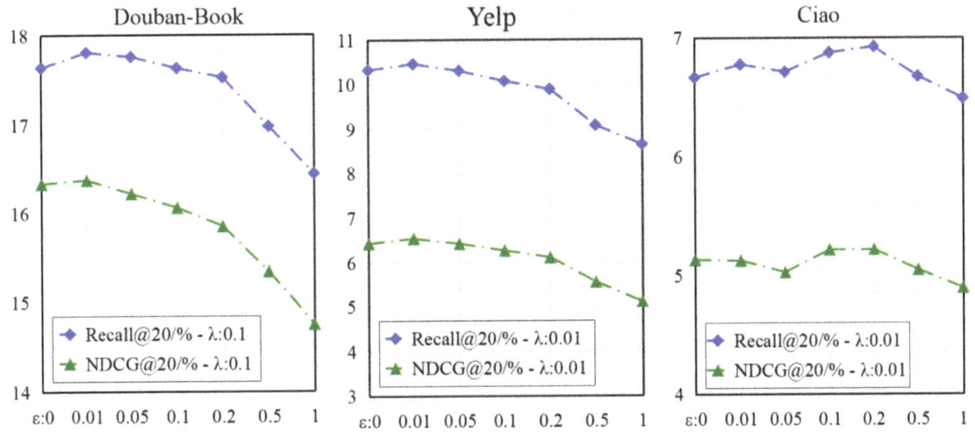

Figure 8. The influence of ε.

4.4.2. Layer Selection for Contrasting

For the recommendations using cross-layer contrastive learning, it is necessary to choose one of the embeddings of one layer and the final embedding for the comparative learning. To explore the optimal model depth and compare the choice of layers, we stacked graph convolution depths from one to five layers. As shown in Table 6, the performance was the best when the model had a depth of two layers. The performance of GAFSRec dropped for all datasets as the number of layers increased. The reason may have been that, as the depth increased, it became more likely to encounter the problem of over-smoothing. In particular, the performance was best only when the first layer of embeddings was learned in contrast to the final embedding. Therefore, the model under consideration could directly select the first layer to compare with the final embedding, making it possible to achieve better results.

Table 6. Influence of the depth of GCLSRec.

Datasets		Douban-Book		Yelp		Ciao	
Metric		Recall@20	NDCG@20	Recall@20	NDCG@20	Recall@20	NDCG@20
One layer	1	17.570%	15.866%	12.739%	8.127%	6.647%	4.982%
Two layers	1	**17.806%**	**16.370%**	**13.002%**	**8.272%**	**6.922%**	**5.219%**
	2	17.001%	15.660%	12.280%	8.153%	6.675%	5.080%
Three layers	1	17.557%	15.915%	12.729%	8.137%	6.550%	4.986%
	2	17.392%	15.832%	12.671%	8.090%	6.494%	5.006%
	3	16.170%	14.745%	12.202%	7.846%	6.456%	4.941%
Four layers	1	17.035%	15.432%	12.427%	7.966%	6.303%	4.877%
	2	16.900%	15.417%	12.385%	7.888%	6.217%	4.856%
	3	15.835%	14.482%	11.860%	7.624%	5.998%	4.634%
	4	15.276%	13.807%	11.211%	7.084%	6.159%	4.783%
Five layers	1	16.761%	15.037%	12.057%	7.676%	6.129%	4.714%
	2	16.724%	15.137%	11.985%	7.636%	6.035%	4.685%
	3	15.687%	14.210%	11.490%	7.324%	5.911%	4.703%
	4	15.132%	13.685%	10.962%	6.952%	5.924%	4.664%
	5	14.121%	12.245%	10.634%	6.640%	5.957%	4.623%

5. Conclusions

In this paper, we proposed a graph-contrastive social recommendation model (GAFS-Rec) for ranking predictions. To fully mine high-order user relationships, the social graph was divided into multiple views, which were modeled using a hypergraph encoder to improve social recommendations. In particular, we presented a method of cross-layer comparative learning to help maintain the consistency of user preferences. Our experiments showed that implicit users outperformed datasets with sparse explicit social relations, and GAFSRec outperformed state-of-the-art baselines using three real datasets. Here, we only considered incorporating the trust relationships between friends in a social network into the recommendations. In the real world, however, a social graph with attributes could better reflect the relationships between users and products. Therefore, exploring social recommendations with attributes will be our next research direction.

Author Contributions: Conceptualization, N.X.; methodology, N.X. and X.M.; software, L.W.; validation, H.L.; formal analysis, X.T.; investigation, H.L.; resources, L.W.; data curation, X.M.; writing—original draft preparation, X.M.; writing—review and editing, X.T.; visualization, H.L.; supervision, N.X.; funding acquisition, N.X. All authors have read and agreed to the published version of the manuscript.

Funding: This research was funded by the Natural Science Foundation of Chongqing Province of China, grant number CSTB2022NSCQ-MSX0786, and the Science and Technology Research Project of Chongqing Education Commission, grant number KJQN202001118.

Institutional Review Board Statement: Not applicable.

Informed Consent Statement: Informed consent was obtained from all subjects involved in the study.

Data Availability Statement: The datasets used and analyzed during the present study are available from the corresponding author upon reasonable request.

Conflicts of Interest: The authors declare no conflict of interest.

References

1. McPherson, M.; Smith-Lovin, L.; Cook, J.M. Birds of a feather: Homophily in social networks. *Annu. Rev. Sociol.* **2001**, *27*, 415–444. [CrossRef]
2. Cialdini, R.B.; Goldstein, N.J. Social influence: Compliance and conformity. *Annu. Rev. Psychol.* **2004**, *55*, 591–621. [CrossRef] [PubMed]
3. Sarwar, B.M. *Sparsity, Scalability, and Distribution in Recommender Systems*; University of Minnesota: Minneapolis, MN, USA, 2001.
4. Chen, J.; Dong, H.; Wang, X.; Feng, F.; Wang, M.; He, X. Bias and debias in recommender system: A survey and future directions. *ACM Trans. Inf. Syst.* **2020**, *41*, 1–39. [CrossRef]
5. Tang, J.; Gao, H.; Liu, H. mTrust: Discerning multi-faceted trust in a connected world. In Proceedings of the Fifth ACM International Conference on Web Search and Data Mining, Seattle, WA, USA, 8–12 February 2012; pp. 93–102.
6. Gao, C.; Lei, W.; Chen, J.; Wang, S.; He, X.; Li, S.; Li, B.; Zhang, Y.; Jiang, P. CIRS: Bursting Filter Bubbles by Counterfactual Interactive Recommender System. *arXiv* **2022**, arXiv:2204.01266.
7. Wang, X.; He, X.; Wang, M.; Feng, F.; Chua, T.-S. Neural graph collaborative filtering. In Proceedings of the 42nd International ACM SIGIR Conference on Research and Development in Information Retrieval, Paris, France, 21–25 July 2019; pp. 165–174.
8. Wu, J.; Wang, X.; Feng, F.; He, X.; Chen, L.; Lian, J.; Xie, X. Self-supervised graph learning for recommendation. In Proceedings of the 44th International ACM SIGIR Conference on Research and Development in Information Retrieval, Virtual Event, Canada, 11–15 July 2021; pp. 726–735.
9. Yu, J.; Yin, H.; Li, J.; Wang, Q.; Hung, N.Q.V.; Zhang, X. Self-supervised multi-channel hypergraph convolutional network for social recommendation. In Proceedings of the Web Conference 2021, Ljubljana, Slovenia, 19–23 April 2021; pp. 413–424.
10. Yu, J.; Yin, H.; Xia, X.; Chen, T.; Cui, L.; Nguyen, Q.V.H. Are graph augmentations necessary? simple graph contrastive learning for recommendation. In Proceedings of the 45th International ACM SIGIR Conference on Research and Development in Information Retrieval, Madrid, Spain, 11–15 July 2022; pp. 1294–1303.
11. Xia, L.; Xu, Y.; Huang, C.; Dai, P.; Bo, L. Graph meta network for multi-behavior recommendation. In Proceedings of the 44th International ACM SIGIR Conference on Research and Development in Information Retrieval, Virtual Event, Canada, 11–15 July 2021; pp. 757–766.
12. Huang, C.; Chen, J.; Xia, L.; Xu, Y.; Dai, P.; Chen, Y.; Bo, L.; Zhao, J.; Huang, J.X. Graph-enhanced multi-task learning of multi-level transition dynamics for session-based recommendation. *Proc. AAAI Conf. Artif. Intell.* **2021**, *35*, 4123–4130. [CrossRef]
13. Wang, J.; Ding, K.; Hong, L.; Liu, H.; Caverlee, J. Next-item recommendation with sequential hypergraphs. In Proceedings of the 43rd International ACM SIGIR Conference on Research and Development in Information Retrieval, Virtual Event, China, 25–30 July 2020; pp. 1101–1110.
14. Xia, L.; Huang, C.; Xu, Y.; Zhao, J.; Yin, D.; Huang, J. Hypergraph contrastive collaborative filtering. In Proceedings of the 45th International ACM SIGIR Conference on Research and Development in Information Retrieval, Madrid, Spain, 11–15 July 2022; pp. 70–79.
15. Creswell, A.; White, T.; Dumoulin, V.; Arulkumaran, K.; Sengupta, B.; Bharath, A.A. Generative adversarial networks: An overview. *IEEE Signal Process. Mag.* **2018**, *35*, 53–65. [CrossRef]
16. Kingma, D.P.; Welling, M. Auto-encoding variational bayes. *arXiv* **2013**, arXiv:1312.6114.
17. Yu, J.; Yin, H.; Xia, X.; Chen, T.; Li, J.; Huang, Z. Self-supervised learning for recommender systems: A survey. *arXiv* **2022**, arXiv:2203.15876.
18. Zhao, H.; Xu, X.; Song, Y.; Lee, D.L.; Chen, Z.; Gao, H. Ranking users in social networks with motif-based pagerank. *IEEE Trans. Knowl. Data Eng.* **2019**, *33*, 2179–2192. [CrossRef]
19. Page, L.; Brin, S.; Motwani, R.; Winograd, T. *The PageRank Citation Ranking: Bringing Order to the Web*; Stanford InfoLab: Stanford, CA, USA, 1999.
20. He, X.; Deng, K.; Wang, X.; Li, Y.; Zhang, Y.; Wang, M. Lightgcn: Simplifying and powering graph convolution network for recommendation. In Proceedings of the 43rd International ACM SIGIR Conference on Research and Development in Information Retrieval, Virtual Event, China, 25–30 July 2020; pp. 639–648.
21. Feng, Y.; You, H.; Zhang, Z.; Ji, R.; Gao, Y. Hypergraph neural networks. *Proc. AAAI Conf. Artif. Intell.* **2019**, *33*, 3558–3565. [CrossRef]
22. Vaswani, A.; Shazeer, N.; Parmar, N.; Uszkoreit, J.; Jones, L.; Gomez, A.N.; Kaiser, Ł.; Polosukhin, I. Attention is all you need. *Adv. Neural Inf. Process. Syst.* **2017**, *30*, 6000–6010.
23. Rendle, S.; Freudenthaler, C.; Gantner, Z.; Schmidt-Thieme, L.B. Bayesian personalized ranking from implicit feedback. In Proceedings of the Uncertainty in Artificial Intelligence, Montreal, QC, Canada, 18–21 June 2009; pp. 452–461.
24. Van den Oord, A.; Li, Y.; Vinyals, O. Representation learning with contrastive predictive coding. *arXiv* **2018**, arXiv:1807.03748.
25. Yu, J.; Yin, H.; Gao, M.; Xia, X.; Zhang, X.; Viet Hung, N.Q. Socially-aware self-supervised tri-training for recommendation. In Proceedings of the 27th ACM SIGKDD Conference on Knowledge Discovery & Data Mining, Virtual Event, Singapore, 14–18 August 2021; pp. 2084–2092.
26. Yin, H.; Wang, Q.; Zheng, K.; Li, Z.; Yang, J.; Zhou, X. Social influence-based group representation learning for group recommendation. In Proceedings of the 2019 IEEE 35th International Conference on Data Engineering (ICDE), Macao, China, 8–11 April 2019; pp. 566–577.
27. Tao, Y.; Li, Y.; Zhang, S.; Hou, Z.; Wu, Z. Revisiting Graph based Social Recommendation: A Distillation Enhanced Social Graph Network. In Proceedings of the ACM Web Conference 2022, Lyon, France, 25–29 April 2022; pp. 2830–2838.

28. Zhao, T.; McAuley, J.; King, I. Leveraging social connections to improve personalized ranking for collaborative filtering. In Proceedings of the 23rd ACM International Conference on Information and Knowledge Management, Shanghai, China, 3–7 November 2014; pp. 261–270.
29. Wu, L.; Li, J.; Sun, P.; Hong, R.; Ge, Y.; Wang, M. Diffnet++: A neural influence and interest diffusion network for social recommendation. *IEEE Trans. Knowl. Data Eng.* **2020**, *34*, 4753–4766. [CrossRef]
30. Lee, D.; Kang, S.; Ju, H.; Park, C.; Yu, H. Bootstrapping user and item representations for one-class collaborative filtering. In Proceedings of the 44th International ACM SIGIR Conference on Research and Development in Information Retrieval, Virtual Event, Canada, 11–15 July 2021; pp. 317–326.
31. Lin, Z.; Tian, C.; Hou, Y.; Zhao, W.X. Improving Graph Collaborative Filtering with Neighborhood-Enriched Contrastive Learning. In Proceedings of the ACM Web Conference 2022, Lyon, France, 25–29 April 2022; pp. 2320–2329.

applied sciences

MDPI

Article

Deep-Learning Based Algorithm for Detecting Targets in Infrared Images

Lifeng Yang [1], Shengzong Liu [2,*] and Yiqi Zhao [3,*]

1 School of Optical and Communication Engineering, Yunnan Open University, Kunming 650000, China; yanglifeng@ynou.edu.cn
2 School of Information Technology and Management, Hunan University of Finance and Economics, Changsha 410000, China
3 Computer Science and Engineering, Central South University, Changsha 410000, China
* Correspondence: lsz@hufe.edu.cn (S.L.); zhaoyiqi@csu.edu.cn (Y.Z.)

Abstract: Infrared image target detection technology has been one of the essential research topics in computer vision, which has promoted the development of automatic driving, infrared guidance, infrared surveillance, and other fields. However, traditional target detection algorithms for infrared images have difficulty adapting to the target's multiscale characteristics. In addition, the accuracy of the detection algorithm is significantly reduced when the target is occluded. The corresponding solutions are proposed in this paper to solve these two problems. The final experiments show that this paper's infrared image target detection model improves significantly.

Keywords: infrared image; deep learning; neural network; target detection; transfer learning; multiscale characteristics; context analysis

Citation: Yang, L.; Liu, S.; Zhao, Y. Deep-Learning Based Algorithm for Detecting Targets in Infrared Images. *Appl. Sci.* **2022**, *12*, 3322. https://doi.org/10.3390/app12073322

Academic Editor: Giacomo Fiumara

Received: 18 January 2022
Accepted: 23 March 2022
Published: 24 March 2022

1. Introduction

Infrared image target detection identifies and labels each target class from an infrared image containing multiple targets. Infrared images consist of information about the thermal radiation emitted by the target and are not susceptible to environmental influences. Therefore, infrared images have advantages over visible images in low-visibility environments, such as night scenes, haze, rain, snow, and dust. In recent years, IoT technologies such as nighttime intrusion warning systems have cited infrared images based on this advantage [1].

Target detection algorithms, in general, can be divided into two categories: traditional target detection algorithms based on image processing and machine learning and new target detection algorithms based on deep learning. Traditional infrared image target detection algorithms include edge detection, module matching, Hough transform, etc. Some target detection algorithms use edges, contours, and textures for target detection. Dalal et al. proposed using gradient direction histograms to detect HOG features of pedestrians [2]. They divided the image and obtained the directional histogram of the gradient edges of each pixel point in each region. The combined directional histogram was used as a feature representation for each area. Papageorgiou et al. proposed using Haar wavelet features for target detection, calculating the pixel values in adjacent rectangles obtained from the detection window and their differences and then using the differences to classify each region in the image [3]. Wu et al. proposed to detect pedestrians using Edgelet features and obtained high target detection performance [4]. Traditional target detection algorithms extract features manually for images. These features rely on a priori knowledge and have limited expressiveness, limiting the accuracy of target detection algorithms.

In recent years, with the rapid development of deep learning, many deep-learning algorithms have been applied to the field of computer vision. Deep-learning-based target detection algorithms have been proposed one after another. Compared with traditional target detection algorithms that use manual feature extraction, deep-learning-based target detection algorithms can self-extract features, which do not require a priori knowledge and have the more expressive power of the extracted features. This is more beneficial to improve the performance of target detection models. In 2014, Girshick applied the regional convolutional neural network to target detection and proposed the R-CNN model [5]. This model is an essential milestone in deep-learning-based target detection algorithms. R-CNN first uses the Selective Search method to extract about 2000 candidate regions, then uses CNN to remove features from the stretched candidate regions, and finally uses support vector machine SVM to classify these features and box regression. In 2015, Girshick proposed a faster Fast R-CNN based on R-CNN [6]. Unlike the computational process of R-CNN, Fast R-CNN first convolves the whole image to get the feature map and then combines the two steps of candidate region classification and frame regression for training so that the computation speed is faster.

Neither R-CNN nor Fast R-CNN solves the problem of relying on the selective search algorithm in the candidate region generation phase, which causes a very time-consuming pain, so Ren et al. proposed the Faster R-CNN model. Faster R-CNN introduces a Region Proposal Network (RPN), which extracts candidate regions directly on the feature map output from the convolutional neural network, significantly improving the detection speed of the target detection model. Then, Bell proposed the ION model based on the Faster R-CNN model [7]. This model uses spatial recurrent neural networks to combine contextual features and the output of the features from different convolutional layers and uses them as multiscale features for target detection.

The above studies mainly focus on target detection in visible images. However, deep learning in target detection research of infrared images is not yet common. Inspired by the idea of transfer learning, this paper migrates the target detection algorithm on visible images to the infrared image target detection field. Firstly, we propose a target detection model CMF Net to solve the problem of the existence of target multiscale features. The CMF Net model is based on the VGG16 network (a convolutional neural network) and uses two multiscale feature extraction mechanisms for image feature extraction and fusion. This makes the final feature map input from the backbone network to the classification network contain low-level visual features that facilitate target localization and high-level semantic features that enable target recognition. Secondly, to solve the problem of low detection accuracy of the algorithm when the target is occluded, we propose the CMF-3DLSTM model. The model improves the classification network into a 3D long- and short-term memory network based on the CMF Net model. We use an attention mechanism to assign weights to the contextual features extracted in different dimensions. Finally, target detection features include multiscale features and contextual features to achieve the fusion of spatio-temporal features.

The rest of this paper is organized as follows: Section 2 summarizes the infrared image target detection algorithm-related work. Section 3 introduces the details of the CMF Net model. Section 4 introduces the structure and details of the CMF-3DLSTM model in detail. Section 5 describes the design and results of relevant experiments. Section 6 summarizes the work of this paper.

2. Related Work

2.1. *Target Detection Framework Based on Deep Learning*

Target detection aims to locate and identify each target instance using a bounding box. Traditional target detection algorithms include edge detection [8], module matching [9], Hough transform [10], etc. These target detection algorithms use edges, contours, and textures for target detection. These features rely on a priori knowledge and have limited expressiveness, limiting the accuracy of target detection algorithms. With the rapid development of deep learning, the field of computer vision has achieved remarkable success in target detection tasks using deep-learning algorithms [11]. Deep-learning-based target detection frameworks have also been proposed one after another [12,13].

Target detection frameworks based on deep learning mainly fall into two categories: two-level detection framework and single-level detection framework [14]. The two-level detection framework includes a pre-processing step for region recommendations. That is, candidate regions are selected and then classified. Such representatives include Faster R-CNN and Mask R-CNN [15], etc. They adopt the R-CNN proposed by Girshick et al. [16] as the target suggestion method [17], which significantly reduces the amount of calculation compared with the traditional method. The conventional method usually uses superpixel, edge [18], and shape to score Windows, containing objects to generate region suggestion boxes. Deepbox [19] used a lightweight ConvNet model for training, which rearranged the regional suggestion boxes generated by the Edge box. Compared with R-CNN, these traditional methods often require a more extensive calculation. The single-stage detection framework adopts the regression method to directly regress the position and type of the target from the feature graph. The grid method or convolution of different scales is used to operate the feature graph to obtain the position and classification information of the target directly. Such representatives include YOLO and SSD. Generally speaking, the two-stage detection frame has higher accuracy, and the single-pole detection frame is faster. In order to improve the performance of CMF-3DLSTM, we use an infrared image target detection model based on multiscale feature fusion and context analysis proposed in this paper, and we adopt the target detection framework of Faster R-CNN as its basic framework.

2.2. *Transfer Learning*

The main idea of transfer learning is to transfer labeled data or knowledge structures from related domains to accomplish or improve the learning of the target domain or task. One of the main assumptions in traditional machine learning algorithms is that training and test data must be in the same feature space and have the same distribution. Transfer learning relaxes the basic assumption that training and test data may be in different feature spaces or follow other data distribution [20]. Specifically for the target detection task of this paper, the task of the source domain is defined as the target detection based on the sizeable visible dataset ImageNet, and the task of the target domain is defined as the target detection based on the small infrared dataset FLIR. Since both infrared imaging and visible imaging are similar, they collect target information for imaging through optical systems. The migration learning approach can be used to initialize the parameters of the infrared image target detection model with the pre-trained model on the visible image dataset. Eventually, the model can be fine-tuned and trained using the infrared image dataset.

2.3. *Cross-Layer Connection Mechanism*

The cross-layer connection mechanism is a classical idea of direct routing from the lower to the higher, ignoring the middle layer. The specific details of the cross-layer connection method vary in different models. A cross-layer connection mechanism is proposed in this paper to solve the problem of multiscale feature detection in images. This method implements two multiscale feature extraction mechanisms and feature fusion mechanisms, which can adapt to the multiscale features of the target and improve the target detection performance of the model. The cross-layer connection mechanism used in this paper is closest to the pedestrian target detection method [21]. In contrast, the two multiscale feature extraction mechanisms proposed in this paper use parameter sharing to process the feature images output from the first, third, and fifth convolution layers in different ways. To keep the resolution consistent, we took the resolution of the feature graphs output by the third convolution layer and the fifth convolution layer as the benchmark, adjusted the resolution of the feature graphs output by other convolution layers, and finally realized the cross-layer connection.

3. CMF Net

This section introduces the CMF Net target detection model in detail. Its innovation lies in using multiscale feature extraction mechanisms of parameter sharing to extract multiscale feature information and carry out feature fusion.

3.1. Network Structure of CMF Net

The network structure of CMF Net is shown in Figure 1, which consists of four parts: backbone network, region proposal network RPN, ROI pooling layer, and classification network. The first part is the backbone network, which mainly adopts the migration learning method, two multiscale feature extraction mechanisms, and a feature fusion mechanism. The output feature map contains both low-level visual features and high-level semantic features. The second part is the region proposal network RPN [22], which is mainly used to extract the region proposal frames containing the target for the feature map output from the backbone network and filter out about 300 high-scoring regions proposal frames. The third part is the ROI pooling layer, mainly used to map ROI regions to convolutional regions and pool them into feature maps of fixed size. The fourth part is the classification network, which is used mainly for target location correction and classification recognition after mapping the ROI pooling layer and achieving target detection.

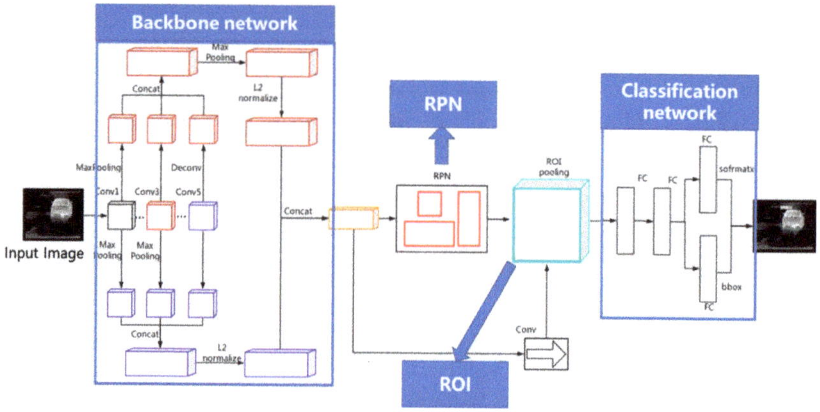

Figure 1. CMF Net architecture.

3.2. Multiscale Feature Extraction I

In the vgg16 network, the visual features extracted from the lower convolutional layer play an important role in the target location, while the semantic features extracted from the higher convolutional layer play an important role in target recognition. The multiscale feature fusion method can retain the low-level visual and high-level semantic features and avoid extracting redundant features from the two adjacent convolutional layers. The first multiscale feature fusion method is shown in Figure 2. Considering that the intermediate convolutional layer contains visual and semantic features, which combine the two, it is essential for target detection. Therefore, the feature map output by the third convolutional layer is retained completely, and the resolution of the feature map is taken as the benchmark. The feature map extracted from the first convolutional layer is divided into two pools, and the feature map extracted from the fifth convolutional layer is deconvoluted [23]. This can further study the feature map output of the first layer and the fifth volume layer, resolve the feature maps of the low-, middle-, and high-volume outputs that are adjusted to the same level, and, finally, connect them to achieve feature fusion.

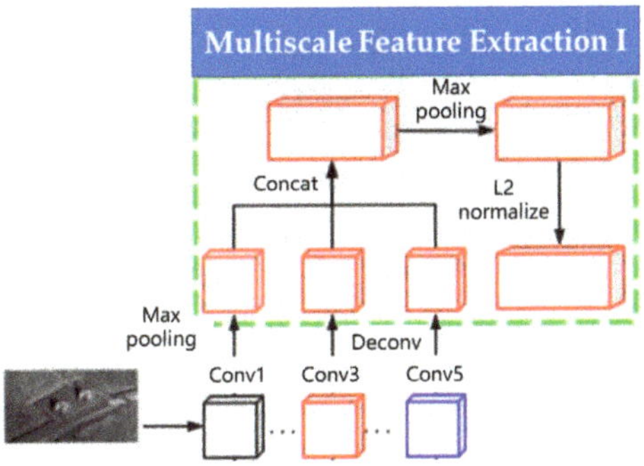

Figure 2. Multiscale feature extraction I.

3.3. Multiscale Feature Extraction II

In the first multiscale feature extraction mechanism, the feature graph output from the third convolution layer is retained completely. Then, based on the resolution of the feature images output from the third layer, the feature images output from the first convolution layer are pooled, and the feature images output from the fifth convolution layer is deconvolution processed. This makes the feature images output by the first, third, and fifth convolution layers maintain the exact resolution. However, this processing method loses the high-level semantic features learned by the fifth convolution layer to some extent, which affects the accuracy of the target detection model. Therefore, a second multiscale feature extraction method is proposed, whose structure is shown in Figure 3. The feature map output from the fifth convolutional layer is retained completely, and the resolution of the feature map is taken as the benchmark. The feature map extracted from the first convolutional layer is pooled twice. The feature map extracted from the third convolutional layer is pooled once. The resolution of the feature map output from the first, third, and fifth layers is adjusted to be the same and connected to achieve feature fusion.

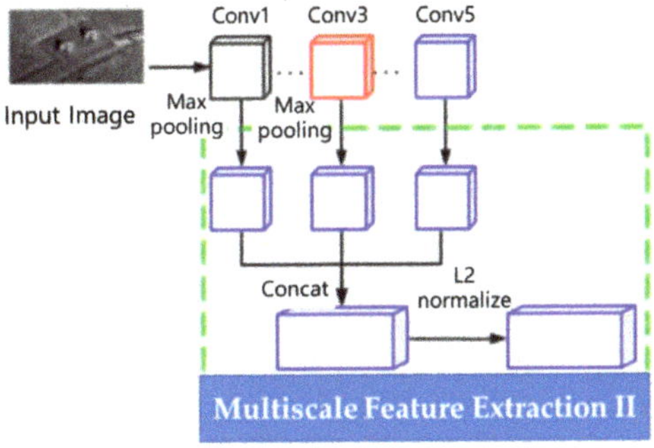

Figure 3. Multiscale feature extraction II.

3.4. Feature Fusion Strategy

Two multiscale feature extraction methods use different strategies to extract multiscale feature maps. The final output of the feature map has its unique advantages. The first multiscale feature extraction method ultimately retains the mixed features of target location and target recognition extracted from the middle convolutional layer. Still, it loses some high-level semantic features and affects target recognition. The second multiscale feature extraction method preserves the semantic features of a high-level convolutional layer but loses some mixed features, which affects the target location. Therefore, it is necessary to fuse the two kinds of feature maps so that the final output feature map simultaneously contains rich mixed features and semantic features.

The final feature maps obtained by the multiscale feature extraction mechanism suffer from inconsistent resolution and inconsistent amplitude of feature values. Therefore, these feature maps cannot be directly fused with features. The feature fusion method proposed in this paper makes the feature maps output by the first multiscale feature extraction mechanism consistent with the resolution of the feature maps output by the second multiscale feature extraction mechanism through a pooling process. Secondly, L2 normalization is performed in the feature maps outputted by the two feature extraction methods, so that the amplitudes of feature values in the two feature maps are consistent. Then, we connect two feature maps to get a feature map containing rich visual features, mixed features, and semantic features. Finally, we input it to RPN to extract ROI information.

We assign a binary class label to each box (including the target or excluding the target). We set a binary class label (include target or not) to each box. We assign a positive title to a box with an IoU threshold higher than 0.7 with any ground truth box and then assign a negative label to a box with an IoU threshold lower than 0.3 with all ground truth boxes. Our goal is to minimize a multitask loss function.

$$L(k,k^*,t,t^*) = L_{cls}(k,k^*) + \lambda L_{reg}(t,t^*) \tag{1}$$

L_{cls} is the classification loss, L_{reg} is the coordinate regression loss of the box with a positive label assigned. k^* and k are true to label and predicted labels separately, respectively. $L_{reg}(t,t^*) = R(t-t^*)$ where R is the smoothed loss function defined in Faster R-CNN. We express the coordinates of the positive box as $t = (t_x, t_y, t_w, t_h)$ and the coordinates of the predicted box as $t^* = \left(t_x^*, t_y^*, t_w^*, t_h^*\right)$.

$$\begin{aligned} t_x &= (G_x - P_x)/P_w \cdot t_y = (G_y - P_y)/P_h \\ t_w &= \log(G_w - P_w) \cdot t_h = \log(G_h - P_h) \end{aligned} \tag{2}$$

where $P^i = (P_x, P_y, P_w, P_h)$ specifies the coordinates of the center point of the predicted box. G^i specifies the coordinates of the center point of the positive box.

The RPN module first resamples the unbalanced sample set of positive and negative samples, using the oversampling method in random sampling to obtain more sample first data balance by randomly repeating examples from a small number of class sample sets. Then, the gradient descent method is used for training, and the classification loss error L_{cls} and regression loss error L_{reg} are back-propagated to update the model parameters until the RPN module converges. The parameters for the training of the RPN module are set as shown in Table 1.

Table 1. RPN training parameters list.

Description	Value
The Anchor scale	32, 64, 128
MiniBatch Quantity	256
PRN foreground–background ratio	1:1
IOU threshold used by NMS for RPN training	0.3
IOU thresholds used by NMS for RPN prediction	0.7

4. CMF-3DLSTM

The infrared image target detection model CMF Net based on multiscale feature fusion proposed in Section 3 adopts two multiscale feature extraction mechanisms and feature fusion methods for the final output feature graph of the backbone network. It adapts to the multiscale characteristics of the target. It can be regarded as the feature fusion of spatial dimension, which is of great help to improve the performance of infrared image target detection. CMF Net can achieve better target detection performance when the background environment is the relatively simple spacing between targets. However, the problem with

CMF Net is that it is easy to misjudge when multiple targets are close together, overlapping, and semantically confusing, as shown in Figure 4.

Figure 4. CMF Net target detection results.

This section proposes CMF-3DLSTM, an infrared image target detection model based on spatio-temporal feature fusion and attention mechanism. This model first inherits the multiscale feature fusion strategy of CMF Net to achieve feature fusion in the spatial dimension. Then, the model is based on 3DLSTM, which extracts contextual information along with the positive and negative directions of each dimension from the length, width, and height dimensions of the 3D feature map. Meanwhile, the model uses an attention mechanism to assign weights to the contextual features extracted in various dimensions and directions. CMF-3DLSTM effectively improves target detection performance in complex situations such as multiple targets approaching each other, overlapping each other, and semantic confusion.

4.1. Network Structure of CMF-3DLSTM

CMF-3DLSTM target detection model includes four modules, namely trunk network, regional proposal network, ROI pooling layer, and classification network, as shown in Figure 5.

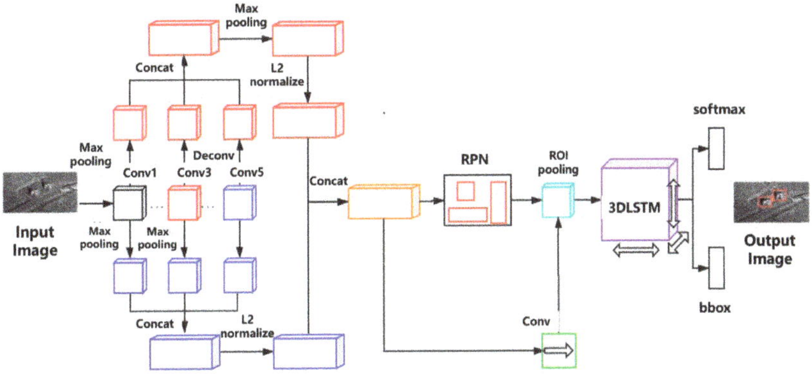

Figure 5. Structure diagram of CMF-3DLSTM model.

The trunk network still adopts a 16-layer VGG16 network. The model file saved after pre-training the target detection model Faster R-CNN on the visible light domain dataset Image Net is used for parameter initialization in CMF-3DLSTM utilizing the idea of transfer learning, and two multiscale feature extraction mechanisms are still used for multiscale feature extraction and feature fusion. The process of generating candidate boxes for regional proposal network RPN is unchanged. It still extracts and screens regional proposal boxes that may contain targets from the feature graph output by the trunk network, and about

300 high-scoring regional proposal boxes are screened. The ROI pooling layer mainly generates a fixed-size feature map based on region proposal mapping of candidate box generated by RPN network recommendation for subsequent classification and regression. The classification network primarily uses the multiscale feature map processed by ROI pooling layer mapping and uses the 3D six-way long- and short-term memory network 3DLSTM, constructed based on BI-LSTM, to extract context information. At the same time, the attention mechanism is used to assign different weights to the context features extracted from other dimensions and directions to achieve the fusion of spatio-temporal features. Finally, the input is given to the classification and location layers for target classification recognition and position correction.

4.2. Context Information Extraction Network

Figure 6 shows a 3D long- and short-term memory network, which can extract context information. The 3DLSTM network firstly transforms the 3D feature image into the 2D feature image. Then, each row in the two-dimensional feature graph is regarded as a vector or a sequence, and each column in the two-dimensional feature graph is considered to be a time step. Finally, the context information of feature map extraction is transformed into the extraction of vector or sequence relations.

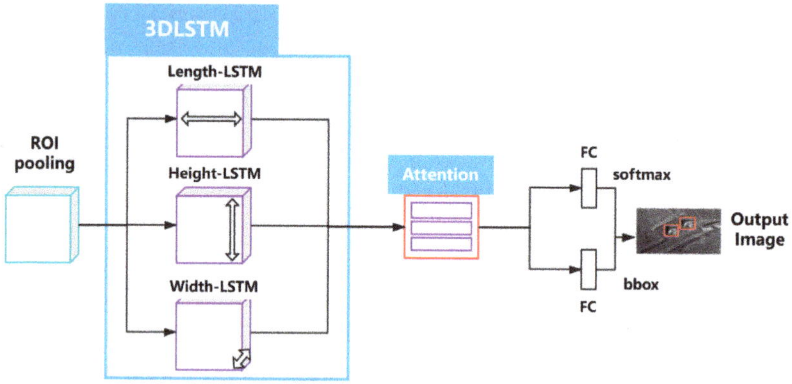

Figure 6. 3DLSTM network structure diagram.

We fix the feature map's length, width, and height separately and stretch the other two directions. In this way, the shape of the 3D feature map can be transformed into a 2D feature map. This two-bit information is then input into Bi-LSTM to extract contextual information along the fixed direction of the original feature map. Taking the length direction of the fixed feature map as an example, the 3D feature map becomes a 2D feature map (length, width × height) after transformation, and the specific transformation process is shown in Figure 7.

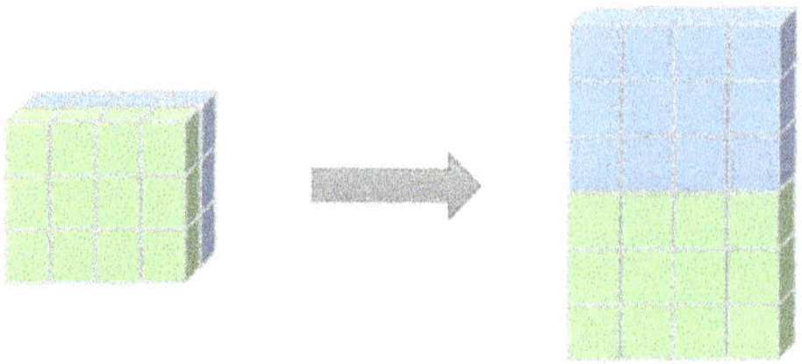

Figure 7. 3D feature-length expansion diagram.

Finally, the features generated are connected. At this time, the feature graph input to the classification regression network contains the context information extracted from the length, width, and height of the original feature graph, so the network is called 3DLSTM. Compared with RNN, LSTM, and BI-LSTM, which extract 2D context information, this network is more conducive to improving the performance of the target detection model.

4.3. Attentional Mechanism

Attention mechanisms are generally used in natural language processing. As the length of text sequence increases in practical applications, the more advanced information in the series is lost more seriously, leading to a significant decline in model performance. A common solution is to input text sequences in both sequential and reverse order or LSTM. Although the two methods can improve the model performance to a certain extent, it is still difficult to effectively solve the problem of a too-long sequence.

The final output size of the feature graph of 3DLSTM is 3256, which adopts bi-LSTM to output vectors with the length of 256 from the three dimensions of length, width, and height of the 3D feature graph, respectively. Each element in the vector represents the neuron's output under a time step. Such a feature map size has the problem of a too-long sequence. Moreover, the weights of the three vectors are different, and the weight of each element in each vector should also be different. Therefore, an attention mechanism is adopted that can selectively screen out a small amount of important information from a large amount of information and focus on this vital information.

A source in attention consists of a series of key–value pairs. The weight coefficient of each key corresponding to value is obtained by calculating the similarity of each key in input vector query and source. Formula (3) is the calculation of the similarity between the query and the key. Formula (4) determines the weight coefficient of each key corresponding to value by the $Soft_{max}$ function.

$$Sim_i(Query, Key_i) = \frac{Query \cdot Key_i}{\|Query\| \cdot \|Key_i\|} \tag{3}$$

$$a_i = Soft_{max}(Sim_i) = \frac{e^{Sim_i}}{\sum_{j=1}^{N} e^{Sim_j}} \tag{4}$$

The weighted sum of values obtains the final attention value according to these weight coefficients. The calculation formula of attention value is as follows:

$$Attention(Query, Source) = \sum_{i=1}^{N} a_i \cdot Value_i \tag{5}$$

Using the attention mechanism, you can assign different weights to the context information collected by 3DLSTM in different directions and to the different elements in the context information in each direction. This enables the target detection model CMF-3DLSTM to pay more attention to the salient features of the target, thus improving the target detection performance of CMF-3DLSTM.

4.4. Model Training Strategy

CMF-3DLSTM uses the same training methods of pre-training migration and model fine-tuning as CMF Net, but the training strategy of the classification module is different. In the classification module, 3DLSTM is based on three bidirectional long- and short-term memory networks, while the network structure of Bi-LSTM is based on LSTM. Therefore, the ultimate goal of 3DLSTM is the same as that of LSTM, which is to minimize a loss function $L(t)$. The specific calculation formula is as follows:

$$L = \sum_{t=1}^{T} l(t) \tag{6}$$

where t represents the current moment, T represents the total time step, and $l(t)$ represents the loss function at the current moment. The calculation formula of $l(t)$ is as follows:

$$l(t) = f(h(t), y(t)) = \|h(t) - y(t)\|^2 \tag{7}$$

$h(t)$ represents the hidden layer output at the current time t, and $y(t)$ represents the output layer output at the present time t. To minimize $l(t)$ loss function, the 3DLSTM network is trained by the gradient descent method. When the error is propagated back, the chain derivative method is used to update the model weight parameters. The specific calculation formula is as follows:

$$\frac{\partial L}{\partial w} = \sum_{t=1}^{T} \sum_{i=1}^{M} \frac{\partial L}{\partial h_i(t)} \cdot \frac{\partial h_i(t)}{\partial w} \tag{8}$$

i represents the memory unit of the hidden layer. M is the number of memory units. w represents the model weight parameter. $h_i(t)$ represents the output of the memory unit in the hidden layer at the current time t. After calculating the gradient of weight parameter w of all models, 3DLSTM uses the gradient descent method to update the parameters iteratively. Finally, it minimizes the loss function L to achieve the purpose of training the classification module.

For infrared image target detection model CMF-3DLSTM, a joint optimization 10-step training process is designed in this paper, as shown in Algorithm 1.

Algorithm 1: CMF-3DLSTM training process

Input: Infrared image dataset.
Output: Target detection model CMF-3DLSTM.
Step 1: Initialize the network parameters in Step2 and Step3 using the pre-training model on the VOC2007 dataset.
Step 2: Use the first multiscale feature extraction mechanism to extract feature information.
Step 3: Using the second multiscale feature extraction mechanism to extract feature information.
Step 4: CMF Net is used to carry out feature fusion for the feature information extracted by Step 2 and Step 3.
Step 5: Train the RPN network to generate the proposals using the characteristic information obtained from Step 4.
Step 6: Implement ROI Pooling of Step 5 and adjust them to the same size.
Step 7: The 3DLSTM network is used to extract the context information of ROI in Step 6.
Step 8: The attention mechanism is used to assign weight to the output of the features by Step 7.
Step 9: Classification layer and regression layer are used for target detection for the output of the features by Ste p8.
Step10: The unified network of Step 5 and Step 9 joint training is taken as the final model.

The training parameters of CMF-3DLSTM, CMF Net, and Faster R-CNN are shown in Table 2, in which Faster R-CNN has 136,708,989 training parameters, and CMF Net has 152,048,765 training parameters. The number of training parameters of CMF-3DLSTMA is 40,761,469, which drops to the level of 10 million and dramatically reduces the space complexity of the algorithm. However, the 3DLSTM network and attention mechanism introduced in the classification module of CMF-3DLSTM is more complex than the fully connected layer in the classification module of CMF Net, thus causing an increase in time complexity. The CMF Net model performs target detection at a speed of about 0.87 s/pc on a machine with a graphics card configuration of GeForce GTX 1080. In the same experimental environment, the Faster R-CNN performs target detection at about 0.75 s/pc, and the CMF-3DLSTMA has a reduced target detection speed of about three s/pc.

Table 2. Experimental environment.

Model	Backbone Network	RPN	ROI Pooling	Classification Network	Total
Faster R-CNN	14,714,688	2,382,893	0	11,961,1408	136,708,989
CMF Net	17,077,824	7,691,309	0	127,279,632	152,048,765
CMF-3DLSTM	17,077,824	7,691,309	0	15,992,336	40,761,469

5. Experiment and Analysis of Experimental Results

5.1. Description of Dataset

We have trained and evaluated our model on FLIR, a public infrared driving image dataset, and achieved excellent results. The introduction of the dataset is shown in Table 3.

Table 3. Dataset specifications.

Content	Synced annotated thermal imagery and non-annotated RGB imagery for reference. Camera centerlines approximately 2 inches apart and collimated to minimize parallax
Images	>10 K from short video segments and random image samples.
Image Capture Refresh Rate	Recorded at 30Hz. Dataset sequences sampled at 2 frames/s or 1 frame/s. Video annotations were performed at 30 frames/s recording.
	10,228 total frames and 9214 frames with bounding boxes.
Frame Annotation Label Totals	1. Person (28,151); 2. Car (46,692); 3. Bicycle (4457); 4. Dog (240); 5. Other vehicle (2228).
Driving Conditions	Day (60%) and night (40%) driving on Santa Barbara, CA area streets and highways from November to May with clear to overcast weather.
Dataset File Format	1. Thermal—14-bit TIFF (no AGC); 2. Thermal—8-bit JPEG (AGC applied) w/o bounding boxes embedded in images; 3. Thermal—8-bit JPEG (AGC applied) with bounding boxes embedded in images for viewing purposes; 4. RGB—8-bit JPEG; 5. Annotations: JSON (MSCOCO format).

5.2. Description of Evaluation

In this paper, the target detection performance of the method on the FLIR infrared driving image dataset is evaluated from mAP (mean average precision), which is widely used as a standard measure in previous target detection research. The calculation formula is as follows:

$$Precision = \frac{TP}{TP + FP} \tag{9}$$

$$MAP(Q) = \frac{1}{|Q|} \sum_{j=1}^{|Q|} \frac{1}{m_j} \sum_{k=1}^{m_j} Precision\left(R_{jk}\right) \tag{10}$$

where TP indicates that the prediction is true and the label is true, FP indicates that the prediction is true and the label is false. Q is the set of target categories to be detected, m_j is the number of pictures of all categories corresponding to Q_j, R_{jk} is the set of all pictures in the returned result until picture k is found. That is to say, the corresponding precision is calculated in this set.

5.3. Experimental Analysis

The training set of Infrared image dataset FLIR contains 7860 IR images, and the test set has 1360 Infrared images. To facilitate the experiment, we converted the image annotation files of the Infrared image from JSON format (MSCOCO format) to XML format (VOC2007 format).

We conducted a total of four experiments. The first set of experiments tested the performance of the Faster R-CNN target detection model under various network layer combinations. The second group of experiments analyzed the performance comparison between CMF Net (Faster R-CNN model that adopts two multiscale feature extraction mechanisms and carries out feature fusion) and those using two multiscale feature extraction mechanisms alone. The third group of experiments analyzed the performance comparison between CMF Net and other target detection models. The fourth group of experiments analyzed the performance comparison between CMF-3DLSTM (using 3DLSTM network to replace the full connection layer in CMF Net) and other target detection models.

(1) *Experiment I*: The performance of the target detection model depends mainly on whether the feature map contains rich features or not. To investigate which network layers and network layer combinations can make the model the best performance, we conduct seven sets of tests based on the Faster R-CNN target detection model. The final target detection performance of the feature maps output by convolutional layer 1 (single 1), convolutional layer 3 (single 2), and convolutional layer 5 (single 3) are first tested separately. Then, the target detection is performed for the feature maps output by the convolutional layer combination 1+2+3 (Group 1) and 3+4+5 (Group 2), respectively. Finally, the target detection is performed for the feature maps output by the convolutional layer combination 1+3+5 with two different multiscale feature extraction mechanisms (Group 3 and Group 4). The experimental results are shown in Table 4.

Table 4. Results of combining different convolutional layers.

Layers	Single 1	Single 2	Single 3	Group 1	Group 2	Group 3	Group 4
mAP	0.514	0.605	0.583	0.567	0.618	**0.636**	**0.661**

Experimental results show that the target detection model of convolution layer combination 1+3+5 with two different multiscale feature extraction mechanisms (Group 3 and Group 4) has better detection performance on FLIR. Therefore, the proposed infrared image target detection model uses a 1+3+5 convolution layer combination.

(2) *Experiment II*: To verify the performance of the two multiscale feature extraction mechanisms and CMF Net, we carried out three experiments, and the experimental results are shown in Figure 8. We found that our target detection model CMF Net has a great improvement.

The mAP of CMF Net improved about 6.8% and 4.4% compared to the first multiscale feature extraction mechanism and the second multiscale feature extraction mechanism, respectively. Although the target detection accuracy of CMF Net decreased in the bicycle category, it improved by 6.1% and 24% in the car and person categories, respectively, compared to the first multiscale feature extraction mechanism, and 13.2% and 9.3%, respectively, compared to the second multiscale feature extraction mechanism, CMF Net's accuracy only in the bicycle target. The accuracy of CMF Net is reduced by 9.6% compared to the first multiscale feature extraction mechanism and 9.3% compared to the second multiscale feature extraction mechanism. This experimental result illustrates the importance of feature fusion based on two multiscale feature extraction mechanisms compared to one multiscale feature extraction mechanism alone to improve the performance of the target detection model.

(3) *Experiment III*: To fully prove the correctness of our multiscale feature extraction strategy, we still adopted the idea of transfer learning to migrate the pre-training networks of Faster R-CNN, YOLO, and SSD, which are currently popular in the

visible light domain, to FLIR infrared driving image dataset, continue training until the model converges.

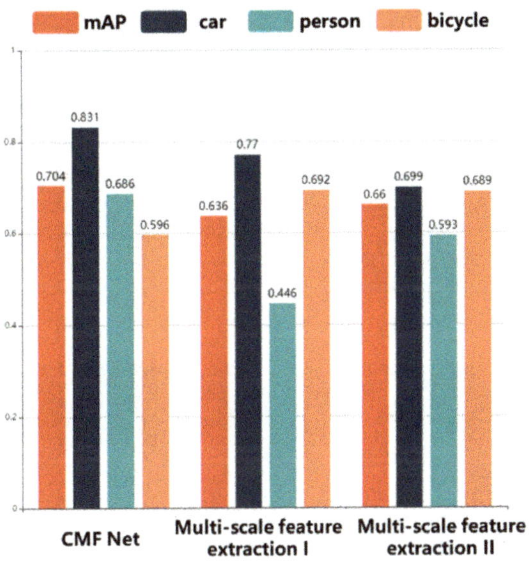

Figure 8. Performance comparison between CMF Net and two multiscale feature extraction mechanisms.

The experimental results are shown in Figure 9 We evaluate the performance of CMF Net on the test set of FLIR infrared driving image dataset. Using the above methods, we get about 71% of mAP by CMF Net, 58% by Faster R-CNN, 65% by YOLO, and 54% by SSD.

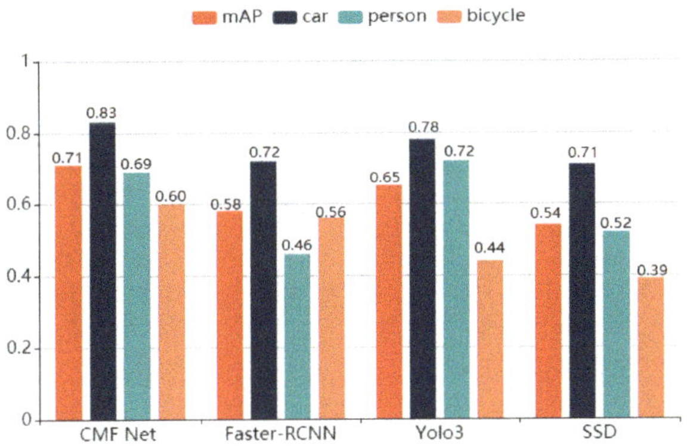

Figure 9. Performance comparison between CMF Net and other target detection models.

Compared with other common target detection model's mAP, our multiscale feature fusion model CMF Net achieved significant improvement in accuracy, about 13 percentage points higher than Faster R-CNN's mAP, about 6 percentage points higher than YOLO's mAP, and about 17 percentage points higher than SSD's mAP.

In FLIR infrared driving image dataset, due to the different shooting distances, the size of the car, person, and bicycle targets in the infrared image is different, which has significant multiscale characteristics. Our model adopts two multiscale feature extraction mechanisms and two-level feature fusion methods, which makes the final output of the backbone network contains rich visual features and semantic features, so the detection accuracy of car, person, and bicycle is far higher than the other three networks. The accuracy of CMF Net on car target is 11%, 5%, and 12% higher than Faster R-CNN, YOLO, and SSD, respectively; the accuracy on person target is 23% and 17% higher than Faster R-CNN and SSD, respectively, and the accuracy on bicycle target is 4%, 16%, and 21% higher than Faster R-CNN, YOLO, and SSD respectively. Compared with YOLO, the accuracy of the personal target is reduced by 3%. The importance of the combination of two multiscale feature extraction mechanisms and two-level feature fusion methods is fully proved.

As shown in Figure 10, the target detection result of CMF Net is on the left, and the target detection result of Faster R-CNN is on the right. The scenes on the left and right are the same, with vehicles and pedestrians appearing on the street at different scales. Faster R-CNN detected most targets in the image well but failed to detect pedestrians appearing at a small scale in the middle of the image. CMF Net can adapt to the multiscale characteristics of the target because it adopts two multiscale feature extraction mechanisms. Therefore, the pedestrians on a small scale can be successfully identified and positioned correctly.

Figure 10. Comparison of target detection results between CMF Net and Faster R-CNN.

(4) *Experiment IV*: CMF Net, a target detection model based on multiscale feature fusion, has a problem: it is easy to cause misjudgment in the complex situation of multi-target detection. In particular, it is challenging to detect CMF Net effectively when multiple targets are close to or even overlapping each other. It is imperative to use the contextual information around the target effectively. This paper proposes an infrared image target detection model CMF-3DLSTM based on multiscale feature fusion and context analysis. CMF-3DLSTM is inherited from CMF Net. The difference between CMF-3DLSTM and CMF Net is that it replaces the complete connection layer of the classification regression network with a 3D long- and short-term memory network. Context information can be extracted based on multiscale feature fusion. CMF-3DLSTM improved target detection performance by about 2.9% on the infrared image dataset FLIR compared to CMF Net's mAP.

The experimental results are shown in Figure 11. The target detection model CMF-3DLSTM and other target detection models are evaluated on the test set of FLIR. Using the above methods, CMF-3DLSTM obtained about 73.3% mAP, while the mAP on CMF Net was about 70.4%, the mAP on Faster R-CNN was about 68.7%, the mAP on YOLO3 was approximately 64.8%, and the mAP on SSD was about 60.8%.

Although the CMF-3DLSTM model does not achieve optimal detection results for car, person, and bicycle alone, the average detection accuracy is more important in complex situations where multiple targets are nearby or even overlap or obscure each other. The target detection results of CMF-3DLSTM on the infrared image dataset FLIR are illustrated in Figure 12.

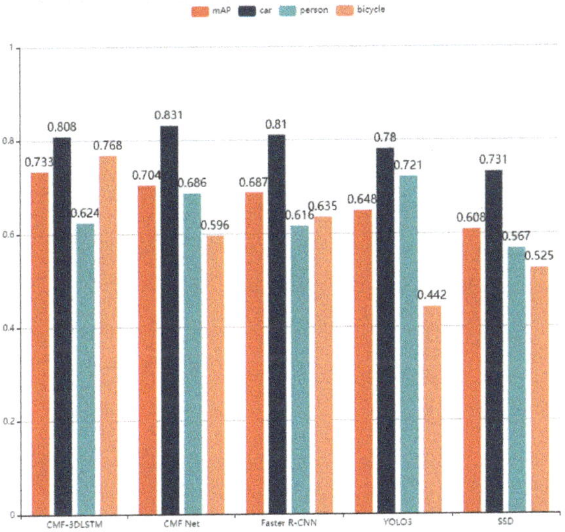

Figure 11. Performance comparison between CMF-3DLSTM and other target detection models.

Figure 12. Comparison of target detection results between CMF-3DLSTM and CMF Net.

6. Conclusions

This paper transfers the target detection model in the visible light domain to the infrared environment by transfer learning. In our work, we creatively proposed the method of feature fusion and multiscale feature extraction with two shared features. We obtained the network architecture of CMF Net that makes full use of multiscale feature fusion information for target detection. Through multiscale feature fusion, rich visual and semantic features can be obtained to improve the accuracy of target detection and adapt to the multiscale characteristics of the target to be detected. To improve the target detection performance of the model in complex scenes such as mutual occlusion and overlapping of multiple targets, we constructed a 3D long- and short-term memory network based on CMF Net to extract context information and finally realized the CMF-3DLSTM model. Compared with Faster R-CNN, YOLO3, and SSD, CMF-3DLSTM achieves higher target detection performance on infrared image dataset FLIR. This proves the importance of constructing an infrared image target detection model based on multiscale feature fusion and context analysis.

However, we still need to make further improvements work. The target detection model proposed in this paper improves the accuracy, reduces the number of parameters, and decreases the spatial complexity compared with models such as Faster-RCNN, but it increases the time complexity, making it challenging to meet the requirements of real-time detection. We need to optimize the network structure further to improve the model's real-time detection capability. At the same time, we need to train and evaluate the model on more infrared image datasets with different scenes to meet the requirements of other scenes in practical applications.

Author Contributions: Data curation, L.Y.; Investigation, S.L.; Methodology, L.Y. and Y.Z.; Resources, S.L.; Supervision, S.L.; Validation, Y.Z.; Writing—review & editing, L.Y., S.L. and Y.Z. All authors have read and agreed to the published version of the manuscript.

Funding: This research received no external funding.

Institutional Review Board Statement: Not applicable.

Informed Consent Statement: Not applicable.

Conflicts of Interest: The authors declare no conflict of interest.

References

1. Tan, P.; Mao, K.; Zhou, S. Image Target Detection Algorithm of Smart City Management Cases. *IEEE Access* **2020**, *8*, 163357–163364. [CrossRef]
2. Dalal, N.; Triggs, B. Histograms of Oriented Gradients for Human Detection. In Proceedings of the Computer Vision and Pattern Recognition, San Diego, CA, USA, 20–26 June 2005; pp. 886–893. [CrossRef]
3. Papageorgiou, C.; Poggio, T. A Trainable System for Object Detection. *Int. J. Comput. Vis.* **2000**, *38*, 15–33. [CrossRef]
4. Wu, B.; Nevatia, R. Detection of multiple, partially occluded humans in a single image by Bayesian combination of edgelet part detectors. In Proceedings of the Tenth IEEE International Conference on Computer Vision (ICCV'2005), Beijing, China, 17–21 October 2005; Volume 1. [CrossRef]
5. Girshick, R.; Donahue, J.; Darrell, T.; Malik, J. Rich Feature Hierarchies for Accurate Object Detection and Semantic Segmentation. In Proceedings of the IEEE Conference on Computer Vision and Pattern Recognition, Columbus, OH, USA, 23–28 June 2014; pp. 580–587. [CrossRef]
6. Girshick, R. Fast r-cnn. In Proceedings of the IEEE International Conference on Computer Vision, Santiago, Chile, 11–18 December 2015.
7. Bell, S.; Lawrence Zitnick, C.; Bala, K.; Girshick, R. Inside-outside net: Detecting objects in context with skip pooling and recurrent neural networks. In Proceedings of the IEEE Conference on Computer Vision and Pattern Recognition, Las Vegas, NV, USA, 27–30 June 2016; pp. 2874–2883.
8. Enkelmann, W.; Struck, G.; Geisler, J. ROMA—A system for model-based analysis of road markings. In Proceedings of the Intelligent Vehicles '95. Symposium, Detroit, MI, USA, 25–28 September 1995. [CrossRef]
9. Otsuka, Y.; Muramatsu, S.; Takenaga, H.; Kobayashi, Y.; Monj, T. Multitype lane markers recognition using local edge direction. In Proceedings of the Intelligent Vehicle Symposium, Versaille, France, 17–21 June 2002. [CrossRef]
10. Kluge, K.; Lakshmanan, S. A deformable-template approach to lane detection. In Proceedings of the Intelligent Vehicles '95. Symposium, Detroit, MI, USA, 25–26 September 2002. [CrossRef]
11. Krizhevsky, A.; Sutskever, I.; Hinton, G.E. Imagenet classification with deep convolutional neural networks. *Commun. ACM* **2012**, *60*, 84–90. [CrossRef]
12. Sermanet, P.; Eigen, D.; Zhang, X.; Mathieu, M.; Fergus, R.; LeCun, Y. Overfeat: Integrated recognition, localization and detection using convolutional networks. *arXiv* **2013**, arXiv:1312.6229.
13. Zhang, Y.; Sohn, K.; Villegas, R.; Pan, G.; Lee, H. Improving object detection with deep convolutional networks via Bayesian optimization and structured prediction. In Proceedings of the IEEE Conference on Computer Vision and Pattern Recognition, Boston, MA, USA, 7–12 June 2015; pp. 249–258. [CrossRef]
14. Liu, L.; Ouyang, W.; Wang, X.; Fieguth, P.; Chen, J.; Liu, X.; Pietikäinen, M. Deep learning for generic object detection: A survey. *Int. J. Comp. Vis.* **2020**, *128*, 261–318. [CrossRef]
15. He, K.; Gkioxari, G.; Dollár, P.; Girshick, R. Mask r-cnn. In Proceedings of the IEEE International Conference on Computer Vision, Venice, Italy, 22–29 October 2017.
16. Cheng, M.M.; Zhang, Z.; Lin, W.Y.; Torr, P. BING: Binarized normed gradients for objectness estimation at 300 fps. In Proceedings of the IEEE Conference on Computer Vision and Pattern Recognition, Columbus, OH, USA, 23–28 June 2014.
17. Felzenszwalb, P.F.; Girshick, R.B.; McAllester, D.; Ramanan, D. Object Detection with Discriminatively Trained Part-Based Models. *IEEE Trans. Pattern Anal. Mach. Intell.* **2009**, *32*, 1627–1645. [CrossRef] [PubMed]
18. Zitnick, C.L.; Dollár, P. Edge boxes: Locating object proposals from edges. In *European Conference on Computer Vision*; Springer: Cham, Switzerland, 2014.
19. Kuo, W.; Hariharan, B.; Malik, J. Deepbox: Learning objectness with convolutional networks. In Proceedings of the IEEE International Conference on Computer Vision, Las Condes, Chile, 11–18 December 2015.
20. Yosinski, J.; Clune, J.; Bengio, Y.; Lipson, H. How transferable are features in deep neural networks? *arXiv* **2014**, arXiv:1411.1792.
21. Sermanet, P.; Kavukcuoglu, K.; Chintala, S.; LeCun, Y. Pedestrian detection with unsupervised multi-stage feature learning. In Proceedings of the IEEE Conference on Computer Vision and Pattern Recognition, Portland, OR, USA, 23–28 June 2013.

22. Long, J.; Shelhamer, E.; Darrell, T. Fully convolutional networks for semantic segmentation. In Proceedings of the IEEE Conference on Computer Vision and Pattern Recognition, Boston, MA, USA, 7–12 June 2015.
23. Wang, C.; Shi, J.; Yang, X.; Zhou, Y.; Wei, S.; Li, L.; Zhang, X. Geospatial object detection via deconvolutional region proposal network. *IEEE J. Sel. Top. Appl. Earth Obs. Remote. Sens.* **2019**, *12*, 3014–3027. [CrossRef]

applied sciences

MDPI

Article

5G Price Competition with Social Equilibrium Optimality for Social Networks

Yuhao Feng [1], Shenpeng Song [1], Wenzhe Xu [1,*] and Huijia Li [2,*]

[1] School of Science, Beijing University of Posts and Telecommunications, Beijing 100876, China
[2] College of Information and Electrical Engineering, China Agricultural University, Beijing 100083, China
* Correspondence: wenzhexu@bupt.edu.cn (W.X.); lihuijia0808@gmail.com (H.L.)

Abstract: Due to the leaps of progress in the 5G telecommunication industry, commodity pricing and consumer choice are frequently subject to change and competition in the search for optimal supply and demand. We here utilize a two-stage extensive game with complete information to mathematically describe user-supplier interactions on a social network. Firstly, an example of how to apply our model in a practical 5G wireless system is shown. Then we build a prototype that offers multiple services to users and provides different outputs for suppliers, where in addition, the user and supplier quantities are independently distributed. Secondly, we then consider a scenario in which we wish to maximize social welfare and determine if there is a perfect answer. We seek the subgame perfect Nash equilibrium and show that it exists, and also show that when both sides reach it, social welfare likewise reaches its maximum. Finally, we provide numerical results that corroborate the efficacy of our approach on a practical example in the 5G background.

Keywords: social network; game theory; provider competition; 5G wireless production; equilibrium

Citation: Feng, Y.; Song, S.; Xu, W.; Li, H. 5G Price Competition with Social Equilibrium Optimality for Social Networks. *Appl. Sci.* **2022**, *12*, 8798. https://doi.org/10.3390/app12178798

Academic Editors: Giacomo Fiumara, Xiaoyang Liu, Annamaria Ficara and Pasquale De Meo

Received: 9 August 2022
Accepted: 30 August 2022
Published: 1 September 2022

1. Introduction

Because of the telecommunication industry's irregularity, wireless consumers have complete freedom in selecting providers to achieve the greatest future tradeoff. Public Wi-Fi connections are a well-known example, where users may connect to any Wi-Fi provider for free but are charged for the time they spend connected. Despite the fact that the majority of users prefer to connect to free public Wi-Fi, there are still many users who are willing to pay for a premium service [1]. In this paper, we focus on the Wireless Service Providers (WSPs) in the 5th Generation Mobile Communication Technology (5G) who offer specific limited resources, such as a wireless frequency band, time slots, or transmission power. 5G is a new generation of broadband mobile communication technology that has high-speed rates, minimal latency, and a strong connection, making it superior to previous generations. How providers set commodity pricing and how users pick a source and commodity quantities is an important and fascinating issue. Suppliers are supposed to give different degrees of service to consumers, and users are aware of the difference for in-depth analysis, to monitor each interaction for each characteristic. As a result, each user is thought to have their own utility functions.

We study the widely used linear pricing schemes in the literature (see [2,3]). This spurs many ideas: the current TCP protocol can be explained as usage-based pricing methods that solve the problem of maximizing network utility [2]. Many researchers in the related literature look at resource supply and interaction through the lens of price strategy and game theory. The related research of wireless settings generally is classified as follows: the majorization-based allocation of one supplier's resource (see [4–9]), theoretical study of the game between one supplier's buyers (see [10–13]), competition between suppliers in the name of users(see [14,15]), and the price competition between suppliers (see [16–24]). Additionally, one work studies a three-tier system for a particular utility function, and the model is similar to ours [20]. The work we are interested in [22] uses evolutionary game theory to study multi-buyer, multi-seller dynamics in a cognitive radio setting. Then finally, the price competition of multihop wireless networks is studied in [23,24]. The work [25] by Chen inspired us to design and prove the decentralized algorithm. Nevertheless, our work has many significant differences. First, we used a less-rigorous precondition to prove our convergence. Second, our research shows that there are a finite (rather than infinite) number of globally optimum solutions. Third, with our work, consumers are free to use any resource quantity they choose. Finally, current research only focused on a single OFDM cell in resource allocation optimization, while we prefer to investigate NOMA or uRLLC with a high-speed network and low delay in the 5G background [26].

In this study, we explain the user-supplier interaction in the 5G wireless system using a two-stage extended game with comprehensive information (see [27]). To understand how to apply our model in a practical 5G wireless system, we take a 5G popular technology as a specific example. Then we explain how the two-stage works. Suppliers set their commodity pricing in the first stage, and consumers select the amount and supplier in the second stage. A user may choose the less costly commodity with poor service or the more expensive commodity with superior service. Based on users' responses to suppliers' prices, the suppliers take advantage and maximize profits. With this in mind, we first create a prototype that provides consumers with a variety of services and distinct outputs for providers. A multistage game model is utilized to describe the user-supplier relationship, and the user and supplier quantities are independently distributed. Next, we consider a social welfare maximization situation and determine that there must be an optimal solution. At that point, we move to the supplier competition game, which generates a decentralized algorithm that gradually finds equilibrium. The flowchart of the proposed algorithm is shown in Figure 1. Users make decisions based merely on the suppliers' set price; meanwhile, suppliers determine the pricing based on demand (the user's want). Finally, we present numerical results that demonstrate the efficacy of the suggested approach.

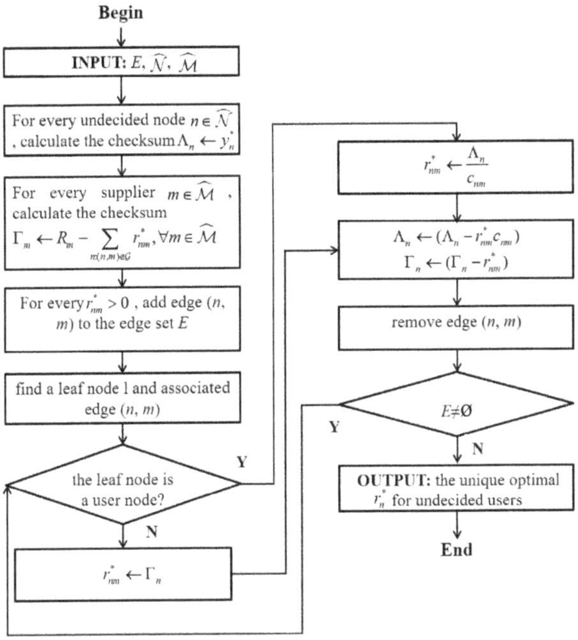

Figure 1. The flowchart of the proposed framework.

Our model aims to address the lack of a strong strategy to address the mismatch between supply and demand in the current 5G market, as well as the insufficient structure of market, which can have extremely negative effects. The model provides the user with a perspective on how to select a supplier, also provides a perspective for suppliers on the needs of users. It targets welfare maximization and provides an efficient way of managing supply and demand-side constraints. At the same time, the model helps to motivate market participants to make decisions that are most beneficial to the remaining economic agents.

2. The Model

Let us start with the features of 5G. In this section, 5G wireless networks will surpass the mobile Internet. In addition to increasing data rates compared to today's 4G and 4.5G (LTE Advanced), new IoT and key communication examples will require new ways to improve performance. For example, "low latency" is about providing real-time interactivity for services that use the cloud: this is crucial to the success of self-driving cars for example. In addition, low power consumption enables networked objects to run for months or years without human assistance.

To better understand our research, we explain some notions here. As we set about formulating our problem, initially, we assume there are two sets: $\mathcal{M} = \{1, \ldots, M\}$ represent the 5G wireless suppliers and $\mathcal{N} = \{1, \ldots, N\}$ represent the 5G wireless users. Supplier $m \in \mathcal{M}$ provides a R_m unit commodity to the users to maximize its return. User $n \in \mathcal{N}$ buys commodities from one or more suppliers to maximize its payoff. We assume that each user utilizes orthogonal resources, there is no interference between them, and meanwhile, the communication can be upward or backward. We simplify the interaction to be a multileader-follower game (see [28,29]), with suppliers leading the way and users following. In a relatively static network environment, channel gains are almost constant and also, public information is known to both sides. For example, every supplier gathers its respective channel information on every user and then applies it to all users. Section 4 assumes that our decentralized algorithm yields the same outcome as the supplier competition game.

2.1. Supplier Competition Game

There are two stages in the supplier competition game. Each supplier claims its price in the first stage. In the price vector $b = [b_1, \ldots, b_M]$, b_m represents the price for the unit commodity that supplier m charges. In addition, every user $n \in \mathcal{N}$ chooses a demand from different suppliers, depicted by vector $r_n = [r_{n1}, \ldots, r_{nM}]$. Then we use a vector to depict the overall demand: $r = [r_1, \ldots, r_N]$.

In the second stage, as prices b have already been set, user n selects its demand r_n to maximize its payoff based on the price. We define the payoff as a utility after subtracting expenses:

$$v_n(r_n, b) = u_n\left(\sum_{m=1}^{M} r_{nm} c_{nm}\right) - \sum_{m=1}^{M} b_m r_{nm}. \tag{1}$$

In this equation, c_{nm} is the offset of channel quality between provider m and user n (see Example 1 and Assumption 2). Here, u_n is the utility function, which is concave and increases with quantity. We can see that the utility function is based on the term $\sum_{m=1}^{M} r_{nm} c_{nm}$, which is the amount of service that users acquire, and also the function of commodity uses. In the first stage, after taking into account the resource constraint $\sum_{n=1}^{N} r_{nm} \leqslant R_m$, and in the second stage, after factoring users' demand, then supplier m sets the price to b_m to maximize its return $b_m \sum_{n=1}^{N} r_{nm}$. We assume linear pricing, with each user facing the same price.

In this model, any user can buy a commodity from more than one supplier simultaneously. In other words, for users, n, more than one r_{nm} $(m \in M)$ can be positive. It may be reasonable only if the user's device has several wireless interfaces. Interestingly, for most users ($N - M$ at least), the optimal strategy is to select one or no supplier.

In the following, a specific example shows how to apply our model in a practical 5G wireless system.

Example 1 (NOMA). *Non-Orthogonal Multiple Access (NOMA) is a popular technology to improve the efficiency of the 5G spectrum, with low latency, low signaling cost, and attenuation resistance (see [30,31]). W_m, $m \in \mathcal{M}$ are the non-orthogonal frequency bands on which wireless providers operate. r_{nm} is the portion of time user n can transmit exclusively on supplier m's frequency band, in which $\sum_{n \in N_m} r_{nm} = 1$, $m \in \mathcal{M}$ is the constraint. We assume a peak power constraint P_n exists as well for each user. Then we define c_{nm} as $W_m \log\left(1 + \frac{P_n |h_{nm}|^2}{\sigma_{nm}^2 W_m}\right)$ by Shannon's theorem, in which the channel's Gaussian noise variance is σ_{nm}^2 between provider m and user n, while h_{nm} is channel gain, channel gain describes the transmission capability characteristics of the channel itself. The payoff for the user, then, is the remaining utility after subtracting payment for the service, $v_n = u_n(\sum_{m=1}^{M} r_{nm} c_{nm}) - \sum_{m=1}^{M} b_m r_{nm}$.*

Similarly, our model applies when suppliers sell bandwidth of ultra-reliable low latency communication (uRLLC) tones to users who face a maximum power constraint [32]. For example, c_{nm}—the offset factor in Example 1—not only represents channel capability but essentially any aspect of the channel capacity's increasing function. Compared with other previous technologies, 5G has a more significant channel gain. According to the Shannon formula, when the channel gain h_{nm} increases, the channel capacity c_{nm} will increase to achieve an extremely low delay.

Although the payment for the 5G service is high, the significant improvement of quality of service(QoS): utility function u_n, is enough to offset the payment, and the final income is significantly higher than that of time division multiple access(TDMA), and it can also meet the requirements of low delay and high spectrum efficiency in modern times.

Finally, we find that this problem resembles a generalized network flow setting's multipath routing problem. A user parallels a source, similar to how a supplier corresponds to a link. Fortunately, there is one fundamental similarity: the multipath routing problem is equal-weighted, which applies to our model and does not hold in the TDMA model.

2.2. Assumptions about the Model

To focus on the problem of social welfare optimization, we make hypotheses, ignoring some unnecessary factors in the supplier competition game. Here, we outline our model assumptions.

Assumption 1. *$u_n(y)$ is increasing, differentiable, and strictly concave in y for each user $n \in \mathcal{N}$. In the network literature, this is how resilient data applications typically are modeled.*

Assumption 2. *We draw offset c_{nm} for the channel's quality from continuous, different probability distributions. c_{nm} are independent of each other, and evidently, different c_{nm} cannot be equal. c_{nm} indicates that a user will have different results if it buys the same commodity quantity from different suppliers.*

As Example 1 shows, c_{nm} is a function of h_{nm}, which is the channel gain between a supplier and user. Given that h_{nm} is drawn from the independent continuous-probability distributions, these assumptions can be fulfilled. In the next section, we study a related socially optimal resource-allocation problem to analyze the supplier competition game. Additionally, we show the solution based on a user's unique demand. In Section 4, we return to the supplier competition game. We find that the socially optimal, unique solution resembles the supplier competition game's unique equilibrium. Here, even when suppliers and users are selfish, the game remains as efficient as previously.

3. Social Welfare Optimization

3.1. Maximizing Social Welfare

In the following, we study social welfare, a problem where we maximize the sum of payoffs for both users and suppliers. We show that the solution is unique based on the user's demand. As users pay for advanced 5G resources and give money to suppliers, the payments between users and suppliers offset one another. Therefore, to maximize social welfare, we need to maximize users' utility functions. We define the social welfare maximization problem as a function of service acquired by users, which is ultimately inherent to users' interests.

Definition 1. *Let $y = [y_1, \dots, y_N]$ be the vector of services acquired, where the service acquired by user n, $y_n = \sum_{m=1}^{M} r_{nm}c_{nm}$ acts as a function of $r_n = [r_{n1}, \dots, r_{nM}]$, the demand for resources by user n.*

Then we define the social welfare optimization problem (SWO) as

$$\begin{aligned} SWO : \max u(y) &= \sum_{n=1}^{N} u_n(y_n) + c \\ s.t. \sum_{n=1}^{N} r_{nm} &= R_m, m \in \mathcal{M} \\ \sum_{m=1}^{M} r_{nm}c_{nm} &= y_n, n \in \mathcal{N} \\ over\ r_{nm}, y_n &\geq 0,\ \forall n \in \mathcal{N}, m \in \mathcal{M}. \end{aligned} \quad (2)$$

Here, c is a variable that denotes the unpredictable change, but for simplicity, we set $c = 0$. Two variables comprise the SWO: the service-acquired vector y and the demand vector r. In fact, y is uniquely determined by r. So y is a function of variable r. Then we can write y as $y(r)$. For brevity, we write $u(y(r))$ as $u(r)$.

3.2. Socially Optimal Demand Vector r^*'s Uniqueness

However, it is interesting that $u_n(\cdot)$s fail to be strictly concave to the demand vector r_n, as is the case of SWO to r. As we all know, a maximization problem that is not strictly concave may have more than one global optimal solution (see [33,34]). To get more than one solution of vector r^* in SWO, we simply modify c_{nm}s, R_ms, and $u_n(\cdot)$s to some value. For example, if c_{nm} is constant and the same for different n, m as it is for R_m, we can get a non-unique maximizer of SWO. However, as we showed previously, c_{nm}s are independent random variables from continuous distributions, and the probability of that case occurring is zero (see Assumption 2).

As we also learn in Lemma 1, no two maximizing demand vectors can exist in SWO that possess the same nonzero components. If two maximizing demand vectors combine, the result is still a maximizing demand vector. Finally, with the previous intermediate result, we see that maximizing demand vectors yields no convex combinations possessing different nonzero components, and this contradicts Lemma 1. So, we can use this to prove the primary finding of this section (Theorem 1).

Next, we define a demand vector r_n's support set.

Definition 2. *User n's support set is composed of suppliers from which user n's demand is strictly positive:*

$$\hat{\mathcal{M}}_n(r_n) = \{m \in \mathcal{M} : r_{nm} > 0.\}$$

Given demand vector r, we define the support sets' ordered collection $\hat{\mathcal{M}}_1, \hat{\mathcal{M}}_1, \dots, \hat{\mathcal{M}}_N$ as $\{\hat{\mathcal{M}}_n\}_{n=1}^{N}$.

Lemma 1. *If r^* is SWO's maximizing demand vector containing the corresponding collection of support sets of $\{\hat{\mathcal{M}}_n\}_{n=1}^{N}$, then it is almost true that r^* can be a unique maximizing demand vector compared to $\{\hat{\mathcal{M}}_n\}_{n=1}^{N}$.*

Proof. Equation (2) holds for the maximizing demand vector r^*, and r^* is uniquely constructed from $\{\hat{\mathcal{M}}_n\}_{n=1}^{N}$. □

Then two categories exist for the users: decided and undecided.

We define that the decided users are those who buy from one supplier ($|\hat{M}_n| = 1$), while the undecided users are those who buy from more than one supplier ($|\hat{M}_n| > 1$). In fact, some users will buy nothing. Without loss of generality, these users are defined as decided users. $y_n^* = \sum_{m=n}^{M} r_{nm}^* c_{nm}$ holds for all users. If user n is a decided user—which means that he only buys from supplier m and buys nothing from other suppliers—we can reduce the equation to $y_n^* = r_{n\bar{m}}^* c_{n\bar{m}}$, because other terms are zero. Then the unique demand vector that corresponds is $r_n^* = [0 \ldots 0, \frac{y_n^*}{c_{n\bar{m}}}, 0 \ldots 0]$.

Theorem 1. *There is a unique maximizing solution b^* with probability 1 in SWO. There are no multiple maximizing demand vectors, and the convex combination of SWO for maximizing demand vectors retains the same support.*

Proof. Suppose that more than one SWO optimal demand vector exists. Two of them are r' and r^*. We learn from Lemma 1 that r^* and r' almost certainly have distinct support sets $\{\hat{M}_n^*\}_{n=1}^N$ and $\{\hat{M}_n'\}_{n=1}^N$. Then, let $r^\lambda = \lambda r^* + \bar{\lambda} r', \lambda \in (0,1), \bar{\lambda} = 1 - \lambda$. If $y_n^* = \sum_{m=1}^{M} r_{nm}^* c_{nm} = \sum_{m=1}^{M} r_{nm}' c_{nm}$ and $\sum_{m=1}^{M} r_{nm}^\lambda c_{nm} = \lambda \sum_{m=1}^{M} r_{nm}^* c_{nm} + \bar{\lambda} \sum_{m=1}^{M} r_{nm}' c_{nm} = y_n^*$, then we can say that r^λ is an SWO maximizing solution for each $\lambda \in (0,1)$. Next, we can say that support set $\hat{M}_n^\lambda(r^\lambda) = \{m \in \mathcal{M} : r_{nm}^\lambda = \lambda r_{nm}^* + \bar{\lambda} r_{nm}' > 0\}$ when user n is $\hat{M}_n^\lambda = \hat{M}_n^* \cup \hat{M}_n'$, for every ($\lambda \in (0,1)$). Note that the support sets $\left(\{\hat{M}_n^\lambda\}_{n=1}^N\right)$, in particular, are the same for all ($\lambda \in (0,1)$). If two maximizing demand vectors exist with different support sets of SWO, then the convex combinations of SWO for two maximizing demand vectors retain the same support. This contradicts Lemma 1. □

We prove the uniqueness and existence of a Lagrange multiplier vector b^* based on an SWO's optimal demand vector r^* [35]. In the following, we explain how the supplier competition game's unique equilibrium is (r^*, b^*).

4. Game Analysis

So far, with the multileader-follower supplier competition game, we showed that the equilibrium is existing and unique, which is compared to the Lagrange multipliers and SWO's unique optimal solution. Now, here we explain that the Lagrange multipliers are prices announced by suppliers. Furthermore, in this equilibrium, there are no more than $M - 1$ undecided users.

The equilibrium concept is interpreted as follows [27]:

Definition 3. *Say that we have a subgame perfect equilibrium (SPE) with a price demand tuple (b^*, r^*), in which no participant would like to change at any stage of the game. Moreover, given the price b^*, every user maximizes their payoff. Given users' demand, $r^*(b^*)$ and other participants' price, every supplier maximizes its return.*

The equilibrium is solved by backward induction. In Stage 2, the users' equilibrium strategy, users choose the best amount of resource $r^*(b)$ based on the vector of prices b. The function of Stage 2 is used to substitute the terms in Stage 1, the suppliers' equilibrium strategy, resulting in equilibrium price b^*. According to BGR decoding, $r^*(b)$ is uniquely determined by equilibrium price b^*.

4.1. Users' Equilibrium Strategy

Taking every user's decision into account, we can solve the problem of user payoff maximization (UPM):

$$UPM : \max_{r_n \geqslant 0} v_n = \max_{r_n \geqslant 0} u_n\left(\sum_{m=1}^{M} r_{nm} c_{nm}\right) - \sum_{m=1}^{M} b_m r_{nm}. \tag{3}$$

Lemma 2. *Regarding the UPM problem, with each maximizer r_n, $\sum_{m=1} c_{nm} r_{nm} = y_n^*$, for a unique nonnegative value of y_n^*. Furthermore, for any m such that $r_{nm} > 0$, $\frac{b_m}{c_{nm}} = \min_{k \in M} \frac{b_k}{c_{ik}}$.*

Proof. We can easily verify that Slater's conditions are satisfied via UPM [36]. The following are the Karush–Kuhn–Tucker (KKT) conditions required for an optimal solution $r_n \geq 0$ of UPM of user n:

$$u_n'(y_n)c_{nm} \leq b_m, m \in \mathcal{M} \tag{4}$$

$$r_{nm}(u_n'(y_n)c_{nm} - b_m) = 0, m \in \mathcal{M}, \tag{5}$$

$$\text{where } y_n = \sum_{m=1}^{M} r_{nm}c_{nm}, r_n \geq 0. \tag{6}$$

Here, (4) implies that $u_n'(y_n) \leq \varphi$, where $\varphi = \min_{k \in \mathcal{M}} \frac{b_k}{c_{nk}}$. Based on user n's utility function, two scenarios are possible: $u_n'(0) < \varphi$ and $u_n'(0) \geq \varphi$.

For the first scenario, $u_n'(0)c_{nm} - b_m < 0$, so $c_{nm}u_n'(y_n) - b_m < 0$ for all $m \in \mathcal{M}$ since, by Assumption 1, $u_n'(\cdot)$ is a marginal utility-with a strictly decreasing function. So, keeping (5) in mind, $r_{nm} = 0$ for all $m \in \mathcal{M}$. Then $r_n = 0$, and with (6), we see $y_n^* = 0$. So, Equations (4)–(6) hold for the $y_n^* = 0$ unique value.

With the second scenario, $u_n'(0) \geq \varphi$. However, keeping in mind that $u_n'(\cdot)$ dwindles to zero (Assumption 1), a unique $\hat{y}_n \geq 0$ exists, such that $u_n'(\hat{y}_n) = \varphi$. First, we make sure r_n exists, such that Equations (4)–(6) hold with $y_n = \hat{y}_n$. We find that Equation (4) holds, because $u_n'(\hat{y}_n) = \varphi \leq b_m/c_{nm}$ for all $m \in \mathcal{M}$. Then, with (5), we remember that for any m such that $b_m/c_{nm} > \varphi = u_n'(\hat{y}_n)$ there is $r_{nm} = 0$. For any other m, $b_m/c_{nm} = \varphi = u_n'(\hat{y}_n)$, so, when it comes to (5), r_{nm} can take any non-negative value. In particular, so that (6) holds for the set $\{m \in \mathcal{M} : b_m/c_{nm} = \varphi\}$, it is possible to choose r_{nm}'s.

We provide the last part of the lemma by noting that r_{nm} is positive only when $b_m/c_{nm} = \varphi$. It remains to be seen whether $\hat{y}_n$ is the only value of y_n for which r_n satisfies Equations (4)–(6). We can say that for any $y_n < \hat{y}_n, u_n'(y_n) > \varphi$, which violates (4) for $m \in \arg\min b_k/c_{nk}$. Then, for each $y_n > \hat{y}_n, u_n(y_n) < \varphi$, which means that $u_n(y_n)c_{nm} - b_m < 0$ for every $m \in \mathcal{M}$. Equation (5) implies, then, that $r_{nm} = 0$ for every $m \in \mathcal{M}$, meaning that $y_n = 0$; this is contradictory to $y_n > \hat{y}_n > 0$. The unique searched value y_n^* is thus $\hat{y}_n$. □

Definition 4. *Each supplier $m \in \mathcal{M}$ with $\frac{b_m}{c_{nm}} = \min_{k \in \mathcal{M}} \frac{b_k}{c_{ik}}$ is included in user n's preference set $\mathcal{M}_n(b)$ for price vector b.*

According to Lemma 2 and Section 2, we can divide users into decided and undecided users based on the preference sets' cardinality. The support sets in Section 2 are quite similar to the preference sets. However, unlike support sets where users buy resources from suppliers, it is just possible for a user to request a resource from suppliers in the preference set. Evidently, the support set acts as a subset for the preference set: $\hat{\mathcal{M}}_n(r(b)) \subset \mathcal{M}_n(b)$. Knowing this, we set about using the preference sets to construct a BGR so that there are on-loops with probability 1.

We define the Lagrange multipliers b^* as prices. Every user knows this information, so it is not difficult to calculate the preference sets of other users and construct the BGR in comparison. Undecided users can determine their unique demand vector using a BGR decoding algorithm. For this, we consider all the demand vectors at a specific time and consider the equality of supply and demand. Then we find the uniqueness of the demand from BGR decoding. Although an infinite amount of best responses exist under prices b^*, the supply and demand will balance only if the demands are found by BGR decoding. Later, we prove that it is the supplier competition's unique SPE.

4.2. Suppliers' Equilibrium Strategy

The user's utility functions determine the suppliers' optimal choice of prices. A utility function u_n can be characterized by its coefficient of relative risk aversion [37], i.e., $k_{RRA}^n = -\frac{yu_n''(y)}{u_n'(y)}$. This quantity characterizes the relationship between price and user demand.

Assumption 3. *Relative risk-aversion coefficient.* $k_{RRA}^n < 1, \forall n \in N$.

Some utility functions satisfy Assumption 3—for example, $\log(1+y)$ and the $\varphi - fair$ utility function s $\frac{y^{1-\varphi}}{1-\varphi}$, for $\varphi \in (0,1)$ [38]. To maximize the return, a monopoly will sell all the resources R_m. Once a supplier decreases the price, the users' demand substantially increases, resulting in the supplier earning more than before. Thus, the supplier will lower the price until total supply and total demand are equal.

Theorem 2. *In keeping with Assumption 3, SPE as a price vector tuple meeting KKT conditions. We constitute the supplier competition game's SPE using the Lagrange multiplier vector b^* and SWO's unique socially optimal demand vector r^*.*

Proof. Suppose $b = [b_1, b_2, \ldots, b_M]$ is the price that suppliers charge. As defined in Equation (3), every user faces a local maximization problem $\text{UPM}_n(b)$. Given Assumption 3, we further remark that r is an SPE of the supplier competition game only if each supplier's supply equals demand, n.e., $\sum_{n=1}^{N} r_{nm} = R_m$ for every $m \in \mathcal{M}$. So, we consider the SPE as a price vector tuple meeting KKT conditions. Moreover, these meet the KKT conditions for any vector tuple b, r to be the SWO's maximizing solution. We, therefore, designed a formal equivalence between the maximizing demand vector and SPE of the supplier competition game and the SWO problem (r^*, b^*)'s Lagrangian multipliers. From this, we deduce that (r^*, b^*) form the supplier competition game's unique SPE. □

That social efficiency is not reduced by suppliers' competition results in users' utility functions being strictly concave and the users' demand is relatively elastic. Therefore, if the price decreases a bit; demand will increase so much that the return is more than before. If suppliers set the price different from the optimal price b^*, the supply and demand are unequal. According to Theorem 2, we define the supplier competition game's unique SPE (b^*, r^*) as the equilibrium.

5. The Algorithm

Here, we provide a continuous-time algorithm, in which all the variables are functions of time. For brevity, we write $r_{nm}(t)$ and $b_m(t)$ as r_{nm} and b_m, respectively. Then we write their time derivatives $\frac{\partial r_{nm}}{\partial t}$ and $\frac{\partial b_m}{\partial t}$ as $\dot{r_{nm}}$ and $\dot{b_m}$. Note that r^* is the SWO's unique maximizer, while the corresponding Lagrange multiplier vector is b^*. According to Theorem 2, we know that the supplier competition game's unique SPE is (b^*, r^*), and its values are invariant.

Given the demand vector $r_n(t)$, we write user N's marginal utility according to r_{nm} as $\psi_{nm}(t)$ or simply ψ_{nm}.

$$\psi_{nm} = \frac{\partial u_n(r_n)}{\partial r_{nm}} = c_{nm} \frac{\partial u_n(y)}{\partial y}\Big|_{y=y_n=\sum_{m=1}^{M} r_{nm}c_{nm}}. \tag{7}$$

Here, we denote $\psi_{nm}(t)$'s value evaluated at r_n^* as ψ_{nm}^*. Then we define column vectors: $\nabla u_n(r_n) = [\psi_{n1}, \ldots, \psi_{nM}]^T$ and $\nabla u_n(r_n^*) = [\psi_{n1}^*, \ldots, \psi_{nM}^*]^T$.

Next, $(y)^+ = \max(0, y)$ is defined, so that

$$(y)_x^+ = \begin{cases} y & x > 0 \\ (y)^+ & x \leq 0. \end{cases}$$

In the following, the standard primal-dual variable update algorithm is motivated by the work in [25]:

$$\dot{r}_{nm} = k^r_{nm}(\psi_{nm} - b_m)^+_{r_{nm}}, n \in \mathcal{N}, m \in \mathcal{M} \quad (8)$$

$$\dot{b}_m = k^b_m(\sum_{n=1}^{N} r_{nm} - R_m)^+_{b_m}, m \in \mathcal{M}. \quad (9)$$

Here, k^r_{nm} and k^b_m are the constants that represent update rates. It is ensured that a variable of interest (r_{nm} or b_m) will not turn negative when it is zero, even if the update's direction is negative. We define the tuple $(r(t), b(t))$ controlled by Equations (8) and (9) as the differential equations' solution trajectory. Users only need to be given the prices that suppliers request. The providers do not need to be given other suppliers' demands of the users, except for that of their resources. Only user n needs to know $c_{nm}, m \in \mathcal{M}$.

The procedure of bipartite graph representation is as follows. First of all, for every undecided node $n \in \hat{\mathcal{N}}$, calculate the checksum $\Lambda_n \leftarrow y^*_n$. Then, for every supplier $m \in \hat{\mathcal{M}}$ calculate the checksum $\Gamma_m \leftarrow R_m - \sum_{n:(n,m)\notin \mathcal{G}} r^*_{nm}, \forall m \in \hat{\mathcal{M}}$. Next, for every $r^*_{nm} > 0$, add edge (n, m) to the edge set E. And we have two steps in the loop. Step 1, Find a leaf node l and associated edge (n, m), if the leaf node is a user node, then $r^*_{nm} \leftarrow \frac{\Lambda_n}{c_{nm}}$, else $r^*_{nm} \leftarrow \Gamma_n$. Step 2, Let $\Lambda_n \leftarrow (\Lambda_n - r^*_{nm}c_{nm})$ and $\Gamma_n \leftarrow (\Gamma_n - r^*_{nm})$, remove edge (n, m). Keep doing these two steps until $E \in \emptyset$.

To find the unique optimal r^*_n for undecided users, the algorithm provides detailed procedures. Here, E, $(\hat{\mathcal{N}})$, and $(\hat{\mathcal{M}})$ are sets of edges, user nodes, and supplier nodes separately. Using the algorithm, we can find the demand of undecided users. Because the probability that a BGR has no loops is 1, suppose that the BGR is an unrooted tree. Then an uncomplicated iterative algorithm can remove a node with a single associated edge, which we define as a leaf node and its incoming edge at each iteration. First, a leaf node is found in the BGR. Second, the demand of the leaf node's incoming line is given from BGR Feature (1) or (2). Third, the parent node's check-sum is updated with this value. Finally, the leaf node and incoming edge are removed. This process is one iteration. Iterate until no edges exist in the graph.

We can run the step finding the demand of undecided users because the probability that a BGR has no loops is 1, and we suppose that the BGR has no loops. However, in the last iteration, only one supplier node m plus one user node n exist, connected via an edge with value r^*_{nm}. Λ_n and Γ_m are their check-sums, which satisfy $\Lambda_n = \Gamma_m c_{nm}$, because $\Lambda_n = r^*_{nm}c_{nm}$ and $\Gamma_m = r^*_{nm}$. At last, undecided users' unique demand is given by the algorithm.

6. Numerical Results

We need to expand the settings to get numerical results. In Example 1, the fraction of time restricted to the 5G wireless supplier's frequency band is the resource that is being sold, i.e., $R_m = 1$ for $m \in \mathcal{M}$. Let W_m = 700 MHz, $m \in \mathcal{M}$. W_m means the 5G wireless suppliers' bandwidth. $a_n \log(\sum_{m=1}^{M} r_{nm}c_{nm} + 1)$ is user n's utility function, in which the spectral efficiency c_{nm} from the Shannon formula $\frac{1}{2}Wlog(1 + \frac{E_b/N_0}{W}|h_{nm}|^2)$, a_n is the "willingness to pay" factor, which we assume is the same among users, r_{nm} is allocated time fraction, and E_b/N_0 is the transmit power divided by thermal noise.

Suppose that the coding choices and modulation are perfect, with a continuum of values supplying a steady communication rate. Users are placed, then, uniformly in field that is 500 × 500 square meters. We only provide these parameters to elucidate our point; we can change the numbers and the theory still applies to the parameters. Let us think of an example in the 5G background with 5 suppliers and 20 users. In Figures 2 and 3, the equilibrium prices are represented as dashed lines. The competition among users can influence the equilibrium. Notice that supplier d offers a higher price than supplier a. Therefore, although supplier d can provide better resource quality, the user's choice will also be affected by the equilibrium prices. Meanwhile, supplier b has the most buyers, its price is the highest, as Figures 2 and 3 shows.

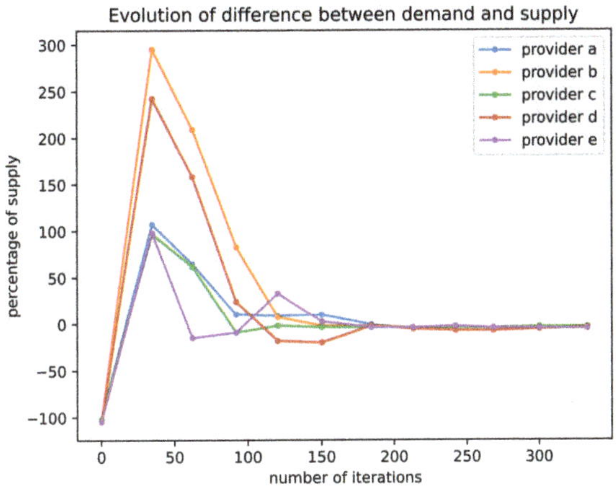

Figure 2. Evolution of difference between demand and supply.

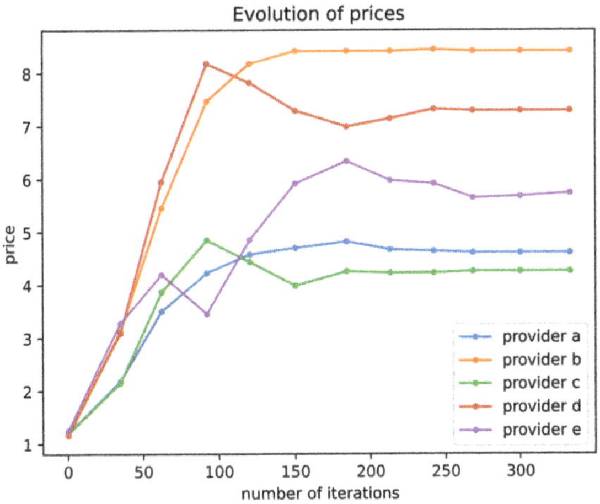

Figure 3. Evolution of prices. The chart shows that when the equilibrium state is reached, the price of supplier *d* is approximately 1.5 times higher than that of supplier *a*.

As for the convergence time in the discrete-time version: Suppose there are only five suppliers, then the number of users increases from 20 to 100. At each condition, we repeat the experiments 2000 times with random locations of suppliers and users. That way, we can obtain the average convergence speed and plot it. We define convergence as the number of iterations after which εR_m is larger than the gap between demand and supply. For different ε, Table 1 shows the average convergence time. Generally, if $\varepsilon = 10^{-3}$, the time to convergence is 2000–4000. If $\varepsilon = 10^{-4}$, the time to convergence is 3000–6000.

Table 1. Average time to convergence for different ε. When εR_m is larger than the gap between demand and supply, we define the number of iterations as convergence. In comparison, the smaller the parameter ε, the larger the gap between supply and demand, so more iterations are required. For instance, if $\varepsilon = 10^{-3}$, the time to convergence is 2000–4000 in most cases, if $\varepsilon = 10^{-7}$, the time to convergence is 6000–12,000.

	Number of Users	$\varepsilon = 10^{-3}$	$\varepsilon = 10^{-4}$	$\varepsilon = 10^{-5}$	$\varepsilon = 10^{-6}$	$\varepsilon = 10^{-7}$
Number of iterations	20	2022.4	3105.7	4209.9	5314.0	6209.9
	30	2068.8	3006.3	4068.8	5110.5	6110.5
	40	2407.0	3802.9	5011.2	6302.9	7532.0
	50	2786.8	4203.5	5807.7	7016.0	8203.5
	60	3145.8	4833.3	6312.5	8020.8	9520.8
	70	3525.6	5129.8	7025.6	8733.9	10,067.3
	80	3988.8	5509.6	7530.4	9509.6	10,926.3
	90	4243.6	6035.3	8056.1	10,076.9	11,931.1
	100	4435.9	6456.8	8435.9	10,540.1	12,519.3

In Table 2, we change the number of suppliers and see how the average convergence time changes. Now, let $\varepsilon = 10^{-4}$, and let it be constant. The update rates decide the convergence time: with low rates, the variables are likely to stabilize, and they will not take long to converge. In contrast, with high rates, the variables may converge rapidly. Based on Section 4's theoretical analysis, to obtain the algorithm's global convergence, let us distribute update variables randomly. Generally, the algorithm will iterate many times to converge if the ratio of users per supplier is too low or too large.

Table 2. Average time to convergence for different numbers of suppliers when $\varepsilon = 10^{-3}$. The chart shows that when the ratio of suppliers to users is too high or low, the algorithm will iterate many times to converge. For instance, nearly 7000 iterations are required for 9 suppliers and 20 users. When the number of users increases gradually, the number of users is approximately positively correlated with the average convergence time.

	Number of Users	5 Suppliers	7 Suppliers	9 Suppliers
Number of iterations	20	6814.7	4199.3	3199.3
	30	3947.5	3493.0	3045.4
	40	3493.0	3793.7	3793.7
	50	3849.6	4150.3	4255.2
	60	4402.0	4604.8	4807.6
	70	4807.6	5101.3	5304.1
	80	5108.3	5437.0	5611.8
	90	5611.8	5723.7	6108.3
	100	6059.4	6213.2	6562.9

At last, Table 3 presents the average convergence time when there are five suppliers with the standard variation. If the number of users is not 20, it does not affect the convergence time variance. If the number of users per supplier is under 4, update rates significantly impact the algorithm. In Table 2, we find that demands and prices vibrate and converge slowly in such instances.

Table 3. Including standard deviation in the average time to convergence.

	Number of Users	Mean	Standard Deviation
Number of iterations	20	3098.0	2397.7
	30	3053.3	899.1
	40	3804.0	1095.1
	50	4208.9	853.1
	60	4798.3	1002.9
	70	5306.9	904.9
	80	5608.1	1089.3
	90	6013.0	1037.5
	100	6510.1	1198.8

7. Conclusions

This paper considers the competition between a random number of 5G wireless providers to attract users with different channel gains and willingness to pay. In this study, we utilized a two-stage wireless provider game to simulate the interaction in this work, and we proved the convergence and unique equilibrium. In the provider competition, our findings show that there is only one socially optimum resource allocation. At equilibrium, there are some undecided users. There are also fewer undecided users than providers. Finally, we designed a decentralized algorithm that uses only regional information to converge to the equilibrium demand vectors and price.

Author Contributions: Funding acquisition, H.L.; Methodology, Y.F., S.S., W.X. and H.L.; Software, Y.F. and S.S.; Supervision, W.X. and H.L.; Validation, H.L.; Writing—original draft, Y.F.; Writing—review & editing, W.X. All authors have read and agreed to the published version of the manuscript.

Funding: This research was funded by National Natural Science Foundation of China grant number 71871233 and Fundamental Research Funds for the Central Universities of China grant number 22RC35.

Data Availability Statement: Not applicable.

Acknowledgments: We are grateful to the anonymous reviewers for their valuable suggestions.

Conflicts of Interest: The authors declare no conflict of interest. The funders had no role in the design of the study; in the collection, analyses, or interpretation of data; in the writing of the manuscript; or in the decision to publish the results.

References

1. Hamed, S.; Wong, V.W.; Huang, J. An incentive framework for mobile data offloading market under price competition. *IEEE Trans. Mob. Comput.* **2017**, *16*, 2983–2999.
2. Gössl, F.; Rasch, A. Collusion under different pricing schemes. *J. Econ. Manag. Strategy* **2020**, *29*, 910–931. [CrossRef]
3. Zhang, M.; Yang, L.; Gong, X.; He, S.; Zhang, J. Wireless service pricing competition under network effect, congestion effect, and bounded rationality. *IEEE Trans. Veh. Technol.* **2018**, *67*, 7497–7507. [CrossRef]
4. Tang, X.; Ren, P.; Han, Z. Hierarchical competition as equilibrium program with equilibrium constraints towards security-enhanced wireless networks. *IEEE J. Sel. Areas Commun.* **2018**, *36*, 1564–1578. [CrossRef]
5. Miao, Z.; Wang, Y.; Han, Z. A supplier-firm-buyer framework for computation and content resource assignment in wireless virtual networks. *IEEE Trans. Wirel. Commun.* **2019**, *18*, 4116–4128. [CrossRef]
6. Zhang, M.; Gao, L.; Huang, J.; Honig, M.L. Hybrid pricing for mobile collaborative Internet access. *IEEE/ACM Trans. Netw.* **2019**, *27*, 986–999. [CrossRef]
7. Yin, S.; Zhao, Y.; Li, L. Resource allocation and basestation placement in cellular networks with wireless powered UAVs. *IEEE Trans. Veh. Technol.* **2018**, *68*, 1050–1055. [CrossRef]
8. Wu, C.; Yan, B.; Yu, R.; Huang, Z.; Yu, B.; Chen, N. An intelligent resource dynamic allocation method for UAV wireless mobile network which supports QoS. *Comput. Commun.* **2020**, *152*, 46–53. [CrossRef]
9. Jiang, L.; Parekh, S.; Walrand, J. Base station association game in multi-cell wireless networks. In Proceedings of the Wireless Communications and Networking Conference, Las Vegas, NV, USA, 31 March–3 April 2008; pp. 1616–1621.
10. Adlakha, S.; Johari, R.; Goldsmith, A.J. Competition in wireless systems via Bayesian interference games. *arXiv* **2007**, arXiv:0709.0516.

11. Etkin, R.; Parekh, A.; Tse, D. Spectrum sharing for unlicensed bands. In Proceedings of the First IEEE International Symposium on New Frontiers in Dynamic Spectrum Access Networks, DySPAN, Baltimore, MD, USA, 8–11 November 2005; pp. 251–258.
12. Huang, J.; Berry, R.A.; Honig, M.L. Distributed interference compensation for wireless networks. *IEEE J. Sel. Areas Commun.* **2006**, *24*, 1074–1084. [CrossRef]
13. Kalathil, D.; Jain, R. Spectrum sharing through contracts. In Proceedings of the 2010 IEEE Symposium on New Frontiers in Dynamic Spectrum (DySPAN), Singapore, 6–9 April 2010.
14. Zhou, C.; Honig, M.L.; Jordan, S. Utility-based power control for a two-cell CDMA data network. *IEEE Trans. Wirel. Commun.* **2005**, *4*, 2764–2776. [CrossRef]
15. Grokop, L.; Tse, D.N. Spectrum sharing between wireless networks. In Proceedings of the 27th Conference on Computer Communications, INFOCOM, Phoenix, AZ, USA, 13–18 April 2008; pp. 201–205.
16. Zemlianov, A.; de Veciana, G. Cooperation and decision-making in a wireless multi-provider setting. In Proceedings of the 24th Annual Joint Conference of the IEEE Computer and Communications Societies, INFOCOM, Miami, FL, USA, 13–17 March 2005; Volume 1, pp. 386–397.
17. Sengupta, S.; Chatterjee, M.; Ganguly, S. An economic framework for spectrum allocation and service pricing with competitive wireless service providers. In Proceedings of the 2nd IEEE International Symposium on New Frontiers in Dynamic Spectrum Access Networks, DySPAN, Dublin, Ireland, 17–20 April 2007; pp. 89–98.
18. Jia, J.; Zhang, Q. Competitions and dynamics of duopoly wireless service providers in dynamic spectrum market. In Proceedings of the 9th ACM International Symposium on Mobile Ad Hoc Networking and Computing, MobiHoc, Hong Kong, China, 26–30 May 2008; Association for Computing Machinery: New York, NY, USA, 2008; pp. 313–322.
19. Ileri, O.; Samardzija, D.; Sizer, T.; Mandayam, N.B. Demand responsive pricing and competitive spectrum allocation via a spectrum server. In Proceedings of the First IEEE International Symposium on New Frontiers in Dynamic Spectrum Access Networks, DySPAN, Baltimore, MD, USA, 8–11 November 2005; pp. 194–202.
20. Acharya, J.; Yates, R.D. Service provider competition and pricing for dynamic spectrum allocation. In Proceedings of the 2009 International Conference on Game Theory for Networks, Istanbul, Turkey, 13–15 May 2009; pp. 190–198.
21. Inaltekin, H.; Wexler, T.; Wicker, S.B. A duopoly pricing game for wireless IP services. In Proceedings of the 2007 4th Annual IEEE Communications Society Conference on Sensor, Mesh and Ad Hoc Communications and Networks, SECON, San Diego, CA, USA, 18–21 June 2007; pp. 600–609.
22. Niyato, D.; Hossain, E.; Han, Z. Dynamics of multiple-seller and multiplebuyer spectrum trading in cognitive radio networks: A game-theoretic modeling approach. *IEEE Trans. Mob. Comput.* **2009**, *8*, 1009–1022. [CrossRef]
23. Xi, Y.; Yeh, E.M. Pricing, competition, and routing for selfish and strategic nodes in multi-hop relay networks. In Proceedings of the IEEE INFOCOM 2008-The 27th Conference on Computer Communications, Phoenix, AZ, USA, 13–18 April 2008; pp. 1463–1471.
24. Gao, L.; Wang, X. A game approach for multi-channel allocation in multihop wireless networks. In Proceedings of the 9th ACM International Symposium on Mobile Ad Hoc Networking and Computing, MobiHoc, Hong Kong, China, 26–30 May 2008; ACM: New York, NY, USA, 2008; pp. 303–312.
25. Chen, M.; Huang, J. Optimal resource allocation for OFDM uplink communication: A primal-dual approach. In Proceedings of the 2008 42nd Annual Conference on Information Sciences and Systems, CISS, Princeton, NJ, USA, 19–21 March 2008; pp. 926–931.
26. Tusha, A.; Dogan, S.; Arslan, H. A hybrid downlink noma with ofdm and ofdm-im for beyond 5g wireless networks. *IEEE Signal Process. Lett.* **2020**, *27*, 491–495. [CrossRef]
27. Fudenberg, D.; Tirole, J. *Game Theory*; MIT Press: Cambridge, MA, USA, 1991.
28. Leyffer, S.; Munson, T. Solving multi-leader-common-follower games. *Optim. Methods Softw.* **2010**, *25*, 601–623. [CrossRef]
29. Wang, J.; Chiu, D.; Lui, J. A game-theoretic analysis of the implications of overlay network traffic on ISP peering. *Comput. Netw.* **2008**, *52*, 2961–2974. [CrossRef]
30. Yang, G.; Xu, X.; Liang, Y.-C. Intelligent reflecting surface assisted non-orthogonal multiple access. In Proceedings of the 2020 IEEE Wireless Communications and Networking Conference (WCNC), Seoul, Korea, 25–28 May 2020; pp. 1–6.
31. Anand, A.; de Veciana, G.; Shakkottai, S. Joint scheduling of urllc and embb traffic in 5g wireless networks. *IEEE/ACM Trans. Netw.* **2020**, *28*, 477–490. [CrossRef]
32. Nguyen, H.V.; Kim, H.M.; Kang, G.-M.; Nguyen, K.-H.; Bui, V.-P.; Shin, O.-S. A survey on non-orthogonal multiple access: From the perspective of spectral efficiency and energy efficiency. *Energies* **2020**, *13*, 4106. [CrossRef]
33. Lin, X.; Shroff, N.B. Utility maximization for communication networks with multipath routing. *IEEE Trans. Autom. Control* **2006**, *51*, 766–781. [CrossRef]
34. Voice, T. Stability of Congestion Control Algorithms with Multi-Path Routing and Linear Stochastic Modelling of Congestion Control. Ph.D. Dissertation, University of Cambridge, Cambridge, UK, May 2006.
35. Bertsekas, D.P. *Nonlinear Programming*; Athena Scientific: Belmont, MA, USA, 1999.
36. Mathur, S.; Sankaranarayanan, L.; Mandayam, N.B. Coalitional games in receiver cooperation for spectrum sharing. In Proceedings of the 2006 40th Annual Conference on Information Sciences and Systems, CISS, Princeton, NJ, USA, 22–24 March 2006; pp. 949–954.

37. Mas-Colell, A.; Whinston, M.D.; Green, J.R. *Microconomic Theory*; Oxford University Press: Oxford, UK, 1995.
38. Rad, A.; Huang, J.; Chiang, M.; Wong, V. Utility-optimal random access without message passing. *IEEE Trans. Wirel. Commun.* **2009**, *8*, 1073–1079. [CrossRef]

applied sciences

MDPI

Article

A Modeling Approach for Measuring the Performance of a Human-AI Collaborative Process

Ganesh Sankaran [1,2,*], Marco A. Palomino [1,*], Martin Knahl [2] and Guido Siestrup [2]

[1] School of Engineering, Computing and Mathematics, University of Plymouth, Plymouth PL4 8AA, UK
[2] Business Information Systems, Hochschule Furtwangen University, 78120 Furtwangen, Germany
* Correspondence: ganesh.sankaran@plymouth.ac.uk (G.S.); marco.palomino@plymouth.ac.uk (M.A.P.)

Abstract: Despite the unabated growth of algorithmic decision-making in organizations, there is a growing consensus that numerous situations will continue to require humans in the loop. However, the blending of a formal machine and bounded human rationality also amplifies the risk of what is known as local rationality. Therefore, it is crucial, especially in a data-abundant environment that characterizes algorithmic decision-making, to devise means to assess performance holistically. In this paper, we propose a simulation-based model to address the current lack of research on quantifying algorithmic interventions in a broader organizational context. Our approach allows the combining of causal modeling and data science algorithms to represent decision settings involving a mix of machine and human rationality to measure performance. As a testbed, we consider the case of a fictitious company trying to improve its forecasting process with the help of a machine learning approach. The example demonstrates that a myopic assessment obscures problems that only a broader framing reveals. It highlights the value of a systems view since the effects of the interplay between human and algorithmic decisions can be largely unintuitive. Such a simulation-based approach can be an effective tool in efforts to delineate roles for humans and algorithms in hybrid contexts.

Keywords: machine learning; system dynamics; simulation modeling; algorithmic decision-making; bounded rationality; supply chain planning

Citation: Sankaran, G.; Palomino, M.A.; Knahl, M.; Siestrup, G. A Modeling Approach for Measuring the Performance of a Human-AI Collaborative Process. *Appl. Sci.* **2022**, *12*, 11642. https://doi.org/10.3390/app122211642

Academic Editor: Mayank Kejriwal

Received: 14 October 2022
Accepted: 12 November 2022
Published: 16 November 2022

1. Introduction

The phenomenal growth of AI in recent years, especially machine learning (ML), a self-improving subfield of AI, has cemented its status as a general-purpose technology [1], like the steam engine or electricity of the past. Therefore, and unsurprisingly, it is also at the center of a strident debate about its impact across multiple dimensions (e.g., economic, social, and ethical) [2], with two very noticeable camps emerging: the optimists and the pessimists [3]. The former camp primarily extolls the virtues (current or anticipated) of ML benefiting all of humanity. The latter, however, warns us about technological sophistication outstripping our ability to reason about its unintended consequences.

Less noticeable, but increasingly gaining traction, is a third camp composed of pragmatists. While acknowledging AI's staggering achievements (thus refuting ardent pessimists), they point out that much progress is still ahead of us and call attention to mounting evidence that should give pause to unchecked optimism. In this view, numerous examples of brittleness (for instance, in the face of adversarial ML) [4,5], poor out-of-distribution performance [6], challenges with explainability [7] (compounded by regulatory pressures [8]), and poor adoption [9] must count as evidence. (On the last point, a recent study has shown that the adoption rate of AI in organizations in the US is less than 7% [10].)

Pragmatism about ML's status and prospects promotes recognition that autonomy is not a viable goal in numerous situations, particularly in open-ended problems (where there is uncertainty about relevant variables, and the effects of causes tend to be distant in space and time). It leads to advocacy for humans in the loop [11]. Of course, given the field's dynamism, the nature of human–ML collaboration must naturally evolve, as well. Therefore, the inevitability of roles for humans in complex decision-making situations coupled with the fast-paced nature of technological change elicits a nuanced view of automation. On this account, a picture of automation antithetical to a simplistic either/or dichotomy [12] emerges.

It is a picture of persistent tension caused by task interdependencies, which are apt to change over time, giving rise to spatial and temporal dynamism. For example, Shestakofsky's [13] empirical work shows that automating a task impacts adjacent tasks in that these (say, previously manual tasks) might benefit from augmentation. Furthermore, the trajectory of these changes heavily depends on the organizational context. Therefore, what further crystallizes is an argument that rejects technological determinism [14] and places importance on context. One where besides the apparent technical aspects, gross behavioral elements such as social relations and politics play substantial roles—a view that accords with the economic theory of complementarities [15]. It holds that studying technological adoption benefits from viewing the human–technology ensemble as a sociotechnical system embedded in an organization, creating a system of complements, a more formal notion of the intuitive idea of synergy.

Although the literature on complementarity illuminates how organizational value derives from the interactions between the embedded technology and the surrounding organizational and broader environmental factors [16], there is a gap when the technology in question is ML [17]. The autonomy that ML affords, albeit partial, represents a break from traditional IT that predominates the discourse about the impact of technology on value creation and capture. In particular, ML's role transcends a mere tool and can assume various other roles, such as those of assistant, peer, and manager [18], depending on context/maturity [19]. A profound consequence of this, plainly stated, is that the ML agents (the technology) now contribute to organizational learning, the object of which is organizational mental models that drive behavior (and create value, or not). Puranam [20] points out an unprecedented dynamic in the history of the technology-driven complementarities that this produces. ML agents can now make the same decisions as humans. So, through aggregation (the wisdom of crowds effect [21]), organizations could generate performance superior to what humans or ML can achieve working alone. Since organizational mental models are the storehouse of creativity, this further implies that, jointly, not just improving existing ways but entirely new ways of doing things (the realm of strategy) open up [22].

The modest premises discussed (self-learning ML, the importance of humans in the loop, and new forms of complementarity that ML affords) combine to yield enormous implications for organizational performance. The challenge of achieving the desired level of performance transforms into a coordinating coalition of human and ML agents that explore the performance landscape in search of tall peaks. Since in the real world of organizational problem-solving, payoffs and the menu of choices are uncertain [23] (as opposed to the closed world of games)—what Hogarth terms "wicked" problems [24]—the exploration has to contend with a "rugged landscape" [25] (i.e., the risk of local maxima).

Knudsen and Srikanth [26] observe that prior work on the normative question of exploring the terrain—or the problem space [27]—in search of satisfactory solutions assumes the organization as a "unitary actor". They note that researchers have scarcely attended to the collaborative aspects (in particular, the role of mutual learning). For instance, the issue of second-guessing arises when there are multiple agents, which can lead to dysfunctional behavior such as, to use their phrasing, "joint myopia" (or *local* rationality [28]) or "mutual confusion" (a result of misperceiving the causes of positive or negative payoffs). Although further complexities arise when humans and ML team up [17], a standout dimension is the inability of ML to fully imbibe tacit knowledge [29] that is crucial to solving many complex tasks.

This is a topic that opens up several avenues for research. However, they fall into two broad categories. The first is research that focuses on the usability aspects of the technology itself. For instance, a burgeoning field known as explanatory AI [30] tries to make ML models less opaque by endowing them with the ability to answer "why" questions, that is, why a specific result or counterfactual questions such as "what would have happened had the input been different?" (in short, an ability to "introspect" their "beliefs"). The second concerns itself with the appropriate use of ML to maximize value—organizational design questions such as the division of labor between humans and ML, and ideal learning configurations [20] fall in this category.

A prerequisite to fruitful research pursuit in either category is the ability to evaluate the combined rationality [31] sufficiently broadly to elucidate the contribution of ML (and, by extension, data) in the context of a longer means-end chain (connecting behavior to business value). It is to this that our work seeks to contribute. We adopt a simulation-based approach (using system dynamics) for the evaluation model. System dynamics is particularly amenable to investigating emergent properties of interdependent actions since it emphasizes dynamic complexity [32] (e.g., due to feedback, a core component of learning) more than component-level complexity. Specifically, our contributions are two-fold:

- We complement the conceptual literature on human–ML teaming that, by necessity (as it caters to various types of organizations), provides general guidelines on effectively structuring collaborations. Our modeling framework allows quantification of the blending of *algorithmic* ML rationality and *bounded* human rationality. We test our approach using an imaginary case of a company trying to improve its supply chain planning process.
- We complement existing work on explanatory AI in terms of framing "why" questions. Concretely, two metrics generally evaluate ML's explanations: interpretability and completeness [33]. Our model provides the organizational problem-solving context (shedding light on the landscape of choices human and ML agents navigate) that must inform the selection of relevant "why" questions.

The structure of the remainder of the paper is as follows. In Section 2, we discuss conceptual frameworks that provide guidelines for human–ML role separation, from which we draw insights that inform the theoretical base for the quantitative framework. In Section 3, we justify our design choices in the framework. Specifically, we explain why choosing a systems approach to modeling best fits the design requirements outlined at the end of Section 2. In Section 4, we describe the details of the framework and run tests using synthetic data to validate our central claim about the risk of local rationality. Finally, in Section 5, we discuss the implications of our findings and comment on what they have to say about related work in this area.

2. Related Work

Various qualitative approaches in the literature suggesting the creation of a human–ML coalition adopt as a guide the insight that ML suffers from what Marcus terms "pointillistic" intelligence [6]. Therefore, in this reading, the overarching brief for humans is to serve as orchestrators in the group such that it can exhibit "general collective intelligence" [18]; Malone describes this group of human strategists and ML tacticians as superminds. Kasparov has written about the strategy/tactics distinction [34] in the context of chess, which serves as a valuable proxy for any intellectual endeavor [35]. It stems from acknowledging that although the cognitive architecture of humans predisposes them to poor performance (compared to ML) on memory and information processing, it allows them to excel in long-term planning, crucial for convergent thinking or the process that results in choosing from among alternatives. (The importance of strategy to decision-making is why Malone recommends that we consider putting computers in the group rather than putting humans in the loop as the mantra for creating effective coalitions.)

Despite the heterogeneity in the details informed by diverse philosophical and intellectual commitments, these approaches share a similar strategy for delineating human and ML roles. They rely on noticing that tasks, seen through the lens of tractability, fall along a spectrum, with some resembling games—fictions of the human mind—or are "game-like", while others, closer to life itself, are "life-like". Game-like tasks are more agreeable to a closed formulation as they have more of the following properties. The rules are well specified and require minimal background knowledge, feedback is unambiguous, feedback loops are short, and behavior is observable. From the perspective of objective attainment, such properties contribute to the connections between the means and the end being neither tenuous nor uncertain (unlike in life-like tasks where the structure is a "tangled web" [36]). It also implies much less difficulty in agreeing on the "best" means for a given end, further aiding a closed formulation.

In contrast, several factors complicate modeling efforts for life-like tasks where social aspects dominate, the value-ladenness of means/ends scuttles efforts to find the "best" option, and poor or absent feedback contributes to flawed mental models. In short, game-like tasks represent a "kind" environment, whereas life-like tasks inhabit a "wicked" environment [24]. A sensible strategy that the approaches often adopt is carefully choosing dimensions that allow the ordering of tasks along the game-like/life-like spectrum, suggesting the appropriate blending of algorithmic and human rationality. In this way, the dimensions proposed include open/closed [19], weak/severe (according to risk) [19], social/asocial [37], creativity/optimization [37], low-dexterity/high-dexterity [37], decision space specificity, size, decision-making transparency, speed, and reproducibility [21], abstraction, intuition/prediction, simulation [38], and thinking/feeling [39].

Although the conceptual models provide a means to assess tasks according to their suitability for ML, they suffer from a critical drawback. Since the recommendations must be broadly applicable, the frameworks have an "objective" bias regarding the problem (that human–ML teams must solve), which yields a disinterested observer or experimenter's eye view of the problem. However, one can scarcely begin to solve real-world problems as posed. A wealth of research in cognitive science supports the importance of framing or problem representation, emphasizing the complexity reduction aspects of problem-solving that make otherwise intractable problems solvable. In their seminal paper on human problem solving, Newell and Simon [27] draw a distinction between the objective problem, the "task environment", in their phrasing, and the problem representation (namely, the "problem space"). The transformation process from the former to the latter is a function of problem complexity. Most problems of interest in organizational decision-making—the consumers of the conceptual frameworks—elude optimal solutions requiring significant simplification efforts. (Simon introduced the term "satisficing" to denote the finding of inexact but satisfactory solutions [40].)

The contrast between the (unreasonable) expectations of optimally solving problems and the reality of searching for a suitable representation that yields good-enough solutions mirrors the contrasting philosophies of the economic man and bounded rationality in cognitive psychology. Several streams of research in organizational theory have explored the implications of bounded rationality in decision-making. They include Klein's naturalistic decision-making [41], Galbraith's organizational information-processing theory [42], Nelson and Winter's evolutionary theory of economic change [43], and Gigerenzer's ecological rationality [44]. These efforts outline structures and tactics that constitute organizational adaptations to the challenges of their task environments. In unison, they reject the idea of an infinitely malleable organization that takes the shape of the problem it is trying to solve, thereby advocating subjectivity (that bounded rationality inevitably entails).

Among the concepts that underpin problem-simplification approaches, the notion of hierarchy stands out as a unifying construct serving as a conceptual glue—in Simon's words, "[h]ierarchy [...] is one of the central structural schemes that the architect of complexity uses" [40]. Hierarchy captures the essence of the near-universal technique of breaking down a complex problem into simpler parts that complex systems embrace. The concept of hierarchical planning systems [45], widespread in supply chain management, vividly illustrates the divide-and-conquer approach inherent in the hierarchical notion. In hierarchical planning's most straightforward formulation, the system stratifies decisions into strategic, tactical, and operational. As a problem passes through the stages, it undergoes the progressive addition of constraints that transform a relatively open problem into a closed one. It results in the sequential imbuing of subjectivity, simultaneously simplifying and providing context to the problem for the organization.

The preceding discussion highlights the importance of an organization-specific problem formulation for evaluating the pairing of humans and ML.

Since such an evaluation typically precedes implementation, it must be quantitative and, given that organizational decisions are context-rich, sufficiently broad in scope, enabling a holistic assessment. To the best of our knowledge, such a quantitative simulation-based model to evaluate the blending of formal/ML and substantive/human rationality in a holistic context is currently lacking, a gap this paper seeks to redress.

3. Design of a Quantitative Model

The desirable traits outlined in the previous section (that an evaluation framework must possess) to assess collaborative human–ML decisions are consistent with findings from the business value literature that deals with the value of technology investments or, more generally, information. A fundamental result from this stream of research, yet one that is often overlooked, is that value does not come from mere investments but derives from proper use [46], reinforcing the importance of quantifying any intervention.

More detailed empirical work [16] on the mechanics of value creation, especially in the resource-based view tradition, recognizes the role of firm-specific resource configuration (erecting "resource position barriers" [47]) in establishing sustained competitive advantage. A resource in this formulation is broad and encompasses such factors as business processes, policies, and culture. This expansive view contrasts with the classical economics definition of resources restricted to only labor, land, and capital. In such an integrated view of value, an assemblage of resources, writ large, mediates technology's performance impact on business outcomes.

An analogous notion to firm-specific resource configuration is the concept of complementarity [48] in organizational economics. In addition to giving quantitative rigor to the hypothesis of synergy behind specific resource configurations, research on complementarities also shows the futility of simplistic ideas of "best practice", a fallacy because business performance is a function of a highly subjective, tenuous mix of internal and external variables. There is substantial empirical [49–51] and anecdotal evidence [48] pointing to the precarity of a desirable system of complements. An organization might suffer significant unintended consequences due to relatively minor changes (also revealing the naivety behind blind imitation). As a result, the metaphor of moving along a rugged landscape (the ruggedness a function of industry dynamism and competition [52]) aptly describes an organization's gradual and tentative attempts to improve its business performance. It aligns with the evolutionary model where the landscape has many local maxima that make finding "good enough" solutions ("satisficing") the only sensible approach.

With empirical support for the subjectivity of technological impact on organizational performance lending credence to intuition from various theoretical bases (chiefly bounded rationality, systems theory, and organizational information processing theory), one can make the implications for an evaluation model—previously mentioned requirements—more precise.

Quantitative. The relative nascency of ML (the technology under consideration here) and the novelty of combining machine and human intelligence further compound the trial-and-error nature of finding an appropriate means of embedding for technology in existing organizational assets. Consequently, it becomes essential to quantify the benefits of competing options, giving rise to the requirement of "quantitative" modeling. Here, the definition of the term quantitative follows from Bertrand and Fransoo [53], which translates to the need for basing the model on a set of variables with "causal relationships" between them.

Simulative. From a modeling perspective, the diversity of paths to value (subjectivity in action) presents the challenge of abstracting from the details while still capturing the richness of context, which plays a pivotal role in determining outcomes. A measure of the appropriateness of a model's level of abstraction is its ability to predict real-world performance. More technically, the model must be "empirical",, again adopting Bertrand and Fransoo's terminology. The alternative approach, called "axiomatic", focuses on better understanding the problem structure (relationship between variables in the model); the objective here is not about achieving correspondence with reality.

The chosen term simulatively performs a double duty: in addition to denoting explanatory power, it forecloses the option of closed-form mathematical formulation, which, in line with the philosophical commitment to bounded rationality, is infeasible given the combinatorial complexity of even moderately sized problems where performance is context-sensitive. In a critique of the predilection for mathematical solutions to closed-form simplifications of real-world problems in the operations research field, Ackoff has cautioned that they tend to be "mathematically sophisticated but contextually naive" [54].

Holistic. A synthetic outlook is a requirement implicit in the term empirical, seen in combination with the premise of subjective problem-framing. However, given its importance, it is a point that bears articulation. The opposite of synthesis is analysis, a reasonable approach to answering mechanistic "how" questions [54]. However, searching for causal explanations of performance requires answering "why" questions, justifying the "holistic" imperative.

Collaborative. The decision-making process must accommodate algorithmic rationality and human judgment or substantive rationality that highlights the (uniquely) human capability for value-rational decisions.

Consistent with evidence from studies about ML's impact on labor [55] that hold that the appropriate unit of analysis is at a task level (rather than at a job level, which is too coarse), the (human–ML) role distinction is likely to be task- and organization-specific. Despite the specificity, a pattern likely to repeat, in agreement with the conceptual frameworks discussed earlier, is the preference for judgment in open contexts and algorithms in relatively closed contexts. Consequently, a challenge—and the raison d'être for such a model—is identifying if what appears to be rational in a limited or local setting [28] remains so when considered globally and does not devolve into dysfunctionality [56]. The need for systemic evaluation narrows the field of candidate paradigms, with system dynamics, a technique created by Jay Forrester [32], emerging as the best choice upon further consideration.

System dynamics buys into a core tenet of complex systems by recognizing that the thrust while modeling must be on the interactions between the components rather than the intricacies of their inner workings. This perspective, inspired by cybernetics, holds that the information flows or feedback are at the heart of learning, influencing our (or organizational) mental models, which manifest as behavior [57]. Noting the pervasiveness of feedback loops (often passing unnoticed) in explaining behavior, Powers goes so far as to say, "... it is as invisible as the air we breathe. Quite literally, it is behavior" [58]. A powerful tool in the system dynamics toolbox, the causal loop diagram, operationalizes this way of thinking about behavior where it is "one of the causes of the same behavior" [58] by depicting a system of cause-and-effect variables in a closed loop.

Therefore, this illustration technique is a common design artifact before implementation in a tool such as Vensim [59,60] that finds extensive use in industry and academia for its straightforward interface and simulation and reporting capabilities. It is also the tool used in the experiment described in the next section.

Causal loop diagrams that visualize a system's feedback structure turn up another vital property of complexity. The individual causal links are simple enough but collectively produce complex emergent behavior, epitomizing the wisdom of complexity theory that Simon articulates thus: "complexity, correctly viewed, is only a mask for simplicity" [40]. In the modeling process, this property has the beneficial effect of simplification. Since the scope boundary is drawn more broadly (compared to alternative approaches)—in line with holistic thinking—the emergent nature of complexity has an overall offsetting effect in the modeling effort.

The discussion about system dynamics has shown that the paradigm meets the qualitative, simulative, and holistic criteria. However, regarding collaboration that requires the mixing of judgment and algorithmic reckoning, a tool such as Vensim does not natively support incorporating ML methods. Here, a Python library, PySD, developed by Houghton and Siegel [60], addresses the gap and allows the infusion of data science techniques into system dynamics models, thus satisfying the collaborative criterion.

PySD enables the bridging of causal modeling (the backbone of system dynamics) and the ever-growing field of data science. It opens doors to exploiting the natural synergy between the fields: the former premised on the tenet that structure drives behavior, and the latter rich in techniques that allow both the modeling of more sophisticated behaviors and their analysis (which can inform improved models).

Despite the potential for embedding ML agents in system dynamics models, the overarching principle that the presence of structural elements such as feedback, delays, and stocks means that one cannot reliably predict the overall dynamic behavior of a system still holds. Thus, such systems' "dynamic complexity" [32] renders analytical solutions infeasible, providing further impetus to simulation-based approaches.

The importance of the structure noted above stems from taking a firm stance (which PySD implicitly does) related to the epistemological question of whether knowledge can be model-free. There have been claims that with big data, we have entered a new paradigm where data can speak for themselves [61], a claim that contradicts the core of the scientific method. However, Pearl [62] and numerous others (e.g., [63,64]) argue that meaning relies on a structure one cannot build from data alone. Trending issues in ML around out-of-distribution performance and explainability further support the position that to progress from merely observing correlations to attributing causes, one has to, in Pearl's words, climb the "ladder of causation" [65]. It requires translating mental representations, the infrastructure humans use so effectively, into formal models that, in conjunction with data, can make understanding possible.

4. Experiment

This section introduces a small-scale experiment to test the viability of the main ideas in the proposed modeling framework (the ML code, system dynamics simulation files, and data are available on GitHub under: https://anonymous.4open.science/r/aicollab-model-C108) For evaluating human-AI collaborative decision-making.

It focuses on the importance of a holistic problem-solving approach that is more resistant to the potentially distracting effect [28] of superior information processing in that such an approach is wary of immediately visible improvements local in time and space, masking unintended consequences that may be quite distant (due to delays and complex feedback structures).

Concretely, improvements in the forecasting process—the result of a machine learning algorithm replacing a judgmental process—represent the "visible" local improvement in the experiment. However, this unearths a suboptimal decision routine in the production process that results in overall underwhelming performance (especially given the magnitude of improvement in forecasting accuracy when viewed narrowly).

4.1. Problem Context

The experiment involves a fictitious company, Acme, attempting to improve its product sales and returns-forecasting process. The company uses a simple first-order exponential smoothing process—a simple but surprisingly hard-to-beat procedure [66]—and would like to evaluate the benefits of implementing an advanced machine learning algorithm, especially for returns. The basic assumption that returns forecasting can benefit from an algorithm more sophisticated than Acme's current univariate (or single-variable) forecasting method is well founded. Specifically, more sophistication afforded by a multivariate approach might improve forecasting accuracy by using additional (leading) indicators, such as historical sales in the case of forecasting returns.

Although a simple comparison of forecast accuracy between the two approaches is a reasonable starting point for evaluating the potential benefits, it is often insufficient. The insufficiency stems from the forecasting process being just one among several processes in end-to-end process chains that encompass forward (material flow from suppliers towards customers) and returns flows (where the customer becomes the supplier for the post-consumer product [67]). Therefore, it is critical to check the local intervention (the forecasting process in the case of this experiment) for unintended global consequences. At Acme, besides the planned machine learning model for forecasting, most other decisions are assumed to be based on rules of thumb or heuristics. Therefore, a systemic assessment (checking if the locally rational algorithmic component translates globally, given that there is a mix of algorithmic and human rationality at this level) of the comingling of human and algorithmic decisions entails modeling the relevant parts of the adjacent production and order-fulfillment processes.

4.2. Data

The experiment (see Figure 1 for experimental protocol) uses a seasonal time series from the M forecasting competition [68] to generate a synthetic sales dataset by first decomposing it (into trend-cycle, seasonal, and remainder components) and subsequently constructing samples with similar demand characteristics.

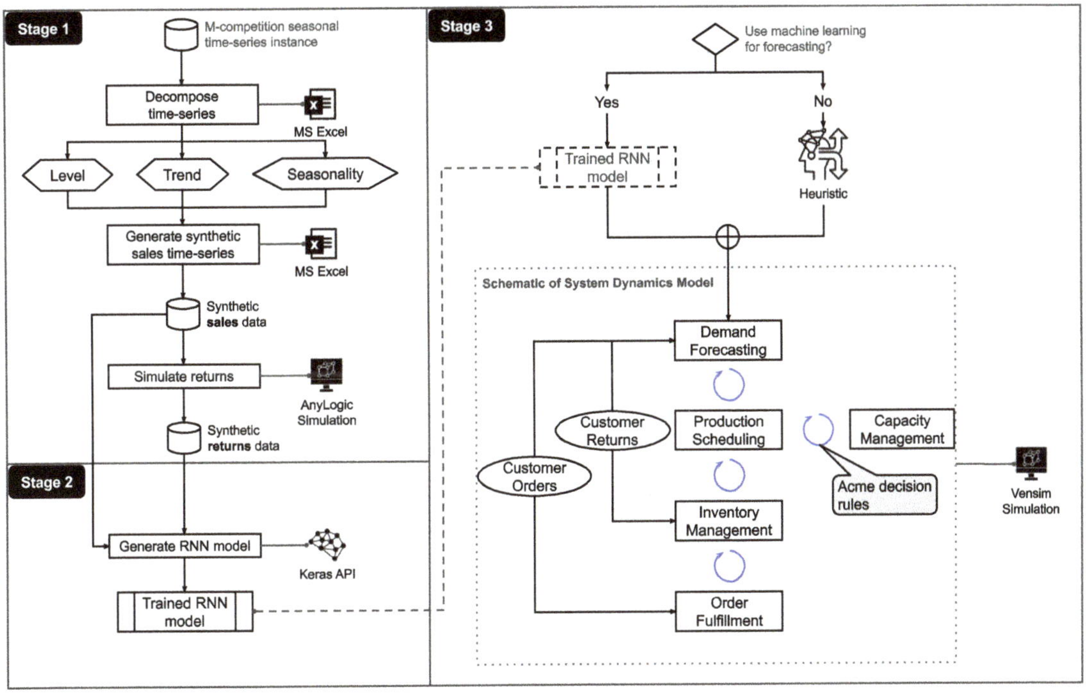

Figure 1. Experimental protocol consists of three stages: generation of synthetic data, creation of the RNN model for predictions, and incorporation of the RNN model in the system dynamics causal model for simulating scenarios for heuristics (human)–ML collaboration.

Regarding the returns time series, the assumption is that Acme has three classes of customers with distinct returns characteristics or profiles (where a profile is a specific combination of the mean and standard deviation of returns that follows a normal distribution). A discrete event-simulation model built in AnyLogic [69] on this assumption generates the requisite returns data. It first spawns customer "agents" (based on the sales data) and sorts them randomly into three groups, assigning them their corresponding returns profiles. After a specified time offset, the model simulates an agent generating a product return per its profile—that is, a sample value, which stands for the number of units returned, is drawn from a normal distribution with the profile's mean and standard deviation (see Figure 2).

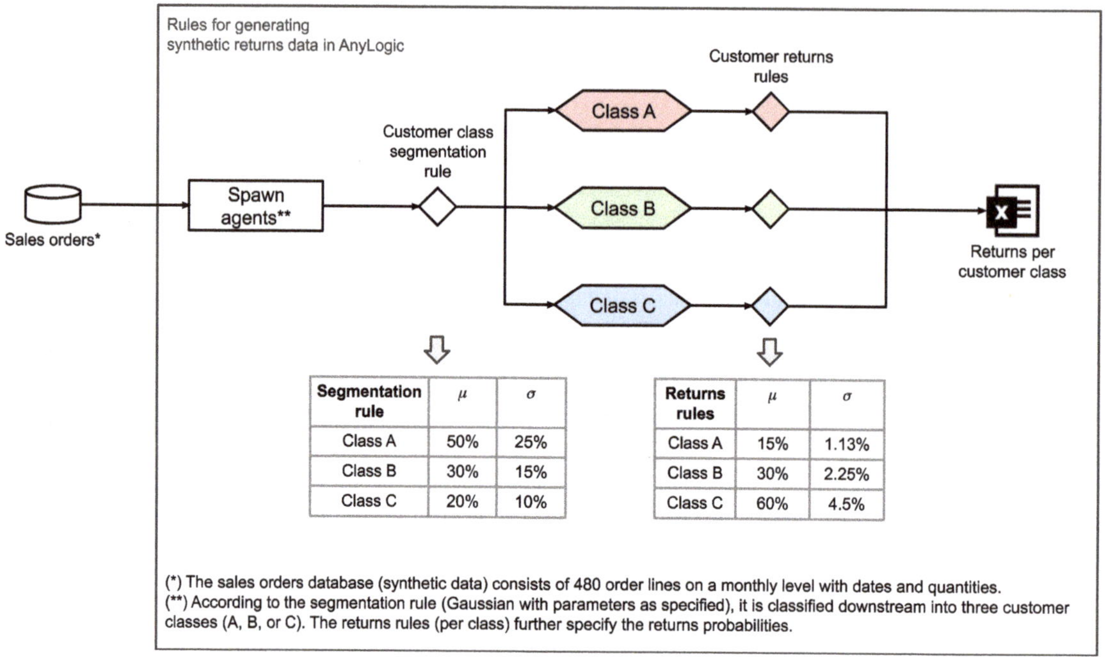

Figure 2. Dataset generation for product returns.

The synthetic data generation process outputs a file consisting of 480 months of monthly sales (broken down by customer group) and returns data. Before model generation, the data undergo a further important processing step designed to tackle the problem of poor generalization common in ML.

Poor generalization performance, or the phenomenon of overfitting, is when model performance deteriorates severely on unseen data. It typically happens when the model learns spurious correlations or memorizes inputs [70] to return good training performance that does not translate to good performance in the real world. The usual recommendation to avoid overfitting is to split the data into train, test, and validation sets [71], which the experiment adopts by splitting the data according to a 60/10/30 ratio. Although several specific techniques exist to perform the split, the experiment takes a simple holdout validation approach. In this variant, the training set determines model parameters, the validation set helps fine-tune those parameters, and the test set forms the basis for the final performance evaluation. Given the relatively simple nature of the original seasonal sales data with a limited number of possible features and the synthetic data generation approach (affording finer control over the noise, thus placing modest demands on sample size), there are no grounds for more complex treatments such as K-fold and iterated K-fold validation more suited to feature-rich/data-sparse contexts [70].

4.3. Main Components

4.3.1. Modeling Judgmental Forecast

System dynamics offers various techniques for modeling simple rules that characterize human decisions in most contexts [57]. One such representation is the so-called anchor-and-adjust [72], which produces an effect similar to the first-order exponential smoothing procedure, which serves as the experiment's current-state judgmental process for forecasting sales and returns. The term anchor-and-adjust alludes to a well-known fact from psychological research that humans, when tasked with estimation, "anchor" on an initial value and adjust it according to the cues they receive [73]. Despite the apparent simple-mindedness of the procedure, there is abundant empirical evidence [72] that supports its use by decision-makers in contexts where the sheer number of influencing factors make satisficing rational in intention.

In the experiment's forecasting process, the initial value or anchor is a historical average of sales or returns. The value undergoes continuous adjustments upon receiving informational feedback (actual sales or returns orders). The adjustment rate, which depends on empirical details regarding such factors as feedback delays experienced by an organization, is set to three months in Acme's case. A delay of three months roughly corresponds to setting the smoothing constant to 0.3 when using the exponential smoothing procedure. Although it is possible to search (for instance, using a grid search technique) for a more optimal value, it is not essential given that the experimental objective only relies on the claim that ML represents any improvement over a heuristic approach. In other words, the qualification criterion for local rationality is that ML is somewhat better than the extant approach. Furthermore, as we will also see, the magnitude of the difference in accuracy between the two approaches renders any fine-tuning effort of the delay parameter moot.

4.3.2. Modeling ML Forecast

As mentioned earlier, the primary focus of the ML method in Acme's context is improving returns forecast accuracy. Since sales (split by the three customer groups) serve as an early indicator for future returns (except, potentially, historical returns), the intuition is apparent behind using an ML algorithm that can learn how the two relate without explicit instructions. More technically, an ML model can learn, from training samples (consisting of input/output pairs), the transformation from the input (set of early indicators) to the output (future returns) that minimizes prediction errors. This description corresponds to a supervised learning regime.

However, the requirements for sequential data (in this context, time series) are slightly more stringent—the architecture must be capable of maintaining temporal ordering. From this perspective, there are two basic ML architectures: feedforward networks that flatten the inputs, hence their lack of means to carry forward information meaningfully, and architectures with a feedback loop. A recurrent neural network (RNN) is an architecture that falls into the latter type, which the evaluation framework uses. An RNN can use its memory about earlier periods in a time-series setting and combine it with the current period while making a prediction. One can best imagine the process by "unrolling (the network) through time" [74]; that is, imagining the network processing each of the periods sequentially. In the simplest case of a network of a single artificial neuron, it receives as input both the current period value and the output of the previous period (usually initialized to zero at the start). Thus, at any given period, the additional input—the output of the previous period—is akin to the memory of the entire past, which influences predictions.

For the RNN implementation, the framework uses the Keras API, which offers convenient routines for training deep learning models [70]. In Keras, there are three types of RNN available: SimpleRNN, long short-term memory (LSTM), and the gated recurrent unit (GRU). Each type shares the basic idea of carrying over information when processing sequential information such as time series. The crucial difference between RNN and both LSTM and GRU lies in the latters' relative ability to handle long sequences. Since the backpropagation procedure has to deal with a significantly deeper network, given the unrolling through time, SimpleRNN (the vanilla implementation) suffers from a debilitating memory loss problem. It is a problem that LSTM and GRU specifically address, partly by being more discerning about what to retain and what to forget [74]. In the experiment, the sales model uses LSTM given the long historical sales horizon (36 months since there are seasonal effects). For the returns scenario, the experiment uses GRU as it performed better during the parameter tuning phase.

Figures 3 and 4 below provide a schematic representation of the ML (using RNN for illustrative purposes) and heuristic approaches.

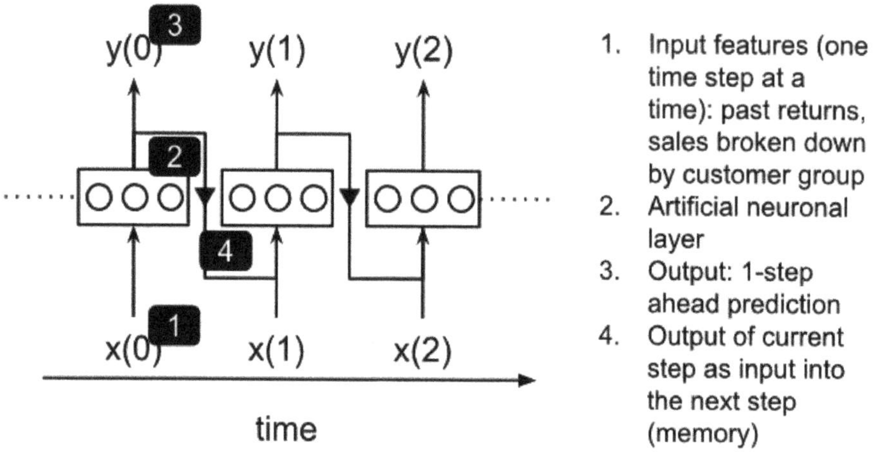

Figure 3. Schematic representation of the unrolling through time in the RNN architecture.

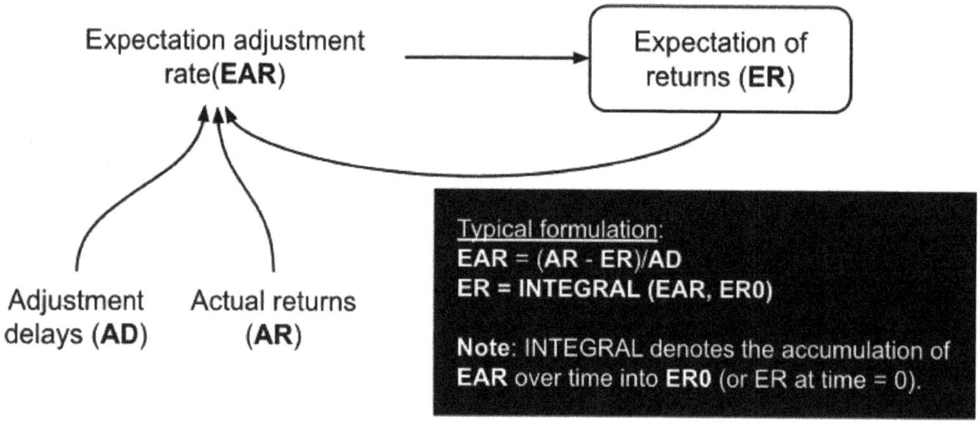

Figure 4. An illustration of how the anchor-and-adjust heuristic works.

4.4. Execution and Results

4.4.1. Comparing Stand-Alone Forecasting Performance

The first stage of the experiment is a straightforward comparison of the forecast accuracy of the RNN and heuristic approaches. The RNN sales- and returns-forecasting models have the following main parameters (see Table 1): a single hidden layer (the former with 225 units and the latter with 150 units), a dropout rate of 10%, a mean squared error (MSE) loss function, and 150 epochs of training. As noted earlier, the sales model uses LSTM and a returns model GRU. (Including dropouts is another effective means to avoid overfitting as the dropping out of units from the network with a certain probability (rate) leads to more robust overall learning since it trains the elements to be more self-reliant and discourages excessive reliance on specific inputs.)

Table 1. RNN model parameters.

Parameter	Description	Value	
		Sales	**Returns**
Historical periods	Actual historical sales or returns horizon to use for forecasting.	36 months	1 month
Forecast periods	Forecast horizon.	1 month	1 month
RNN layer	There are three built-in layers in Keras: SimpleRNN, GRU, and LSTM (the latter two support longer time-series sequences; we describe the rationale in the text).	LSTM	GRU
Optimizer	Gradient method used.	Adam	Adam
Layers	Depth of the neural network.	1	1
Number of units	Artificial neurons per layer.	225	150
Dropout	Regularization parameter (described in the text).	10%	10%
Epochs	Training iterations.	150	150

After model fit, an evaluation of the model to assess overfitting (Figure 5) shows the converging training and validation loss curves, which indicates a robust fit. The canonical overfitting behavior is when training loss decreases while the validation loss increases—the divergence is predictive of poor generalizability.

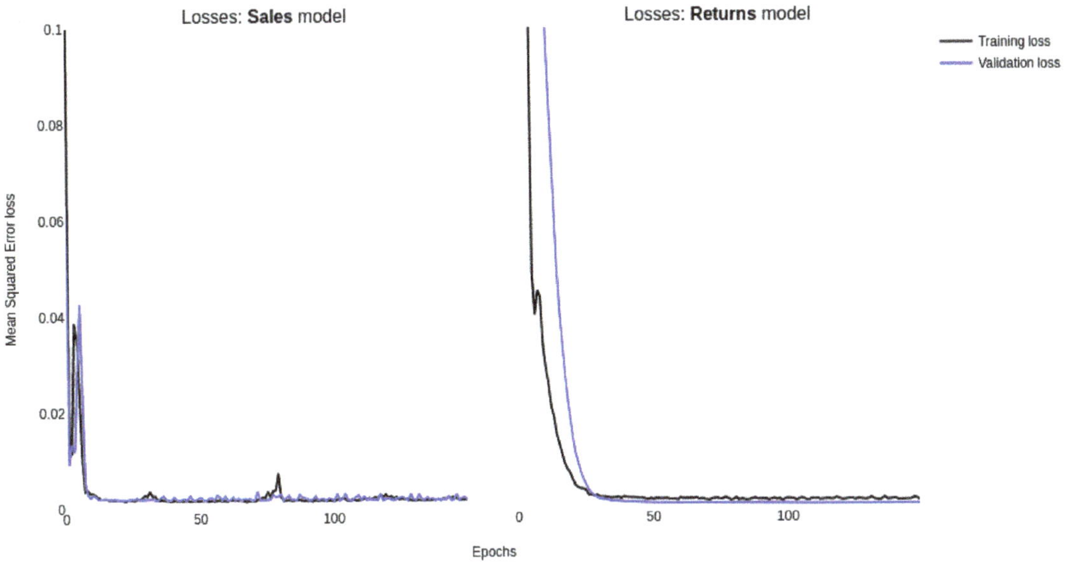

Figure 5. Loss curves for the sales and returns models.

Carrying out a partial model test to verify intended rationality [56] by embedding the forecasting routine in the system dynamics model shows that the accuracy (measured using the mean squared error metric) of RNN is 87.4% better for sales (21.4 compared to 2.7) and 81% better for returns (5.8 compared to 1.1). Figure 6 below shows the system dynamics model for returns; the sales model follows the same structure. Table A1 in Appendix A provides a complete list of the parameters for the system dynamics model.

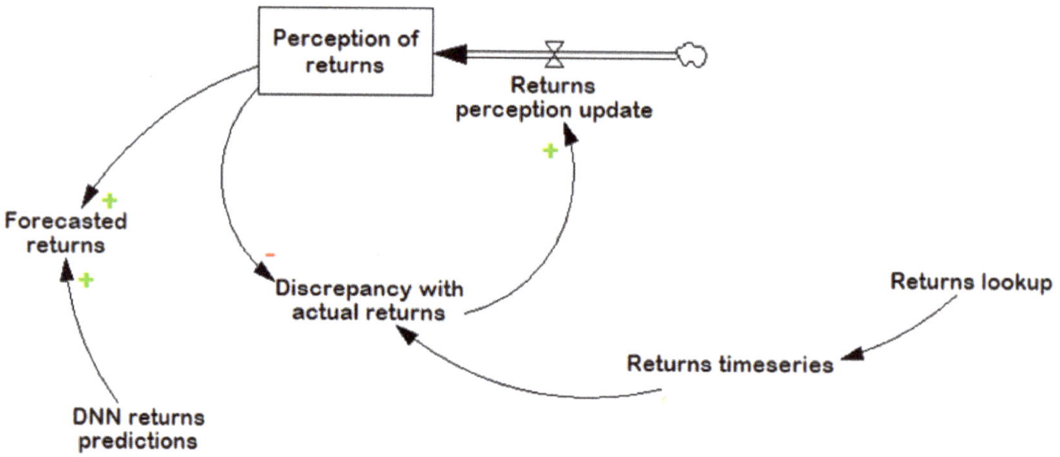

Figure 6. The partial system dynamics model for predicting returns.

The graphs below (Figures 7 and 8) compare the RNN and heuristic predictions against the actual sales and returns orders. At first glance, a more significant improvement in sales-forecasting accuracy with RNN might be surprising. However, this is because the sales time series shows seasonality, but the anchor-and-adjust heuristic does not account for seasonal factors, providing a satisfactory explanation.

As an additional sanity check, a comparison of the accuracy of the heuristic to a simple exponential smoothing procedure using the Statsmodels library [75] (with a smoothing equivalent to a delay of three months, as described earlier) shows that the MSEs are roughly the same (Figure 9). It confirms the magnitude of the local improvement indicated earlier.

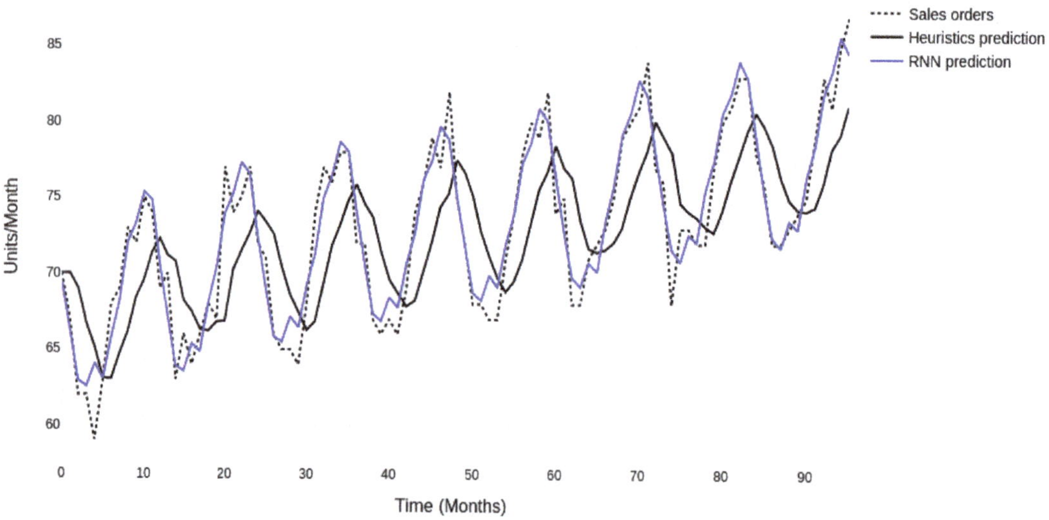

Figure 7. Comparing heuristic and RNN sales predictions.

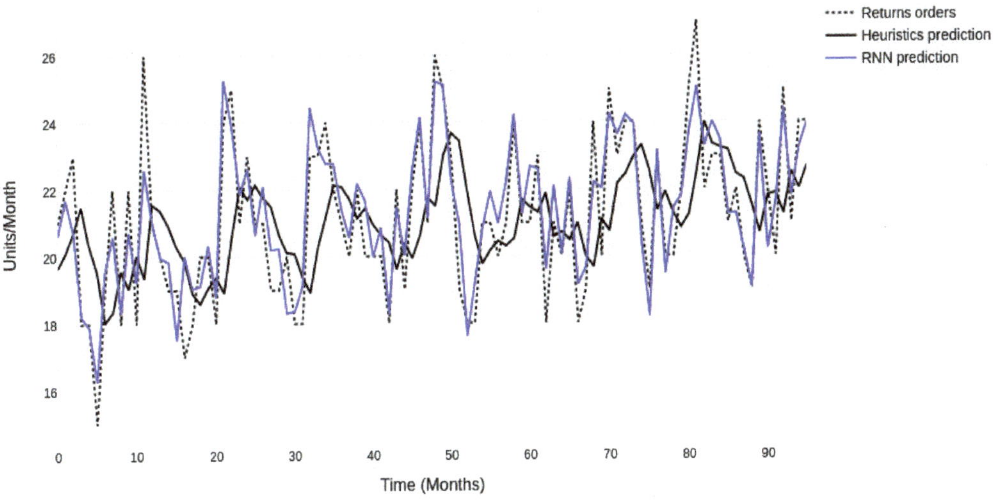

Figure 8. Comparing heuristic and RNN returns predictions.

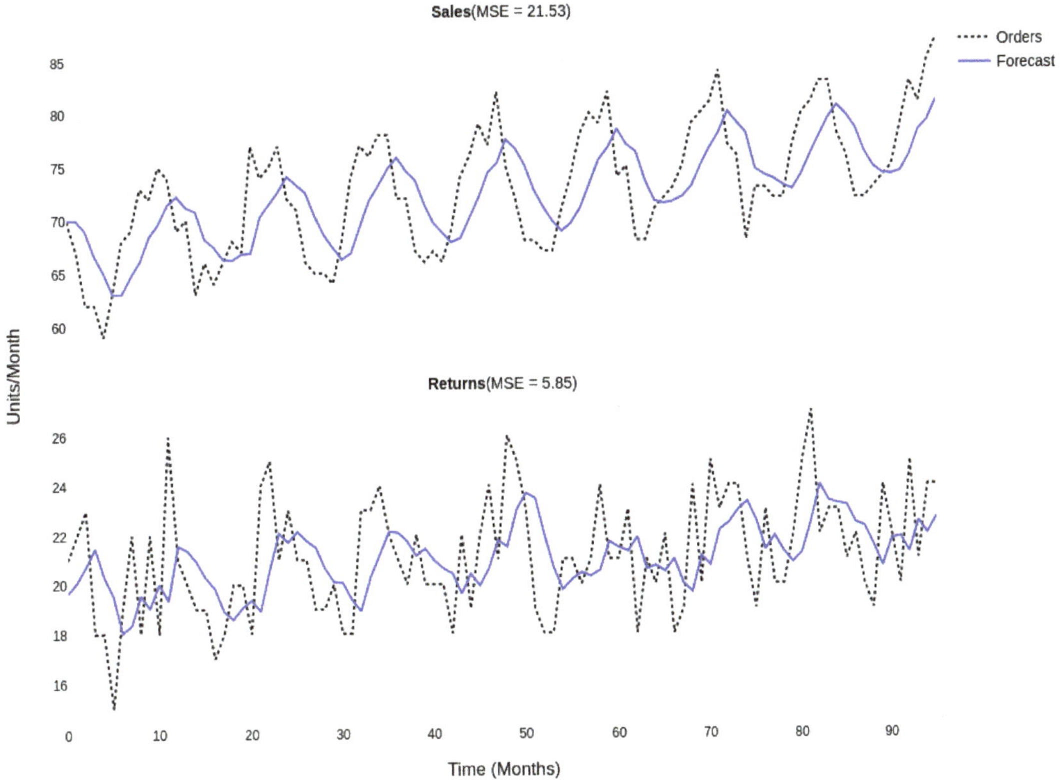

Figure 9. An exponential smoothing approach to forecasting sales and returns.

4.4.2. Comparing Overall System Performance

The second and most crucial stage involves evaluating the impact of the forecasting intervention (use of RNN) in its proper context by including relevant aspects of Acme's overall forecasting, order fulfillment, and extended production processes. A decision parameter in the whole-model simulation (Figure 10a) determines the source of predictions (for sales and returns)—heuristics (or judgmental) or RNN. If set, PySD substitutes a hook in the model with a function that makes online predictions that integrate with the rest of the model (Figure 10b). Otherwise, judgmental or heuristic predictions take effect (Figure 10c). Acme follows a make-to-stock strategy, implying that it fulfills orders from inventory. Here (order fulfillment sector in the figure below), the system dynamics model includes simple rules to satisfy orders as they come in and to ascertain backorders and lost sales if there is a shortage—in other words, when the forecast is inaccurate. If the delivery lead time exceeds the goal, delivery pressure (a function of backlog) builds up, resulting in lost sales if the delay exceeds the tolerance limit (Figure 10d). Production orders are simply the difference between forecasted sales and returns in the production sector. For simplicity, the experiment assumes a negligible production lead time (a few days) relative to the planning periodicity (of months). At the end of the month, the forecasts generate production orders, assumed to be available as inventory at the start of the following period.

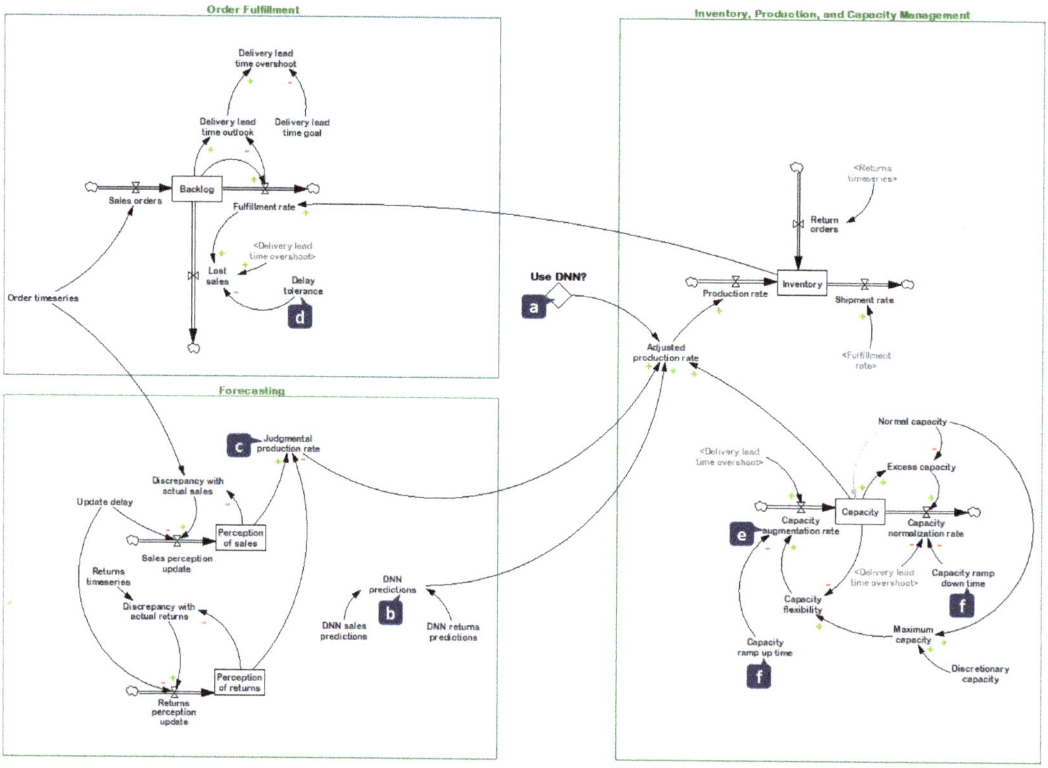

Figure 10. Whole simulation model. (**a**): Choice of forecasting procedure. (**b**): RNN forecast—a "hook" that is programmatically substituted with online predictions via PySD. (**c**): Judgmental forecast. (**d**): Delay tolerance. (**e**): Capacity adjustment under delivery pressure. (**f**): Capacity ramp-up and ramp-down delays.

The decision rules discussed thus far are operational (in their evolutionary theory of economic change, Nelson and Winter use the term "operating characteristics" to characterize such rules [43]). However, there is a routine in the capacity management sector that is at a higher level, typically considered tactical by supply chains. It is a routine that calibrates the available capacity and is a crucial determinant of overall performance. The routine is responsible for augmenting capacity under delivery pressure (Figure 10e). Augmentation increases the capacity at a rate defined by the ramp-up delay (Figure 10f) to a maximum capacity determined by the available discretionary capacity. On the other hand, capacity normalization happens as the pressure eases. The equation for delivery lead time overshoot captures the easing of delivery pressure. This is the difference between the goal delivery lead time and the delivery lead time outlook (the ratio between the backlog and the current clearance rate). As the overshoot moves close to zero—all other things being equal—the normalization rule down-regulates the available capacity until it reaches the standard available capacity.

As all the rules in the whole-model simulation, except the source of predictions, are the same, verifying the effect of improved forecast accuracy on the overall performance is allowed. In an ideal scenario—perfect forecast accuracy—the lost sales are zero, and the average inventory equals half of the average production orders (since the planned production is available at the start of the period and the consumption of inventory by sales is assumed to proceed at a constant rate). Thus, lost sales and average inventory are the outcome metrics—closer to actual business performance—that provide a window into how well the process metric (forecast accuracy) translates to improved performance, seen holistically.

Base Case

As the first step of the whole-model simulation, setting the capacity profile to two months each for ramp-down and ramp-up and a 20% discretionary capacity, the results show a 39% improvement in lost sales (RNN over heuristics) performance and a 6% improvement in inventory (see Figures 11 and 12, and Table 2).

Table 2. Outcome metrics summary.

					Case B vs. Case A	
	Metric (in Units)	**Heuristics (H)**	**RNN (R)**	**R vs. H**	**Heuristics**	**RNN**
(A) Base case: 20% discretionary capacity, quick ramp-up and ramp-down (1)	Lost Sales	29.06	17.85	**−39%**	**N/A**	**N/A**
	Inventory	39.46	37.17	**−6%**	**N/A**	**N/A**
(B) After heuristic adjustment: 20% discretionary capacity; quick ramp-up and slow ramp-down (2)	Lost Sales	26.10	10.27	**−61%**	**−10.2%**	**−42.5%**
	Inventory	39.52	37.54	**−5%**	**0.2%**	**1.0%**

Notes: (1) Quick ramp-up and ramp-down: Capacity ramp-up lasts two months. Capacity normalization when delivery pressure eases also lasts two months. (2) Quick ramp-up and slow ramp-down: Capacity ramp-up lasts two months. However, capacity normalization when delivery pressure eases lasts four months.

At first glance, the results seem to live up to the promise of the forecast accuracy gains of RNN over heuristics. However, studying the graphs gives pause as it suggests that there are further improvements to be made. Focusing on the lost sales and capacity subplots and comparing the RNN and heuristics graphs, one sees that the capacity profile in the case of RNN has significantly more spikes. The lost sales in the case of RNN are also much more densely clustered compared to heuristics. This behavior results from RNN's superior ability to capture the peaks and troughs in customer demands (one can see this by comparing the production rate curves). In particular, the inability of heuristics to anticipate the troughs results in excess inventory. The inventory build-up obviates the need for sustained additional capacity in the case of heuristics—thus, the capacity availability curve is smoother.

On the other hand, the much-improved forecast accuracy of RNN translates to the production rate closely chasing actual demands, thereby leading to a leaner inventory profile. An additional consequence of the better anticipation of lows is that capacity seems to normalize too quickly during periods with "rugged" peaks. This suggests a simple adjustment (an increase) to the capacity ramp-down delay. Intuitively, a slower ramp-down should allow the provisioning of some buffer capacity to clear the backlog, even as the production rate continues to roughly trace the sharp turns in the demands.

Heuristic Adjustment

After adjusting the ramp-down to four months (up from two months), the results show a 61% improvement in lost sales (RNN over heuristics) performance and a 5% improvement in inventory. This represents a 42.5% improvement in lost sales (with a slight 1% degradation in inventory performance) for RNN over the base case (see Figures 13 and 14, and Table 2).

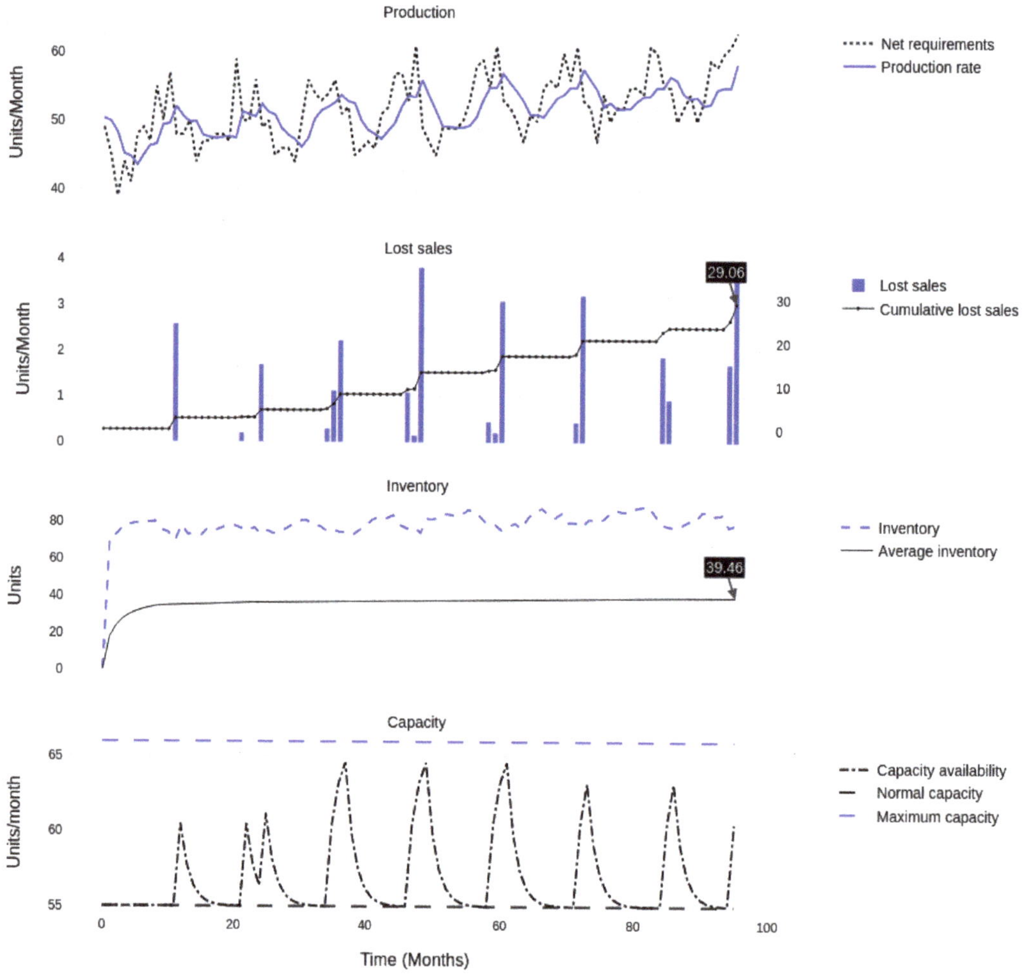

Figure 11. Whole-model simulation results: heuristics with discretionary capacity; quick ramp-up and ramp-down.

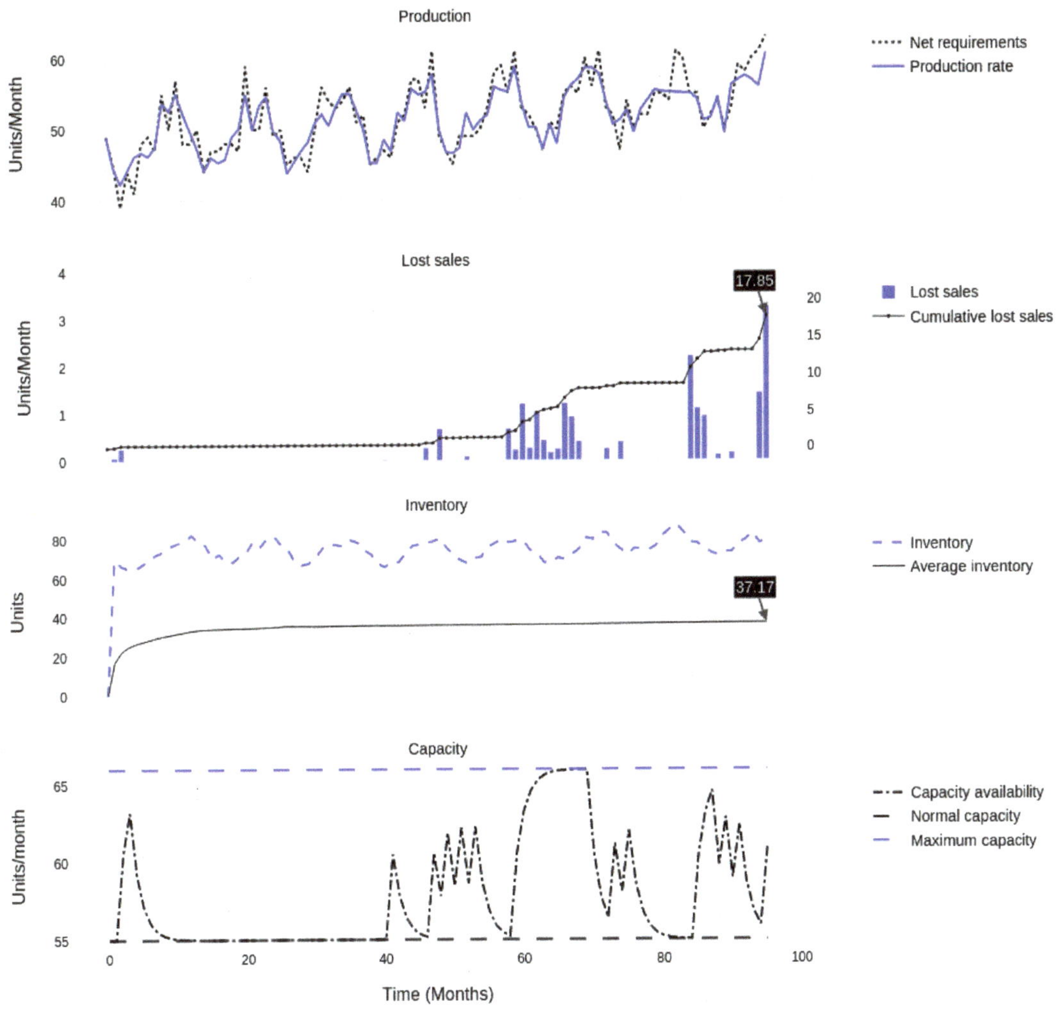

Figure 12. Whole-model simulation results: RNN with discretionary capacity; quick ramp-up and ramp-down.

As intuitively hypothesized, the improvement in the case of heuristics over the previous scenario is minor in comparison (10% improvement in lost sales and a 0.2% degradation in inventory performance) to RNN, given its tendency to build excess inventory. Furthermore, as the capacity utilization for RNN is only slightly more than heuristics (3% more; 57.8 units/month versus 56.1 units/month), the nearly cost-neutral rule adjustment projects substantial overall gains.

A graph that overlays lost sales and the capacity profiles in the second scenario for RNN (see Figure 15) confirms our intuition regarding why RNN benefits disproportionately from this rule change. The capacity availability profile in the second case has fewer spikes owing to the more gradual ramp-down, which allows for additional buffer capacity (relative to case A) for clearing the backlog, resulting in less dense clustering of lost sales than before.

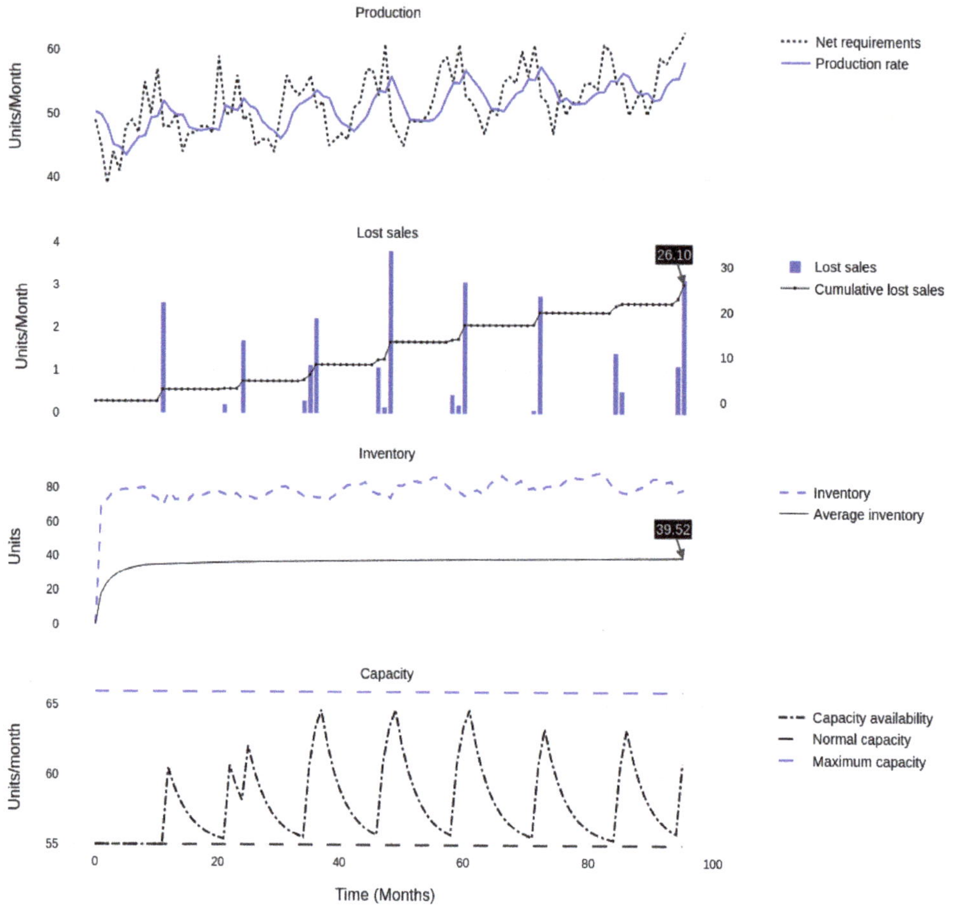

Figure 13. Whole-model simulation results: heuristics with discretionary capacity; quick ramp-up and slow ramp-down.

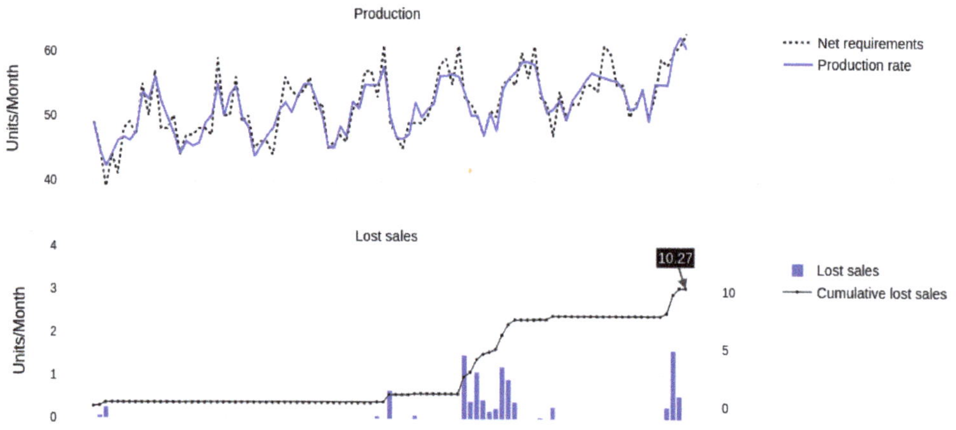

Figure 14. *Cont.*

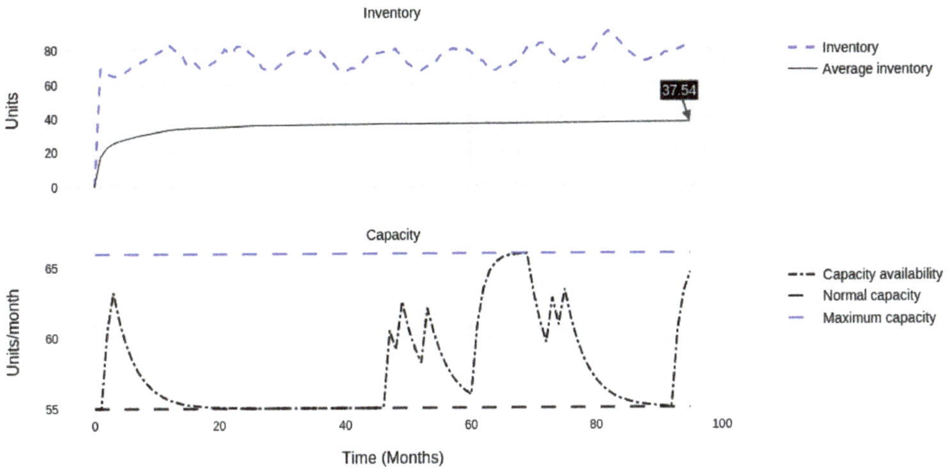

Figure 14. Whole-model simulation results: RNN with discretionary capacity; quick ramp-up and slow ramp-down.

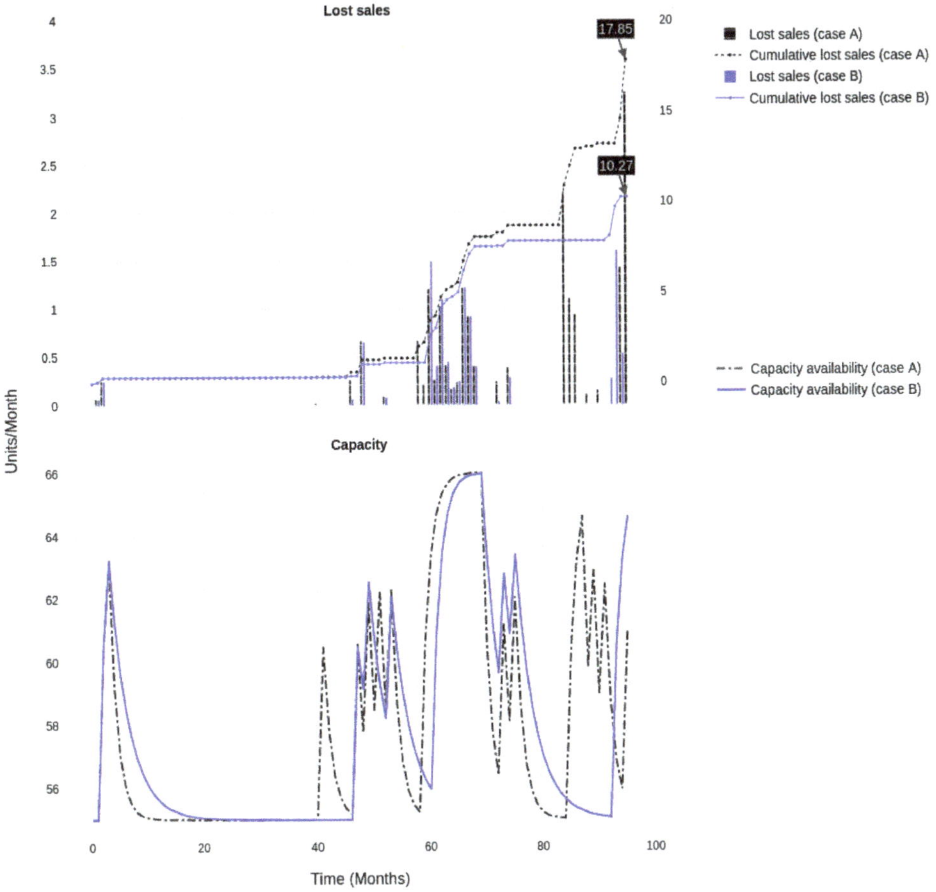

Figure 15. Comparing quick and slow ramp-down along lost sales and capacity measures.

5. Discussion and Conclusions

Despite the rapid pace of progress in ML, there are concerns about the disconnect between innovation and adoption [9] and between investments and value (e.g., [76,77]). A growing body of work reflecting on the state of ML notes the overemphasis of technology over transformation (e.g., [78,79]) and worker substitution over augmentation [15]. Consequently, there is neglect in studying the unique ways by which humans and ML can jointly unlock significantly more value (e.g., [14,22,80]). We contribute to the conversation by adopting the view that ML is a technological asset that combines in an organizational-specific manner with other assets, chiefly personnel, to generate value. In the following paragraphs, we discuss the primary insights from the simulations performed using our proposed quantitative model that is suitably subjective (and holistic) in its conceptualization of the value-generation process. We also note the implications of these insights and how they relate to other works of a more conceptual/abstract nature in this area.

Our simulations have highlighted that, although procedurally rational, local process improvements (measured via *process metrics*) do not automatically translate to commensurate overall benefits (measured via *outcome metrics*). The system dynamics approach provides an elegant way to confirm the rationale of the improvement (in this case, the ML intervention) through partial model tests before proceeding to whole-model simulations to check for unintended global consequences. Although, as noted earlier, the idea of partial-model tests is not new [56], employing the idea when the improvement comes from ML is novel. It assumes greater significance in light of recent work [17] on organizational learning (using an abstract agent-based modeling approach) in a human–ML collaborative context that shows that ML strongly influences the classic explore–exploit trade-off [81]. Specifically, since ML agents do not subscribe to preexisting organizational mental models, they tend to facilitate nimbler exploration of the performance landscape. However, this also amplifies the type of risk our experiments illustrate (entering an organization into operating regimes with an increased likelihood of untested decision routines or operating characteristics that might produce dysfunctional global outcomes). The new dynamics caused by the introduction of ML forecasting in our experiment underscores this point.

Before ML, the incumbent heuristic method was slow to react to peaks and troughs in actual sales and returns (because of the inertia inherent in the anchor-and-adjust heuristic, current perceptions change only slowly). More pertinent to the earlier point regarding exploration, ML predictions (from the forecasting process) that are closer to actual values readily expose the inadequacy of the adjacent capacity management process. For instance, some high values that ML predicts are more than the standard available capacity, and the rules for using additional discretionary capacity suffer from latency, leading to poor order fulfillment performance. The example confirms the folk wisdom in manufacturing, supported by rigorous research, that saving time at a non-bottleneck resource is a mirage [82]. Translated to the experiment, the forecast improvement beyond a point collides with the capacity bottleneck, which limits the performance (unless addressed). This example reinforces the point about broadening the scope of analysis—organizational decision-making involves complex feedback loops that make it unrealistic to anticipate high-level outcomes accurately.

In addition, the approach taken in the experiment to alleviate the problem demonstrates the importance of complementarity between decision pairs. Concretely, the improvement took the form of reducing the delay in using the discretionary capacity. In general, ML increases the "clock speed" [83] of an organization, and the decision structures must keep pace, for example, through decentralization that tends to reduce the number of levels a decision has to pass through (reducing delays).

A further implication of bottlenecks preventing subsystem improvements cascading to the system level—discovered through a synthetic rather than an analytical view of performance—is how it provides a valuable frame for questions about the value of data. In case additional data (costly to acquire and process) push the system to an operating point that surfaces limiting constraints fixed in the short term (e.g., physical assets or lead times), it puts a cap on benefits. This, in turn, helps ascertain the value of data collection efforts. From a more technical standpoint, a systems lens strengthens the argument for a reasonable statistical baseline before attempting ML methods that usually require many predictors and complex nonlinear relationships (between predictors and the target variable) for their superior performance [84].

Treating data as instrumental to value (and not valuable in themselves) is a position that follows naturally from the causal modeling approach that is the bedrock of system dynamics simulation. Thus, the importance given to the data-generating process aligns with the position of the causal inference research community (gaining wider acceptance) that espouses the need for good explanations. In Pearl's words, "empiricism should be balanced with the principles of model-based science" [62]. One can surmise the upshot of this from our simulations. By situating the forecasting process in the context of the end-to-end order-to-delivery process chain, the model makes prioritizing aspects of the explanation possible. For instance, one can focus on predictions that most impinge outcomes and pose "why" questions (see Figure 16) to understand if they are representative or a product of anomalous inputs. In this way, the proposed modeling approach contributes to the explanatory AI work by identifying what the "completeness" criterion (for evaluating explanations) must entail.

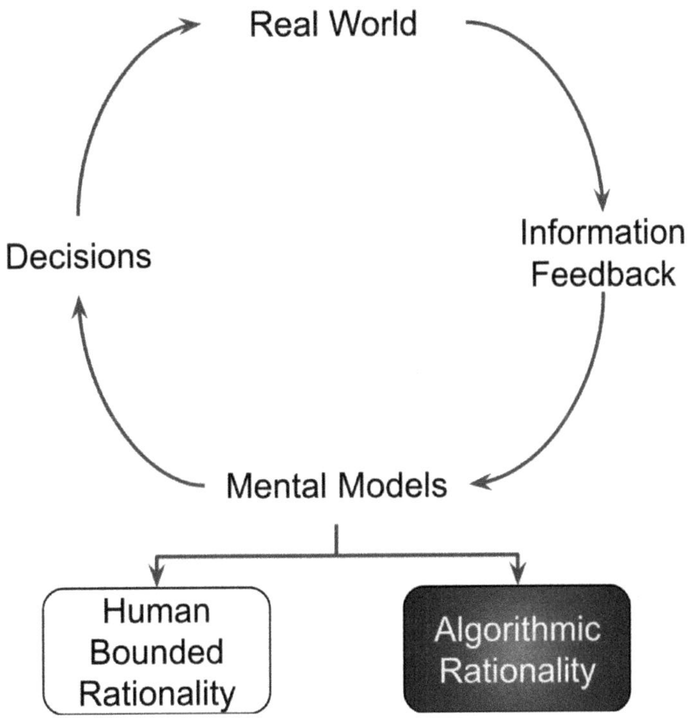

Figure 16. Causal modeling engenders asking relevant "why" questions to make algorithmic rationality less opaque.

Author Contributions: Conceptualization, G.S. (Ganesh Sankaran); Methodology, G.S. (Ganesh Sankaran); Software, G.S. (Ganesh Sankaran); Validation, G.S. (Ganesh Sankaran) and M.A.P.; Formal analysis, G.S. (Ganesh Sankaran); Investigation, G.S. (Ganesh Sankaran); Data curation, G.S. (Ganesh Sankaran); Writing—original draft, G.S. (Ganesh Sankaran); Writing—review & editing, G.S. (Ganesh Sankaran), M.A.P. and G.S. (Guido Siestrup); Visualization, G.S. (Ganesh Sankaran); Supervision, M.A.P., M.K. and G.S. (Guido Siestrup); Project administration, M.A.P.; Funding acquisition, G.S. (Guido Siestrup). All authors have read and agreed to the published version of the manuscript.

Funding: This research received no external funding.

Institutional Review Board Statement: Not applicable.

Informed Consent Statement: Not applicable.

Data Availability Statement: The ML code, system dynamics simulation files, and data are available on GitHub under: https://anonymous.4open.science/r/aicollab-model-C108.

Conflicts of Interest: The authors declare no conflict of interest.

Credits: The "Excel" icon on pages 8 and 9 is by Gleb Khorunzhiy, the "Mind" icon on page 8 is by Med Marki, the "Decision" icon on page 8 is by Template, and the "Simulation Computer" icon on page 8 is by Ian Rahmadi Kurniawan; they were all sourced from thenounproject.com.

Appendix A

Table A1. System dynamics parameters.

Variable	Equation or Value	Units
Adjusted production rate	IF THEN ELSE ("Use DNN?" = 1, MIN (Capacity, DNN predictions), MIN (Capacity, Judgmental production rate))	Pcs/Month
Backlog	INTEG (Sales orders − Fulfillment rate − Lost sales, 0)	Pcs
Capacity	INTEG (Capacity augmentation rate − Capacity normalization rate, Normal capacity)	Pcs/Month
Capacity augmentation rate	IF THEN ELSE (Delivery lead time overshoot > 0, Capacity flexibility/Capacity ramp-up time, 0)	Pcs/(Month × Month)
Capacity flexibility	Maximum capacity − Capacity	Pcs/Month
Capacity normalization rate	IF THEN ELSE (Delivery lead time overshoot > 0, 0, Excess capacity/Capacity ramp-down time)	Pcs/(Month × Month)
Capacity ramp-down time	2 (base case); 4 (heuristic adjustment)	Month
Capacity ramp-up time	2	Month
Delay tolerance	2	Month
Delivery lead time goal	1	Month
Delivery lead time outlook	IF THEN ELSE (Backlog = 0, 1, Backlog/Fulfillment rate)	Month
Delivery lead time overshoot	MAX(0, Delivery lead time outlook − Delivery lead time goal)	Month
Discrepancy with actual returns	Returns time series (Time/One month) − Perception of returns	Pcs/Month
Discrepancy with actual sales	Order time series (Time/One month) − Perception of sales	Pcs/Month
Discretionary capacity	0.2	Dmnl
DNN predictions	DNN sales predictions (Time/One month) − DNN returns predictions (Time/One month)	Pcs/Month
DNN returns predictions	The result of RNN returns forecast is programmatically fed.	Pcs/Month
DNN sales predictions	The result of RNN sales forecast is programmatically fed.	Pcs/Month
Excess capacity	MAX(0, Capacity − Normal capacity)	Pcs/Month
FINAL TIME	96	Month
Fulfillment rate	MIN(Backlog, Inventory)/One month	Pcs/Month
INITIAL TIME	1	Month
Inventory	INTEG (Production rate + Return orders − Shipment rate, 0)	Pcs
Judgmental production rate	Perception of sales − Perception of returns	Pcs/Month
Lost sales	(Delivery lead time overshoot × Fulfillment rate)/Delay tolerance	Pcs/Month
Maximum capacity	Normal capacity × (1 + Discretionary capacity)	Pcs/Month
Net requirements	Sales orders − Return orders	Pcs/Month
Normal capacity	55	Pcs/Month
One month	1	Month
Order time series	Test dataset for actual customer orders.	Pcs/Month

Table A1. *Cont.*

Variable	Equation or Value	Units
Perception of returns	INTEG (Returns perception update, 19.67) (Note: 19.67 is the initial value; equals the average of last 6 months of returns)	Pcs/Month
Perception of sales	INTEG (Sales perception update, 70) (Note: 70 is the initial value; equals the average of last 6 months of sales)	Pcs/Month
Production rate	Adjusted production rate	Pcs/Month
Return orders	Returns time series (Time/One month)	Pcs/Month
Returns perception update	Discrepancy with actual returns/Update delay	Pcs/(Month × Month)
Returns time series	Test dataset for actual customer returns.	Pcs/Month
Sales orders	Order time series (Time/One month)	Pcs/Month
Sales perception update	Discrepancy with actual sales/Update delay	Pcs/(Month × Month)
Shipment rate	Fulfillment rate	Pcs/Month
Update delay	3	Month
"Use DNN?"	Programmatically set to switch between heuristic forecasting and RNN.	Dmnl

References

1. Brynjolfsson, E.; Mitchell, T. What can Machine Learning Do? Workforce Implications. *Science* **2017**, *358*, 1530–1534. [CrossRef]
2. Dwivedi, Y.K.; Hughes, L.; Ismagilova, E.; Aarts, G.; Coombs, C.; Crick, T.; Duan, Y.; Dwivedi, R.; Edwards, J.; Eirug, A.; et al. Artificial Intelligence (AI): Multidisciplinary perspectives on emerging challenges, opportunities, and agenda for research, practice and policy. *Int. J. Inf. Manag.* **2021**, *57*, 101994. [CrossRef]
3. Makridakis, S. The forthcoming Artificial Intelligence (AI) revolution: Its impact on society and firms. *Futures* **2017**, *90*, 46–60. [CrossRef]
4. Mitchell, M. Why AI Is Harder Than We Think. *arXiv* **2021**, arXiv:2104.12871.
5. Chollet, F. On the Measure of Intelligence. *arXiv* **2019**, arXiv:1911.01547.
6. Marcus, G. The Next Decade in AI: Four Steps Towards Robust Artificial Intelligence. *arXiv* **2020**, arXiv:2002.06177.
7. Doshi-Velez, F.; Kim, B. Towards A Rigorous Science of Interpretable Machine Learning. *arXiv* **2017**, arXiv:1702.08608.
8. Blackman, R.; Ammanath, B. When—and Why—You Should Explain How Your AI Works. *Harvard Business Review*, 31 August 2022. Available online: https://hbr.org/2022/08/when-and-why-you-should-explain-how-your-ai-works(accessed on 28 October 2022).
9. Zolas, N.; Kroff, Z.; Brynjolfsson, E.; McElheran, K.; Beede, D.N.; Buffington, C.; Goldschlag, N.; Foster, L.; Dinlersoz, E. *Advanced Technologies Adoption and Use by U.S. Firms: Evidence from the Annual Business Survey*; National Bureau of Economic Research: Cambridge, MA, USA, 2020. [CrossRef]
10. Karp, R.; Peterson, A. Find the Right Pace for Your AI Rollout. *Harvard Business Review*, 25 August 2022. Available online: https://hbr.org/2022/08/find-the-right-pace-for-your-ai-rollout(accessed on 28 October 2022).
11. Agrawal, A.; Gans, J.S.; Goldfarb, A. What to Expect from Artificial Intelligence. *MIT Sloan Management Review*, 7 February 2017. Available online: https://sloanreview-mit-edu.plymouth.idm.oclc.org/article/what-to-expect-from-artificial-intelligence/(accessed on 14 September 2021).
12. Raisch, S.; Krakowski, S. Artificial Intelligence and Management: The Automation–Augmentation Paradox. *Acad. Manag. Rev.* **2021**, *46*, 192–210. [CrossRef]
13. Shestakofsky, B. Working Algorithms: Software Automation and the Future of Work. *Work. Occup.* **2017**, *44*, 376–423. [CrossRef]
14. Brynjolfsson, E.; McAfee, A. Will Humans Go the Way of Horses. *Foreign Aff.* **2015**, *94*, 8.
15. Autor, D. *Polanyi's Paradox and the Shape of Employment Growth*; NBER Working Papers 20485; National Bureau of Economic Research: Cambridge, MA, USA, 2014. [CrossRef]
16. Melville, N.; Kraemer, K.; Gurbaxani, V. Review: Information Technology and Organizational Performance: An Integrative Model of IT Business Value. *MIS Q.* **2004**, *28*, 283–322. [CrossRef]
17. Sturm, T.; Gerlach, J.P.; Pumplun, L.; Mesbah, N.; Peters, F.; Tauchert, C.; Nan, N.; Buxmann, P. Coordinating Human and Machine Learning for Effective Organizational Learning. *MIS Q.* **2021**, *45*, 1581–1602. [CrossRef]
18. Malone, T.W. How Human-Computer 'Superminds' Are Redefining the Future of Work. *MIT Sloan Management Review*, 21 May 2018. Available online: https://sloanreview-mit-edu.plymouth.idm.oclc.org/article/how-human-computer-superminds-are-redefining-the-future-of-work/(accessed on 22 September 2021).
19. Elena Revilla, M.J.S.; Simón, C. Designing AI Systems with Human-Machine Teams. *MIT Sloan Management Review*, 18 March 2020. Available online: https://sloanreview.mit.edu/article/designing-ai-systems-with-human-machine-teams/(accessed on 8 September 2021).

20. Puranam, P. Human-AI collaborative decision-making as an organization design problem. *J. Org. Design* **2021**, *10*, 75–80. [CrossRef]
21. Shrestha, Y.R.; Ben-Menahem, S.M.; von Krogh, G. Organizational Decision-Making Structures in the Age of Artificial Intelligence. *Calif. Manag. Rev.* **2019**, *61*, 66–83. [CrossRef]
22. Brynjolfsson, E. The Turing Trap: The Promise & Peril of Human-Like Artificial Intelligence. *Daedalus* **2022**, *151*, 272–287. [CrossRef]
23. Rahmandad, H.; Repenning, N.; Sterman, J. Effects of feedback delay on learning. *Syst. Dyn. Rev.* **2009**, *25*, 309–338. [CrossRef]
24. Hogarth, R.M.; Lejarraga, T.; Soyer, E. The Two Settings of Kind and Wicked Learning Environments. *Curr. Dir. Psychol. Sci.* **2015**, *24*, 379–385. [CrossRef]
25. Ethiraj, S.K.; Levinthal, D. Bounded Rationality and the Search for Organizational Architecture: An Evolutionary Perspective on the Design of Organizations and Their Evolvability. *Adm. Sci. Q.* **2004**, *49*, 404–437. [CrossRef]
26. Knudsen, T.; Srikanth, K. Coordinated Exploration: Organizing Search by Multiple Specialists to Overcome Mutual Confusion and Joint Myopia. *Adm. Sci. Q.* **2013**, *59*, 409–441. [CrossRef]
27. Simon, H.A.; Newell, A. Human problem solving: The state of the theory in 1970. *Am. Psychol.* **1971**, *26*, 145–159. [CrossRef]
28. Glazer, R.; Steckel, J.H.; Winer, R.S. Locally Rational Decision Making: The Distracting Effect of Information on Managerial Performance. *Manag. Sci.* **1992**, *38*, 212–226. [CrossRef]
29. Nonaka, I. The Knowledge-Creating Company. *Harvard Business Review*, 1 July 2007. Available online: https://hbr.org/2007/07/the-knowledge-creating-company(accessed on 6 September 2021).
30. Narayanan, M.; Chen, E.; He, J.; Kim, B.; Gershman, S.; Doshi-Velez, F. How do Humans Understand Explanations from Machine Learning Systems? An Evaluation of the Human-Interpretability of Explanation. *arXiv* **2018**, arXiv:1802.00682.
31. Elgendy, N. Enhancing Collaborative Rationality between Humans and Machines through Data-Driven Decision Evaluation. In Proceedings of the 21st International Conference on Perspectives in Business Informatics Research (BIR), Rostock, Germany, 20–23 September 2022; p. 12.
32. Sterman, J. System Dynamics: Systems Thinking and Modeling for a Complex World. Massachusetts Institute of Technology. Engineering Systems Division, Working Paper, May 2002. Available online: https://dspace.mit.edu/handle/1721.1/102741 (accessed on 2 June 2022).
33. Gilpin, L.H.; Bau, D.; Yuan, B.Z.; Bajwa, A.; Specter, M.; Kagal, L. Explaining Explanations: An Overview of Interpretability of Machine Learning. In Proceedings of the 2018 IEEE 5th International Conference on Data Science and Advanced Analytics (DSAA), Turin, Italy, 1–3 October 2018; pp. 80–89. [CrossRef]
34. Kasparov, G. *Deep Thinking: Where Machine Intelligence Ends and Human Creativity Begins*, 1st ed.; PublicAffairs: New York, NY, USA, 2017.
35. Hassabis, D. Artificial Intelligence: Chess match of the century. *Nature* **2017**, *544*, 7651. [CrossRef]
36. Simon, H.A. *Administrative Behavior*, 4th ed.; Free Press: New York, NY, USA, 1997.
37. Lee, K.-F. *AI Superpowers: China, Silicon Valley, and the New World Order*, 1st ed.; Mariner Books: Boston, MA, USA, 2018.
38. Reeves, M.; Ueda, D. Designing the Machines That Will Design Strategy. *Harvard Business Review*, 18 April 2016. Available online: https://hbr.org/2016/04/welcoming-the-chief-strategy-robot(accessed on 20 September 2021).
39. Huang, M.-H.; Rust, R.; Maksimovic, V. The Feeling Economy: Managing in the Next Generation of Artificial Intelligence (AI). *Calif. Manag. Rev.* **2019**, *61*, 43–65. [CrossRef]
40. Simon, H.A. *The Sciences of the Artificial*, 3rd ed.; The MIT Press: Cambridge, MA, USA, 1996.
41. Klein, G.A. *Sources of Power: 20th Anniversary Edition*, 1st ed.; The MIT Press: Cambridge, MA, USA, 2017.
42. Galbraith, J.R. Organization Design: An Information Processing View. *INFORMS J. Appl. Anal.* **1974**, *4*, 28–36. [CrossRef]
43. Nelson, R.R.; Winter, S.G. Neoclassical vs. Evolutionary Theories of Economic Growth: Critique and Prospectus. *Econ. J.* **1974**, *84*, 886–905. [CrossRef]
44. Gigerenzer, G.; Goldstein, D.G. Reasoning the fast and frugal way: Models of bounded rationality. *Psychol. Rev.* **1996**, *103*, 650–669. [CrossRef]
45. Panchalavarapu, P.R.; De Kok, A.G.; Stephen, C.; Graves, C.S. (Eds.) 2004. Handbooks in Operations Research and Management Science: Supply Chain Management: Design, Coordination and Operation. *Interfaces* **2005**, *35*, 339–341.
46. Devaraj, S.; Kohli, R. Performance Impacts of Information Technology: Is Actual Usage the Missing Link? *Manag. Sci.* **2003**, *49*, 273–289. [CrossRef]
47. Wernerfelt, B. A resource-based view of the firm. *Strateg. Manag. J.* **1984**, *5*, 171–180. [CrossRef]
48. Brynjolfsson, E.; Milgrom, P. 1. Complementarity in Organizations. In *The Handbook of Organizational Economics*; Princeton University Press: Princeton, NJ, USA, 2012; pp. 11–55. [CrossRef]
49. Mithas, S.; Ramasubbu, N.; Sambamurthy, V. How Information Management Capability Influences Firm Performance. *MIS Q.* **2011**, *35*, 237–256. [CrossRef]
50. Brynjolfsson, E.; Hitt, L. Computing Productivity: Firm-Level Evidence. *Rev. Econ. Stat.* **2003**, *85*, 793–808. [CrossRef]
51. Haynes, C.; Palomino, M.A.; Stuart, L.; Viira, D.; Hannon, F.; Crossingham, G.; Tantam, K. Automatic Classification of National Health Service Feedback. *Mathematics* **2022**, *10*, 983. [CrossRef]
52. Melville, N.; Gurbaxani, V.; Kraemer, K. The productivity impact of information technology across competitive regimes: The role of industry concentration and dynamism. *Decis. Support Syst.* **2007**, *43*, 229–242. [CrossRef]

53. Will, J.; Bertrand, M.; Fransoo, J.C. Operations management research methodologies using quantitative modeling. *Int. J. Oper. Prod. Manag.* **2002**, *22*, 241–264. [CrossRef]
54. Ackoff, R.L. The Future of Operational Research is Past. *J. Oper. Res. Soc.* **1979**, *30*, 93–104. [CrossRef]
55. Frank, M.R.; Autor, D.; Bessen, J.E.; Brynjolfsson, E.; Cebrian, M.; Deming, D.J.; Feldman, M.; Groh, M.; Lobo, J.; Moro, E.; et al. Toward understanding the impact of artificial intelligence on labor. *Proc. Natl. Acad. Sci. USA* **2019**, *116*, 6531–6539. [CrossRef]
56. Morecroft, J.D.W. Rationality in the Analysis of Behavioral Simulation Models. *Manag. Sci.* **1985**, *31*, 900–916. [CrossRef]
57. Sterman, J.D. *Business Dynamics*; International Edition; McGraw-Hill Education: Boston, MA, USA, 2000.
58. Powers, W.T. Feedback: Beyond Behaviorism. *Science* **1973**, *179*, 351–356. [CrossRef]
59. Pruyt, E. *Small System Dynamics Models for Big Issues: Triple Jump towards Real-World Complexity*; TU Delft Library: Delft, The Netherlands, 2013.
60. Houghton, J.; Siegel, M. Advanced data analytics for system dynamics models using PySD. In Proceedings of the 33rd International Conference of the System Dynamics Society, Cambridge, MA, USA, 19–23 July 2015.
61. Anderson, C. The End of Theory: The Data Deluge Makes the Scientific Method Obsolete. Wired, 23 June 2008. Available online: https://www.wired.com/2008/06/pb-theory/ (accessed on 26 August 2022).
62. Pearl, J. Radical Empiricism and Machine Learning Research. *J. Causal Inference* **2021**, *9*, 78–82. [CrossRef]
63. Kitchin, R. Big Data, new epistemologies and paradigm shifts. *Big Data Soc.* **2014**, *1*, 2053951714528481. [CrossRef]
64. Bender, E.M.; Gebru, T.; McMillan-Major, A.; Shmitchell, S. On the Dangers of Stochastic Parrots: Can Language Models Be Too Big? 🦜. In Proceedings of the 2021 ACM Conference on Fairness. Accountability, and Transparency, Virtual Event Canada, 3–10 March 2021; pp. 610–623. [CrossRef]
65. Pearl, J. Theoretical Impediments to Machine Learning with Seven Sparks from the Causal Revolution. *arXiv* **2018**, arXiv:1801.04016.
66. Makridakis, S.; Spiliotis, E.; Assimakopoulos, V. Statistical and Machine Learning forecasting methods: Concerns and ways forward. *PLoS ONE* **2018**, *13*, e0194889. [CrossRef] [PubMed]
67. Souza, G.C. Closed-Loop Supply Chains: A Critical Review, and Future Research*. *Decis. Sci.* **2013**, *44*, 7–38. [CrossRef]
68. Makridakis, S.; Hibon, M. The M3-Competition: Results, conclusions and implications. *Int. J. Forecast.* **2000**, *16*, 451–476. [CrossRef]
69. Borshchev, A. Multi-method modelling: AnyLogic. *Discret. Event Simul. Syst. Dyn. Manag. Decis. Mak.* **2014**, *9781118349*, 248–279. [CrossRef]
70. Chollet, F. *Deep Learning with Python*, 2nd ed.; Manning: Shelter Island, New York, USA; Hong Kong, China, 2021.
71. Grus, J. *Data Science from Scratch: First Principles with Python*, 2nd ed.; O'Reilly Media: Sebastopol, CA, USA, 2019.
72. Sterman, J.D. Misperceptions of feedback in dynamic decision making. *Organ. Behav. Hum. Decis. Process.* **1989**, *43*, 301–335. [CrossRef]
73. Kahneman, D.; Slovic, S.P.; Slovic, P.; Tversky, A.; Press, C.U. *Judgment under Uncertainty: Heuristics and Biases*; Cambridge University Press: Cambridge, UK, 1982.
74. Géron, A. *Hands-On Machine Learning with Scikit-Learn, Keras, and TensorFlow: Concepts, Tools, and Techniques to Build Intelligent Systems*, 2nd ed.; O'Reilly Media: Beijing, China; Sebastopol, CA, USA, 2019.
75. Seabold, S.; Perktold, J. Statsmodels: Econometric and Statistical Modeling with Python. In Proceedings of the 9th Python in Science Conference (SciPy), Austin, TX, USA, 28 June–3 July 2010; pp. 92–96. [CrossRef]
76. Tabrizi, B.; Lam, E.; Girard, K.; Irvin, V. Digital Transformation Is Not About Technology. *Harvard Business Review*, 13 March 2019. Available online: https://hbr.org/2019/03/digital-transformation-is-not-about-technology(accessed on 7 September 2021).
77. LaValle, S.; Lesser, E.; Shockley, R.; Hopkins, M.S.; Kruschwitz, N. Big Data, Analytics and the Path from Insights to Value. *MIT Sloan Management Review*. Available online: https://sloanreview.mit.edu/article/big-data-analytics-and-the-path-from-insights-to-value/ (accessed on 11 October 2021).
78. Weill, P.; Woerner, S.L. Is Your Company Ready for a Digital Future? *MIT SMR*, 4 December 2017. Available online: https://sloanreview.mit.edu/article/is-your-company-ready-for-a-digital-future/(accessed on 7 November 2022).
79. Westerman, G.; Bonnet, D.; McAfee, A. *Leading Digital: Turning Technology into Business Transformation*; Harvard Business Press: Bosto, MA, USA, 2014.
80. Case, N. How To Become A Centaur. *J. Des. Sci.* **2018**. [CrossRef]
81. Sutton, R.S.; Barto, A.G. *Reinforcement Learning*, 2nd ed.; An Introduction; MIT Press: Cambridge, MA, USA, 2018.
82. Hopp, W.J.; Spearman, M.L. *Factory Physics*; Reissue Edition; Waveland Pr Inc.: Long Grove, IL, USA, 2011.
83. Galbraith, J.R. Organizational Design Challenges Resulting from Big Data. 10 April 2014. Available online: https://papers.ssrn.com/abstract=2458899 (accessed on 7 November 2022).
84. Clark, S.; Hyndman, R.J.; Pagendam, D.; Ryan, L.M. Modern Strategies for Time Series Regression. *Int. Stat. Rev.* **2020**, *88*, S179–S204. [CrossRef]

applied sciences

MDPI

Article

Leveraging Artificial Intelligence in Blockchain-Based E-Health for Safer Decision Making Framework

Abdulatif Alabdulatif [1,*], Muneerah Al Asqah [2], Tarek Moulahi [2,*] and Salah Zidi [3]

[1] Department of Computer Science, College of Computer, Qassim University, Buraidah 52571, Saudi Arabia
[2] Department of Information Technology, College of Computer, Qassim University, Buraidah 52571, Saudi Arabia; 411207283@qu.edu.sa
[3] ISSIG, University of Gabes, Gabes 6072, Tunisia; salah_zidi@yahoo.fr
* Correspondence: ab.alabdulatif@qu.edu.sa (A.A.); t.moulahi@qu.edu.sa (T.M.)

Abstract: Machine learning-based (ML) systems are becoming the primary means of achieving the highest levels of productivity and effectiveness. Incorporating other advanced technologies, such as the Internet of Things (IoT), or e-Health systems, has made ML the first choice to help automate systems and predict future events. The execution environment of ML is always presenting contrasting types of threats, such as adversarial poisoning of training datasets or model parameters manipulation. Blockchain technology is known as a decentralized network of blocks that symbolizes means of protecting block content integrity and ensuring secure execution of operations.Existing studies partially incorporated Blockchain into the learning process. This paper proposes a more extensive secure way to protect the decision process of the learning model. Using smart contracts, this study executed the model's decision by the reversal engineering of the learning model's decision function from the extracted learning parameters. We deploy Support Vector Machine (SVM) and Multi-Layer Perceptron (MLP) classifiers decision functions on-chain for more comprehensive integration of Blockchain. The effectiveness of this proposed approach is measured by applying a case study of medical records. In a safe environment, SVM prediction scores were found to be higher than MLP. However, MLP had higher time efficiency.

Keywords: blockchain; e-health; machine learning; deep learning; smart contract; decision function

Citation: Alabdulatif, A.; Al Asqah, M.; Moulahi, T.; Zidi, S. Leveraging Artificial Intelligence in Blockchain-Based E-Health for Safer Decision Making Framework. *Appl. Sci.* **2023**, *13*, 1035. https://doi.org/10.3390/app13021035

Academic Editor: Giacomo Fiumara

Received: 2 December 2022
Revised: 8 January 2023
Accepted: 9 January 2023
Published: 12 January 2023

1. Introduction

Present-day lives require people to depend on various types of technology to assist in achieving higher levels of productivity and better operational efficiency. The continuous growth of computer-based technologies has placed them as essential pillars in new world development. Such intelligent technologies include the Internet of Things (IoT) [1] systems such as smart homes and supply chain management systems, spam filtering systems, and many others. As a critical life sector, smart healthcare systems, such as diagnostic systems or e-health decision systems, are another application of these technologies. These applications rely on Machine Learning (ML) models to detect and diagnose diseases and help disease spread prediction, such as the COVID-19 virus. ML uses models that train with heterogeneous data types to progressively develop and learn to make decisions based on their calculated outcomes. These outcomes are the decisions that can help automate routine tasks, detect abnormalities, and predict disease spreads.

Unfortunately, the outcome of this decision can encounter various threats that affect and change its value. These threats can include model parameter manipulation, poisoning attacks, and evasion attacks. The latter two are types of Adversarial Machine Learning (AML) attacks which are manipulative attacks that affect the integrity of ML datasets. A recent study of [2] applied adversarial attacks on six different COVID-19 detection systems with underlying ML Deep Neural Network (DNN) models. The authors showed that the confidence of the DNN model dropped from 91% to 9% on a subject having positive

COVID-19 results when adding random noise of black and white batches in Computed tomography (CT) scan training images. The previous experiment is one of many other examples that proved ML models' susceptibility to AML attacks.

1.1. Motivation and Problem Background

As shown by Figure 1, AML attacks include falsifying data samples to achieve ML model inaccuracy in classifying new data inputs [3]. A typical ML process splits the dataset between two phases, training and testing. A poisoning attack affects the training dataset, while an evasion attack injects carefully crafted samples into the testing dataset. AML research is literature designed to measure AML's impact on ML models to find a way to increase their robustness against such attacks [4]. For instance, works of [5–7] evaluated Neural Networks (NN)-based systems against different types of evasion attacks. The work of [5] performed an evasion attack on a Multi-Layer Perceptron (MLP) Intrusion Detection System (IDS), where they succeeded in dropping the model's accuracy from 99.8% to 29.87%. Moreover, Ref. [6] injected adversarial samples to fool Conventional Neural Networks (CNN) malware detection. Meanwhile, Ref. [7]'s work was successful in tricking a DNN visual recognition model into classifying adversarial inputs as benign.

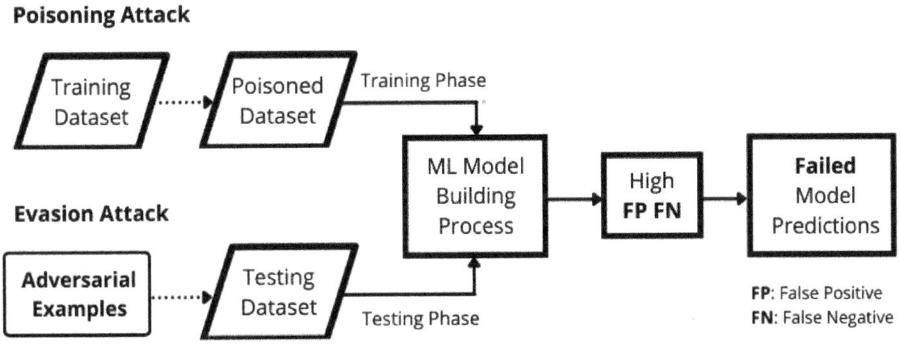

Figure 1. An illustration of poisoning and evasion attacks workflow.

Additionally, Ref. [8]'s work is an example of a poisoning attack that tested the susceptibility of a Support Vector Machine (SVM) spam filtering system by inserting a well-crafted label-flipped malicious sample into the training dataset.

In e-Health applications, the authors of [9] applied a poisoning attack on a LASSO regression ML model trained on a dataset that contained records of 5700 patients that predicted the dosage of Warfarin, an anticoagulant drug. Applying a 20% poisoning attack caused patients' dosages to change by an average of 139.31%. An increased dosage of Warfarin can cause severe bleeding, while a decreased dosage could cause blood clots, leading to heart attacks if the patient has a history of blood clotting [10]. This notable impact of AML attacks on people's health can severely affect other similar e-health systems [4]. A more comprehensive survey of similar studies on AML effects on other ML models, and domains can be found in [3].

The AML field of research can easily state that most types of ML are prone to adversarial attacks since it is impossible to assume that the system's environment is entirely benign. Security researchers are always on the work to deploy robust methods against AML. For example, Blockchain is an emerging technology that uses means of cryptography and decentralization to provide stable, secure, and immutable blocks of records. Multiple blocks are connected together through a hash-based procedure. This hashing procedure and the utilization of other cryptography methods have given Blockchain technology its property of protecting block content integrity [11]. More security researchers incorporate

Blockchain with ML to protect against AML. However, we believe that a more sophisticated kind of integration of the ML decision process is still needed.

1.2. Study Contribution and Novelty

The paper's main contributions are:

- Develop a trust-based AI framework that relies on the integration of Blockchain and ML models;
- Develop an effective method to secure the decision functions of SVM and MLP models by using immutable smart contracts.

1.3. Why Blockchain?

In our paper, the blockchain is used for two reasons:

- To protect the dataset against poisonous attacks. Since the used dataset is securely uploaded to the blockchain instead of publishing it in a shared repository. Indeed, the blockchain will guarantee the integrity of data;
- To protect machine learning techniques against evasion attacks. We perform this goal by embedding model decision functions as smart contracts in the Ethereum blockchain.

The rest of this paper is organized as follows: Section 2 provides a brief review of existing Blockchain-adopted ML research. Moreover, it explains the necessary background details of Blockchain and ML models. The applied system flow is illustrated in Section 3, while Section 4 elaborates on the implementation steps of the proposed system, including the reconstruction of SVM and MLP decision functions. Lastly, Section 5 analyzes and discusses the performance of the proposed system.

2. Background Study

The revolutionized growth in Blockchain technology has encouraged its emergence with ML solutions. This section outlines related works and provides a brief technological preview.

2.1. Related Literature

Several studies cover combining Blockchain with ML to solve security and privacy issues in the literature. We summarize these integrations into three main categories, NN-based integration, partial integration of Blockchain, and Blockchain with Federated Learning. One of the NN-based integration examples is called DeepRing, where authors of [12] designed each NN layer to be presented as a Blockchain block to protect against tampering attacks. Although this integration stood robust against a tampering attack that downgraded a regular CNN performance by 20.71%, its application is limited to NN-based ML models. Other research included the partially separated integration of ML with Blockchain [13]. One example is [14]'s work of combining ML with Blockchain for a more efficient and safer COVID-19 vaccine supply chain. Their solution queried records from the Blockchain to feed them into a separate Long Short Term Memory (LSTM) classifier. The demand forecasting LSTM helped preserve 4% of the vaccine ratio, 6 million vaccine doses.

Most Blockchain-integrated ML solutions focus on employing the technology to protect the privacy of the ML model. Studies with such scope deployed Federated Learning (FL) [15], or as can be known as Decentralized ML (DML) [16], where a centralized server collects and aggregates learning parameters among participating nodes.

The majority of found FL-based integration, such as works of [17–19], relied on off-chain execution of ML training while applying different types of consensus algorithms to manage work among nodes. This partial application of the ML decision process is due to the metered usage and storage of Blockchain, which can result in the prosperous implementation of the whole ML fitting and decision process.

One other noticeable example is the collaborative deep learning framework called DeepChain. Similar to the previously mentioned studies, Ref. [20] designed DeepChain for nodes to train a global model locally and then upload their gradients through a smart contract. They also relied on consensus algorithms for control updating the model's gradients which are averaged and broadcasted again for the next learning iteration. Table 1 summarises existing related work and their limitations.

Table 1. Summary of existing solutions incorporating ML with Blockchain and their limitations.

Ref.	Summary	Limitation
[14]	Combined Blockchain and ML to design a more efficient COVID-19 supply chain system.	Separate integration of ML modules. No focus on security/privacy
[12]	Designed DeepRing, where each NN layer is presented as a Blockchain block to protect against tampering attacks.	Not suitable for other none NN-based ML models
[21]	Developed dynamic malware detection system based on behavioral logs using Deep learning and Blockchain	No consideration to poisoning attacks prevention. No privacy consideration
[22]	Used DanKu protocol to build a malware detection system on Blockchain	Resource wastage due to using DanKu protocol. No consideration to poisoning attacks
[18]	Used FL with Blockchain to mitigate end-point corruption attacks.	The consensus committee is only one member who is considered non-hostile

As shown in Table 1, the studies incorporating ML with Blockchain are scarce. At the same time, most of these integrations focused on collaborative learning and partially included Blockchain in the ML decision process. None of the found literature included the decision process to be performed on-chain. To our knowledge, there is still no ML decision integration with Blockchain. This study proposes the further inclusion of the decision function in smart contracts to achieve a more reliable and secure decision process.

2.2. Technological Background

Since the apparition of bitcoin as an electronic cash system in 2008, researchers and developers have been working to enhance Blockchain systems to be the main pillars of future systems [23].

2.2.1. Blockchain Technology

The term "Blockchain" started to be popularly used to refer to the technology presented in Nakamoto's paper. Although it was considered a breakthrough in technology, concepts of Blockchain, such as cryptography and hashing, were explored way before Bitcoin's publication [24].

Blockchain has many definitions, but it can be defined as the technology that uses block-type data structure to store data, uses consensus algorithms to generate and update the distributed ledger, and uses encryption to ensure security during transmission [25]. To be put in other words, Blockchain is a Peer to Peer (P2P) network, where nodes share a distributed ledger [26].

Blockchain protects the integrity of the ledger content through hashing; each block's hash connects with the previous block, and a small change in any block's content will not go unnoticed [11]. Consensus algorithms are rules and agreements which the P2P network uses to draw verdicts on the new block's validity to the ledger. Nodes calculate a cryptography challenge, called a nonce, to prove the block's validity. Once the nonce is validated, the new block is added. The previous operation is also called mining. Different consensus algorithms rely on various intensives to encourage mining nodes [26].

2.2.2. Smart Contracts

Smart contracts control the ledger's state by dynamically sending specific transactions through execution conditions and logic. When conditions are met, the execution logic is invoked [25]. Smart contracts enabled Blockchain to manipulate digital assets inside the Blockchain. It was first implemented in the Ethereum Blockchain platform in 2015 [26]. Smart contracts can be written in Solidity or Vyper programming languages, and they execute on the Ethereum Virtual Machine (EVM). Smart contracts enabled the development of distributed applications (dApp) [26].

3. Proposed Methodology

This section presents the proposed model followed by experiments approve to achieve ML execution security. Figure 2 shows the main steps to achieve the objective of this study. A smart contract writes the dataset to Blockchain. There are two ways to store the dataset. One way is to store the dataset off-chain while keeping the hash value of the dataset on-chain. In contrast, the whole dataset can be stored on-chain. The choice of on- or off-chain storage depends on the size of the dataset; storing a vast dataset on-chain can be costly. Direct complete on-chain execution of the ML fitting process was not applicable due to the following reasons:

1. EVM is still immature to execute complex ML operations;
2. Storing the large-size of ML libraries is considered costly in terms of deployment and runtime.

The proposed methodology suggests that the ML fitting process be executed off-chain in a local client. Fitted models' parameters will then be extracted to be stored on-chain.

Since the dataset followed a classification problem, this study employed two classifiers, SVM and MLP. SVM is a popular ML classifier that uses influence functions to find a hyperplane that best separates two data classes. On the other hand, MLP is a deep learning model that is a type of feedforward NN that uses a set of nodes organized into multiple layers to draw prediction conclusions.

A smart contract writes the model's parameters to Blockchain. These written parameters are used to reversely construct the decision function of the ML model on-chain which will help to classify a new given datapoint vector.

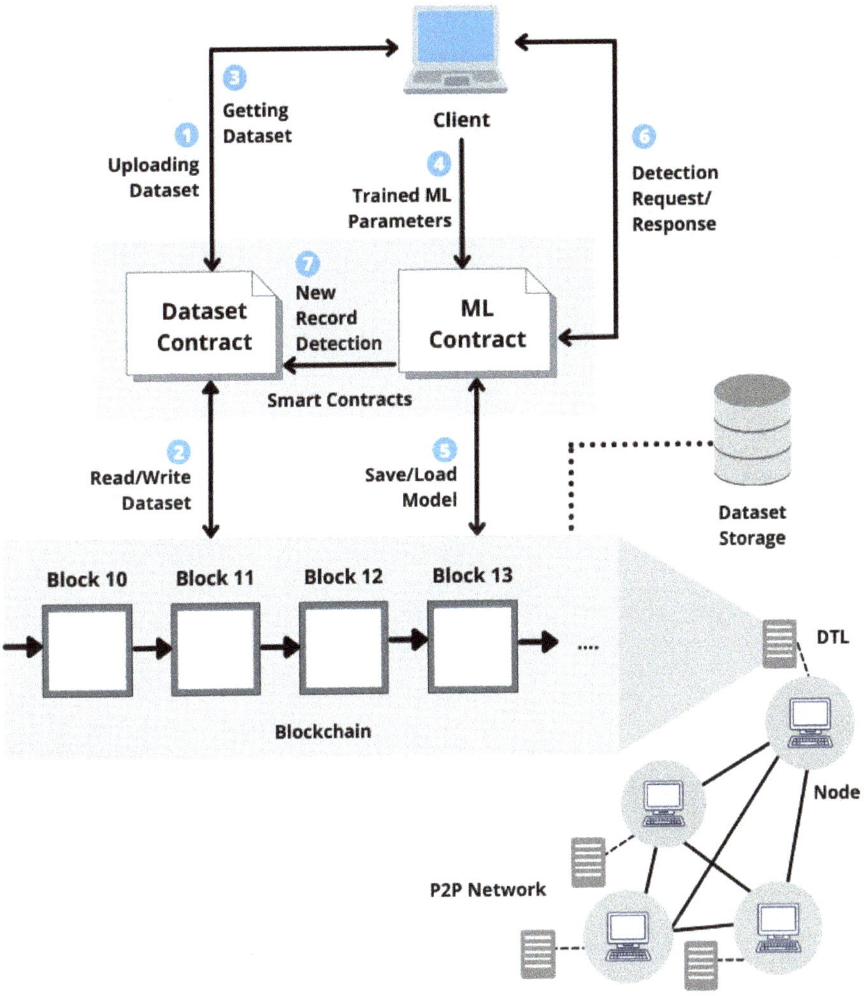

Figure 2. An overview of the proposed model along with entities interaction and workflow.

4. Implementation

This section elaborates on the implementation details of the proposed methodology. The implementation device was an AMD64 3.20 GHz CPU with a 16 GB RAM computer, and the implementation of Blockchain was performed using Ganache [27].

4.1. Experimental Dataset Setup

The dataset used in this experiment is the Pima Indians Diabetes dataset, which contained the measurements of labeled 21 or older 768 females, where 268 records were of females with type 2 diabetes, and the rest were healthy females.

In preparation for ML fitting, two copies of the dataset were prepared. Copy 1 of the dataset was set to be preserved on-chain. On the other hand, an evasion attack of manual label-flipping was implemented on copy 2 of the dataset, where healthy records appeared with diabetes labels and vice versa. Almost 33% of records were manipulated and labeled as poisonous samples, and the rest were labeled as normal records.

4.2. Writing Dataset

In this implementation, the data storage was on-chain storage since the 24 kilobytes size of the experiment dataset was quite manageable. Algorithm 1 shows the pseudo code of writing copy 1 of the dataset procedure, where each record of copy 1 of the dataset was converted to Javascript Object Notation (JSON) string format before writing them to the ledger. The storing of dataset on Blockchain will help to protect its integrity against poisonous attacks. In addition, it makes it available for use with a high safety level.

Algorithm 1 Write dataset to the distributed ledger

```
    Input: file dataset file
1: procedure UPLOADDATASET
2:     for i ← 1 to length of file do
3:         Record ← file[i] in JSON
4:         DatasetProtect.writeRecord(Record)
```

Line 4 in Algorithm 1 is a call to a setter function in the *DatasetHandler* smart contract.

4.3. Reading Dataset

Loading records from the ledger is less expensive and more straightforward. Algorithm 2 shows the steps of the loading procedure.

Algorithm 2 Read dataset from the distributed ledger

```
    Output: recordList[] list of records in JSON
1: procedure LOADDATASET
2:     latestID ← DatasetHandler.dCount
3:     Init recordList as Array
4:     for i ← 1 to latestID do
5:         R ← DatasetHandler.readRecord(i)
6:         Push R to recordList[]
7:     return recordList[]
```

Line 2 in Algorithm 2 is a call to a smart contract getter function to obtain the latest record ID value. Line 5 is a call to a getter function inside the smart contract to retrieve the record JSON string. This procedure returns a list of JSON strings representing all the dataset records which then can be converted to any other format for ML model training, a Comma-Seperated Value (CSV) file for instance.

4.4. Parameters Extraction and Preservation

At this stage, an off-chain client loads the dataset copy 2 and starts the ML fitting process to detect poisonous records. In this use case of application, an additional standardizer is implemented to normalize the dataset.

4.4.1. Scalar Parameters

Copy 1 of the dataset is scaled using a standard scalar that uses means and standard deviation values to standardize data to values close to 1 and –1 . The extraction of the fitted scaler produced two vectors with a length of 9, which is the same number as the dataset's features. The two vectors are the *means* for the means values, and the *vars* for the variances. These vectors were set and preserved on-chain for future use.

4.4.2. SVM Parameters

SVM was trained using the radial basis kernel function (RBF), which calculates the Euclidean distance between vectors. After the training process is complete, the following model's training parameters were extracted to be preserved on-chain as shown in Table 2.

Table 2. SVM extracted parameters.

Name	Description	Data Type
support_vectors_	Datapoints defining hyperplane decision boundaries placements	9 × 395 Decimal Matrix
_dual_coef	Weights of support vectors	1 × 395 Decimal Array
_intercept	The bias	Decimal
_gamma	To handle non-linear classification	Decimal

Support vectors were saved individually through a loop similar to the one previously used to store dataset records due to the HyperText Transfer Protocol (HTTP) limitations to pass the sizeable multi-dimensional list to the chain in one transaction.

4.4.3. MLP Parameters

MLP training process included two hidden layers of sizes 5 and 2 over 1000 epochs. The feed-forward deep learning NN relied on a nonlinear activation function, known as the Rectifier Linear Unit activation function (reLU). Likewise, Table 3 shows MLP parameters stored on-chain after the training is complete.

Table 3. MLP extracted parameters.

Name	Description	Data Type
coefs_	Weights of neuron's inputs in three layers, two input layers, and the output layer	5 × 9, 2 × 5, 1 × 2 Decimal Matrices
intercepts_	Biases of each neuron in three layers	1 × 5, 1 × 2, 1 × 1 Decimal Arrays

The learning settings of both SVM and MLP were found to be best in this case of classifying poisonous diabetes data records by balancing performance and avoiding overfitting.

4.5. ML Detection Implementation

This study proposes preserving efficiency by manually deploying decision functions built from both algorithms' previously-stored parameters to carry the detection of new data input on-chain. As mentioned earlier, the entire algorithm learning process on-chain was not cost-efficient.

4.5.1. Scalar Standardization Function

A new data point needs to be scaled first to be classified by the ML model. The scalar applies the standardization function below to the means and variances values acquired in the previous step:

$$z = \frac{x - \mu}{\sigma} \tag{1}$$

where z is the scaled vector, x is the input vector, μ is the mean value, and σ is the standard deviation value, which is the square root of the variance value. Algorithm 3 shows the scaling procedure steps.

Algorithm 3 Standardize new data point vector.

Input:
vector integer input vector,
means means array,
vars variances array
Output:
scaledVector scaled decimal output vector

```
1: procedure SCALEDATA
2:     stds ← sqrt(vars)                                    ▷ call math sqrt() function
3:     for i ← 1 to length of vector do
4:         vector[i] ← vector[i] in Decimal                 ▷ convert integer to decimal format
5:         scaledVector[i] ← (vector[i] − means[i])/stds[i]
```

It was not possible to pass decimal arrays directly to smart contract functions during the implementation of the proposed methodology. For this reason, as seen in line 4 in Algorithm 3, every decimal array was passed as an integer and then converted back to decimal on-chain for further calculations.

4.5.2. SVM Decision Function

An RBF-kernel SVM's decision function returns values close to (–1,1) and is generally described in the math equation below:

$$h^* = (\mathbf{x}^\star \phi(x)) + w_0^\star$$
$$h^* = \sum_{i \in P_S} a_i^\star u_i \cdot K(\mathbf{x}_i - \mathbf{x}) + w_0^\star \tag{2}$$

In the above representation, h^* is the decision function, $a_i^\star$ is the value of the coefficients, u_i is the support vector output of the kernel function K, x is the new data point vector, $\mathbf{x}_i$ is the support vector, and $w_0^\star$.

Since the kernel function of this SVM implementation was the RBF function, it has the mathematical representation as follows:

$$K(\mathbf{x}, \mathbf{x}') = \mathbf{exp}(-\gamma \|\mathbf{x} - \mathbf{x}'\|^2) \tag{3}$$

In the above equation, γ is the gamma value acquired previously. The RBF kernel function returns the product of negative gamma with the Frobenius norm of two input vectors. Function **exp()** is the exponent of Euler number, e.

$$\|\mathbf{x}, \mathbf{x}'\|_F = \sqrt{\sum_{i=1}^{m} \sum_{j=1}^{n} | a_{i,j} |^2} \tag{4}$$

The above math representation shows that the Frobenius norm, F, is the square root of the summation of two input vectors, x, x', squared difference, a.

Smart contracts did not provide complex decimal math libraries support for exp() function, nor did they allow execution for decimal numbers to be the base or the exponent of exponentials. This study came with the workaround to use the Taylor Maclaurin series to calculate the exponential of decimals:

$$f(x) = \sum_{n=0}^{\infty} \frac{x^n}{n!} \tag{5}$$

The above math representation shows that the Taylor Maclaurin series is the summation of a number x raised to power n divided by the factorial of that n.

Taylor Maclaurin series is an approximation calculation, which means that calculating e^x using the summation of an infinite number n iterations will progressively produce a value closer to the actual value.

The following Algorithm 4 shows the steps of calculating the Taylor Maclaurin series in fifty rounds of calculations, as it was better suited for execution efficiency while obtaining more precise values. Algorithm 5 shows the steps of calculating the Frobenius norm of Equation (4).

Algorithm 4 Approximate e^x using Taylor Maclaurin series

```
    Input: x decimal number
    Output: result decimal number
1: procedure TAYLOR
2:     term ← result ← n ← 1
3:     for i ← 1 in range(50) do
4:         term ← (term * x)/n
5:         n ← n + 1
6:         result ← result + term
7:     return result
```

Algorithm 5 Frobenius norm of two vectors

```
    Input:
x decimal support vector,
z decimal input vector
    Output:
sum decimal number
1: procedure FNORM
2:     sum ← 0
3:     for i ← 1 to length of x do
4:         y ← (x[i] − z[i])^2
5:         sum ← y + sum
6:     return sqrt(sum)
```

Equation (3) of the kernel function is calculated by using the Frobenius norm and the Taylor Maclaurin series, as shown by Algorithm 6.

Algorithm 6 RBF kernel of support vector and input vector

```
    Input:
x decimal support vector,
z decimal input vector,
g gamma value
    Output:
y decimal number
1: procedure RBF
2:     norm ← Fnorm(x, z)^2                    ▷ call Fnorm procedure
3:     y ← (g * norm) * (− − 1)
4:     return Taylor(y)                        ▷ call Taylor procedure
```

It is worth noting that the square root step in Algorithm 5 line 7 cancels the squaring step in Algorithm 6, line 2. For this reason, these steps were omitted in the smart contract code implementation.

Algorithm 7 applies Equation (2), where it takes an integer vector input and returns a value > 0 or <0. If the output of this function is larger than zero, a positive number, it has a label of class 1. Alternatively, if the output is less than zero, a negative number, it has a label of class 0.

Algorithm 7 SVM decision function

```
Input:
x integer input vector,
sv support vectors matrix,
df dual coefficients array,
incpt intercept decimal value,
g gamma decimal value
Output:
v decimal number
procedure SVMDECFUN
    z ← ScaledData(x)                          ▷ call ScaleData procedure
    init rbfList[] as Array
    for i ← 1 to length of df do
        rbf ← RBF(sv[i], z, g)                 ▷ call RBF procedure
        push rbf to rbfList[]
    sum ← 0
    for i ← 1 to length of df do
        y ← df[i] * rbfList[i]
        sum ← sum + y
    v ← (sum + incpt) * (− − 1)
    return Taylor(v)
```

4.5.3. MLP Decision Function

MLP has an input layer of 9-dimensions, and two hidden layers of 5 and 2 dimensions, and since it is solving a binary classification problem, it has a 1-dimension output layer.

The decision function of MLP concludes a series of addition and multiplication to classify an input. In this calculation, each hidden neuron's value equals the linear summation of all previous layer's neurons' values multiplied by their coefficients, or the weights between the neuron's layer and the last layer. An additional value of intercept, or bias, is added to this summation:

$$\begin{aligned} h_i^{(1)} &= \phi(\sum_j x_j w_{i,j} + b_i^{(1)}) \\ h_i^{(2)} &= \phi(\sum_j h_j^{(1)} w_{i,j} + b_i^{(2)}) \\ y_i &= \phi(\sum_j h_j^{(2)} w_{i,j} + b_i^{(3)}) \end{aligned} \tag{6}$$

Equation (6) calculates MLP decision function where h_i^n is the neuron i value in the nth layer. This implementation includes two layers and a final output layer with one neuron, y_i, which gives the final summation value. $\phi()$ is the nonlinear activation function that calculates the neuron's value by a weighted sum, where x_j is the input features vector, h_j^n is the neurons' values at layer $i-1$, $w_{i,j}$ is the weight, and b_i^n is the intercept of neuron i at the nth layer.

$$f(x) = max(0, x) \tag{7}$$

This MLP implementation follows a Rectifier Linear Unit activation function (reLu), which, as illustrated by Equation (7), returns the max between 0 and the weighted sum, x, of a neuron.

Algorithm 8 shows the steps in calculating the Equation (6) decision function of this paper's implementation of MLP by using a three-level loop. It is worth noting that the current development of smart contracts did not allow for multi-loop applications. For this reason, the inner loops were applied and called separate functions inside the contract.

Algorithm 8 MLP decision function

```
Input:
x integer input vector,
w weight matrices,
b biases array
Output:
v decimal number
procedure MLPDecFun
    z ← ScaleData(x)                                  ▷ ScaleData procedure
    init A[] as Array
    init B[] as Array
    for i ← 0 to i ← 2 do
        if i ← 1 then
            for j ← 0 to j ← 4 do
                xSum ← 0
                for k ← 0 to k ← 8 do
                    xw ← z[k] * w[i][k][j]
                    xSum ← xSum + xw
                a ← xSum + b[i][j]
                a ← max(0,a)
                push a to A[]
        if i ← 1 then
            for j ← 0 to j ← 1 do
                aSum ← 0
                for k ← 0 to k ← 4 do
                    aw ← A[k] * w[i][k][j]
                    aSum ← aSum + aw
                b ← aSum + b[i][j]
                b ← max(0,b)
                push b to B[]
        if i ← 2 then
            bSum ← 0
            for k ← 0 to k ← 1 do
                bw ← B[k] * w[i][k]
                bSum ← bSum + bw
            y ← bSum + b[i]
    y ← max(0, y)                                     ▷ last reLU function
    return y
```

Eventually, the MLP decision function returns a decimal value of 0 or >0. If the output is greater than zero, the classification label is 1; otherwise, it is 0.

The number of iterations in each loop is related to the number of neurons in each layer of MLP, and Algorithm 8 was tailored according to this paper's MLP implementation; a different implementation should follow different specifications accordingly.

5. Results and Discussion

The evaluation and analysis of the proposed system's execution steps are divided by analyzing the experimental results and the decision function execution measurements.

After applying the proposed methodology, several results were obtained to measure the proposed system's efficiency.

5.1. ML Detection Performance

As previously mentioned, this study applied two classification models to develop a poisonous record detection system. Table 4 shows SVM and MLP performance measurement metrics.

Table 4. ML Classifiers Performance.

Classifier	Accuracy	Precision	Recall	F1-Score
SVM	**0.81**	0.86	0.72	0.74
MLP	0.71	0.67	0.61	0.62

Figure 3 shows the Receiver Operating Characteristics (ROC) curve, which shows the relation between the true positive rates (TPR) and the True Negative Rates (TNR) of the two classifiers. The two classifiers' performances show that SVM achieved higher accuracy scores of 81% compared to MLP's 71% accuracy score. SVM also obtained better TPR scores than those MLP.

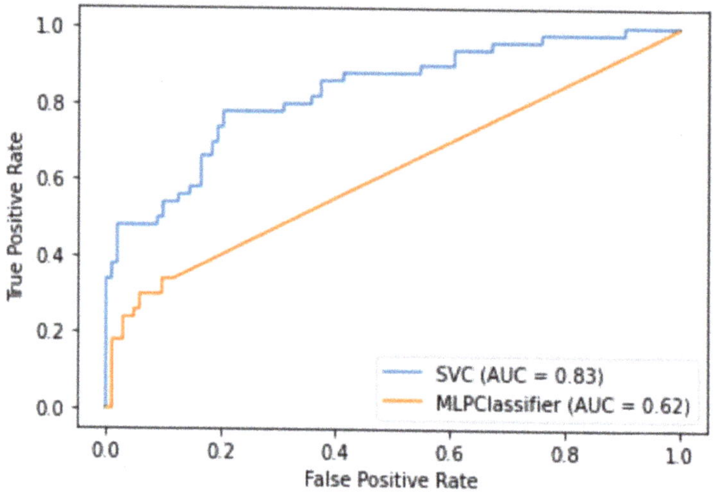

Figure 3. ROC curve for SVM and MLP classifiers.

5.2. *Smart Contract Performance*

This paper employs Blockchain technology through smart contracts. This subsection measures the performance and cost of such employment.

Table 5 shows details of the deployed smart contracts, including the deployment cost in gas. ML models of SVM and MLP were separately implemented to evaluate their performance better. Table 4 shows that the SVM contract has the most expensive deployment.

Table 5. Smart Contract Deployment.

Smart Contract	Language	Gas Cost (gwei)
Dataset Handler	Solidity	357,364
SVM Model Handler	Vyper	3,320,656
MLP Model Handler	Vyper	1,046,866

There was no clear way to measure the performance of smart contract functions regarding the CPU performance of the EVM. For this case, this study chooses to follow the judgment of each procedure's performance based on the elapsed time taken to complete each operation.

Table 6 shows the elapsed time for each read and write operation in the proposed methodology. It is worth noting that these measurements were taken on a local Blockchain network, with each transaction mined instantly to avoid any additional time latency.

Table 6. Elapsed time for methods' executions.

Procedure	Method	Run Time (s)	Expected (s)		Time Avg (s)
			Runs	Time	
Upload Dataset	writeRecord	0.57844	768	444.24 [1]	934.95 [2]
	uploadDataset	1425.66631	-		
Load Dataset	readRecord	0.16492	768	129.6246 [1]	114.41 [2]
	loadDataset	99.1903	-		
Write Scalar Parameters	setScalar	20.05179	-		20.05
Write SVM Parameters	setSupportVector	1.68084	395	663.9318 [1]	634.5 [2]
	setSupportVectors	605.0704	-		
	setSVM	0.47049	-		0.47049
	setDualCoef	46.240966	-		46.240966
					Total = 681.21
Write MLP Parameters	setFirstWeights	40.60692	-		40.60692
	setSecondWeights	11.84798	-		11.84798
	setThirdWeights	2.745	-		2.745
	setBiases	7.25123	-		7.25123
					Total = 62.45

[1] Calculated expected elapsed time by single run time × number of items' runs. [2] Averaged time between calculated expected and actual time of operation execution.

The elapsed time analysis showed that the write operation with smart contract setter functions has low time efficiency than the getter functions. The longest time was to set SVM's support vectors since writing a single support vector takes about two seconds to complete.

As mentioned before, the smart contract did not allow for the direct passing of decimal vectors; vectors were sent as integers and then converted back to decimals. These additional conversion steps could be the reason for the setting support vector procedure's low time efficiency; a more thorough CPU analysis could determine the cause for such latency.

5.3. Decision Function Performance

As mentioned before, this study chose to lower the implementation cost and only deploy the ML model's decision functions (DFs). Table 7 shows SVM and MLP performance details, including the execution cost and the elapsed time with average of 10 runs.

Table 7. DF performance details.

DF	Execution Cost (gwei)	Elapsed Avg (s)
SVM DF	16,495,436	8.4415
MLP DF	316,215	0.1914

A similar script was applied to time the completion of each deployed classifier prediction to calculate the smart contract DF. Figure 4 shows the elapsed time in seconds of both SVM and MLP in comparison with the built-in functions executed on the test machine.

MLP performed better than SVM regarding deployed DF time-efficiency and cost-efficiency. While obtaining a classification with the SVM's smart contract DF takes almost 10 s, it takes less than a second to obtain a classification with MLP. However, both classifiers' DFs fell behind in comparison with the client test machine's performance, which could be because of the humble EVM abilities to execute complex math methods in contrast with the test machine's abilities.

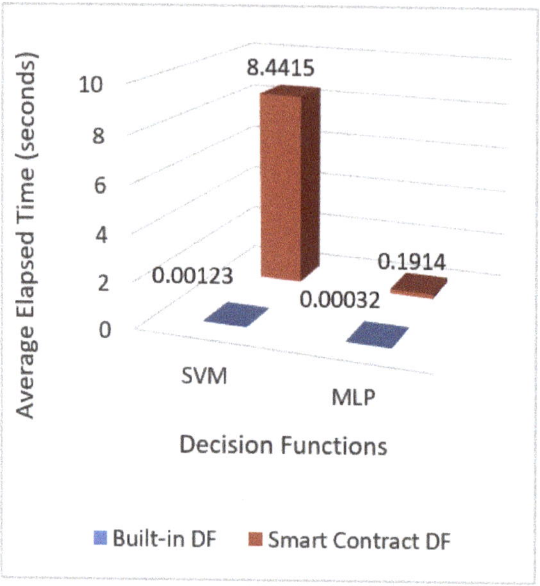

Figure 4. Elapsed time of SVM and MLP DFs.

5.4. *Overall Performance Comparison*

This section discusses the proposed methodology's overall performance by examining three essential perspectives, security, cost efficiency, and run-time efficiency.

- Security Perspective: No system can be 100% secured, and smart contracts are not an exception to that rule. Smart contracts can contain numerous security vulnerabilities: re-entrancy, unhandled exceptions, Integer Overflow, and unrestricted action [28]. Many reports and literature have discussed and studied such vulnerabilities. For instance, a study of [29] that evaluated 1.2 million smart contracts found that only 4% of real-world smart contracts are affected by the integer bug vulnerability.
  Authors of [28] argued that, even if the smart contract contained reported vulnerabilities, it does not mean it can be exploited in practice. They analyzed 23,327 vulnerable contracts and found that only 1.98% are exploited. By remarking on this, the proposed smart contract-based system can preserve the safety of execution by 96–98%.
- Cost Efficiency Perspective: The proposed methodology of integrating ML into Blockchain by only deploying the decision function of the ML model preserves the cost of storing large ML libraries and the execution cost of training the ML model entirely on-chain. Training the ML model off-chain and only deploying the decision process on-chain enhances the ML-integrated Blockchain cost efficiency.
- Run-time Efficiency Perspective: The run-time of the proposed system depends on EVM execution abilities. EVM is a run-time virtual machine with limited resources, which causes it to take longer to execute basic programming procedures, such as loops. Although the proposed system can preserve higher security and better cost efficiency, it has low run-time efficiency, which is believed to be increased with EVM development.

6. Conclusions

The proposed methodology in this paper provides a more exhaustive and efficient way to integrate AI abilities with Blockchain. In fact, it is more important to secure the process of using ML than improve ML itself. Blockchain provides the means of hashing to ensure this process of ML model decision integrity. This study proposed the flow of preserving trained models' gradients on-chain and reverse-engineering the decision function of SVM and MLP models on-chain.

During implementation, SVM proved to be more accurate than MLP in poisonous records detection. However, MLP achieved a higher time efficiency. Joining Blockchain to ML is very useful in sensitive domains like e-Health, which affects human life. This joining provides a safe environment for ML techniques for decision-making.

For future work, a more improved ML training procedure will increase the detection performance of the system. In addition, a layer of consensus could be added to control the uploading of training parameters. The obstacles faced by this study were primarily because of the immature EVM execution of complex math. When EVM becomes more efficient, the application of the proposed methodology will achieve better performance. Additionally, this integration of smart decisions gives Blockchain and smart contracts the AI ability to intelligently classify and detect, which is applicable to various Blockchain scenarios besides poisoning attack detection.

In our framework, the role of blockchain is to secure the decision process by using smart contracts. As for the consensus mechanisms among the connected nodes, it can be future work to further strengthen the learning process.

Finally, this study presented a prototype for the future incorporation of ML with Blockchain to take both technologies further in their evolution by securing the decision process.

Author Contributions: M.A.A., T.M. and S.Z. contributed to the conceptualization, methodology and writing—original draft. T.M. and A.A. contributed to methodology, project administration, visualization, and writing—review and editing. All authors have read and agreed to the published version of the manuscript.

Funding: This research was funded by Deputyship for Research and Innovation, Ministry of Education, Saudi Arabia Grant No. QU-IF-04-01-28436.

Data Availability Statement: The dataset used to support the findings of this study has been deposited in the website of kaggle repository (https://www.kaggle.com/uciml/pima-indians-diabetes-database, accessed on 8 January 2023).

Acknowledgments: The authors extend their appreciation to the Deputyship for Research and Innovation, Ministry of Education, Saudi Arabia for funding this research work through the project number (QU-IF-04-01-28436). The authors also thank Qassim University for technical support.

Conflicts of Interest: The authors declare no conflict of interest.

References

1. Alfrhan, A.; Moulahi, T.; Alabdulatif, A. Comparative study on hash functions for lightweight blockchain in Internet of Things (IoT). *Blockchain Res. Appl.* **2021**, *2*, 100036. [CrossRef]
2. Rahman, A.; Hossain, M.S.; Alrajeh, N.A.; Alsolami, F. Adversarial Examples—Security Threats to COVID-19 Deep Learning Systems in Medical IoT Devices. *IEEE Internet Things J.* **2021**, *8*, 9603–9610. [CrossRef]
3. Pitropakis, N.; Panaousis, E.; Giannetsos, T.; Anastasiadis, E.; Loukas, G. A taxonomy and survey of attacks against machine learning. *Comput. Sci. Rev.* **2019**, *34*, 100199. [CrossRef]
4. Ren, K.; Zheng, T.; Qin, Z.; Liu, X. Adversarial Attacks and Defenses in Deep Learning. *Engineering* **2020**, *6*, 346–360. [CrossRef]
5. Ayub, M.A.; Johnson, W.A.; Talbert, D.A.; Siraj, A. Model Evasion Attack on Intrusion Detection Systems using Adversarial Machine Learning. In Proceedings of the 2020 54th Annual Conference on Information Sciences and Systems (CISS), Princeton, NJ, USA, 18–20 March 2020. [CrossRef]
6. Demetrio, L.; Biggio, B.; Lagorio, G.; Roli, F.; Armando, A. Explaining Vulnerabilities of Deep Learning to Adversarial Malware Binaries. *arXiv* **2019**, arXiv:1901.03583.
7. Papernot, N.; Mcdaniel, P.; Jha, S.; Fredrikson, M.; Celik, Z.B.; Swami, A. The Limitations of Deep Learning in Adversarial Settings. In Proceedings of the 2016 IEEE European Symposium on Security and Privacy (EuroS&P), Saarbruecken, Germany, 21–24 March 2016. [CrossRef]
8. Xiao, H.; Biggio, B.; Nelson, B.; Xiao, H.; Eckert, C.; Roli, F. Support vector machines under adversarial label contamination. *Neurocomputing* **2015**, *160*, 53–62. [CrossRef]
9. Jagielski, M.; Oprea, A.; Biggio, B.; Liu, C.; Nita-Rotaru, C.; Li, B. Manipulating Machine Learning: Poisoning Attacks and Countermeasures for Regression Learning. In Proceedings of the 2018 IEEE Symposium on Security and Privacy (SP), San Francisco, CA, USA, 21–23 May 2018. [CrossRef]
10. Tideman, P.A.; Tirimacco, R.; St John, A.; Roberts, G.W. How to manage warfarin therapy. *Aust. Prescr.* **2015**, *38*, 44. [CrossRef] [PubMed]

11. Alharby, M.; Moorsel, A.V. BlockSim: An Extensible Simulation Tool for Blockchain Systems. *Front. Blockchain* **2020**, *3*, 28. [CrossRef]
12. Goel, A.; Agarwal, A.; Vatsa, M.; Singh, R.; Ratha, N. DeepRing: Protecting Deep Neural Network With Blockchain. In Proceedings of the 2019 IEEE/CVF Conference on Computer Vision and Pattern Recognition Workshops (CVPRW), Long Beach, CA, USA, 16–17 June 2019. [CrossRef]
13. Srivastava, G.; Crichigno, J.; Dhar, S. A light and secure healthcare blockchain for iot medical devices. In Proceedings of the 2019 IEEE Canadian Conference of Electrical and Computer Engineering (CCECE), Edmonton, AB, Canada, 5–8 May 2019; pp. 1–5.
14. Yong, B.; Shen, J.; Liu, X.; Li, F.; Chen, H.; Zhou, Q. An intelligent blockchain-based system for safe vaccine supply and supervision. *Int. J. Inf. Manag.* **2020**, *52*, 102024.
15. Chowdhury, D.; Banerjee, S.; Sannigrahi, M.; Chakraborty, A.; Das, A.; Dey, A.; Dwivedi, A.D. Federated learning based COVID-19 detection. *Expert Syst.* **2022**, e13173. [CrossRef]
16. Kim, H.; Kim, S.H.; Hwang, J.Y.; Seo, C. Efficient Privacy-Preserving Machine Learning for Blockchain Network. *IEEE Access* **2019**, *7*, 136481–136495. [CrossRef]
17. Preuveneers, D.; Rimmer, V.; Tsingenopoulos, I.; Spooren, J.; Joosen, W.; Ilie-Zudor, E. Chained Anomaly Detection Models for Federated Learning: An Intrusion Detection Case Study. *Appl. Sci.* **2018**, *8*, 2663. [CrossRef]
18. Sun, Y.; Esaki, H.; Ochiai, H. Blockchain-Based Federated Learning Against End-Point Adversarial Data Corruption. In Proceedings of the 2020 19th IEEE International Conference on Machine Learning and Applications (ICMLA), Miami, FL, USA, 14–17 December 2020. [CrossRef]
19. Qi, Y.; Hossain, M.S.; Nie, J.; Li, X. Privacy-preserving blockchain-based federated learning for traffic flow prediction. *Future Gener. Comput. Syst.* **2021**, *117*, 328–337. [CrossRef]
20. Weng, J.; Weng, J.; Zhang, J.; Li, M.; Zhang, Y.; Luo, W. DeepChain: Auditable and Privacy-Preserving Deep Learning with Blockchain-based Incentive. *IEEE Trans. Dependable Secur. Comput.* **2021**, *18*, 2438–2455. [CrossRef]
21. Jan, S.; Musa, S.; Syed, T.A.; Nauman, M.; Anwar, S.; Tanveer, T.A.; Shah, B. Integrity verification and behavioral classification of a large dataset applications pertaining smart OS via blockchain and generative models. *Expert Syst. J. Knowl. Eng.* **2021**, *38*. [CrossRef]
22. Rana, M.; Gudla, C.; Sung, A.H. Evaluating machine learning models on the ethereum blockchain for android malware detection. In *Intelligent Computing-Proceedings of the Computing Conference*; Springer: Cham, Switzerland, 2019; pp. 446–461.
23. Alsayegh, M.; Moulahi, T.; Alabdulatif, A.; Lorenz, P. Towards Secure Searchable Electronic Health Records Using Consortium Blockchain. *Network* **2022**, *2*, 239–256. [CrossRef]
24. Haber, S.; Stornetta, W.S. How to time-stamp a digital document. *J. Cryptol.* **1991**, *3*, 99–111. [CrossRef]
25. Lu, Y. The blockchain: State-of-the-art and research challenges. *J. Ind. Inf. Integr.* **2019**, *15*, 80–90. [CrossRef]
26. Namasudra, S.; Deka, G.C.; Johri, P.; Hosseinpour, M.; Gandomi, A.H. The Revolution of Blockchain: State-of-the-Art and Research Challenges. *Arch. Comput. Methods Eng.* **2020**, *28*, 1497–1515. [CrossRef]
27. Aladhadh, S.; Alwabli, H.; Moulahi, T.; Al Asqah, M. BChainGuard: A New Framework for Cyberthreats Detection in Blockchain Using Machine Learning. *Appl. Sci.* **2022**, *12*, 12026. [CrossRef]
28. Perez, D.; Livshits, B. Smart Contract Vulnerabilities: Vulnerable Does Not Imply Exploited. In Proceedings of the 30th USENIX Security Symposium (USENIX Security 21). USENIX Association, Virtual, 11–13 August 2021; pp. 1325–1341.
29. Torres, C.F.; Schütte, J.; State, R. Osiris: Hunting for Integer Bugs in Ethereum Smart Contracts. In Proceedings of the 34th Annual Computer Security Applications Conference, San Juan, PR, USA, 3–7 December 2018. [CrossRef]

applied sciences

MDPI

Article

Infrared Small and Moving Target Detection on Account of the Minimization of Non-Convex Spatial-Temporal Tensor Low-Rank Approximation under the Complex Background

Kun Wang [1,2], Defu Jiang [1,*], Lijun Yun [2] and Xiaoyang Liu [3]

1 College of Computer and Information, Hohai University, Nanjing 211000, China
2 School of Information Science and Technology, Yunnan Normal University, Kunming 650500, China
3 Department of Computer Science and Engineering, Chongqing University of Technology, Chongqing 400054, China
* Correspondence: surfer_jiangdf0801@163.com

Featured Application: Authors are encouraged to provide a concise description of the specific application or a potential application of the work. This section is not mandatory.

Abstract: Infrared point-target detection is one of the key technologies in infrared guidance systems. Due to the long observation distance, the point target is often submerged in the background clutter and large noise in the process of atmospheric transmission and scattering, and the signal-to-noise ratio is low. On the other hand, the target in the image appears in the form of fuzzy points, so that the target has no obvious features and texture information. Therefore, scholars have proposed many object detection methods for dimming infrared images, which has become a hot research topic on account of the flow-rank model based on the image patch. However, the result has a high false alarm rate because the most low-rank models based on the image patch do not consider the spatial-temporal characteristics of the infrared sequences. Therefore, we introduce 3D total variation (3D-TV) to regularize the foreground on account of the non-convex rank approximation minimization method, so as to consider the spatial-temporal continuity of the target and effectively suppress the interference caused by dynamic background and target movement on the foreground extraction. Finally, this paper proposes the minimization of the non-convex spatial-temporal tensor low-rank approximation algorithm (MNSTLA) by studying the related algorithms of the point infrared target detection, and the experimental results show strong robustness and a low false alarm rate for the proposed method compared with other advanced algorithms, such as NARM, RIPT, and WSNMSTIPT.

Keywords: complex background; infrared image; MNSTLA; point target detection

Citation: Wang, K.; Jiang, D.; Yun, L.; Liu, X. Infrared Small and Moving Target Detection on Account of the Minimization of Non-Convex Spatial-Temporal Tensor Low-Rank Approximation under the Complex Background. *Appl. Sci.* **2023**, *13*, 1196. https://doi.org/10.3390/app13021196

Academic Editor: Emanuele Carpanzano

Received: 10 December 2022
Revised: 8 January 2023
Accepted: 11 January 2023
Published: 16 January 2023

1. Introduction

The infrared detection system has the advantages of not being affected by light and, therefore, being capable of working at all times of the day [1]; not emitting electromagnetic waves and, therefore, being a system using a non-automatic detection method [2]; and having a strong penetrability and, therefore, being capable of penetrating the covers of dust, clouds, and smoke so as to better identify false camouflage targets, making it an effective supplement or substitute for the traditional visible light detection system and the radar detection system [3]. Therefore, the infrared point and moving target detection on account of the infrared detection system has always been an important topic and hotspot of research.

The infrared images have a low rank feature due to the many repetitive elements in the background, and they have a sparse feature due to the few feature points of infrared points and moving targets [4,5]. In this case, the detection of infrared points and moving targets is transformed into a classification task on account of the good performance of sparse representation in the classification task, which is what the method on of low-rank sparse is concerned about.

A sparse representation-based multispectral image target detection method was first proposed by the US Army Sensor Research Laboratory in 2014 [6]. He adopted the augmented Lagrange multiplier method to perform the optimization on account of the SR theory and the low-rank matrix [7] in 2015 based off the LRSR mode. This method can detect dim and point targets in a background with strong noise but does not have a good background suppression effect.

To overcome the limitations of conventional methods, Gao put forward an IPI (Infrared Patch-Image) model on account of the image segmentation by means of a sliding window, and the method can detect dim and point targets according to the targeted sparse feature of each patch image [8]. Considering the non-local autocorrelation structure for the background, the assumptions of the infrared patch image (IPI) model are in excellent agreement with the true scenario, which rephrases:

$$D_P = B_P + T_P + N_P \tag{1}$$

where D_P, B_P, T_P, and N_P are patch-images corresponding to the original, background, target, and random noise images, which are shown separately. Furthermore, the features of low-rank for the background B, and the target T, which is sparsity.

Dai et al. of Nanjing University introduced a structural prior model into the detection process of infrared points and moving targets, namely WIPI (Weighted Infrared Patch-Image). This method can better preserve the infrared point and moving targets while suppressing the strong edges [9]. Dai proposed an RIPT model. Furthermore, in view of the detection of infrared points and moving targets with insufficient prior information and strong edges [10], the SNN is used to separate the real target from the background by combining the non-local and local spatial priors. In order to solve the problems that the observation values of strong edge information are insufficient and the implicit assumptions do not match, The NIPPS model put forward by Dai, which can detect the residual error in the target image and is used for singular values [11]. As the SNN is not a convex envelope of low-rank background, and in view of the fact that the traditional IPT method only uses spatial information, Sun proposed the WNRIPT model [12].

In order to adapt to different images and solve the problem of images with strong edges, Xiong Bin used adaptive weights and an augmented Lagrange multiplier method [13]. Wang put forward an IPI model-based detection method for infrared point and moving targets, which maintains the spatial correlation among images, constructs a patch image form, and uses the ADMM multiplier method to optimize the solution finding so as to deal with the non-smooth and non-uniform background by the TV-PCP method [14].

Wang used different multi-subspaces for the areas to reduce the interference in each area, combined the APG with the patch coordinate descent method, and used the SMSL method to improve the accuracy of heterogeneous background [15]. However, for the infrared images with a complex background, and especially for those that contain clutter signals, as the noise also has a sparse feature as the target, the false alarm rate will increase. For complex scenarios, Zhang et al. put forward a non-convex rank approximation minimization (NRAM) detection method for infrared points and moving targets, which introduces extra regular terms into the edges [16]. Although the NRAM method has achieved good results in single image frame detection, the false alarm rate of this method is still high in complex and changing scenarios because it does not consider spatial and temporal information.

The above methods only vectorize the infrared image into a matrix, but do not well consider the temporal information. Therefore, many methods on account of the tensor analysis are applied in the IRST system, such as multi-view clustering [17], subspace clustering [18], super-resolution image generation [19], and image video processing [20]. Tensor analysis not only considers the spatial information of image sequences but also the temporal information thereof.

First, to fully exploit the inter-frame correlation between infrared image sequences, considering the time consistency and local spatial smoothness between the consecutive frames of the target, we introduced the spatial-temporal tensor into the NRAM model. To obtain more precise background estimations in the detection of infrared points and moving targets, as there was considerable noise in the infrared scenario, the norm was introduced because, compared with the norm, the norm requires not only sparse columns but also sparse rows, which can better remove the strong edge non-target noise. In order to simplify the computational complexity, we introduced the Frobenius norm. Finally, we proposed a minimization of the non-convex spatial-temporal tensor low-rank approximation algorithm (MNSTLA). The main contributions of the MNSTLA model are:

(1) A non-convex spatial-temporal tensor low-rank approximation minimization method for the detection of infrared points and moving targets in the sequence scenarios was proposed. We introduced 3D-TV regularization into the NRAM model. The 3D-TV constraint on the background is helpful for keeping the image details and removing the noise, so it can achieve better detection performance under complex backgrounds.
(2) The norm is introduced into the detection of IR points and moving targets to better describe the target components. By combining structured sparsity terms, non-target components, especially those with strong edges, can be eliminated.
(3) The ADMM is used to efficiently reduce the computational complexity and solve the low-rank component recovery problem.

The paper is organized as follows: in Section 2, the work related to the MNSTLA method-based detection of infrared dim and point targets is briefly described; in Section 3, the proposed MNSTLA model; in Section 4, the extensive experiments carried out on various sequence scenarios are described to illustrate the efficiency of the MNSTLA model, and the results are evaluated subjectively and objectively; and in Section 5, we give the discussion and conclusion.

2. Related Work

In this section, we first briefly introduce how to construct an image sequence into a spatial-temporal patch tensor model of image tensors. Furthermore, we introduce the 3D-TV regularization model and the tensor kernel norm model, respectively, and model the foreground and background of the sequence image tensor considering both models.

2.1. Spatial-Temporal Patch Tensor Model

Generally speaking, given an image sequence $f_1, f_2, \ldots, f_p \in R^{m \times n}$ and a cube patch tensor $F \in R$, the frames can be obtained by stacking them in time order. The tensor of the IR point target image can be expressed as:

$$D_T = B_T + N_T + T_T \tag{2}$$

where $D_T, B_T, N_T, T_T \in R^{m \times n \times L}$ present the original patch-tensor, background tensor, target-tensor, and noise-tensor. According to the infrared imaging mechanism, the relative motion between the imaging sensor and the target is usually due to small changes at a long distance, such as an early warning system. Therefore, it is generally believed that the backgrounds of different frames change slowly in the whole sequence images, which means that there is a correlation between adjacent sequences [8,21]. For the reason that images containing infrared points and moving targets are considered to be of low rank, the constructed background tensor can also be considered a low-rank tensor. Compared with the matrix model, constructing a tensor model can not only mine the internal relations between data from more angles in the tensor domain but also further improve the capability of target detection by combining the spatial-temporal information.

2.2. Foreground Modeling on Account of 3D-TV Regularization

Total variation (TV) regularization is widely used to detect the sharp edges and corners of images, which can represent the desired spatial smoothness. In this study, we use 3D-TV to leverage spatio-temporal information. Assuming $N \in R^{m \times n \times t}$, we define the 3D-TV norm as:

$$||T||_{3D-TV} = \sum_{m,n,t} TV_{m,n,t}(T) = |T_{m+1,n,t} - T_{m,n,t}| + |T_{m,n+1,t} - T_{m,n,t}| + |T_{m,n,t+1} - T_{m,n,t}| \tag{3}$$

where $T_{m,n,t}$ represents the intensity of the pixels (m, n, t); at the same time, the difference operator along the temporal direction shows that it considers the persistence of the foreground target in time.

We introduced the vector difference operators for the horizontal, vertical and time directions:

$$\begin{cases} V_h||T|| = vec(|T_{m+1,n,t} - T_{m,n,t}|) \\ V_v||T|| = vec(|T_{m,n+1,t} - T_{m,n,t}|) \\ V_t||T|| = vec(|T_{m,n,t+1} - T_{m,n,t}|) \end{cases} \tag{4}$$

Then, the Formula (3) can be rewritten as:

$$||T||_{3D-TV} = ||VT||_1 = ||\mathrm{V_h}T||_1 + ||\mathrm{V_v}T||_1 + ||\mathrm{V_t}T||_1 \tag{5}$$

2.3. Background Modeling on Account of the Tensor Nuclear Norm

In the TRPCA model [22], the tensor nuclear norm is usually used instead of the rank function to constrain the background. However, the general tensor nuclear norm is used to matrix the tensor, and using the singular value of matrix to define the tensor nuclear norm will destroy the spatial structure of the video, and the degree of approximation to the rank function will be insufficient. On account of the t-product, Lu, et al. [23] an improved tensor nuclear norm is proposed:

$$||B||_{**} = \sum_{i=1}^{r} S(i,i,1) \tag{6}$$

where $r = rank_t(B)$, $B = U * S * V$. and converted into the nuclear norm of the matrix:

$$||B||_{**} = \frac{1}{N}||bcric(B)||_* = \frac{1}{N}||\overline{B}||_* \tag{7}$$

From the Formulas (6) and (7), we obtain:

$$||B||_{**} = \frac{1}{n_3} \sum_{i=1}^{r} \sum_{j=1}^{n_3} \overline{S}(i,i,j) \tag{8}$$

where $bcric(B)$ represents the patch cyclic matrix of B, and $\overline{B}$ represents the patch diagonal matrix of B.

It can be seen from the Formula (6) that the improved tensor nuclear norm is directly defined by the singular value tensor S, and it can be seen from the patch cyclic matrix and patch diagonal matrix of the Formula (7) that the above tensor nuclear norm is defined on account of the front-side slicing (the third-dimension time). In addition, the improved tensor nuclear norm $||B||_{**}$ is a convex envelope of the average rank in the unit sphere of the tensor spectral norm, which has a better approximation to the rank function [23] on account of the above considerations; this paper uses the above tensor nuclear norm to perform low-rank constraining on the background, which strengthens the low rank of the background.

3. Methods

The spatial-temporal infrared patch-tensor model is described as:

$$f_D = f_B + f_T + f_N \tag{9}$$

where f_D, f_B, f_T, and f_N represent the original, background, target, and noise images, respectively. As shown in Figure 1, each image frame is split into small image patches, and all the small image patches of consecutive L frames are superimposed into the 3D patch-tensor. Therefore, the above formula can be rewritten into a tensor form as shown in Formula (2) in Section 2.2.

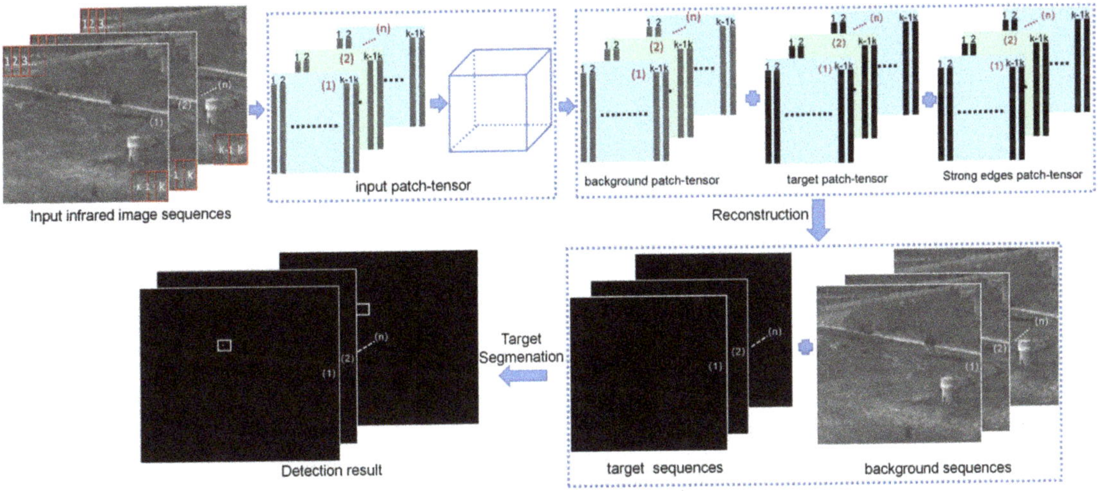

Figure 1. Flow Chart of the MNSTLA Method.

In the WNRIPT model, the problem of point target detection is expressed as:

$$B.T = \underset{B.T}{min} ||B||_{W_B,*} + \lambda ||W_T \odot T|| \tag{10}$$

where $\left|\left|B\right|\right|_{W_B,*} = \frac{1}{L} \sum_{i=1}^{r} \sum_{j=1}^{L} W_B(i,i,j)\overline{S}(i,i,j)$.

In order to further improve the performance and efficiency of point target detection, the 3D-TV regularization is introduced into the spatial-temporal tensor model, and its expression is:

$$B.T.N = \underset{B.T.N}{arg\ min} ||B||_{W_B,*} + \lambda_1 ||V(B)||_{3D-TV} + \lambda_2 ||T||_1 + \lambda_3 ||N||_F^2$$

$$s.t.\ F = B + T + N \tag{11}$$

where $k \times || * ||_{3D-TV}$ is the norm of 3D-TV, and λ_1, λ_2, and λ_3 represent the regularization parameters of the 3D-TV term, target component, and noise component.

As the Frobenius norm [24,25] has a good noise suppression effect, the Frobenius norm term is further introduced:

$$B.T.N = \underset{B.T.N}{arg\ min} ||B||_{W_B,*} + \lambda_1 ||V(B)||_1 + \lambda_2 ||T||_1 + \lambda_3 ||N||_F^2$$

$$s.t.\ F = B + T + \mathrm{N} \tag{12}$$

In this model, the 3D-TV regularization term is introduced, which can fully capture the spatial-temporal information of infrared sequence images, so it is expected to achieve better performance.

3.1. Low Rank and Sparse Frame Model

Different values of singularities in the conventional convex kernel norm solve the imbalance penalty. Due to the equal treatment mechanism, if singular values are far from 1, the nuclear norm will have a considerable deviation. Each time the nuclear norm weight is determined, additional SVD will appear [26], which increases the running time of the method. Zhao proposed the γ norma which is a new rank of non-convex function [27]. The γ norm is unitarily invariant. The γ norm is almost in agreement with the true rank (γ = 0.002), and the heuristic of the log-det performs poorly at minimal singular values [28], in particular when the value is close to 0; the γ norm of the matrix B is described as:

$$||B||_{\gamma} = \sum_{i} \frac{(1+\gamma)\sigma_i(B)}{\gamma+\sigma_i(B)} \tag{13}$$

For the reason the l_0 norm is NP-hard, the l_1 norm [29] assigns the same weight to each single element. Therefore, many other methods use the l_1-norm to characterize the sparsity of the target patch-image [30–32], and the target T with the l_1-norm is described as follows:

$$||T||_1 = \sum_{i,j} W_{ij}|T_{ij} \tag{14}$$

where $W_{ij} = C/\left(\left|T_{ij}\right| + \varepsilon_T\right)$ is an element at position (*i, j*), C is a compromise constant; moreover, ε_T is a small positive number.

Infrared images also have a lot of strong edge noise, which makes many advanced methods [33–35] leave residual errors in the target image. The strong edge E is linearly sparse relative to the whole image, and each line (i.e., line vector) is described by the vector l_2 norm, $w_i = \sqrt{\sum_j \left|E_{i,j}\right|^2}$, that is, the vector $w = [w_1, w, \ldots, w_d]^T$, and then the whole matrix E needs to be described by the norm. Therefore, the l_1 norm is used to describe w, that is, the $l_{2,1}$ norm of the strong edge E:

$$||E||_{2,1} = ||w||_1 = \sum_{i=1}^{d} \sqrt{\sum_{j=1}^{n} \left|E_{i,j}\right|^2} \tag{15}$$

According to the foregoing discussion, the patch-tensor model for the infrared image sequences is proposed on account of the minimization of the non-convex spatial-temporal tensor low-rank approximation algorithm (MNSTLA), that is, Formula (10) is redefined as:

$$B.T.E = \underset{B.T.E}{arg\ min} ||B||_{\gamma,*} + \lambda_1||L||_{\gamma} + \lambda_2||T||_1 + \lambda_3||E||_{2,1}$$
$$s.t.\ D = B + T + E \tag{16}$$

3.2. Solution Finding of MNSTLA Model

The optimization method based on the ADMM is used to work out Formula (16). Formula (16) can be rewritten as an augmented Lagrange function:

$$L(D, B, T, E, L, Z, Y, \mu)$$
$$= ||Z||_{\gamma,*} + \lambda_1||L||_{\gamma} + \lambda_2||T||_{w,1} + \langle Y_1, Z - B\rangle + \langle Y_2, L - V(B)\rangle + \langle Y_3, D - B - T - E\rangle$$
$$+ \tfrac{\mu}{2}\left(||Z - B||_F^2 + ||L - V(B)||_F^2 + ||D - B - T - E||_F^2\right) + \lambda_3||E||_{2,1}$$

$$s.t.\ D = B + T + E, Z = B, L = V(Z) \tag{17}$$

where Y_*, μ are an Lagrange multiplier and a positive penalty scalar, $\langle * \rangle$ represents the inner product, and $||*||_F$ is the norm for Frobenius.

The ADMM method is used to iteratively update the Z and L by the Formula (17), respectively:

$$Z^{k+1} = \underset{Z}{arg\ min} ||Z||_{\gamma,*} + \frac{\mu^k}{2}||Z - B^k + \frac{{Y_1}^k}{\mu^k}||_F^2 \tag{18}$$

$$L^{k+1} = \underset{L}{arg\ min} ||L||_{\gamma} + \frac{\mu^k}{2}||L - V(B^k) - \frac{{Y_1}^k}{\mu^k}||_F^2 \tag{19}$$

Find their solutions by t-SVD [20] operation and unit contraction operator, respectively:

$$Z^{k+1} = D_{W/\mu^k}(B^k - \frac{Y_1^k}{\mu^k}) \tag{20}$$

$$L^{k+1} = Th_{\lambda_1/\mu^k}\left(V\left(B^k\right) - \frac{{Y_2}^k}{\mu^k}\right) \tag{21}$$

where D(*) represent the t-SVD operation and Th(*) represent the unit contraction operator.

Extract the term containing B from the Formula (17):

$$B^{k+1} = \frac{\mu^k}{2}(||D - B - T^k - E^k + \frac{Y_1^k}{\mu^k}||_F^2 + ||Z^{k+1} - B + \frac{Y_2^k}{\mu^k}||_F^2 + ||L^{k+1} - V(B) + \frac{Y_3^k}{\mu^k}||_F^2) \tag{22}$$

The Formula (22) is equivalent to the following linear equations:

$$(2I + V(B))B^{k+1} = D - T^k - E^k + \frac{Y_1^k}{\mu^k} + Z^k + \frac{Y_2^k}{\mu^k} + V^T(V_B^k + \frac{Y_3^k}{\mu^k}) \tag{23}$$

The closed form of the Formula (23) can be obtained by 3D Fast Fourier Transform:

$$B^{k+1} = ifftn(\frac{fftn(D - T^k - E^k + \frac{Y_1^k}{\mu^k} + Z^k + \frac{Y_2^k}{\mu^k} + V^T(V_B^k + \frac{Y_3^k}{\mu^k}))}{2\mu^k I + \mu^k|fftn(V(B)|^2}) \tag{24}$$

where $fftn$ is the fast 3D Fourier transform and $ifftn$ is the inverse transform of the $fftn$.

Variables T and E are corrected:

$$T^{k+1} = \underset{T}{arg\ min} \lambda_2||T||_{W,1} + \frac{\mu^k}{2}||D - B^{k+1} - T - E^k + \frac{Y_3^k}{\mu^k}||_F^2 \tag{25}$$

$$E^{k+1} = \underset{E}{arg\ min} \lambda_3||E||_{2,1} + \frac{\mu^k}{2}\left|\left|D - B^{k+1} - T^{k+1} - E\right|\right|_F^2 \tag{26}$$

By using the element-by-element shrinkage operation method in references [29,36], we obtain:

$$T^{k+1} = Th_{\lambda W/\mu^k}(D - B^{k+1} - E^k - \frac{Y_3^k}{\mu^k}) \tag{27}$$

$$E^{k+1} = \frac{\mu^k(D - B^{k+1} - T^{k+1} - \frac{Y^k}{\mu^k}) + Y_3^k}{\mu^k + 2\lambda_3} \tag{28}$$

3.3. The Processing of the MNSTLA

The steps of the MNSTLA model (Algorithm 1):

Algorithm 1: The Minimization of Non-Convex Spatial-Temporal Tensor Low-Rank Approximation Algorithm(MNSTLA)

Input: Input the $f_1, f_2, \ldots, f_p \in R^{m\times n}$, $\lambda_1, \lambda_2, \lambda_3, L$ and $tol = 10^{-7}$
Initialize: Original patchtensor$D \in R^{m\times n\times L}$, $B^0 = T^0 = E^0 = Y_1{}^k = Y_2{}^k = Y_3{}^k = 0, \mu = 1e-2$
ADMM for solving the Equation (17)
while
(1) Fix the others and update and $L$by (20) and (21) Z^{k+1}, L^{k+1}
(2) Fix the others and update B by (24) B^{k+1}
(3) Fix the others and update $T$by (25) T^{k+1}
(4) Fix the others and update E by (26) E^{k+1}
(5) Check the convergence conditions $\frac{||D-B^{k+1}-T^{k+1}||_F^2}{||D||_F^2} \leq tol$
(6) Update $k = k + 1$.
Output B^{k+1}, T^{k+1}

The flow chart of the MNSTLA model is shown in Figure 1.
The specific detection steps are as follows:

(1) The original infrared image sequences $f_1, f_2, \ldots, f_p \in R^{m\times n}$ are sequentially arranged by n_3 adjacent frames and are converted into several patch-tensor tensors $D \in R^{m\times n\times L}$.
(2) The original patch-tensor is decomposed into the target patch-tensor T, background patch-tensor B, and structural noise (strong edge) patch-tensor E by using the method 1.
(3) The target image I_T and the background image I_B are reconstructed by inverse operation.
(4) In the last step, we segment the target using the adaptive threshold [8]:

$$t_{seg} = mean(C) + \lambda \times std(C) \tag{29}$$

where $mean(C)$ is the mean value of the reconstructed confidence map, $std(C)$ is the standard deviation, and λ is a constant.

4. Experiment and Analysis of Experimental Results

Where $mean(C)$ is the mean value of the reconstructed confidence map, $std(C)$ is the standard deviation, and λ is a constant.

4.1. Data Set and Evaluation Indicators

4.1.1. Test Data Set

In the experiment, the "A data set for infrared detection and tracking of dim-small aircraft targets underground/air background [37]" collected by Hui Bingwei et al. was used. The sensors used for data acquisition were refrigerated medium-wave infrared cameras with a resolution of 256 × 256 pixels.

There are 22 data scenarios in this dataset. The 22 image sequences of data 1–data 22 of this data set data are described and shown in Table 1:

Table 1. Detailed Description of 22 Real Scenarios.

Data	No. Frame	Scenario Description
data1	399	Close range, single target, sky background
data2	599	Close range, two targets, sky background, cross flight
data3	100	Close range, single target, air-ground interface background, the target enters the field of view again after leaving the field of view.
data4	399	Close range, two targets, sky background, cross flight
data5	3000	Long range, single target, ground background, long time
data6	399	From near to far, single target, ground background
data7	399	From near to far, single target, ground background
data8	399	From far to near, single target, ground background
data9	399	From near to far, single target, ground background
data10	401	Target from near to far, single target, ground-air interface background
data11	745	Target from far to near, single target, ground background
data12	1500	Target from far to near, single target, target mid-course maneuver, ground background
data13	763	Target from near to far, single target, dim target, ground background
data14	1462	Target from near to far, single target, ground background, target interfered by ground vehicles
data15	751	Single target, target maneuver, ground background
data16	499	Target from far to near, single target, extended target, target maneuver, ground background
data17	500	Target from near to far, single target, dim target, ground background
data18	500	Target from far to near, single target, ground background
data19	1599	Single target, target maneuver, ground background
data20	400	Single target, target maneuver, air-ground background
data21	500	Long range, single target, ground background
data22	500	Target from far to near, single target, ground background

As can be seen from the above table, data1–data 4 all have a sky background. As they have a single background and large targets as shown by Figure 2, they are not suitable for our set conditions and are not used.

Figure 2. Data1–data 4 Sequence Images.

Six sequences of data 6, data10, data13, data14, data17 and data 22 were selected from data 5–data 22 as the sequence images of our experiment. As shown by Figure 3a–f, they are six representative images in the six sequences of the selected six data sets, namely, data 6, data 10, data 13, data 14, data 17, and data 22. The point-target is in the white boxes.

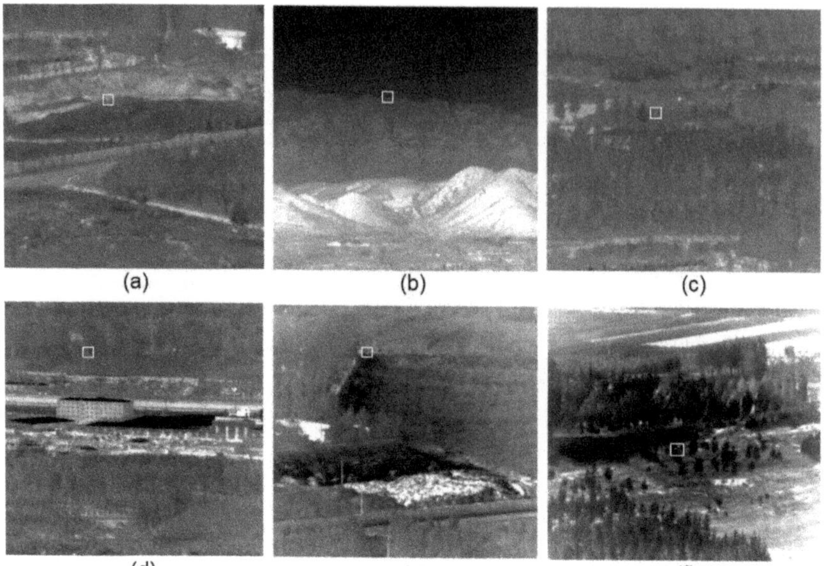

Figure 3. Six Infrared Image Sequences Selected.

4.1.2. Evaluation Indicators

The performance of dim object detection methods is generally evaluated using three criteria: background suppression, target enhancement, and detection accuracy.

(1) Background suppression factor (*BSF*) [9]:

The *BSF* is defined as follows:

$$BSF = \frac{\delta_{out}}{\delta_{in}} \tag{30}$$

where δ_{out} and δ_{in} represent the local background standard deviation around the target of the output image and the original image.

(2) Local contrast gain (LCG)

The *SCRG* represents the signal and noise ratios (*SCR*) before and after processing:

$$SCRG = \frac{SCR_{out}}{SCR_{in}} \tag{31}$$

In which the *SCR* uses the same expression as in reference [38]:

$$SCR = \frac{|\mu_t - \mu_b|}{\delta_b} \tag{32}$$

where μ_t , μ_b and δ_b represent the average gray values of the targets in the image.

In this paper, both *BSF* and *SCRG* need the determination of the background range around the target. Figure 4 shows the background around the target calculated in this paper, where d takes 20.

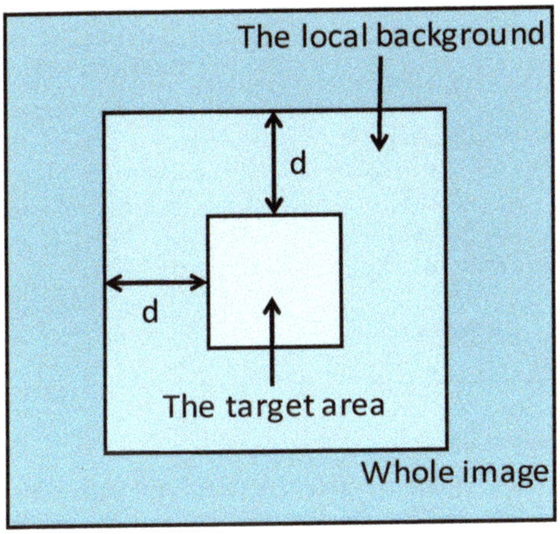

Figure 4. Local Background around the point targets in the Infrared Image.

For the reason the δ_b is close to zero in the Formula (32), it is difficult to evaluate the performance as the SCR approaches infinity. Therefore, we evaluate the performance of the target augmentation using LCG:

$$LCG = \frac{LC_{out}}{LC_{in}} \tag{33}$$

$$LC = \frac{|\mu_t - \mu_b|}{\mu_t + \mu_b} \tag{34}$$

where LC_{out} and LC_{in} represent the local contrast (LC) of the output image and the input image, the μ_t and μ_b the are consistent with those in the Formula (32).

(3) Receiver operating characteristic curve (ROC)

In order to further compare the methods, the ROC curve is used to evaluate the methods which can be used to select the best category judgment model and abandon the sub-optimal model. When judging the category, the ROC curve can give a correct evaluation without being limited by cost or benefit.

All the samples, which is actually the target but is wrongly judged. It is defined as follows:

$$P_d = \frac{N_{true}}{N_{act}} \tag{35}$$

$$P_f = \frac{N_{false}}{N_{img}} \tag{36}$$

where N_{true}, N_{act}, N_{false} and N_{img} represent the number of really detected targets, the actual targets, the falsely detected targets and the frames, respectively.

4.2. Parameter Setting

We quote the values of μ, γ, and C in reference [16], which are the penalty factor $\mu = c\sqrt{min(m,n)}$, where c = 3, γ = 0.002, and C = 2.5, where m and n are the length and width of patch images, respectively. References [39–41] all made a detailed analysis of the frame number L, and we also take its value and the frame number L = 3. For details, please refer to these references.

In order to better verify the advancement of the MNSTLA method, we will compare it with seven advanced methods, including the Top-Hat method. Table 2 lists the parameter settings for these methods.

Table 2. The parameters for the 7 tested methods.

Methods	Parameter Setting
Top-Hat	Structure size: 3 × 3, structure shape: square
PSTNN	Sliding step : 40, $\lambda = 0.6/\sqrt{max(n_1, n_2) * n_3}$, patch size: 40 × 40, $\varepsilon = 1 \times 10^{-7}$
IPI	Patch size : 50 × 50, sliding step : 10, $\lambda = 1/\sqrt{min(m,n)}$, $\varepsilon = 10^{-7}$
RIPT	Patch size : 30 × 30, $\lambda = L/\sqrt{min(m,n)}$, sliding step: 10, $L = 0.7$, h = 1, $\varepsilon = 10^{-7}$
WSNMSTIPT	Patch size : 30 × 30, sliding step : 30 $L = 6$, p = 0.8, $\lambda = 1/\sqrt{max(n_1, n_2) * n_3}$
NRAM	Patch size : 50 × 50, sliding step : 10, $\lambda = 1/\sqrt{min(m,n)}, \mu 0 = 3\sqrt{min(m,n)}$, $\gamma = 0.002, C = \sqrt{min(m,n)}/2.5$, $\varepsilon = 10^{-7}$
MNSTLA	Patch size : 50 × 50, sliding step : 10, $\gamma = 0.002$, $\mu = c\sqrt{min(m,n)}$ where $c = 3$, $L = 3$. C = 2.5, $\varepsilon = 1 \times 10^{-7}$

4.3. Subjective Evaluation in Different Scenes

In this sub-section, we give the detection results of six infrared image sequences. The method proposed herein is compared with six related advanced methods, namely Top-Hat [41], IPI [9], PSTNN [22], IPT [23], WSNMSTIPT [24], and NRAM [16]. For the convenience of observing the results, the experimental results obtained and the three-dimensional grid diagrams generated by all the test methods in different scenarios are given intuitively in Figures 5–10.

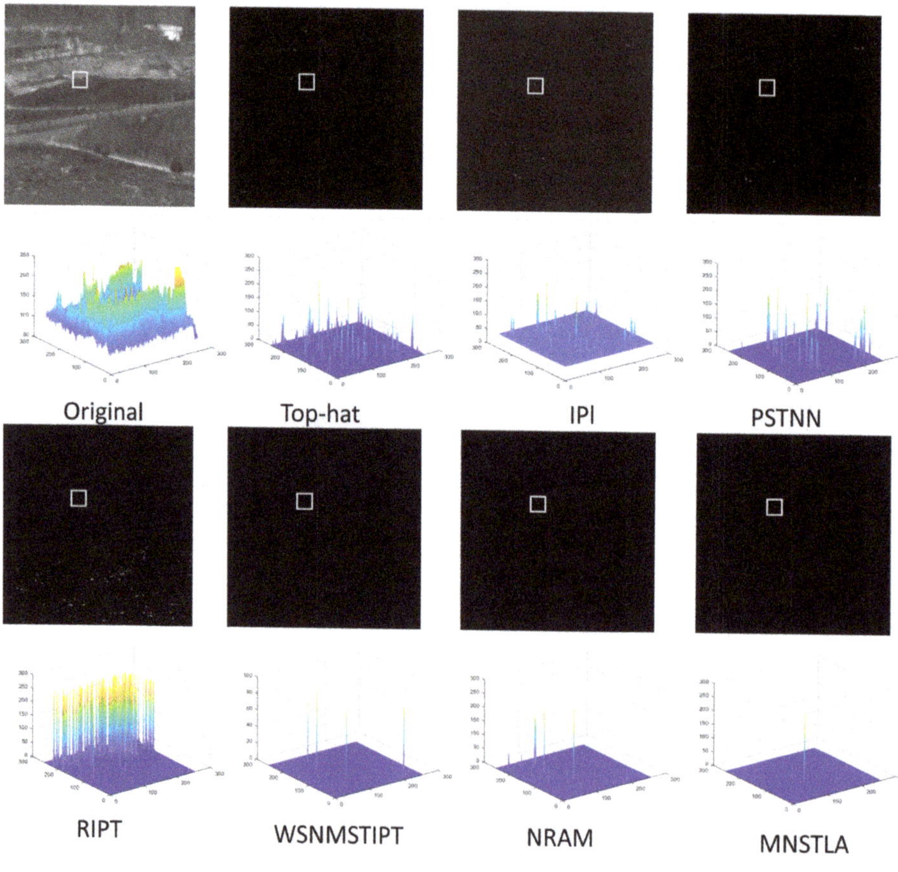

Figure 5. Infrared Sequence (a) Original image and Detection Results and the 3d grid diagrams.

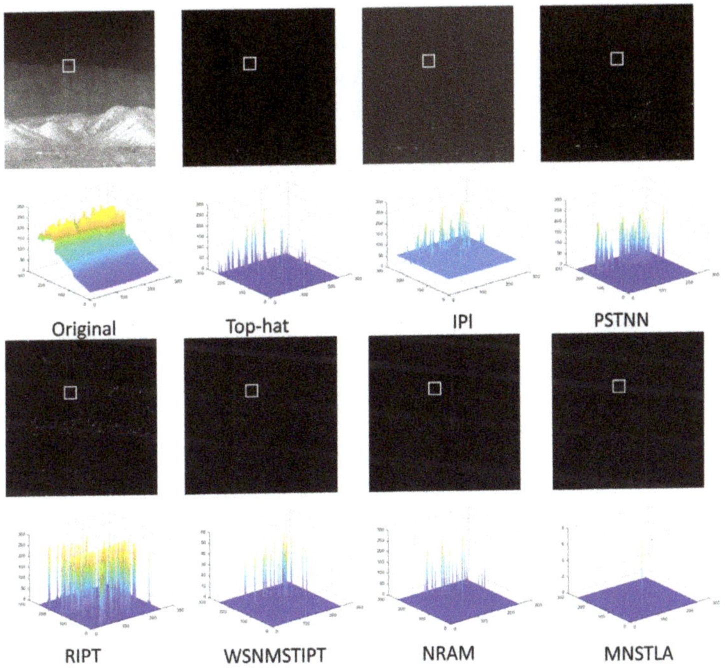

Figure 6. Infrared Sequence (b) Original image and Detection Results and the 3d grid diagrams.

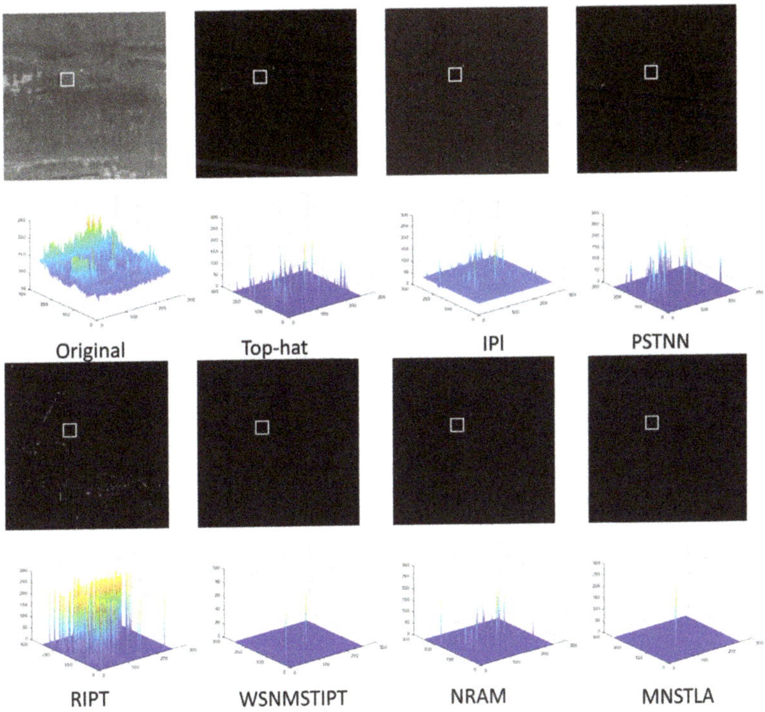

Figure 7. Infrared Sequence (c) Original image and Detection Results and the 3d grid diagrams.

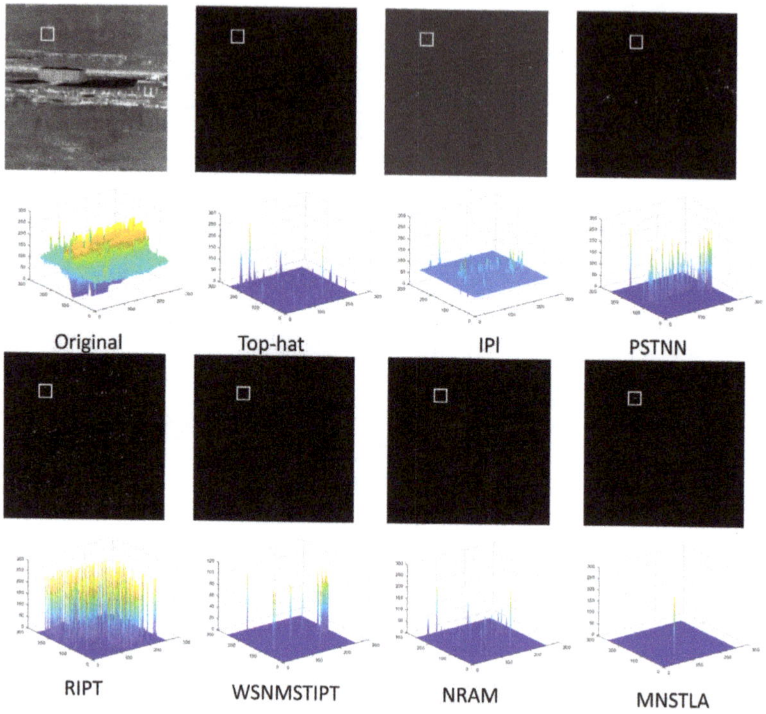

Figure 8. Infrared Sequence (d) Original image and Detection Results and the 3d grid diagrams.

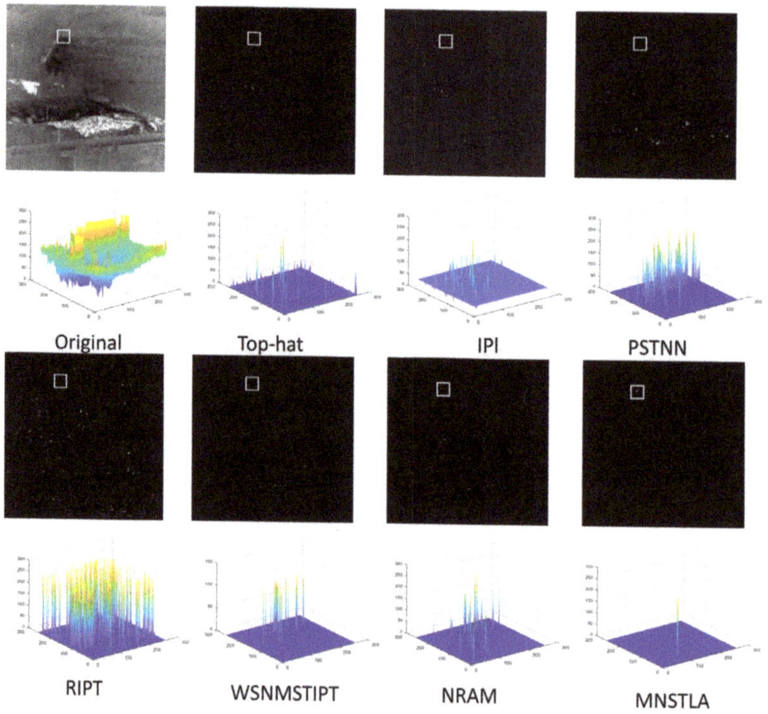

Figure 9. Infrared Sequence (e) Original image and Detection Results and the 3d grid diagrams.

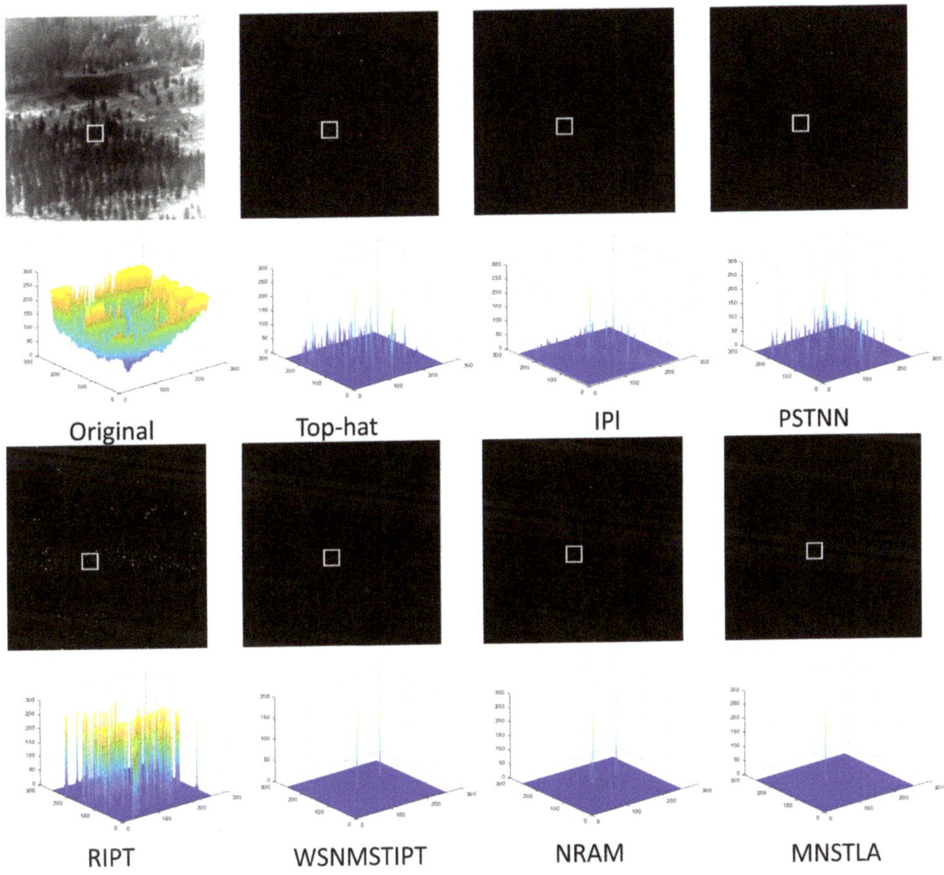

Figure 10. Infrared Sequence (f) Original image and Detection Results and the 3d grid diagrams.

It can be seen from Figures 5–10 that the RIPT model has the worst detection efficiency; the Top-Hat and PSTNN methods do enhance the targets, but edges and noise are introduced, which is mainly due to the assumption of fixed structural elements and a smooth background. Undoubtedly, among all the results from the test methods, the Top-Hat and PSTNN methods have the worst performance. This is because this contrast mechanism is not suitable for complex backgrounds. The IPI method is slightly better than the Top-Hat and PSTNN methods. The WSNMSTIPT models are on account of the IPI model and refer to the spatial-temporal information. Compared with the IPI model, although their false alarm rates are effectively reduced, not only do the images with dim targets selected from data sets 13 and 17 (corresponding to Figures 7 and 9) lose their targets, but also the images selected from the data sets with complex backgrounds lose their targets. Compared with the WSNMSTIPT models, the NRAM model does not consider the spatial-temporal information; it constructs the target-patches and background-patches according to the sparse feature of infrared target images. It can be seen from Figures 5–10 that, compared with the IPI model, the NRAM method not only effectively reduces the false alarm rate but also effectively enhances the strong edges. Therefore, the potential target points are also enhanced, and a better detection rate is achieved compared with the IPI model. The MNSTLA model proposed herein constructs, on account of the NRAM model and the spatial-temporal information, a spatial-temporal tensor model of infrared dim moving targets that fully considers the correlation between the frames of infrared dim moving targets and can further reduce the false alarm rate and improve the detection efficiency of infrared dim moving targets.

4.4. Objective Evaluation for Different Scenes

We evaluate the performance of the MNSTLA model using the LCG and the BSF. The experimental results of the six actual sequences (Figures 5 and 6) are shown in Table 3. It can be seen that the method presented here can achieve the best values.

Table 3. Average Values of BSF and LCG of the Six Infrared Sequence Images Obtained by the Methods.

Methods	a		b		c		d		e		f	
	BSF	LCG	BSF	LCG	BSF	LCG	BSF	LCG	BSF	LCG	BSF	LCG
Top-Hat	7.73	5.94	3.28	6.76	7.86	1.67	9.66	7.53	10.25	3.64	7.34	3.45
PSTNN	3.85	1.23	3.86	8.20	4.16	1.18	3.67	2.43	4.14	3.16	3.14	2.99
IPI	3.35	1.70	2.30	5.65	3.45	1.06	3.19	3.18	5.61	2.37	2.02	1.94
RIPT	0.92	3.11	0.72	3.16	1.76	1.29	1.62	2.01	1.26	1.29	0.56	1.93
WSNMSTIPT	5.16	6.22	2.08	22.35	4.26	2.36	5.08	2.86	3.46	4.16	3.29	3.38
NRAM	26.45	1.235	23.74	6.39	7.08	1.68	18.16	16.18	9.31	2.17	10.67	4.86
MNSTLA	**61.25**	**8.353**	**36.29**	**26.58**	**63.42**	**6.98**	**39.61**	**7.69**	**54.36**	**5.93**	**53.17**	**5.29**

Table 3 shows the average BSF and LCG of different methods on the six infrared image sequences. The Top-Hat and PSTNN methods have the lowest BSF and LCG values, and the corresponding background suppression capability is the worst. The IPI, RIPT, and WSNMSTIPT models have achieved good results in the six infrared image sequences, among which the RIPT and WSNMSTIPT models are slightly better than the IPI models in terms of performance; the NRAM model obtained a higher BSF value in the first sequence, but compared with the RIPT and WSNMSTIPT model, its background suppression ability is still not ideal; the MNSTLA model proposed herein achieved the highest BSF value on all six infrared image sequences, which means the robustness and efficiency of background suppression are better. In terms of LCG, this method has the highest LCG value and the best target enhancement of the six image sequences. From the evaluation results, it can be seen that the LCG and BSF values of the MNSTLA model proposed herein are much higher than those of other methods, indicating that it has great advantages in object enhancement and that the signal-to-noise ratio of images is improved effectively.

In order to compare the above optimization methods more objectively, the comparison of the ROC curves of the sequences 1–6 is shown in Figure 11. It is found in the study that the RIPT was the worst performer and that the Top-Hat method and the PSTNN method are not satisfactory. The IPI model achieved good results on the six infrared image sequences, and the WSNMSTIPT methods are slightly better than the IPI model in terms of performance. The detection rate of the NRAM model is not as high as that of the WSNMSTIPT models, and this is because the NRAM model does not consider the temporal-spatial information. Finally, under the same false alarm ratio, the MNSTLA model proposed herein achieved the highest detection probability, which means that the proposed MNSTLA model has better performance than that of any of the other models.

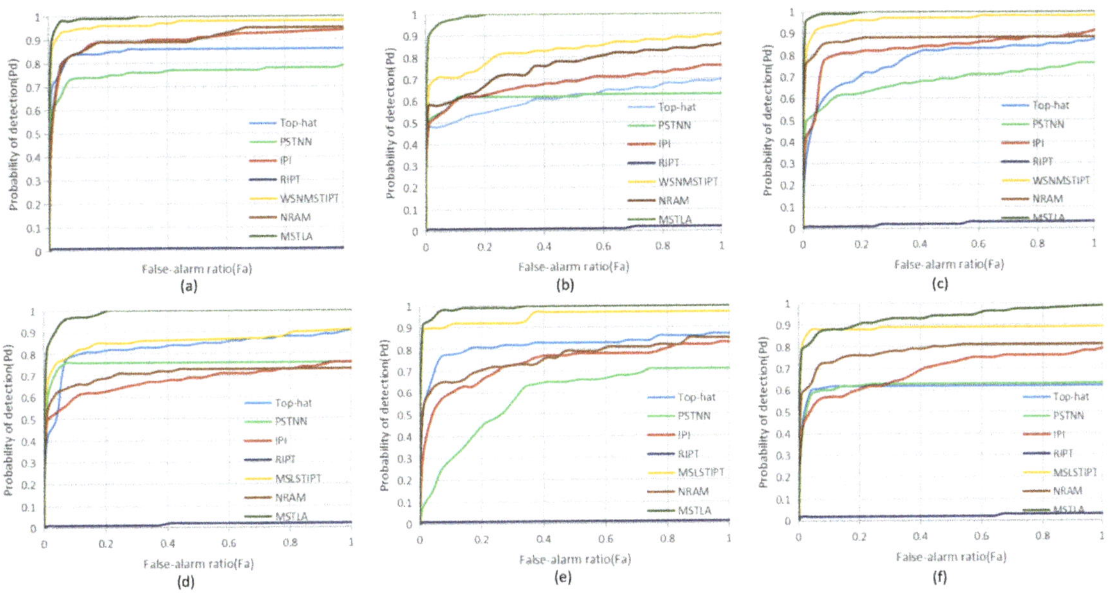

Figure 11. This is a figure. Schemes follow the same formatting. ROC curves of Six Image Sequences (**a–f**) Detected by Different Methods.

5. Discussion

The non-local auto-correlation on account of the infrared background and the target's sparsity has been extensively employed in the field of infrared tiny target detection. When the infrared image is homogeneous, a classical IPI effectively represents low-rank patch-background matrices using the nuclear norm. Larger solitary values really hold more information and visual detail. In other words, the complex infrared image is too complicated for the nuclear standard to handle, resulting in residual error and a blurry backdrop after reconstruction because of the rich details.

Currently, the majority of approaches concentrate on the priori backdrop and target, but this does not effectively separate the target from the background. In order to address the residual performance issue, RIPT proposes the structure tensor. The case of a poor signal-to-noise ratio, which leads to a lack of structure information and then target loss, is ignored by RIPT in complicated scenes. The NRAM model, on account of the IPI model, introduces a tighter rank proxy.

Based on the NRAM model, this article initially constrains the low-rank background using the tensor kernel norm rather than the rank function. The proposed MNSTLA model and other cutting-edge techniques can effectively suppress the interference caused by dynamic background and object moving on the foreground extraction and also show good performance in background suppression and object enhancement, according to qualitative and quantitative comparisons.

6. Conclusions

The robustness and effectiveness of a detection method for infrared point and moving targets are of great importance to the requirements of the early warning system. However, it is difficult to detect infrared dim and point targets, especially the point and moving targets. Therefore, we proposed a detection method using the minimization of a non-convex spatial-temporal tensor low-rank approximation for infrared points and moving targets. Our method introduces the concept of a spatial-temporal tensor on the basis of the non-convex rank approximation method. The experimental results on the real sequence data sets in different scenes illustrate that this method is robust and effective in detecting infrared points and moving targets, and is less affected by background changes and poor image quality.

By the above discussion, while the MNSTLA model has a lower false alarm rate, the comparison is based on single target detection. However, in the IRST system, for multi-target detection of infrared sequence images or infrared videos, the spatial and temporal information is extremely crucial to improve the detection rate of dim and point targets and reduce the false alarm rate. Therefore, constructing a model that can simultaneously use the spatial-temporal information of infrared image sequences for multi-target detection is the focus of our further research. Therefore, we will consider combining the spatial-temporal information with the existing method in the follow-up research in the hopes of realizing the multi-target detection, improving the efficiency of target detection, and reducing the false alarm rate.

Author Contributions: Data curation, K.W.; Investigation, K.W.; Methodology, K.W.; Supervision, D.J. and X.L.; Validation, K.W.; Writing—review & editing, K.W. and D.J. and L.Y. All authors have read and agreed to the published version of the manuscript.

Funding: This research was funded by Youth Project of Applied Basic Research Project, grant number 2013FD016.

Institutional Review Board Statement: Not applicable.

Informed Consent Statement: Not applicable.

Data Availability Statement: No new data were created or analyzed in this study. Data sharing is not applicable to this article.

Conflicts of Interest: The authors declare no conflict of interest.

References

1. Huan, K.W.; Pang, B.; Shi, X.G.; Zhao, Q.Y.; Shi, N.N. Research on Performance Testing and Evaluation of Infrared Imaging System. *Infrared Laser Eng.* **2008**, *6*, 482–486.
2. Wang, G.H.; Mao, S.Z.; He, Y. A Survey of Radar and Infrared Data Fusion. *Fire Control. Command. Control.* **2002**, *27*, 4.
3. Yang, L.; Sun, Q.; Wang, J.; Guo, B.; Li, C. Design of long-wave infrared continuous zoom optical system. *Infrared Laser Eng.* **2012**, *41*, 99–100.
4. Zhou, X.; Yang, C.; Yu, W. Moving Object Detection by Detecting Contiguous Outliers in the Low-Rank Representation. *IEEE Trans. Pattern Anal. Mach. Intell.* **2013**, *35*, 597–610. [CrossRef] [PubMed]
5. Zhi, G.; Cheong, L.-F.; Wang, Y.-X. Block-Sparse RPCA for Salient Motion Detection. *IEEE Trans. Pattern Anal. Mach. Intell.* **2014**, *36*, 1975–1987.
6. Chen, C.; Li, H.; Wei, Y.; Xia, T.; Tang, Y.Y. A Local Contrast Method for Small Infrared Target Detection. *IEEE Trans. Geosci. Remote Sen.* **2013**, *52*, 574–581. [CrossRef]
7. He, Y.J.; Li, M.; Zhang, J.L.; An, Q. Small infrared target detection on account based on flow-rank and sparse representation. *Infrared Phys. Technol.* **2015**, *68*, 98–109. [CrossRef]
8. Gao, C.; Meng, D.; Yang, Y.; Wang, Y.; Zhou, X.; Hauptmann, A.G. Infrared Patch-Image Model for point target Detection in a Single Image. *IEEE Trans. Image Process.* **2013**, *22*, 4996–5009. [CrossRef]
9. Dai, Y.; Wu, Y.; Song, Y. Infrared point target and background separation via column-wise weighted robust principal component analysis. *Infrared Phys. Technol.* **2016**, *77*, 421–430. [CrossRef]
10. Dai, Y.; Wu, Y. Reweighted Infrared Patch-Tensor Model With Both Nonlocal and Local Priors for Single-Frame point target Detection. *IEEE J. Sel. Top. Appl. Earth Obs. Remote Sens.* **2017**, *10*, 3752–3767. [CrossRef]
11. Dai, Y.; Wu, Y.; Song, Y.; Guo, J. Non-negative infrared patch-image model: Robust target-background separation via partial sum minimization of singular values. *Infrared Phys. Technol.* **2017**, *81*, 182–194. [CrossRef]
12. Sun, Y.; Yang, J.; Long, Y.; Shang, Z.; An, W. Infrared patch tensor model with weighted tensor nuclear norm for point target detection in a single frame. *IEEE Access* **2018**, *6*, 76140–76152. [CrossRef]
13. Bin, X.; Xinhan, H.; Min, W. Infrared dim point target detection on account of adaptive target image recovery. *J. Huazhong Univ. Sci. Technol. (Nat. Sci. Ed.)* **2017**, *45*, 25–30.
14. Wang, X.Y.; Peng, Z.; Kong, D.; Zhang, P.; He, Y. Infrared dim target detection on account of total variation regularization and principal component pursuit. *Image Vis. Comput.* **2017**, *63*, 1–9. [CrossRef]
15. Wang, X.Y.; Peng, Z.; Kong, D.; He, Y. Infrared dim and small target detection on account of stable multisubspace learning in heterogeneous scene. *IEEE Trans. Geosci. Remote Sens.* **2017**, *55*, 5481–5493. [CrossRef]

16. Zhang, L.D.; Peng, L.; Zhang, T.; Cao, S.; Peng, Z. Infrared small target detection via non-convex rank approximation minimization joint $l_{2,1}$ norm. *Remote Sens.* **2018**, *10*, 1821. [CrossRef]
17. Wu, J.; Lin, Z.; Zha, H. Essential tensor learning for multi-view spectral clustering. *IEEE Trans. Image Process.* **2019**, *28*, 5910–5922. [CrossRef]
18. Zhang, J.; Li, X.; Jing, P.; Liu, J.; Su, Y. Low-rank regularized heterogeneous tensor decomposition for subspace clustering. *IEEE Signal Process. Lett.* **2018**, *25*, 333–337. [CrossRef]
19. Jing, P.; Guan, W.; Bai, X.; Guo, H.; Su, Y. Single image super-resolution via low-rank tensor representation and hierarchical dictionary learning. *Multimed. Tools Appl.* **2020**, *79*, 11767–11785. [CrossRef]
20. Zhou, P.; Lu, C.; Feng, J.; Lin, Z.; Yan, S. Tensor low-rank representation for data recovery and clustering. *IEEE Trans. Pattern Anal. Mach. Intell.* **2021**, *43*, 1718–1732. [CrossRef]
21. Gao, C.; Wang, L.; Xiao, Y.; Zhao, Q.; Meng, D. Infrared small-dim target detection on account of Markov random field guided noise modeling. *Pattern Recognit.* **2018**, *76*, 463–475. [CrossRef]
22. Chen, L.X.; Liu, J.L.; Wang, X.W. Foreground detection with weighted Schatten-p norm and 3D total variation. *J. Comput. Appl.* **2019**, *39*, 1170–1175.
23. Lu, C.; Feng, J.; Chen, Y.; Liu, W.; Lin, Z.; Yan, S. Tensor robust principal component analysis with a new tensor nuclear norm. *IEEE Trans. Pattern Anal. Mach. Intell.* **2020**, *42*, 925–938. [CrossRef] [PubMed]
24. Sun, Y.; Yang, J.; Li, M.; An, W. Infrared point target detection via spatial–temporal infrared patch-tensor model and weighted schatten p-norm minimization. *Infrared Phys. Technol.* **2019**, *102*, 103050. [CrossRef]
25. Wang, Y.; Peng, J.; Zhao, Q.; Leung, Y.; Zhao, X.; Meng, D. Hyperspectral image restoration via total variation regularized low-rank tensor decomposition. *IEEE J. Sel. Top. Appl. Earth Obs. Remote Sens.* **2017**, *11*, 1227–1243. [CrossRef]
26. Gu, S.H.; Zhang, L.; Zuo, W.M.; Feng, X.C. Weighted Nuclear Norm Minimization with Application to Image Denoising. In Proceedings of the 2014 IEEE Conference on Computer Vision and Pattern Recognition, Columbus, OH, USA, 23–28 June 2014; pp. 2862–2869.
27. Kang, Z.; Peng, C.; Cheng, Q. Robust PCA via Nonconvex Rank Approximation. In Proceedings of the 2015 IEEE International Conference on Data Mining (ICDM), Atlantic City, NJ, USA, 14–17 November 2015; pp. 211–220.
28. Fazel, M.; Hindi, H.; Boyd, S.P. Log-det heuristic for matrix rank minimization with applications to Hankel and Euclidean distance matrices. In Proceedings of the 2003 American Control Conference, Denver, CO, USA, 4–6 June 2003; pp. 2156–2162.
29. Guo, J.; Wu, Y.Q.; Dai, Y.M. Point target detection on account of reweighted infrared patch-image model. *IEEE Image Process.* **2018**, *12*, 70–79. [CrossRef]
30. Wright, J.; Ganesh, A.; Rao, S.; Peng, Y.; Ma, Y. Robust principal component analysis: Exact recovery of corrupted low-rank matrices via convex optimization. In Proceedings of the Advances in Neural Information Processing Systems, Vancouver, BC, Canada, 7–10 December 2009; pp. 2080–2088.
31. Peng, Y.; Suo, J.; Dai, Q.; Xu, W. Reweighted low-rank matrix recovery and its application in image restoration. *IEEE Trans. Cybern.* **2014**, *44*, 2418–2430. [CrossRef]
32. Liu, Z.S.; Li, J.C.; Li, G.; Bai, J.C.; Liu, X.N. A New Model for Sparse and Low-Rank Matrix Decomposition. *J. Appl. Anal. Comput.* **2017**, *7*, 600–616.
33. Zhao, Y.; Pan, H.; Du, C.; Peng, Y.; Zheng, Y. Bilateral two-dimensional least mean square filter for infrared point target detection. *Infrared Phys. Technol.* **2014**, *65*, 17–23. [CrossRef]
34. Bae, T.W.; Zhang, F.; Kweon, I.S. Edge directional 2D LMS filter for infrared point target detection. *Infrared Phys. Technol.* **2012**, *55*, 137–145.
35. Bae, T.W.; Kim, Y.C.; Ahn, S.H.; Sohng, K.I. A novel Two-Dimensional LMS (TDLMS) using sub-sampling mask and step-size index for point target detection. *IEICE Electron. Express* **2010**, *7*, 112–117. [CrossRef]
36. Yuan, M.; Lin, Y. Model selection and estimation in regression with grouped variables. *J. R. Stat. Soc. Stat. Methodol. Ser. B* **2006**, *68*, 49–67. [CrossRef]
37. Hui, B.; Song, Z.; Fan, H.; Zhing, P.; Hu, W.; Zhang, X.; Ling, J.; Su, H.; Jin, W.; Jang, Y.; et al. A dataset for infrared detection and tracking of dim-small aircraft targets underground/air background. *China Sci. Data* **2020**, *5*, 286–297. [CrossRef]
38. Gao, C.; Zhang, T.; Li, Q. Small infrared target detection using sparse ring representation. *IEEE Aerosp. Electron. Syst. Mag.* **2012**, *27*, 21–30.
39. Sun, Y.; Yang, J.; Long, Y.; An, W. Infrared point target Detection Via Spatial-Temporal Total Variation Regularization and Weighted Tensor Nuclear Norm. *IEEE Access* **2019**, *7*, 56667–56682. [CrossRef]
40. Sun, Y.; Yang, J.; An, W. Infrared Dim and point target Detection via Multiple Subspace Learning and Spatial-Temporal Patch-Tensor Model. *IEEE Trans. Geosci. Remote Sens.* **2021**, *59*, 3737–3752. [CrossRef]
41. Rivest, J.-F.; Fortin, R. Detection of dim targets in digital infrared imagery by morphological image processing. *Opt. Eng.* **1996**, *35*, 1886–1893. [CrossRef]

applied sciences

MDPI

Article

Multi-View Gait Recognition Based on a Siamese Vision Transformer

Yanchen Yang [1], Lijun Yun [1,2,*], Ruoyu Li [1], Feiyan Cheng [1] and Kun Wang [1]

1 College of Information, Yunnan Normal University, Kunming 650000, China
2 Yunnan Key Laboratory of Optoelectronic Information Technology, Kunming 650000, China
* Correspondence: yunlijun@ynnu.edu.cn

Abstract: Although the vision transformer has been used in gait recognition, its application in multi-view gait recognition remains limited. Different views significantly affect the accuracy with which the characteristics of gait contour are extracted and identified. To address this issue, this paper proposes a Siamese mobile vision transformer (SMViT). This model not only focuses on the local characteristics of the human gait space, but also considers the characteristics of long-distance attention associations, which can extract multi-dimensional step status characteristics. In addition, it describes how different perspectives affect the gait characteristics and generates reliable features of perspective–relationship factors. The average recognition rate of SMViT for the CASIA B dataset reached 96.4%. The experimental results show that SMViT can attain a state-of-the-art performance when compared to advanced step-recognition models, such as GaitGAN, Multi_view GAN and Posegait.

Keywords: multi-view gait recognition; Siamese neural network; vision transformer; view-feature conversion; gradual view

Citation: Yang, Y.; Yun, L.; Li, R.; Cheng, F.; Wang, K. Multi-View Gait Recognition Based on a Siamese Vision Transformer. *Appl. Sci.* **2023**, *13*, 2273. https://doi.org/10.3390/app13042273

Academic Editor: Krzysztof Koszela

Received: 29 November 2022
Revised: 1 February 2023
Accepted: 3 February 2023
Published: 10 February 2023

1. Introduction

The identification of human individuals based on their gait, alongside a range of biological features, including facial features, fingerprints, and irises, has the benefits of being a long-range, non-intrusive, and passive mode of identification [1]. In addition, as the security facilities of urban and public places are gradually improved, monitoring facilities, such as cameras are ubiquitous, facilitating the use of basic, low-resolution instruments of which the identified target is unaware [2]. Personality traits determine one's identity. This has led to the widespread use of deep-learning-based gait-recognition technology in modern society [3], particularly in criminal investigations and public security; this technique has significant potential for future applications [4]. To sum up, the fact that gait recognition allows for the undetectable identification of individuals means that it has obvious advantages in anti-terrorism and in fugitive tracking. Therefore, we believe that this research is of great significance to the long-term interests of society.

To achieve a reliable identification of people in public spaces, it is necessary to overcome the problem of variability in pedestrian behaviors in such environments through the collection and identification of pedestrian gait information from multiple views [5,6]. Formal gait recognition uses 90° gait features that provide the most salient and comprehensive details of human posture as experimental data. The rationale for this method is that the gait characteristics in other views overlap due to the perspective problems of human physical characteristics, with the result that the contour characteristics are not effectively rendered. This is also one of the complications of multi-view gait recognition. Moreover, in practical terms, in order to preserve the advantages of passive identification, it is crucial not to establish a fixed walking position and camera viewpoint for the pedestrian [7]. This problem needs to be solved urgently.

In the task of multi-view gait recognition, when the angle of view moves from 90° to 0° and 180°, the contour of the human body is affected by the shooting angle, and some of the gait feature information is lost. This significantly impacts the extraction of the gait contour characteristics. In response to this issue, this paper uses a Siamese neural network to calculate the posture relationship between the two views and calculate the characteristic conversion factor. Under the premise of retaining identity information, the useful high-dimensional intensive characteristics of the network are strengthened to make its high-dimensional features clearer, and the effect of the loss of gait characteristics on recognition accuracy is lessened. The SMViT model constructed using this concept can obtain higher recognition accuracy in a non-90° multi-view; moreover, this model is more robust.

In summary, this paper makes the following contributions:

(1) It designs a reasonable and novel gait-view conversion method, which can deal with the problem of multi-view gait;
(2) It constructs the SMViT model, and uses the view characteristic relation module to calculate the association between multi-view gait characteristics;
(3) We develop a gradually moving view-training strategy that can raise the model's robustness while raising the recognition rate for less precise gait-view data.

The structure of this paper is as follows. The technologies related to gait recognition are introduced in Section 2. Then, the SMViT is constructed and the gradual-moving-view training method is explained in Section 3. In Section 4, experimentation with the CASIA B gait dataset [8] is employed to explore the models and methods that are presented in this paper. Finally, in Section 5, we summarize the research contained herein and consider the future directions of gait recognition technology.

2. Related Work

At present, there are many methods of solving the problem of multi-view gait recognition. Some researchers adopt the method of constructing a three-dimensional model and use the close cooperation of multiple cameras to construct a three-dimensional model of pedestrian movement, so as to weaken the influence of multiple perspectives, clothing, and other factors. Bodor et al. proposed combining arbitrary views taken by multiple cameras to construct appropriate images that match the training view for pedestrian gait recognition [9]. Ariyanto et al. constructed a correlation energy map between their proposed generative gait model and the data, and adopted a dynamic programming method to select possible paths and extract gait kinematic trajectory features, proposing that the extracted features were inherent to the 3D data [10]. In addition, Tome et al. set out a comprehensive approach that combines the probability information related to 3D human poses with convolutional neural networks (CNNs), and introduced a unified formula to address the challenge of estimating 3D human poses from a single RGB image [11]. In subsequent research, Weng et al. changed the extraction method of human 3D pose modeling, and proposed a deformable pose ergodic convolution to optimize the convolution kernel of each joint by considering context joints with different weights [12]. However, this method of 3D pose modeling is more complicated to calculate and has high requirements regarding the number of cameras and the shooting environment, so it is difficult to use in application settings with ordinary cameras.

Some scholars used the view transformation model (VTM) to extract the frequency domain features of gait contours by transforming them from different views. For instance, Makihara et al. proposed a gait recognition method based on frequency domain features and view transformation. First, a spatio-temporal contour set of gait characteristics was constructed, and the periodic characteristics of gait were subjected to the Fourier analysis to extract the frequency domain features of pedestrian gait; the multi-view training set was used to calculate the view transformation model [13]. In this method, the Spatio-temporal gait images in the gait cycle are usually first fused into a gait energy image (GEI), which is a Spatio-temporal gait representation method first proposed by Han et al. [14]. Kusakunniran et al. combined the gait energy image (GEI) with the view transition model (VTM), and used a linear discriminant analysis (LDA) to optimize the feature vectors and improve the performance of VTM [15]. Later, Kusakunniran et al. used a motion clustering method to classify gaits from different views into groups according to their correlation; within each group, a canonical correlation analysis (CCA) was used to further enhance the linear correlation between gaits from different views [16]. In addition, researchers have considered how to perform gait recognition from any view. Hu et al. proposed a viewpoint invariant discriminant projection (ViDP) method to improve the discrimination accuracy of gait features using linear projection [17]. However, most of these methods are realized by domain transformation or singular value decomposition, and the perspective of transformation is complicated.

Others have used an adversarial generative network to normalize multiple views into a common perspective. Zhang et al. proposed a perspective-shifting adversarial generative network (VT-GAN), which can transform gait views across two arbitrary views with only one model [18]. Shi et al. designed GaitGANv1 and GaitGANv2, versions of a gait adversarial generation network, which use GAN as a regressor to generate a standardized side view of a normal gait; this not only prevents the falsification of gait images, but also helps to maintain identity information, and the networks achieved good results in cross-view gait recognition [19,20]. In addition, Wen et al. used GAN to convert gait images with arbitrary decorations and views into normal states of 54°, 90°, and 126°, so as to extract view-invariant features and reduce the loss of feature information caused by view transformation [21]. Focusing on the problem of limited recognition accuracy arising from the lack of gait samples from different views, Chen et al. proposed a multi-view gait generation ad hoc network (MvGGAN) to generate false gait samples to expand the dataset and improve recognition accuracy [22]. However, the ability of this adversarial generative network structure to accurately undertake recognition tasks from the same perspective is easily affected by decorative features, such as clothes and backpacks, resulting in limited recognition accuracy.

3. SMViT and the Gradually Moving-View Training Method

3.1. Model Structure

In order to solve the problem of multi-perspective situations, this paper uses the Siamese neural network as a design basis and calculates the correlation between the characteristics of different views and uses this as the basis for the conversion of the characteristics of the view. When there are few specimens, a Siamese neural network can extract and learn the links between two groups of photos [23]. ViT is advantageous for the extraction of multi-scale features because of its robust strength and resistance to interference from mistaken samples [24,25]. In this paper, a two-channel Siamese module (Conv and MViT, CM Block) of convolution is constructed to extract the characteristics of multi-view gait contour features. The specific model structure is shown in Figure 1.

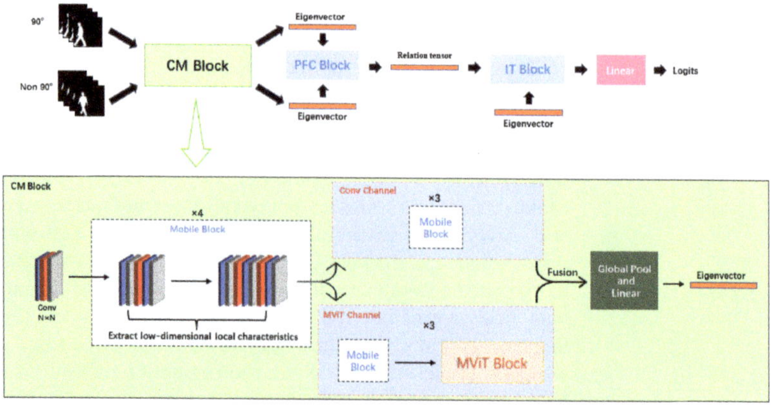

Figure 1. Structure diagram of SMViT.

In order to extract the gait information from various viewpoints, two feature extraction networks are used in the Siamese network module described in this paper. Convolution channels are used inside each module to obtain the contour's high-dimensional local features. Furthermore, we utilize the Mobile ViT channel to create high-dimensional states that are indicative long-distance attention characteristics of the current view.

In addition, the mobile view transformation (MVT) module is used to extract the view characteristics and tensors, meaning that the advantages of convolution and ViT are retained in the extraction of the gait contour features. This module is based on the Mobile ViT model, and incorporates the convolution with the transformer. Local processing is replaced with deeper global treatment in the convolution. In an effective receiving domain, we model long-distance non-local dependencies. The model has smaller parameters and produces ideal experimental results. The specific details of the module are shown in Figure 2.

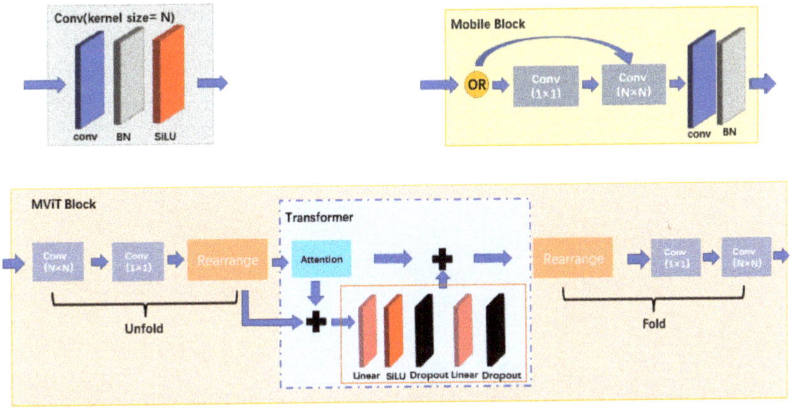

Figure 2. Details of the module.

As shown in Figure 2, there are two types of the conv block module (Conv Block), $N \times N$ convolution and point-by-point convolution, according to the difference in the convolution kernel size. This module consists of a convolution layer and a batch normalization layer. Mobile Block takes MobileNet as the essential conceptual basis [26–28] and controls the network's depth and the number of parameters by constructing depth-separable convolution. In addition, due to the differences in the processing methods and content of the feature extraction between the convolution and the ViT, the format conversion needs to be carried out before and after the transformer module in order to control the data processing format. $SiLU$ is used as the activation function in each module, as shown in Equation (1), and the global average pooling layer is used in the pooling layer, as shown in Equation (2), where x means the input matrix and x_w represents the operation area of the pooling layer:

$$SiLU(x) = x \cdot Sigmoid(x) \tag{1}$$

$$Pooling(x_w) = Avg(x_w) \tag{2}$$

The transformer module absorbs some of the advantages of the convolutional computing and retains its characteristic processing capabilities in the space-perception domain. By dividing a large global receptive field into different patches in a non-overlapping way, $P = Wh$, where w and h are the width and height of the patches, respectively. Then, the transformer is used to encode the relationship between the patches. Specifically, the self-attention mechanism calculates the scaled dot-product attention by constructing the query vector Q, the value vector V, and the key vector K, as shown in Equations (3)–(5):

$$f(Q,K) = \frac{QK^T}{\sqrt{d_k}} \tag{3}$$

$$X = softmax(f(Q,K)) \tag{4}$$

$$Attention(X,V) = X \times V \tag{5}$$

In this case, the module's computation cost of multi-head self-attention is $O(N^2Pd)$. Compared with the traditional ViT, the calculation cost $O(N^2d)$ is increased but its speed in practical applications is faster [29].

3.2. Perspective Feature Conversion Block and Inverse Transformation Block

The inverse transformation block (IT block) and the perspective feature conversion block (PFC block) are designed concurrently. The former is used to calculate the characteristics of the two perspectives obtained by the Siamese network, and the relation tensor is taken as the view conversion factor as shown in Formula (6). The latter is used to convert the high-dimensional characteristics between the two views as shown in Formula (7). Among them, x and y are two view-cornering characteristics and N is the capacity of the target view set. The process of calculating the gait characteristics from different perspectives in the PFC block and IT block are shown in Figure 3.

$$IT(x,y) = x + PFC(x,y) \tag{6}$$

$$PFC(x,y) = \frac{\sum_{i=1}^{N}(x_i - y_i)}{N} \tag{7}$$

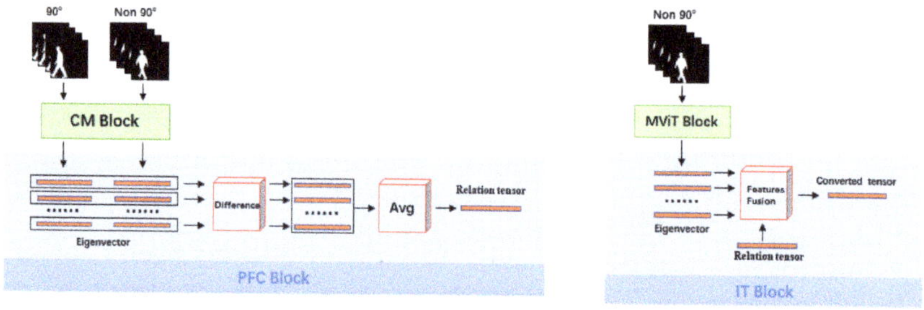

Figure 3. PFC block and IT block.

3.3. Gradually Moving-View Training Method

To develop the multi-view gait recognition model SMViT, this article designs a brand new, suitable, multi-view gait-recognition method; that is, the gradually moving-view training method. The training strategy of this method differs between the perspective feature relationship calculation module and the classification module.

In the characteristic view relationship calculation module, SMViT uses convolution and dual-channel VIT to extract the gait characteristics. To better calculate the difference between the 90°- and non-90°-perspective characteristics, the same pre-training weights are used to calculate the characteristics tensor of the perspective. Because the purpose of this module is to calculate the perspective characteristic tensor, not the classification, the pre-training weight of the last two layers of the network module needs to be eliminated. The specific training steps are shown in Algorithm 1.

Algorithm 1: Process of training the characteristic view relationship calculation module.

Input: CASIA B and CASIA C gait datasets.
Step 1: First, supplement the dual-channel view feature relationship calculation module as a complete classification model and conduct pre-training in the CASIA C dataset.
Step 2: Freeze the pre-training parameters to remove the weight of the final classification layer.
Step 3: Load the parameters from step 2 to the dual-channel perspective characteristic relationship calculation module.
Step 4: Use the module obtained in step 3 to extract the dual-channel features of 90° and non-90° CASIA B gait data.
Step 5: Calculate the relationship tensor between the two perspectives obtained in step 4 with the PFC block.
Step 6: Store the characteristic relationship between the two views obtained in step 5, and hand it over to the classification module.
Output: The characteristic relationship tensor between the two views.

In the classification module, the tensors of the gait characteristics of the different views after conversion should be identified and classified. Therefore, starting with the weight of the 90° model with high accuracy, training is undertaken in two directions of 0° and 180°. That is, a training weight of 90° is used as the initial weight of the model when training 72° and 108° gait data. When training at 54° and 126°, the training weights of, respectively, 72° and 108° are loaded and so on. This part uses a cross-entropy loss function as the method of loss calculation as shown in formula 8. The specific training steps are shown in Algorithm 2.

$$Loss(output, class) = weight_{[class]}(-output_{[class]} + \log(\sum_j e^{output_j})) \quad (8)$$

Here, $output$ is the prediction result, $class$ is the actual label of this sample, and $output_{[class]}$ represents the element of the $class$ position in $output$, that is, the predicted value of the real classification. Finally, $weight_{[class]}$ is a weight parameter.

Algorithm 2: Classification Module Training Process

Input: CASIA B gait dataset.
Step 1: First, the 90° gait data are transformed and recognized (at this time, there is no change in the characteristics of the perspective), and the parameter weight is saved.
Step 2: The weight parameters obtained in step 1 are loaded to the classification module, the gait dataset (such as 72° and 108°) of the adjacent perspective is trained, and the parameter weight is saved.
Step 3: The characteristic relationship tensor between the two perspectives is matched and the parameter weights obtained in the previous step are loaded into the model.
Step 4: The trained perspective weight is loaded to the model, the gait dataset of adjacent non-90° perspectives is trained, and the weight parameters are saved.
Step 5: The classification layer and the regression layer are used to identify and classify the characteristic tensor of the view.
Step 6: Push in two directions (90°→0° and 90°→180°), and repeat steps 3, 4, and 5.
Output: The gait recognition model SMViT.

4. Experiment and Analysis

4.1. Experimental Data

The CASIA B dataset is a large dataset that is widely used in multi-view gait recognition tasks. It consists of 124 subjects (31 women and 93 men) [19]. The gait images of each subject in the three different states of normal walking (NM), walking with a bag (BG), and walking with a coat (CL) were collected [30] from 11 points of view, from 0° to 180° (with an interval of 18° for close views) as shown in Figure 4.

Figure 4. CASIA B multi-view gait dataset [8].

4.2. Experimental Design

In multi-view gait recognition tasks, due to the offset of the perspective, some human gait contour characteristics are lost and the recognition accuracy is reduced. This experiment designated 90° as a high-precision standard perspective. The remaining 10 views of the gait characteristics are calculated using the standard view relationship. The mutual verification between the perspectives is not considered; only the gait recognition accuracy inside the perspective is calculated. Additionally, a simple perspective conversion factor group is established to transform the view feature tensor with less feature information into a 90° feature tensor with more useful and distinct feature information.

In order to compare the results with the comparison model for the same data, we directly used the gait contour data provided by the CASIA B gait database. In addition, this allowed us to be more attuned to the uncertainty caused by people's attire and walking speeds, and other aspects of practical application scenarios. From the same perspective, we ignore slight differences in dress, walking speeds, and other personal features, and divide the overall data into the training set and the verification set according to the 7:3 ratio. There is no crossover between each view, in order to improve the recognition accuracy within each view. The gait data obtained in the actual application scenario may not necessarily contain a complete gait cycle, and the gait characteristics are random. Therefore, this experiment does not adopt the gait-cycle group as the input data. Instead, the gait group with three walking states is scattered at will to ensure that the model's effect is similar to a complex real-world environment.

Setting the initial learning rate to 1×10^{-3}, with Adam as the optimizer, we used the categorical_ The crossentropy multiclass cross-entropy loss function to calculate loss. In this experiment, Pycharm, an efficient Python IDE, was used to write code. The code was tested in Pytorch 1.8 and CUDA 11. The various equipment parameters used in the calculation process are shown in Table 1.

Table 1. Experimental environment.

Environment	Parameter/Version
CPU	*I7-10700K*
GPU	NVIDIA RTX 3060
CUDA	11.0
Pytorch	1.8
Operating System	Win10

4.3. Experimental Results for the CASIA B Dataset

To evaluate the effectiveness of the SMViT model and the view movement method (SMViT_T) proposed in this paper, we used the first 10,000 training loss changes of the two intermediate views as an assessment of the convergence effect. It can be seen that, even in the middle of the view offset, the SMViT model proposed in this paper can still effectively converge and stabilize under the general trend as shown by the blue line in Figures 5 and 6. After the gradually moving-perspective training, not only is the model's drop in loss significantly improved, but the unstable jumping phenomenon of losses is also suppressed to a certain extent as shown by the orange line in Figures 5 and 6.

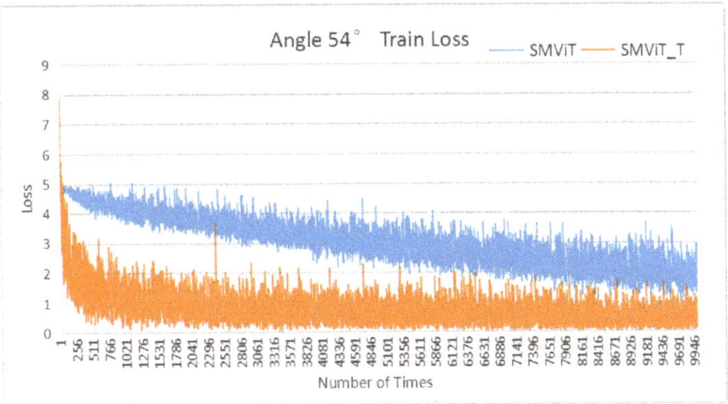

Figure 5. Loss change when the view is 54°.

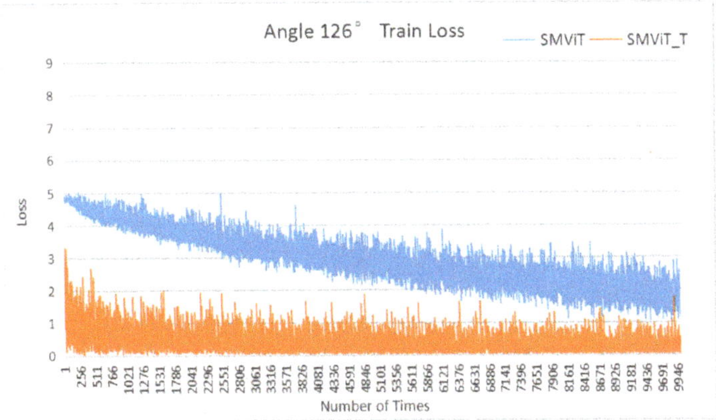

Figure 6. Loss change when the view is 126°.

The experiment was carried out at 10 angles, from 0 to 180 degrees, in order to prove that the model can effectively learn gait characteristics from multiple perspectives and overcome the problems caused by the poor learning effect and fluctuation in the accuracy rate, which are commonplace in multi-perspective gait recognition. Figure 7 shows the effect pictures of the model trained with or without gradually moving-view training for 10 views, but not for the 90° view. It can be observed that, except for the basic model (SMViT_BASE) at a perspective of 36°, there is a small oscillation in accuracy, and the experimental effects for the other views steadily increased. Both the base and T models quickly reach their peak accuracy and then stabilize as shown by the point line in Figure 7. The model proposed in this paper (SMViT_T), with gradually moving-view training, has a high accuracy for all the viewpoints, and there is no significant fluctuation in the recognition rate during the training process. The recognition rate of our model is always higher than that of the base model.

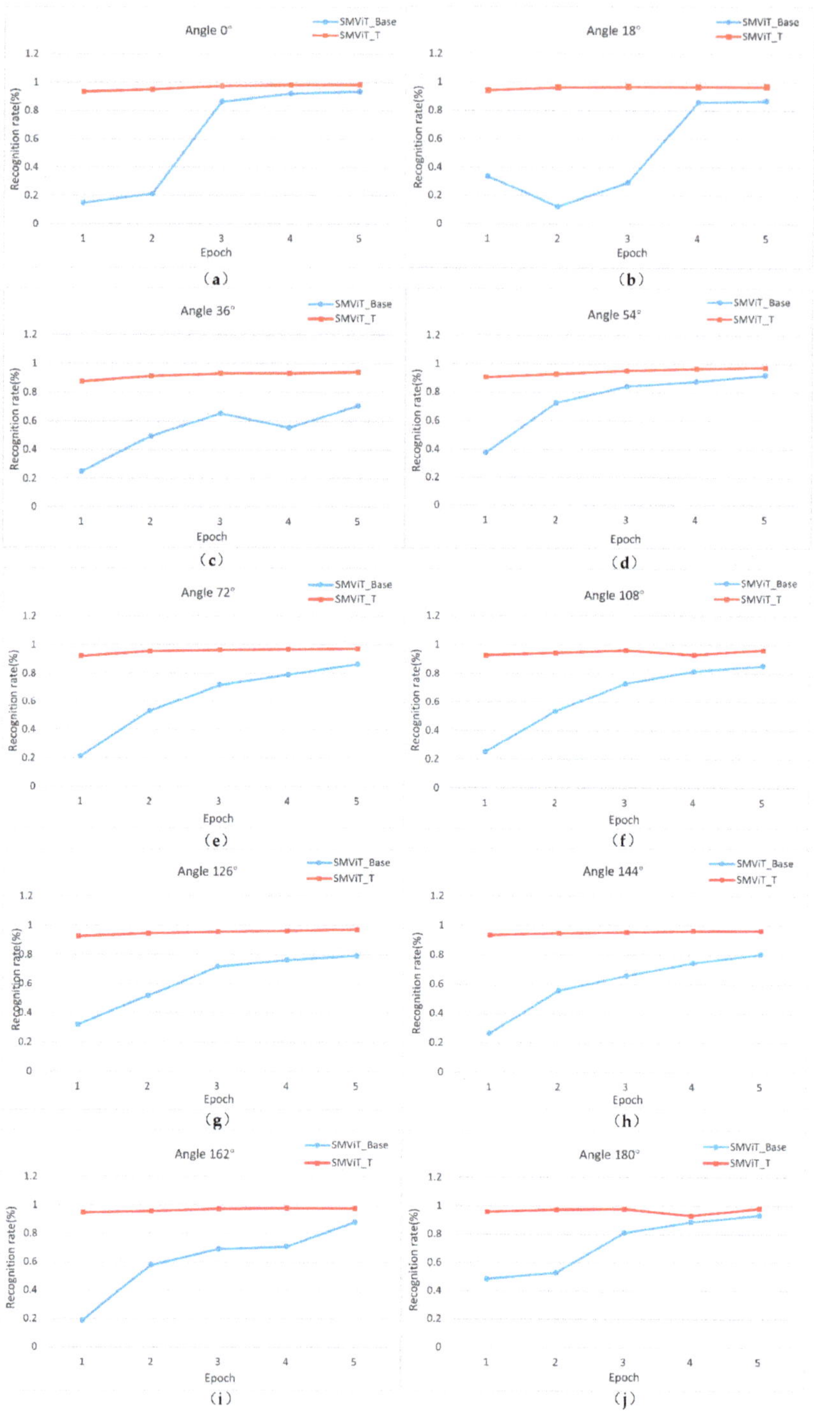

Figure 7. Diagram of the training process for the model proposed in this paper for various views in the CASIA B dataset (In the figure (**a**–**j**) are the experimental results of SMViT base and T in each angle).

4.4. Comparison with the Latest Technology

4.4.1. Ablation Experiment

At present, in many gait recognition studies, walking speed, clothing, and other characteristics are tested separately. As such, it can be considered an idealized experiment where some of the complexity of the data are eliminated by manual classification. This allows the model to obtain good results in NM classification. Other classifications, however, have poor accuracy. We suggest that, in real-world application scenarios, human appearance is highly uncertain. Therefore, the data obtained for different states are trained and verified separately, and the results cannot be used as the possible results of practical applications.

We experimented with a mixture of data for various characteristics to maximize the complexity of the gait data. In addition, our structural design focuses on improving the accuracy of gait recognition from multiple perspectives, rather than using cross-perspective experimental methods.

For the 11 views in the CASIA B dataset, our model (SMViT_Base) was compared with SPAE [31], GaitGAN [19,20], Multi_View GAN [21], Slack Allocation GAN [32], GAN based on U-Net [33], and PoseGait [34] in terms of the internal recognition rate of non-cross-view offset views. That is, without considering cross-verification, verification experiments only considered the multi-view perspective. Due to the limitations of the experimental environment and equipment, we could not effectively restore the experimental results of the multiple comparative models. Therefore, we directly used the experimental results presented in the papers. Other model data, shown in Table 2, were taken from the average value of the three-state gait recognition rates from the same view in the same dataset. It can be seen that, for all the views, the model presented here showed significant improvements when compared to the other gait recognition models. Additionally, the average upgrade index exceeded 20 percentage points. It is verified that, in the task of multi-view recognition with a non-crossing view, the model proposed in this paper is better than the selected comparison model.

Table 2. Precision comparison of CASIA B with the latest technology for each view.

Comparison of Model Accuracy for Each View When Not Crossing Views											
	0°	**18°**	**36°**	**54°**	**72°**	**90°**	**108°**	**126°**	**144°**	**162°**	**180°**
SPAE [31]	0.7419	0.7661	0.7150	0.6989	0.7311	0.6801	0.6854	0.7258	0.7016	0.6881	0.7231
GaitGANv1 [19]	0.6828	0.7123	0.7285	0.7339	0.6962	0.7043	0.7150	0.7285	0.7204	0.7042	0.6828
GaitGANv2 [20]	0.7258	0.7554	0.7150	0.7332	0.7527	0.707	0.6962	0.7392	0.7150	0.7311	0.6989
Multi_View GAN [21]	0.7213	0.7869	0.7814	0.7589	0.7568	0.7131	0.7322	0.7431	0.7431	0.7480	0.7513
Slack Allocation GAN [32]	0.7473	0.7258	0.7258	0.7141	0.7560	0.7336	0.6967	0.7365	0.7277	0.7243	0.7221
GAN based on U-Net [33]	0.7365	0.7715	0.7956	0.7957	0.8521	0.7822	0.8172	0.7956	0.7984	0.7419	0.7580
PoseGait [34]	0.7231	0.7365	0.7688	0.7822	0.7446	0.7473	0.7607	0.7284	0.7553	0.7365	0.6586
SMViT_Base	**0.9802**	**0.9704**	**0.9318**	**0.9805**	**0.9689**	**0.9744**	**0.9668**	**0.9617**	**0.9529**	**0.9451**	**0.9831**

At the same time, the average values for the 11 views of normal walking (NM), walking with a backpack (BG), and walking while wearing a jacket (CL) were compared. For this comparison, the training sets and verification sets of each model were taken from the same view. The proportion of internal training sets and verification sets for each view is 7:3, and the cross-verification of the view is not considered. From Figure 8, it can be seen that the red model, which was proposed in this paper, significantly increased the average value of multi-view mixed recognition rates, which increased by about 20 percentage points.

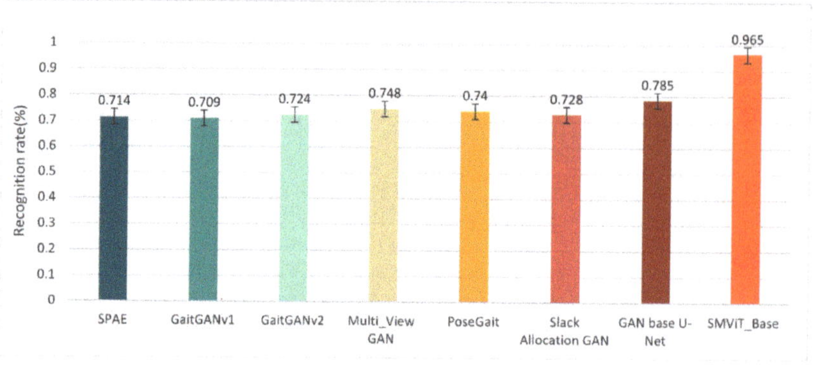

Figure 8. Comparison of the average validation rates of the model from multiple views.

4.4.2. Validation of the Gradually Moving-View Training Strategy

As shown in Figure 7, we initially demonstrated the effectiveness of this strategy by comparing the T model with the base model in SMViT. During the experiment, we found that, with this training strategy, SMViT can still maximize the stability of learning for some perspectives with low-quality data. Taking the 18° gait data as an example, we conducted ablation experiments to verify the effectiveness of the gradually moving-view training method. As shown by the square point line in Figure 9, during the first 15 rounds of training, the SMViT_base model proposed in this paper dropped significantly in the second and ninth rounds and there was a significant saturation of recognition rates. After the experimental analysis, we surmised that the first decline was due to the loss of a number of gait characteristic outlines. At the same time, this model does not use view-mobilization training methods to convert and strengthen the gait characteristics, making the model unable to effectively learn the characteristics, and the accuracy decreases sharply. The second drop is due to the small number of features, which led to the abnormal situation of gradual overfitting in the verification accuracy; this also reflects the improvement in the model's robustness facilitated by the view-transition training method. At around 13 rounds, the base model reaches the upper limit of saturation accuracy but is still about one percentage point lower than the SMViT_T model's recognition accuracy. On the whole, due to the gradually moving-view training strategy of SMViT, the initial recognition rate is about 70 percentage points higher than that of the basic model; our model also maintains a relatively stable level of recognition accuracy. Although the accuracy saturation trend also appeared quickly, the upper limit of the saturation value was about one percentage point higher than that of the base model, and the oscillation amplitude of the validation rate remained below one percentage point.

In this paper, we integrated the design concept of Siamese neural networks and a variant mobile vision transformer model and built a multi-view Siamese ViT gait recognition model: SMViT. At the same time, we designed a gradually moving-view training strategy for multi-view gait recognition, referred to as SMViT_Base and SMViT_T. After conducting a number of experiments on the CASIA B dataset, it was shown that the Siamese feature relationship calculation method can be used to obtain the perspective characteristic conversion factor, which can be used to determine the relationship between different perspective gait characteristics; this effectively improves the accuracy of multi-perspective gait recognition. Our experimental results show that the proposed model can significantly improve the recognition rate when compared with the existing generative multi-view gait recognition methods, without considering cross-view verification. We demonstrated an increase of 20 percentage points in the hybrid recognition rate, without considering the external attire of the pedestrians. Therefore, SMViT expands the gait recognition view while ensuring high accuracy, improving efficient gait recognition in multi-view practical application scenarios.

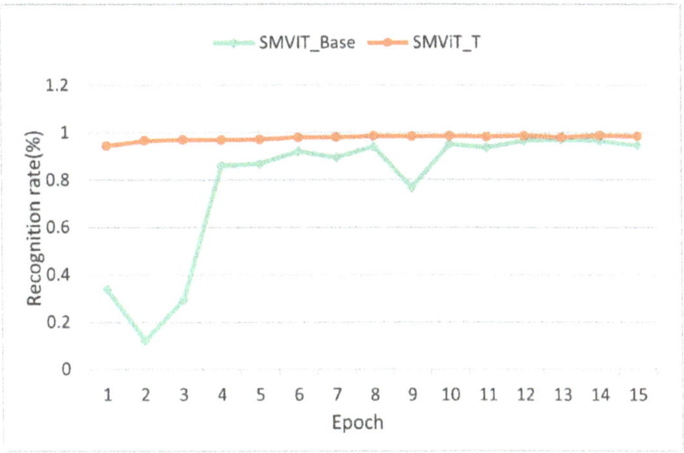

Figure 9. Reliability verification of the gradually moving-view training method at 18°.

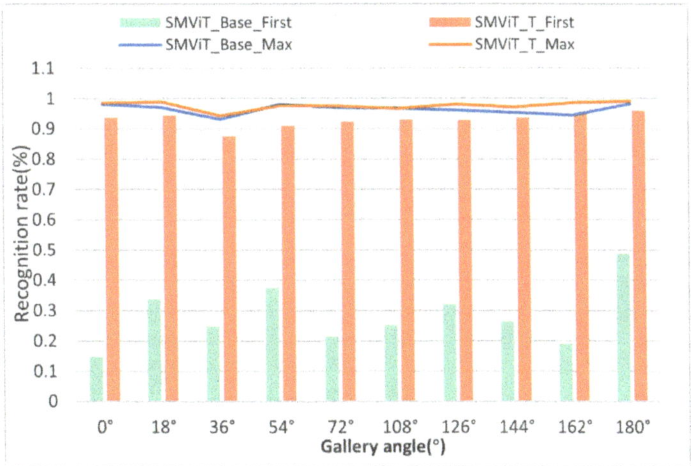

Figure 10. Comparison of the initial validation rate and the maximum validation rate of the proposed model trained with or without gradually moving-view training.

5. Conclusions and Future Prospects

In the future, a more abundant dataset can be used to verify the recognition effect, and a more sophisticated view-feature conversion module can be used to enhance the application scope of SMViT. Additionally, when the visible light intensity is insufficient, the infrared thermal imaging target tracking method can be used to extract the gait contour features, creating the possibility of dealing with more complex and variable natural environments [35] and undertaking the tracking of more obscure gaits [36,37]. We believe that the design of SMViT with multiple covariates will open up new methods for multi-view gait recognition; the vision transformer can also play a role in multi-view gait recognition tasks in complex environments.

Author Contributions: Data curation, L.Y.; investigation, R.L.; methodology, Y.Y. and L.Y.; resources, L.Y.; supervision, L.Y. and K.W.; validation, Y.Y. and R.L.; writing—review and editing, Y.Y., L.Y. and F.C. All authors have read and agreed to the published version of the manuscript.

Funding: This research was funded by the Key Projects of Yunnan Applied Basic Research Plan, grant number 2018FA033.

Institutional Review Board Statement: Not applicable.

Informed Consent Statement: Not applicable.

Data Availability Statement: The datasets analyzed in the current study are available in the CASIA gait database of the Chinese Academy of Sciences: http://www.cbsr.ia.ac.cn/china/Gait%20Databases%20CH.asp (Website viewed on 16 October 2021). The data used in the experiment described in this article come from the CASIA gait database provided by the Institute of Automation of the Chinese Academy of Sciences.

Conflicts of Interest: The authors declare no conflict of interest.

References

1. Singh, J.P.; Jain, S.; Arora, S.; Singh, U.P. Vision-based gait recognition: A survey. *IEEE Access* **2018**, *6*, 70497–70527. [CrossRef]
2. A survey on gait recognition. *ACM Comput. Surv. (CSUR)* **2018**, *51*, 1–35.
3. Sepas-Moghaddam, A.; Etemad, A. Deep gait recognition: A survey. *IEEE Trans. Pattern Anal. Mach. Intell.* **2022**, *45*, 264–284. [CrossRef] [PubMed]
4. Fan, C.; Peng, Y.; Cao, C.; Liu, X.; Hou, S.; Chi, J.; Huang, Y.; Li, Q.; He, Z. Gaitpart: Temporal part-based model for gait recognition. In Proceedings of the IEEE/CVF Conference on Computer Vision and Pattern Recognition, Seattle, WA, USA, 13–19 June 2020; pp. 14225–14233.
5. Zhu, Z.; Guo, X.; Yang, T.; Huang, J.; Deng, J.; Huang, G.; Du, D.; Lu, J.; Zhou, J. Gait recognition in the wild: A benchmark. In Proceedings of the IEEE/CVF International Conference on Computer Vision, 11–17 October 2021; pp. 14789–14799.
6. Chao, H.; He, Y.; Zhang, J.; Feng, J. Gaitset: Regarding gait as a set for cross-view gait recognition. In Proceedings of the AAAI Conference on Artificial Intelligence, Palo Alto, CA, USA, 22 February–1 March 2019; Volume 33, pp. 8126–8133.
7. Asif, M.; Tiwana, M.I.; Khan, U.S.; Ahmad, M.W.; Qureshi, W.S.; Iqbal, J. Human gait recognition subject to different covariate factors in a multi-view environment. *Results Eng.* **2022**, *15*, 100556. [CrossRef]
8. Yu, S.; Tan, D.; Tan, T. A framework for evaluating the effect of view angle, clothing and carrying condition on gait recognition. In Proceedings of the 18th International Conference on Pattern Recognition (ICPR'06), Hong Kong, China, 20–24 August 2006; IEEE: Piscataway, NJ, USA, 2006; Volume 4, pp. 441–444.
9. Bodor, R.; Drenner, A.; Fehr, D.; Masoud, O.; Papanikolopoulos, N. View-independent human motion classification using image-based reconstruction. *Image Vis. Comput.* **2009**, *27*, 1194–1206. [CrossRef]
10. Ariyanto, G.; Nixon, M.S. Model-based 3D gait biometrics. In Proceedings of the 2011 International Joint Conference on Biometrics (IJCB), Washington, DC, USA, 11–13 October 2011; IEEE: Piscataway, NJ, USA, 2011; pp. 1–7.
11. Tome, D.; Russell, C.; Agapito, L. Lifting from the deep: Convolutional 3d pose estimation from a single image. In Proceedings of the IEEE Conference on Computer Vision and Pattern Recognition, Seattle, WA, USA, 14–19 June 2017; pp. 2500–2509.
12. Weng, J.; Liu, M.; Jiang, X.; Yuan, G. Deformable pose traversal convolution for 3D action and gesture recognition. In Proceedings of the European Conference on Computer Vision (ECCV), Munich, Germany, 8–14 September 2018; pp. 136–152.
13. Makihara, Y.; Sagawa, R.; Mukaigawa, Y.; Echigo, T.; Yagi, Y. Gait recognition using a view transformation model in the frequency domain. In Proceedings of the European Conference on Computer Vision, Graz, Austria, 7–13 May 2006; Springer: Berlin/Heidelberg, Germany, 2006; pp. 151–163.
14. Han, J.; Bhanu, B. Individual recognition using gait energy image. *IEEE Trans. Pattern Anal. Mach. Intell.* **2005**, *28*, 316–322. [CrossRef] [PubMed]
15. Kusakunniran, W.; Wu, Q.; Li, H.; Zhang, J. Multiple views gait recognition using view transformation model based on optimized gait energy image. In Proceedings of the 2009 IEEE 12th International Conference on Computer Vision Workshops, ICCV Workshops, Kyoto, Japan, 27 September–4 October 2009; IEEE: Piscataway, NJ, USA, 2009; pp. 1058–1064.
16. Kusakunniran, W.; Wu, Q.; Zhang, J.; Li, H.; Wang, L. Recognizing gaits across views through correlated motion co-clustering. *IEEE Trans. Image Process.* **2013**, *23*, 696–709. [CrossRef] [PubMed]
17. Hu, M.; Wang, Y.; Zhang, Z.; Little, J.J.; Huang, D. View-invariant discriminative projection for multi-view gait-based human identification. *IEEE Trans. Inf. Forensics Secur.* **2013**, *8*, 2034–2045. [CrossRef]
18. Zhang, P.; Wu, Q.; Xu, J. VT-GAN, View transformation GAN for gait recognition across views. In Proceedings of the 2019 International Joint Conference on Neural Networks (IJCNN), Budapest, Hungary, 14–19 July 2019; IEEE: Piscataway, NJ, USA, 2019; pp. 1–8.
19. Yu, S.; Chen, H.; Garcia Reyes, E.B.; Poh, N. Gaitgan: Invariant gait feature extraction using generative adversarial networks. In Proceedings of the IEEE Conference on Computer Vision and Pattern Recognition Workshops, Honolulu, HI, USA, 21–26 July 2017; pp. 30–37.
20. Shiqi, Y.; Chen, H.; Liao, R.; An, W.; García, E.B.; Huang, Y.; Poh, N. GaitGANv2: Invariant gait feature extraction using generative adversarial networks. *Pattern Recognit* **2019**, *87*, 179–189.

21. Wen, J.; Shen, Y.; Yang, J. Multi-view gait recognition based on generative adversarial network. *Neural Process. Lett.* **2022**, *54*, 1855–1877. [CrossRef]
22. Chen, X.; Luo, X.; Weng, J.; Luo, W.; Li, H.; Tian, Q. Multi-view gait image generation for cross-view gait recognition. *IEEE Trans. Image Process.* **2021**, *30*, 3041–3055. [CrossRef] [PubMed]
23. Koch, G.; Zemel, R.; Salakhutdinov, R. Siamese neural networks for one-shot image recognition. In Proceedings of the ICML Deep Learning Workshop, Lille, France, 10–11 July 2015; Volume 2.
24. Chen, X.; Yan, X.; Zheng, F.; Jiang, Y.; Xia, S.-T.; Zhao, Y.; Ji, R. One-shot adversarial attacks on visual tracking with dual attention. In Proceedings of the IEEE/CVF Conference on Computer Vision and Pattern Recognition, Seattle, WA, USA, 13–19 June 2020; pp. 10176–10185.
25. Chen CF, R.; Fan, Q.; Panda, R. CrossViT: Cross-attention multi-scale vision transformer for image classification. In Proceedings of the IEEE/CVF International Conference on Computer Vision, Montreal, QC, Canada, 10–17 October 2021; pp. 357–366.
26. Howard, A.G.; Zhu, M.; Chen, B.; Kalenichenko, D.; Wang, W.; Weyand, T.; Andreetto, M.; Adam, H. Mobilenets: Efficient convolutional neural networks for mobile vision applications. *arXiv* **2017**, arXiv:1704.04861.
27. Sandler, M.; Howard, A.; Zhu, M.; Zhmoginov, A.; Chen, L.-C. Mobilenetv2: Inverted residuals and linear bottlenecks. In Proceedings of the IEEE Conference on Computer Vision and Pattern Recognition, Seattle, WA, USA, 14–19 June 2018; pp. 4510–4520.
28. Howard, A.; Sandler, M.; Chu, G.; Li, Q. Searching for mobilenetv3. In Proceedings of the IEEE/CVF International Conference on Computer Vision, Seoul, Republic of Korea, 27 October–2 November 2019; pp. 1314–1324.
29. Mehta, S.; Rastegari, M. MobileViT: Light-weight, general-purpose, and mobile-friendly vision transformer. *arXiv* **2021**, arXiv:2110.02178.
30. Chai, T.; Li, A.; Zhang, S.; Li, Z.; Wang, Y. Lagrange Motion Analysis and View Embeddings for Improved Gait Recognition. In Proceedings of the IEEE/CVF Conference on Computer Vision and Pattern Recognition, New Orleans, LA, USA, 18–24 June 2022; pp. 20249–20258.
31. Yu, S.; Chen, H.; Wang, Q.; Shen, L.; Huang, Y. Invariant feature extraction for gait recognition using only one uniform model. *Neurocomputing* **2017**, *239*, 81–93. [CrossRef]
32. Gao, J.; Zhang, S.; Guan, X.; Meng, X. Multiview Gait Recognition Based on Slack Allocation Generation Adversarial Network. *Wirel. Commun. Mob. Comput.* **2022**, *2022*, 1648138. [CrossRef]
33. Alvarez IR, T.; Sahonero-Alvarez, G. Cross-view gait recognition based on u-net. In Proceedings of the 2020 International Joint Conference on Neural Networks (IJCNN), Glasgow, UK, 19–24 July 2020; IEEE: Piscataway, NJ, USA, 2020; pp. 1–7.
34. Liao, R.; Yu, S.; An, W.; Huang, Y. A model-based gait recognition method with body pose and human prior knowledge. *Pattern Recognit.* **2020**, *98*, 107069. [CrossRef]
35. Zhao, X.; He, Z.; Zhang, S.; Liang, D. Robust pedestrian detection in thermal infrared imagery using a shape distribution histogram feature and modified sparse representation classification. *Pattern Recognit.* **2015**, *48*, 1947–1960. [CrossRef]
36. Li, L.; Xue, F.; Liang, D.; Chen, X. A Hard Example Mining Approach for Concealed Multi-Object Detection of Active Terahertz Image. *Appl. Sci.* **2021**, *11*, 11241. [CrossRef]
37. Kang, B.; Liang, D.; Ding, W.; Zhou, H.; Zhu, W.-P. Grayscale-thermal tracking via inverse sparse representation-based collaborative encoding. *IEEE Trans. Image Process.* **2019**, *29*, 3401–3415. [CrossRef] [PubMed]

applied sciences

MDPI

Article

Integrating Spherical Fuzzy Sets and the Objective Weights Consideration of Risk Factors for Handling Risk-Ranking Issues

Kuei-Hu Chang

Department of Management Sciences, R.O.C. Military Academy, Kaohsiung 830, Taiwan; evenken2002@yahoo.com.tw

Abstract: Risk assessments and risk prioritizations are crucial aspects of new product design before a product is launched into the market. Risk-ranking issues involve the information that is considered for the evaluation and objective weighting considerations of the evaluation factors that are presented by the data. However, typical risk-ranking methods cannot effectively grasp a comprehensive evaluation of this information and ignore the objective weight considerations of the risk factors, leading to inappropriate evaluation results. For a more accurate ranking result of the failure mode risk, this study proposes a novel, flexible risk-ranking approach that integrates spherical fuzzy sets and the objective weight considerations of the risk factors to process the risk-ranking issues. In the numerical case validation, a new product design risk assessment of electronic equipment was used as a numerically validated case, and the simulation results were compared with the risk priority number (RPN) method, improved risk priority number (IRPN) method, intuitionistic fuzzy weighted average (IFWA) method, and spherical weighted arithmetic average (SWAA) method. The test outcomes that were confirmed showed that the proposed novel, flexible risk-ranking approach could effectively grasp the comprehensive evaluation information and provide a more accurate ranking of the failure mode risk.

Keywords: spherical fuzzy sets; objective weights; risk ranking; risk priority number; artificial intelligence

Citation: Chang, K.-H. Integrating Spherical Fuzzy Sets and the Objective Weights Consideration of Risk Factors for Handling Risk-Ranking Issues. *Appl. Sci.* **2023**, *13*, 4503. https://doi.org/10.3390/app13074503

Academic Editor: Mayank Kejriwal

Received: 12 March 2023
Revised: 25 March 2023
Accepted: 31 March 2023
Published: 2 April 2023

1. Introduction

Risk assessment and risk-ranking issues include multiple evaluation criteria, multiple failure modes, and multiple experts, which can be categorized as multi-criteria decision making (MCDM) problems. The results of the risk assessment and risk-ranking of a product or system directly affect the product quality, profit, and market competitiveness. These risk-ranking problems primarily involve two important issues: the method of evaluating the information processing and the consideration of the risk factor weights. The typical risk priority number (RPN) approach is the most widely applied method for risk assessments and has been adopted by different industry standards, such as QS9000, IATF 16949, MILSTD-1629A, ISO 9001, and IEC 60812 [1]. In the RPN method, the failure risk of the failure mode is ranked using the RPN value, which is obtained by multiplying the three risk factors, severity (*Sev*), occurrence (*Occ*), and detection (*Det*). The RPN method involves simple calculations and, in recent years, has thus been widely applied in various areas, such as hospital radiopharmacy management [2], semiconductor manufacturing [3], robot-assisted rehabilitation processes [4], photovoltaic cell manufacturing [5], power transformer equipment [6], submersible pump risk analyses [7], and high-dose-rate brachytherapy treatments [8]. However, the RPN method is not able to process the uncertainty of the evaluation information [9,10] and ignores the objective weight consideration of the risk factors [4,11], also violating the definition of the measurement scale [12,13].

To process the uncertainty of the evaluation information, Zadeh [14] first presented a fuzzy set for handling the decision making issues in everyday life. The fuzzy set (FS) method applied membership degrees (MD) and non-membership degrees (NMD) to express the content of the evaluation information. The NMD is equal to 1 minus the MD in the FS method. To solve the restriction of the FS, Atanassov [15] proposed an intuitionistic FS to increase the consideration of the indeterminacy degree (ID), which required that the sum of MD, ID, and NMD must be equal to 1. The intuitionistic FS method has the advantage of an ID consideration; therefore, the intuitionistic FS method has recently been used within many different fields, such as stock prediction [16], supplier selection [17], enterprise resource planning systems [18], medical diagnoses [19], risk assessments [20], supply chain management [21], tourist destination selection [22], and so on. Extending the concept of the intuitionistic FS, the picture fuzzy set applied the MD, ID, NMD, and refusal degree to express an expert's opinion [23], and the sum of the MD, ID, and NMD had to be less than or equal to 1. However, in the actual execution of the MCDM problems, sometimes, the sum of the MD and NMD exceeds one. To overcome the restriction of the MD and NMD of the intuitionistic FS, Yager [24] proposed a Pythagorean FS, allowing the sum of the MD and NMD to be greater than 1, but restricting the sum of squares of the MD and NMD to be less than 1. The Pythagorean FS has the advantage of being able to consider the MD, ID, and NMD simultaneously. To fully consider all the possible situations in a decision analysis, Mahmood et al. [25] used a three-dimensional FS mode to propose a spherical FS. A spherical FS allows the sum of the MD, ID, and NMD to be greater than 1, but restricts the sum of the squares of the MD, ID, and NMD to a value of less than 1. The main difference between the spherical FS and Pythagorean FS is that the spherical FS increases the consideration of the refusal degree. In a spherical FS, decision makers can specify the MD, ID, and NMD values [26]. Currently, the spherical FS is being widely used in many different areas, such as vehicle model selection [27], the construction of Fangcang shelter hospitals [28], community epidemic prevention [29], medical diagnoses [30], waste management [31], green supply chain management [32], and performance evaluation [33,34].

Another key issue in risk assessments is the objective weight consideration of the evaluation factors, which affects the accuracy of the risk assessment results. However, the traditional RPN method only considers the subjective assessment of the experts in the risk assessment process, ignoring the objectivity of the research data, which leads to incorrect assessment results [35]. Scholars have also used different calculation methods to deal with the objective weights of the MCDM problems. For example, Liang et al. [36] used the structural entropy weight approach to calculate the indicator weights of the index and then combined the fuzzy technique for order of preference with a similarity to ideal solution (TOPSIS) model, structural entropy weight approach, and cloud inference, in order to process the risk assessments of urban polyethylene gas pipelines. Likewise, Paramanik et al. [37] applied the criteria importance through an intercriteria correlation (CRITIC) approach to obtain the objective weights of the evaluation criteria, and then combined the linear programming technique for a multidimensional analysis of preference and the best–worst approach to process the web service selection problems. Earlier, Barukab et al. [38] combined the spherical FS, entropy measures, and fuzzy TOPSIS methods to process the group decision making problems for a robot selection. Recently, Chang [39] reported the use of the combined compromise solution (CoCoSo) approach and subjective–objective weights consideration to process the supplier selection problems.

To fully solve the limitations of these typical risk assessment methods, considering the information and weights, a novel flexible approach that integrates the spherical FS and objective-weight-considering factors is proposed in this study to process the risk-ranking issues. The proposed novel, flexible risk-ranking approach uses the spherical FS to fully grasp the fuzzy, intuitionistic fuzzy, and spherical fuzzy information that is provided by experts. The proposed approach also uses the preference selection index (PSI) to probe the objective weights of the evaluation factors that are presented by the data itself.

The remainder of this paper is organized as follows. In Section 2, some of the basic concepts, definitions, and algorithm rules of the RPN method, spherical weighted arithmetic average (SWAA) method, and PSI method are presented and briefly reviewed. In Section 3, a novel, flexible risk-ranking approach that integrates the SWAA and PSI methods is proposed. Section 4 presents a risk assessment numerical example of a new electronic equipment product design and compares the calculation results of the RPN, improved risk priority number (IRPN) method, intuitionistic fuzzy weighted average (IFWA) method, SWAA method, and proposed method. Section 5 presents the conclusions and future research directions.

2. Preliminaries

Here, we briefly review some of the basic definitions, concepts, and algorithm rules of the RPN method, SWAA method, and PSI method.

2.1. Risk Priority Number Method

At present, failure mode and effect analysis (FMEA) is the most commonly used risk assessment method by different industries; this method originated in the aerospace industry in the 1950s and has been widely used within different industries since [40]. The FMEA approach uses the *RPN* value to rank the possible failure risks. The *RPN* value is the product of three risk factors with equal weights: severity (*Sev*), occurrence (*Occ*), and detection (*Det*). The *RPN* value is calculated using Equation (1).

$$RPN = Sev \times Occ \times Det \tag{1}$$

The risk factor *Sev* represents the severity of the failure occurrence, *Occ* is the probability of the failure occurrence, and *Det* is the probability that a failure occurrence cannot be detected. These risk factors, *Sev*, *Occ*, and *Det*, use risk assessment ratings of 1–10. The potential failure mode (FM) has a higher RPN value, which means that this FM has a higher risk of failure, and a higher risk priority must be given to prevent the occurrence of such failures.

2.2. Spherical Fuzzy Set Method

The intuitionistic FS is the basis of the spherical FS. The basic principles related to the intuitionistic FS and the calculation rules are described as follows:

Definition 1 [41]. *Assuming that X is the universe of discourse. Then, an intuitionistic FS I in X and the IFWA are expressed as follows:*

$$I = \{x, \mu_I(x), \nu_I(x) | x \in X\} \tag{2}$$

where $\mu_I(x)$ and $\nu_I(x)$ represent the MD and NMD, respectively, and $\mu_I(x)$ and $\nu_I(x) \in [0,1]$ satisfy the condition $\mu_I(x) + \nu_I(x) \leq 1$.

$$IFWA(I_1, I_2, \ldots, I_n) = \left(1 - \prod_{g=1}^{n} (1 - \mu_g(x))^{w_g}, \prod_{g=1}^{n} \nu_g^{w_g}\right) \tag{3}$$

where w_g represents the weight of I_g, $w_g \in [0,1]$ and $\sum_{g=1}^{n} w_g = 1$.

The score value of the intuitionistic FS is defined as follows:

$$Score(I) = \mu_I(x) - \nu_I(x) \tag{4}$$

Mahmood et al. [25] used a three-dimensional FS mode by extending the concepts of the FS, intuitionistic FS, and Pythagorean FS to propose a spherical FS for processing the MCDM problems under uncertain conditions. The basic principles related to the spherical FS and the calculation rules are described as follows.

Definition 2 [42]. *Assuming that X is the universe of discourse, then, a spherical FS S in X is defined as follows:*

$$S = \{x, \mu_S(x), \pi_S(x), \nu_S(x) | x \in X\} \tag{5}$$

where the $\mu_S(x)$, $\pi_S(x)$, and $\nu_S(x)$ represent the MD, ID, and NMD, and $\mu_S(x)$, $\pi_S(x)$, and $\nu_S(x) \in [0,1]$ satisfy the condition $0 \leq (\mu_S(x))^2 + (\pi_S(x))^2 + (\nu_S(x))^2 \leq 1$.

The refusal degree $(R_S(x))$ can be expressed as follows:

$$R_S(x) = \sqrt{1 - (\mu_S(x))^2 - (\pi_S(x))^2 - (\nu_S(x))^2} \tag{6}$$

Definition 3 [42,43]. *Supposing that the $S_1 = \langle \mu_{S_1}(x), \pi_{S_1}(x), \nu_{S_1}(x) \rangle$ and $S_2 = \langle \mu_{S_2}(x), \pi_{S_2}(x), \nu_{S_2}(x) \rangle$ are any two spherical FSs, the basic algorithm rules of the spherical FSs are as follows:*

$$S_1 \oplus S_2 = \left\{ \sqrt{\mu_{S1}^2 + \mu_{S2}^2 - \mu_{S1}^2 \cdot \mu_{S2}^2}, \sqrt{(1 - \mu_{S2}^2) \cdot \pi_{S1}^2 + (1 - \mu_{S1}^2) \cdot \pi_{S2}^2 - \pi_{S1}^2 \cdot \pi_{S2}^2}, \nu_{S1} \cdot \nu_{S2} \right\} \tag{7}$$

$$S_1 \otimes S_2 = \left\{ \mu_{S1} \cdot \mu_{S2}, \sqrt{(1 - \nu_{S2}^2) \cdot \pi_{S1}^2 + (1 - \nu_{S1}^2) \cdot \pi_{S2}^2 - \pi_{S1}^2 \cdot \pi_{S2}^2}, \sqrt{\nu_{S1}^2 + \nu_{S2}^2 - \nu_{S1}^2 \cdot \nu_{S2}^2} \right\} \tag{8}$$

$$kS_1 = \left\{ \sqrt{1 - (1 - \mu_{S1}^2)^k}, \sqrt{(1 - \mu_{S1}^2)^k - (1 - \mu_{S1}^2 - \pi_{S1}^2)^k}, \nu_{S1}^k \right\}; \ k > 0 \tag{9}$$

$$S_1{}^k = \left\{ \mu_{S1}^k, \sqrt{(1 - \nu_{S1}^2)^k - (1 - \nu_{S1}^2 - \pi_{S1}^2)^k}, \sqrt{1 - (1 - \nu_{S1}^2)^k} \right\}; \ k > 0 \tag{10}$$

Definition 4 [43]. *Let $S_g = \langle \mu_S(x), \pi_S(x), \nu_S(x) \rangle$ be the spherical FS and w_g represent the weights of S_g, $w_g \in [0,1]$ and $\sum_{g=1}^n w_g = 1$. The spherical weighted arithmetic average (SWAA) is defined as:*

$$\begin{aligned} SWAA(S_1, S_2, \ldots, S_n) &= \textstyle\sum_{g=1}^n w_g S_g \\ &= \left(\sqrt{1 - \prod_{g=1}^n \left(1 - \mu_g^2\right)^{w_g}}, \sqrt{\prod_{g=1}^n \left(1 - \mu_g^2\right)^{w_g} - \prod_{g=1}^n \left(1 - \mu_g^2 - \pi_g^2\right)^{w_g}}, \prod_{g=1}^n \nu_g^{w_g} \right) \end{aligned} \tag{11}$$

Definition 5 [43]. *Let $S_g = \langle \mu_S(x), \pi_S(x), \nu_S(x) \rangle$ be the spherical FS and w_g represent the weight of S_g, $w_g \in [0,1]$ and $\sum_{g=1}^n w_g = 1$. The spherical weighted geometric average (SWGA) is defined as:*

$$\begin{aligned} SWGA(S_1, S_2, \ldots, S_n) &= \textstyle\prod_{g=1}^n S_g{}^{w_g} \\ &= \left(\prod_{g=1}^n \mu_g^{w_g}, \sqrt{\prod_{g=1}^n \left(1 - \nu_g^2\right)^{w_g} - \prod_{g=1}^n \left(1 - \nu_g^2 - \pi_g^2\right)^{w_g}}, \sqrt{1 - \prod_{g=1}^n \left(1 - \nu_g^2\right)^{w_g}} \right) \end{aligned} \tag{12}$$

Definition 6 [28,43]. *Let $S_g = \langle \mu_S(x), \pi_S(x), \nu_S(x) \rangle$* be the spherical FS, *$\mu_S(x)$, $\pi_S(x)$, and $\nu_S(x) \in [0,1]$, then the score and accuracy values are defined as follows:*

$$Score(S) = (\mu_S - \pi_S)^2 - (\upsilon_S - \pi_S)^2 \tag{13}$$

$$Accuracy(S) = \mu_S^2 + \pi_S^2 + \nu_S^2 \tag{14}$$

Definition 7 [1,28]. *The comparison rules of the two spherical FSs,* $S_1 = \langle \mu_{S1}(x), \pi_{S1}(x), \nu_{S1}(x) \rangle$ and $S_2 = \langle \mu_{S2}(x), \pi_{S2}(x), \nu_{S2}(x) \rangle$, *are defined as follows.*

(1) If $Score(S_1) > Score(S_2)$, then $S_1 > S_2$;
(2) if $Score(S_1) = Score(S_2)$, and $Accuracy(S_1) > Accuracy(S_2)$, then $S_1 > S_2$;
(3) if $Score(S_1) = Score(S_2)$, and $Accuracy(S_1) = Accuracy(S_2)$, then $S_1 = S_2$.

2.3. The Preference Selection Index (PSI) Method

The PSI approach was first introduced by Maniya and Bhatt [44]; in this approach, statistical concepts are used to calculate the overall preference value of the assessment factors and then process the material selection issues. The algorithm program of the PSI approach is as follows:

(1) Create an initial decision matrix, x_{ij}:

The x_{ij} values represent the values of the ith alternative and jth decision criterion. $i = 1,2,\ldots,m$, and $j = 1,2,\ldots,n$.

$$x_{ij} = \begin{bmatrix} x_{11} & x_{12} & \cdots & x_{1n} \\ x_{21} & x_{22} & \cdots & x_{2n} \\ \vdots & \vdots & \ddots & \vdots \\ x_{m1} & x_{m2} & \cdots & x_{mn} \end{bmatrix} \quad (15)$$

(2) The decision matrix is normalized as, N_{ij}:

$$N_{ij} = \frac{x_{ij}}{x_j^{max}}, \text{ for the profit decision criteria} \quad (16)$$

$$N_{ij} = \frac{x_j^{min}}{x_{ij}}, \text{ for the cost decision criteria} \quad (17)$$

(3) The preference variation value PV_j is calculated as:

$$PV_j = \sum_{i=1}^{m} \left(N_{ij} - \overline{N_j}\right)^2, \ \overline{N_j} = \frac{1}{m}\sum_{i=1}^{m} N_{ij} \quad (18)$$

(4) The overall preference value OP_j is calculated as:

$$OP_j = \frac{1 - PV_j}{n - \sum_{j=1}^{n} PV_j} \quad (19)$$

(5) The preference selection value PS_i is calculated as:

$$PS_i = \sum_{j=1}^{n} N_{ij} \times OP_j \quad (20)$$

3. Proposed Novel Flexible Risk-Ranking Approach

Failure risk analysis is a crucial factor in product design and manufacturing processes. FMEA is the most commonly and widely used risk assessment method and is used as a different industry standard. It is a systematic, structured approach to risk assessment and uses RPN values to rank the risks of the FM. In product or system failure risk assessment, two main factors need to be considered: the information for the evaluation and the objective weighting considerations of the risk factors that are presented by the data themselves. However, the RPN method cannot process intuitionistic and spherical fuzzy information, nor does it consider the objective weighting of the risk factors that are presented by the data. Moreover, the calculation mode of an RPN method violates the definition of the measurement scale. To solve the restrictions of the RPN method, this study integrated the spherical FS and an objective weight consideration of the risk factors to process these risk-ranking issues. The proposed method uses the MD, ID, and NMD of the spherical FS to represent the assessment information of the risk factors. Thus, the proposed novel, flexible risk-ranking approach can process fuzzy, intuitionistic fuzzy, and spherical fuzzy information simultaneously and can fully consider various types of information. The proposed novel, flexible risk-ranking approach used the PSI approach to calculate the objective weights of the risk factors and the SWAA method to obtain the aggregation values of the risk factors, which solves the problem of the RPN method violating the definition of the measurement scale.

The proposed method can be broadly divided into eight steps (as shown in Figure 1), as follows.

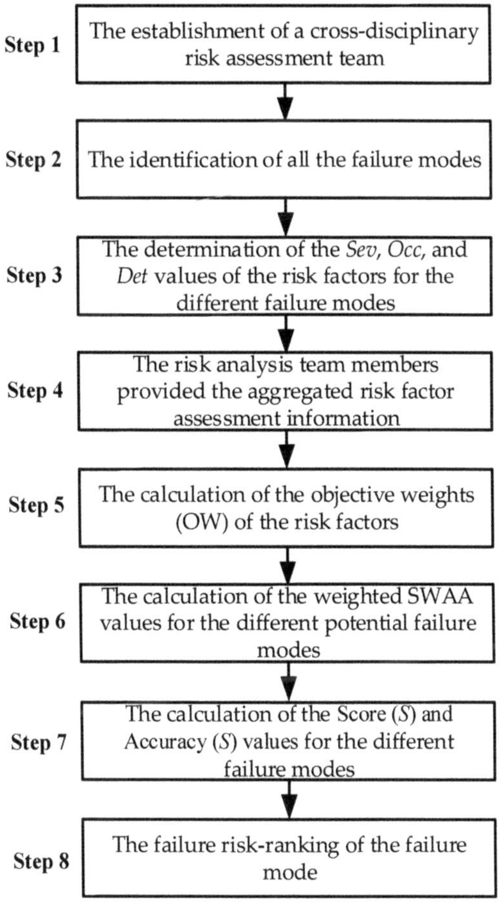

Figure 1. The flowchart of the proposed method.

Step 1. The establishment of a cross-disciplinary risk assessment team.

This was achieved based on their respective professional backgrounds.

Step 2. The identification of all the failure modes.

The risk analysis team members had a discussion to identify all the possible potential FMs based on the risk topic being evaluated.

Step 3. The determination of the *Sev*, *Occ*, and *Det* values of the risk factors for the different failure modes.

The risk analysis team members, according to their professional experience and background, determined the *Sev*, *Occ*, and *Det* values of the risk factors for the different FMs.

Step 4. The risk analysis team members provided the aggregated risk factor assessment information.

Based on the data from Step 3, Equation (11) was used to aggregate the assessment information of the risk factors that were provided by the risk analysis team members.

Step 5. The calculation of the objective weights (OW) of the risk factors.

Based on the data from Step 4, Equations (15)–(18) were used to calculate the preference variation value (PV_j). Then, Equation (19) was used to calculate the overall preference value (OP_j).

Based on the overall preference value (OP_j), Equation (21) was used to calculate the objective weights (OW_j) of the risk factors.

$$OW_j = \frac{(\mu_j - \pi_j)^2 - (v_j - \pi_j)^2}{\sum_{j=1}^{3}\left((\mu_j - \pi_j)^2 - (v_j - \pi_j)^2\right)} \tag{21}$$

Step 6. The calculation of the weighted SWAA values for the different potential failure modes.

Based on the data from Steps 4 and 5, Equation (11) was used to calculate the weighted SWAA values of the different FMs.

Step 7. The calculation of the $Score(S)$ and $Accuracy(S)$ values for the different failure modes.

Based on the data from Step 6, Equations (13) and (14) were used to calculate the $Score(S)$ and $Accuracy(S)$ values of the different FMs, respectively.

Step 8. The failure risk-ranking of the failure mode.

The failure risk of the FM was ranked according to the $Score(S)$ and $Accuracy(S)$ values.

4. Numerical Example

4.1. Case Overview

The completeness of the information considerations and the rationality of the evaluation results of the proposed novel, flexible risk-ranking approach were verified in this study by using the new product design of electronic equipment as a numerically validated case (adapted from [45]). The new product design for electronic equipment requires a risk assessment, avoiding a product failure with limited resources and instantly completing the system design within the specification constraints specified by the customer. The risk analysis assessment team for electronic equipment includes three domain experts (DE1, DE2, and DE3) in engineering and electronic design. The main goal of the risk analysis assessment team is to confirm the possible failure risk items in the product design process of the electronic equipment, correctly sort the risk-ranking of the potential FM, and allocate resources under the limited resources in the best possible way to prevent the occurrence of risks. The relationship between the linguistic terms and spherical fuzzy numbers within the new product design of an electronic equipment case is shown in Table 1, according to which, the domain experts are given these linguistic terms based on the different potential FMs, the results of which are shown in Table 2.

Table 1. Relationship between the linguistic terms and spherical fuzzy numbers.

Linguistic Terms	μ_S	π_S	v_S
Extremely high impact (EH)	0.95	0.10	0.20
Very high impact (VH)	0.85	0.20	0.30
High impact (HI)	0.75	0.30	0.40
Slightly high impact (SH)	0.65	0.40	0.50
Medium impact (MI)	0.55	0.50	0.60
Slightly low impact (SL)	0.45	0.40	0.70
Low impact (LI)	0.35	0.30	0.80
Very low impact (VL)	0.25	0.20	0.85
Extremely low impact (EL)	0.15	0.10	0.90
Extremely very low impact (EV)	0.10	0.10	0.95

Table 2. Linguistic values of potential failure items given by experts. (FM: failure mode).

Items	Potential Failure Mode	*Sev*			*Occ*			*Det*		
		DE1	DE2	DE3	DE1	DE2	DE3	DE1	DE2	DE3
1	Extremely limited launch time (FM1)	MI	SL	SL	SL	SL	LI	SH	SL	SL
2	Customer request changes (FM2)	SL	SL	LI	SL	SL	LI	EH	EH	VH
3	Lack of aesthetic consideration (FM3)	MI	SL	MI	MI	MI	SL	SH	MI	MI
4	Product technical failure (FM4)	HI	SH	HI	VL	VL	SL	VL	SL	VL
5	Design changes at the last minute (FM5)	VH	EH	EH	SL	SL	SL	SL	SL	SL
6	Poor product performance (FM6)	VH	HI	VH	MI	MI	SL	SH	MI	MI
7	Manufacturing is not ready to start (FM7)	SL	SL	LI	SL	LI	LI	SL	SL	LI
8	Insufficient manufacturing capacity (FM8)	MI	SL	MI	LI	VL	VL	MI	SL	MI
9	Long lead times for materials (FM9)	SL	SL	LI	SL	SL	LI	LI	LI	LI
10	Potential market saturation (FM10)	VL	VL	LI	SH	MI	MI	MI	LI	MI
11	Failed test run (FM11)	SL	LI	SL	LI	SL	LI	LI	SL	LI
12	Customer sample failed (FM12)	MI	SH	MI	MI	LI	MI	SH	MI	MI
13	Insufficient stock to start (FM13)	LI	LI	LI	SL	LI	SL	SL	SL	SL
14	Incorrect market analysis (FM14)	VH	HI	VH	MI	MI	SL	VH	HI	VH
15	Unavailability of any new technology for development (FM15)	LI	LI	SL	SL	SL	SL	SL	LI	LI
16	Environmental compliance not considered (FM16)	LI	SL	LI	SL	LI	LI	LI	LI	SL
17	New technologies in the manufacturing process (FM17)	SH	SL	SL	SL	SL	LI	SL	LI	SL
18	Lack of experts to develop products (FM18)	SL	LI	SL	SL	SL	LI	LI	SL	SL
19	Poor quality raw materials (FM19)	EL	VL	EL	EL	EL	LI	VL	EL	EL

4.2. Solution with the Risk Priority Number Approach

The RPN approach [2] uses the RPN value to rank the possible failure risks. The RPN value is the product of the three equal weighted risk factors: *Sev*, *Occ*, and *Det*. The higher the RPN value that is represented, the higher the risk level of the FM, and it must be given a higher risk prevention priority to prevent the occurrence of this FM. However, the RPN method can only handle the MD information of the FM. As shown in Tables 1 and 2, Equation (1) was used to calculate the RPN value of the electronic equipment new product design failure, and the results are expressed in Table 3.

Table 3. The RPN value of the electronic equipment new product design failure.

Items	*Sev*	*Occ*	*Det*	RPN	Rank
1	0.483	0.417	0.517	0.104	7
2	0.417	0.417	0.917	0.159	5
3	0.517	0.517	0.583	0.156	6
4	0.717	0.317	0.317	0.072	12
5	0.917	0.450	0.450	0.186	3
6	0.817	0.517	0.583	0.246	2
7	0.417	0.383	0.417	0.067	13
8	0.517	0.283	0.517	0.076	10
9	0.417	0.417	0.350	0.061	17
10	0.283	0.583	0.483	0.080	9
11	0.417	0.383	0.383	0.061	16
12	0.583	0.483	0.583	0.164	4
13	0.350	0.417	0.450	0.066	15
14	0.817	0.517	0.817	0.345	1
15	0.383	0.450	0.383	0.066	14
16	0.383	0.383	0.383	0.056	18
17	0.517	0.417	0.417	0.090	8
18	0.417	0.417	0.417	0.072	11
19	0.183	0.217	0.183	0.007	19

4.3. Solution with the Improved Risk Priority Number Method

To solve the problem of the RPN method violating the definition of the measurement scale, the improved risk priority number (IRPN) [46] is used as the sum of the *Sev*, *Occ*, and *Det* risk factors to estimate the IRPN value. The IRPN method is the same as the RPN approach and can only process the MD information of the FM. According to Tables 1 and 2, the sum of *Sev*, *Occ*, and *Det* risk factors was used to calculate the IRPN value for the electronic equipment new product design failure, and the results are expressed in Table 4.

Table 4. The IRPN value of the electronic equipment new product design failure.

Items	*Sev*	*Occ*	*Det*	IRPN	Rank
1	0.483	0.417	0.517	1.417	7
2	0.417	0.417	0.917	1.750	4
3	0.517	0.517	0.583	1.617	6
4	0.717	0.317	0.317	1.350	8
5	0.917	0.450	0.450	1.817	3
6	0.817	0.517	0.583	1.917	2
7	0.417	0.383	0.417	1.217	13
8	0.517	0.283	0.517	1.317	11
9	0.417	0.417	0.350	1.183	16
10	0.283	0.583	0.483	1.350	8
11	0.417	0.383	0.383	1.183	16
12	0.583	0.483	0.583	1.650	5
13	0.350	0.417	0.450	1.217	13
14	0.817	0.517	0.817	2.150	1
15	0.383	0.450	0.383	1.217	13
16	0.383	0.383	0.383	1.150	18
17	0.517	0.417	0.417	1.350	8
18	0.417	0.417	0.417	1.250	12
19	0.183	0.217	0.183	0.583	19

4.4. Solution with the Intuitionistic Fuzzy Weighted Average Method

The intuitionistic fuzzy weighted average (IFWA) method [41] can simultaneously consider the MD and NMD in the risk assessment problem of the new product design of the electronic equipment. According to Tables 1 and 2, Equations (3) and (4) were used to calculate the IFWA and score values for the electronic equipment new product design failure, and results are expressed in Table 5.

Table 5. The IFWA value of the electronic equipment new product design failure.

Items	*Sev*	*Occ*	*Det*	IFWA	*Score*(*I*)	Rank
1	(0.486, 0.514)	(0.419, 0.581)	(0.527, 0.473)	(0.479, 0.521)	−0.042	8
2	(0.419, 0.581)	(0.419, 0.581)	(0.928, 0.072)	(0.710, 0.290)	0.420	3
3	(0.519, 0.481)	(0.519, 0.481)	(0.586, 0.414)	(0.542, 0.458)	0.085	6
4	(0.720, 0.280)	(0.324, 0.676)	(0.324, 0.676)	(0.496, 0.504)	−0.008	7
5	(0.928, 0.072)	(0.450, 0.550)	(0.450, 0.550)	(0.721, 0.279)	0.441	2
6	(0.822, 0.178)	(0.519, 0.481)	(0.586, 0.414)	(0.672, 0.328)	0.343	4
7	(0.419, 0.581)	(0.385, 0.615)	(0.419, 0.581)	(0.408, 0.592)	−0.185	13
8	(0.519, 0.481)	(0.285, 0.715)	(0.519, 0.481)	(0.451, 0.549)	−0.098	11
9	(0.419, 0.581)	(0.419, 0.581)	(0.350, 0.650)	(0.397, 0.603)	−0.207	16
10	(0.285, 0.715)	(0.586, 0.414)	(0.491, 0.509)	(0.468, 0.532)	−0.064	9
11	(0.419, 0.581)	(0.385, 0.615)	(0.385, 0.615)	(0.397, 0.603)	−0.207	16
12	(0.586, 0.414)	(0.491, 0.509)	(0.586, 0.414)	(0.557, 0.443)	0.113	5
13	(0.350, 0.650)	(0.419, 0.581)	(0.450, 0.550)	(0.408, 0.582)	−0.185	13
14	(0.822, 0.172)	(0.519, 0.481)	(0.822, 0.178)	(0.752, 0.248)	0.504	1
15	(0.385, 0.615)	(0.450, 0.550)	(0.385, 0.615)	(0.408, 0.592)	−0.185	13
16	(0.385, 0.615)	(0.385, 0.615)	(0.385, 0.615)	(0.385, 0.615)	−0.230	18
17	(0.527, 0.473)	(0.419, 0.581)	(0.419, 0.581)	(0.457, 0.543)	−0.086	10
18	(0.419, 0.581)	(0.419, 0.581)	(0.419, 0.581)	(0.419, 0.581)	−0.163	12
19	(0.185, 0.815)	(0.223, 0.777)	(0.185, 0.815)	(0.198, 0.802)	−0.605	19

4.5. Solution with the Spherical Weighted Arithmetic Average Method

The spherical weighted arithmetic average (SWAA) method [43] can simultaneously consider the MD, ID, and NMD of the new product design of the electronic equipment. As mentioned in Tables 1 and 2, Equation (11) was used to aggregate the evaluation opinions of the different domain experts on the risk factors *Sev*, *Occ*, and *Det*. Then, Equations (11), (13) and (14) were used to calculate the SWAA, score, and accuracy values for the electronic equipment new product design failure, and the results are expressed in Table 6.

Table 6. The SWAA, score, and accuracy values of the electronic equipment new product design failure.

Items	*Sev*	*Occ*	*Det*	SWAA	*Score*(*S*)	*Accuracy*(*S*)	Rank
1	(0.487, 0.443, 0.665)	(0.420, 0.373, 0.732)	(0.533, 0.403, 0.626)	(0.484, 0.409, 0.673)	−0.064	0.854	8
2	(0.420, 0.373, 0.732)	(0.420, 0.373, 0.732)	(0.928, 0.127, 0.229)	(0.739, 0.263, 0.497)	0.172	0.862	3
3	(0.520, 0.475, 0.632)	(0.520, 0.475, 0.632)	(0.587, 0.467, 0.565)	(0.544, 0.473, 0.608)	−0.013	0.890	6
4	(0.721, 0.332, 0.431)	(0.334, 0.296, 0.797)	(0.334, 0.296, 0.797)	(0.526, 0.324, 0.649)	−0.065	0.803	9
5	(0.928, 0.127, 0.229)	(0.450, 0.400, 0.700)	(0.450, 0.400, 0.700)	(0.745, 0.278, 0.482)	0.177	0.865	2
6	(0.823, 0.231, 0.330)	(0.520, 0.475, 0.632)	(0.587, 0.467, 0.565)	(0.681, 0.382, 0.490)	0.078	0.850	4
7	(0.420, 0.373, 0.732)	(0.387, 0.341, 0.765)	(0.420, 0.373, 0.732)	(0.410, 0.363, 0.743)	−0.142	0.851	13
8	(0.520, 0.475, 0.632)	(0.288, 0.240, 0.833)	(0.520, 0.475, 0.632)	(0.461, 0.431, 0.693)	−0.068	0.878	10
9	(0.420, 0.373, 0.732)	(0.420, 0.373, 0.732)	(0.350, 0.300, 0.800)	(0.399, 0.352, 0.754)	−0.159	0.851	16
10	(0.288, 0.240, 0.833)	(0.587, 0.467, 0.565)	(0.497, 0.461, 0.660)	(0.482, 0.424, 0.677)	−0.061	0.871	7
11	(0.420, 0.373, 0.732)	(0.387, 0.341, 0.765)	(0.387, 0.341, 0.765)	(0.399, 0.352, 0.754)	−0.159	0.851	16
12	(0.587, 0.467, 0.565)	(0.497, 0.461, 0.660)	(0.587, 0.467, 0.565)	(0.560, 0.466, 0.595)	−0.008	0.885	5
13	(0.350, 0.300, 0.800)	(0.420, 0.373, 0.732)	(0.450, 0.400, 0.700)	(0.410, 0.363, 0.743)	−0.142	0.851	13
14	(0.823, 0.231, 0.330)	(0.520, 0.475, 0.632)	(0.823, 0.231, 0.330)	(0.759, 0.303, 0.410)	0.197	0.836	1
15	(0.387, 0.341, 0.765)	(0.450, 0.400, 0.700)	(0.387, 0.341, 0.765)	(0.410, 0.363, 0.743)	−0.142	0.851	13
16	(0.387, 0.341, 0.765)	(0.387, 0.341, 0.765)	(0.387, 0.341, 0.765)	(0.387, 0.341, 0.765)	−0.178	0.852	18
17	(0.533, 0.403, 0.626)	(0.420, 0.373, 0.732)	(0.420, 0.373, 0.732)	(0.463, 0.385, 0.695)	−0.090	0.845	11
18	(0.420, 0.373, 0.732)	(0.420, 0.373, 0.732)	(0.420, 0.373, 0.732)	(0.420, 0.373, 0.732)	−0.126	0.852	12
19	(0.190, 0.143, 0.883)	(0.239, 0.199, 0.865)	(0.190, 0.143, 0.883)	(0.208, 0.164, 0.877)	−0.506	0.839	19

4.6. Solution with the Proposed Novel Flexible Risk-Ranking Approach

To solve the restrictions of the typical risk assessment approach in its information processing and objective weighting considerations, the proposed method integrates the spherical FS and considers the objective weights of the risk factors to process the risk-ranking issues. The proposed novel, flexible approach is implemented in eight distinct steps, as described below. The process first must establish a cross-disciplinary risk assessment team, identify all the potential FMs, and determine the *Ser*, *Occ*, and *Det* values of the risk factors for the different potential FMs (Steps 1–3).

Step 4. The risk analysis team members provided the aggregated risk factor assessment information.

Based on Tables 1 and 2, Equation (11) was used to aggregate the evaluation opinions of the different domain experts on the risk factors *Sev*, *Occ*, and *Det*, and the results are expressed in Table 7.

Table 7. The weighted SWAA, score, and accuracy values of the proposed method.

Items	*Sev*	*Occ*	*Det*	Weighted SWAA	$Score(S)$	$Accuracy(S)$	Rank
1	(0.487, 0.443, 0.665)	(0.420, 0.373, 0.732)	(0.533, 0.403, 0.626)	(0.450, 0.403, 0.704)	−0.088	0.860	9
2	(0.420, 0.373, 0.732)	(0.420, 0.373, 0.732)	(0.928, 0.127, 0.229)	(0.476, 0.360, 0.703)	−0.104	0.850	10
3	(0.520, 0.475, 0.632)	(0.520, 0.475, 0.632)	(0.587, 0.467, 0.565)	(0.523, 0.475, 0.629)	−0.021	0.895	5
4	(0.721, 0.332, 0.431)	(0.334, 0.296, 0.797)	(0.334, 0.296, 0.797)	(0.534, 0.325, 0.641)	−0.056	0.802	7
5	(0.928, 0.127, 0.229)	(0.450, 0.400, 0.700)	(0.450, 0.400, 0.700)	(0.755, 0.272, 0.472)	0.194	0.867	1
6	(0.823, 0.231, 0.330)	(0.520, 0.475, 0.632)	(0.587, 0.467, 0.565)	(0.675, 0.378, 0.500)	0.073	0.849	3
7	(0.420, 0.373, 0.732)	(0.387, 0.341, 0.765)	(0.420, 0.373, 0.732)	(0.401, 0.354, 0.752)	−0.156	0.851	15
8	(0.520, 0.475, 0.632)	(0.288, 0.240, 0.833)	(0.520, 0.475, 0.632)	(0.401, 0.375, 0.748)	−0.138	0.861	14
9	(0.420, 0.373, 0.732)	(0.420, 0.373, 0.732)	(0.350, 0.300, 0.800)	(0.418, 0.371, 0.734)	−0.130	0.852	13
10	(0.288, 0.240, 0.833)	(0.587, 0.467, 0.565)	(0.497, 0.461, 0.660)	(0.509, 0.427, 0.651)	−0.044	0.865	6
11	(0.420, 0.373, 0.732)	(0.387, 0.341, 0.765)	(0.387, 0.341, 0.765)	(0.399, 0.353, 0.753)	−0.158	0.851	16
12	(0.587, 0.467, 0.565)	(0.497, 0.461, 0.660)	(0.587, 0.467, 0.565)	(0.535, 0.464, 0.621)	−0.020	0.888	4
13	(0.350, 0.300, 0.800)	(0.420, 0.373, 0.732)	(0.450, 0.400, 0.700)	(0.399, 0.352, 0.754)	−0.159	0.851	17
14	(0.823, 0.231, 0.330)	(0.520, 0.475, 0.632)	(0.823, 0.231, 0.330)	(0.684, 0.370, 0.491)	0.084	0.846	2
15	(0.387, 0.341, 0.765)	(0.450, 0.400, 0.700)	(0.387, 0.341, 0.765)	(0.427, 0.380, 0.725)	−0.117	0.852	11
16	(0.387, 0.341, 0.765)	(0.387, 0.341, 0.765)	(0.387, 0.341, 0.765)	(0.387, 0.341, 0.765)	−0.178	0.852	18
17	(0.533, 0.403, 0.626)	(0.420, 0.373, 0.732)	(0.420, 0.373, 0.732)	(0.465, 0.386, 0.692)	−0.088	0.845	8
18	(0.420, 0.373, 0.732)	(0.420, 0.373, 0.732)	(0.420, 0.373, 0.732)	(0.420, 0.373, 0.732)	−0.126	0.852	12
19	(0.190, 0.143, 0.883)	(0.239, 0.199, 0.865)	(0.190, 0.143, 0.883)	(0.221, 0.180, 0.872)	−0.478	0.842	19

Step 5. The calculation of the objective weights (OW) of the risk factors.

Based on the data from Step 4, Equations (15)–(18) were used to calculate the preference variation value (PV_j), as given below:

$$PV_{Sev} = (0.670, 0.192, 0.607);\ PV_{Occ} = (0.126, 0.106, 0.099);\ PV_{Det} = (0.517, 0.190, 0.464)$$

According to the preference variation value (PV_j), Equation (19) was used to calculate the overall preference value (OP_j), as given below:

$$OP_{Sev} = (0.196, 0.322, 0.215);\ OP_{Occ} = (0.518, 0.356, 0.492);\ OP_{Det} = (0.286, 0.322, 0.293)$$

According to the overall preference value (OP_j), Equation (21) was used to calculate the objective weights (OW_j) of the risk factors, as given below:

$$OW_{Sev} = 0.353;\ OW_{Occ} = 0.612;\ OW_{Det} = 0.035$$

Step 6. The weighted SWAA values for the different potential failure modes were calculated.

Based on the data from Steps 4 and 5, Equation (11) was used to calculate the weighted SWAA values of the different potential FMs; the results are expressed in Table 7.

Step 7. The calculation of the $Score(S)$ and $Accuracy(S)$ values for the different failure modes.

Based on the data from Step 6, Equations (13) and (14) were used to calculate the $Score(S)$ and $Accuracy(S)$ values of the different potential FMs, respectively, and the results are expressed in Table 7.

Step 8. The failure risk-ranking of the failure mode.

According to the $Score(S)$ and $Accuracy(S)$ values, the comparison rules of the spherical FS (Definition 7) were applied to the failure risk-ranking of the potential FM, and the results are expressed in Table 7.

4.7. Comparison between Different Methods

In order to verify the comprehensiveness and effectiveness of the proposed novel, flexible risk-ranking approach in the information processing and weight processing of the risk-ranking problem, Section 4 adopts a risk assessment case of the new product design of electronic equipment to verify and compare its calculation results with the RPN method, IRPN method, IFWA method, and SWAA method. These five calculation methods were calculated using the same input data (Tables 1 and 2). After the calculation, the risk-ranking results of the different calculation methods for the potential FMs are expressed in Table 8 and Figure 2. The main differences in the factors considered by the five different calculation approaches are expressed in Table 9.

Table 8. The risk-ranking results of different calculation methods for potential failure mode.

Items	RPN Method [2]		IRPN Method [46]		IFWA Method [41]		SWAA Method [43]			Proposed Method		
	RPN	Rank	IRPN	Rank	$Score(I)$	Rank	$Score(S)$	$Accuracy(S)$	Rank	$Score(S)$	$Accuracy(S)$	Rank
1	0.104	7	1.417	7	−0.042	8	−0.064	0.854	8	−0.088	0.860	9
2	0.159	5	1.750	4	0.420	3	0.172	0.862	3	−0.104	0.850	10
3	0.156	6	1.617	6	0.085	6	−0.013	0.890	6	−0.021	0.895	5
4	0.072	12	1.350	8	−0.008	7	−0.065	0.803	9	−0.056	0.802	7
5	0.186	3	1.817	3	0.441	2	0.177	0.865	2	0.194	0.867	1
6	0.246	2	1.917	2	0.343	4	0.078	0.850	4	0.073	0.849	3
7	0.067	13	1.217	13	−0.185	13	−0.142	0.851	13	−0.156	0.851	15
8	0.076	10	1.317	11	−0.098	11	−0.068	0.878	10	−0.138	0.861	14
9	0.061	17	1.183	16	−0.207	16	−0.159	0.851	16	−0.130	0.852	13
10	0.080	9	1.350	8	−0.064	9	−0.061	0.871	7	−0.044	0.865	6
11	0.061	16	1.183	16	−0.207	16	−0.159	0.851	16	−0.158	0.851	16
12	0.164	4	1.650	5	0.113	5	−0.008	0.885	5	−0.020	0.888	4
13	0.066	15	1.217	13	−0.185	13	−0.142	0.851	13	−0.159	0.851	17
14	0.345	1	2.150	1	0.504	1	0.197	0.836	1	0.084	0.846	2
15	0.066	14	1.217	13	−0.185	13	−0.142	0.851	13	−0.117	0.852	11
16	0.056	18	1.150	18	−0.230	18	−0.178	0.852	18	−0.178	0.852	18
17	0.090	8	1.350	8	−0.086	10	−0.090	0.845	11	−0.088	0.845	8
18	0.072	11	1.250	12	−0.163	12	−0.126	0.852	12	−0.126	0.852	12
19	0.007	19	0.583	19	−0.605	19	−0.506	0.839	19	−0.478	0.842	19

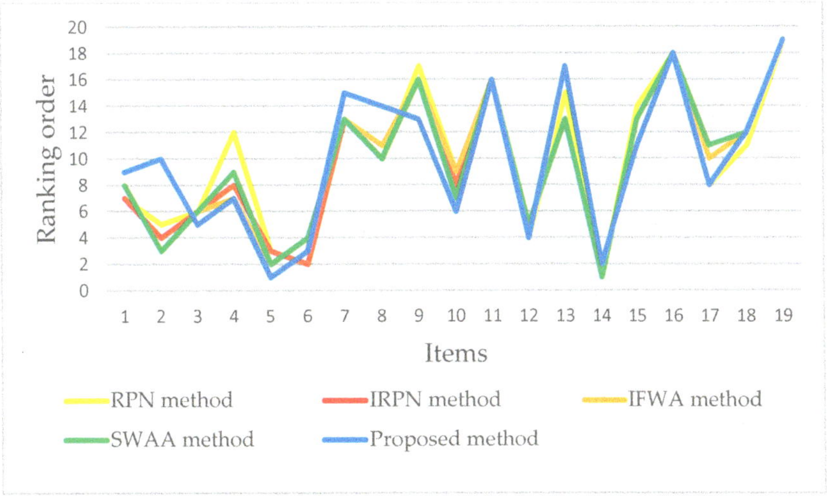

Figure 2. The risk-ranking results of different calculation methods.

Table 9. The main differences in factors considered by different calculation methods.

	Information Consideration		Measurement Scale Consideration	Objective Weight Consideration
	Intuitionistic Fuzzy Information	Spherical Fuzzy Information		
RPN method [2]	No	No	No	No
IRPN method [46]	No	No	Yes	No
IFWA method [41]	Yes	No	Yes	No
SWAA method [43]	Yes	Yes	Yes	No
Proposed method	Yes	Yes	Yes	Yes

According to the contents of Tables 3–9, the primary advantages of the proposed novel, flexible risk-ranking approach over the other calculation methods are as follows. Firstly, its information consideration is an advantage; both the RPN method and IRPN method can only process the MD information of a potential FM, and cannot handle the ID and refusal degree information, while the IFWA method can effectively grasp the intuitionistic fuzzy information that is provided by the experts on the risk factors (MD, ID, and NMD information of a potential FM). However, the IFWA method still cannot effectively deal with the spherical fuzzy information (MD, ID, NMD, and refusal degree information of a potential FM) that is provided by the experts on the risk factors. The SWAA method and the proposed method can simultaneously process the MD, ID, NMD, and refusal degree information of a potential FM and can fully consider various types of information.

Secondly, its measurement scale consideration is also advantageous. The attributes of the data distinguish the different measurement scales. The measurement scale includes the nominal scale, ordinal scale, interval scale, and ratio scale. The data attributes of the risk factors belong to the ordinal scale, and the geometric mean cannot be used for calculation. The RPN method uses the concept of the Ser, Occ, and Det risk factor products for its calculation; this violates the definition of the measurement scale and leads to biased risk-ranking results. The IRPN, IFWA, SWAA, and the proposed novel, flexible risk-ranking approach can fully consider the definition of the data attribute measurement scale and apply a more reasonable calculation mode.

The other advantage is its objective weight consideration. The RPN, IRPN, IFWA, and SWAA methods ignore the objective weighting considerations of the risk factors that are presented by the data, which may lead to distortion of the risk-ranking results. The proposed novel approach used the PSI technique to calculate the objective weights of the different risk factors to truly reflect the significance of the data.

5. Conclusions

For any industry, risk analysis and risk prioritization are key issues. Maximizing the yield rate of products under limited resources will ensure the profitability of the company and the overall customer satisfaction. Risk analysis and risk ranking must be considered as the processing modes of the information evaluation and the relative weight of the risk factors. The lack of a comprehensive evaluation information consideration or ignoring the objective weighting of the risk factors can lead to incorrect evaluation results. However, most of the risk-ranking methods cannot simultaneously handle the comprehensive evaluation information consideration, measurement scale consideration, and relative weight of the risk factors, which causes biased risk-ranking results. This study proposed a novel, flexible risk-ranking approach to obtain rigorous and correct risk-ranking results; here, the spherical FS and objective weight considerations of the risk factors are integrated to process the risk-ranking issues.

The contributions of the proposed novel, flexible risk-ranking method are as follows:

(1) The proposed novel, flexible risk-ranking method can grasp the information on the intuitionistic fuzzy evaluation of the risk factors,

(2) The proposed novel, flexible risk-ranking method can grasp the information on the spherical fuzzy evaluation of the risk factors,

(3) The proposed novel, flexible risk-ranking method considers the measurement scale of the data,
(4) The proposed novel, flexible risk-ranking method considers the relative weights of the risk factors,
(5) The IRPN, IFWA, and SWAA methods are special examples of the proposed novel, flexible risk-ranking method.

In the future, researchers can extend the concept of the proposed method to process different decision making problems such as performance evaluation, supplier selection, reliability evaluation, green energy planning, resource allocation, big data processing, and project management. In addition, future researchers can probe the impact of different subjective and objective weight combinations on their risk-ranking results.

Funding: The authors would like to thank the National Science and Technology Council, Taiwan, for financially supporting this research under Contract No. MOST 110-2410-H-145-001 and MOST 111-2221-E-145-003.

Institutional Review Board Statement: Not applicable.

Informed Consent Statement: Not applicable.

Data Availability Statement: Not applicable.

Conflicts of Interest: The author declares no conflict of interest.

References

1. Chang, K.H. A new emergency-risk-evaluation approach under spherical fuzzy-information environments. *Axioms* **2022**, *11*, 474. [CrossRef]
2. Romero-Zayas, I.; Anon, F.C.; Virosta, M.S.; del Pozo, J.C.; Montero, C.S.; Baizan, A.N.; Fuster, D. Implementation of the failure modes and effects analysis in a hospital radiopharmacy unit. *Rev. Esp. Med. Nucl. Imagen Mol.* **2022**, *41*, 300–310. [CrossRef]
3. Chang, Y.C.; Chang, K.H.; Chen, C.Y. Risk assessment by quantifying and prioritizing 5S activity for semiconductor manufacturing. *Proc. Inst. Mech. Eng. Part B J. Eng. Manuf.* **2013**, *227*, 1874–1887. [CrossRef]
4. Liu, J.W.; Wang, D.J.; Lin, Q.L.; Deng, M.K. Risk assessment based on FMEA combining DEA and cloud model: A case application in robot-assisted rehabilitation. *Expert Syst. Appl.* **2023**, *214*, 119119. [CrossRef]
5. Wen, T.C.; Chung, H.Y.; Chang, K.H.; Li, Z.S. A flexible risk assessment approach integrating subjective and objective weights under uncertainty. *Eng. Appl. Artif. Intell.* **2021**, *103*, 104310. [CrossRef]
6. Jiang, M.L.; Ren, H.P. Risk priority evaluation for power transformer parts based on intuitionistic fuzzy preference selection index method. *Math. Probl. Eng.* **2022**, *2022*, 8366893.
7. Bhattacharjee, P.; Dey, V.; Mandal, U.K. Failure mode and effects analysis (FMEA) using interval number based BWM-MCDM approach: Risk expected value (REV) method. *Soft Comput.* **2022**, *26*, 12667–12688. [CrossRef]
8. Chang, K.H. A novel risk ranking method based on the single valued neutrosophic set. *J. Ind. Manag. Optim.* **2022**, *18*, 2237–2253. [CrossRef]
9. Chang, K.H.; Wen, T.C.; Chung, H.Y. Soft failure mode and effects analysis using the OWG operator and hesitant fuzzy linguistic term sets. *J. Intell. Fuzzy Syst.* **2018**, *34*, 2625–2639. [CrossRef]
10. Chang, K.H.; Chung, H.Y.; Wang, C.N.; Lai, Y.D.; Wu, C.H. A new hybrid Fermatean fuzzy set and entropy method for risk assessment. *Axioms* **2023**, *12*, 58. [CrossRef]
11. Yu, J.X.; Zeng, Q.Z.; Yu, Y.; Wu, S.B.; Ding, H.Y.; Ma, W.T.; Gao, H.T.; Yang, J. Failure mode and effects analysis based on rough cloud model and MULTIMOORA method: Application to single-point mooring system. *Appl. Soft. Comput.* **2023**, *132*, 109841. [CrossRef]
12. Chang, K.H.; Wen, T.C. A novel efficient approach for DFMEA combining 2-tuple and the OWA operator. *Expert Syst. Appl.* **2010**, *37*, 2362–2370. [CrossRef]
13. Song, W.Y.; Ming, X.G.; Wu, Z.Y.; Zhu, B.T. Failure modes and effects analysis using integrated weight-based fuzzy TOPSIS. *Int. J. Comput. Integr. Manuf.* **2013**, *26*, 172–1186. [CrossRef]
14. Zadeh, L.A. Fuzzy sets. *Inf. Control.* **1965**, *8*, 338–353. [CrossRef]
15. Atanassov, K.T. Intuitionistic fuzzy sets. *Fuzzy Sets Syst.* **1986**, *20*, 87–96. [CrossRef]
16. Wang, W.M.; Lin, W.W.; Wen, Y.M.; Lai, X.Z.; Peng, P.; Zhang, Y.; Li, K.Q. An interpretable intuitionistic fuzzy inference model for stock prediction. *Expert Syst. Appl.* **2023**, *213*, 118908. [CrossRef]
17. Chang, K.H. A novel supplier selection method that integrates the intuitionistic fuzzy weighted averaging method and a soft set with imprecise data. *Ann. Oper. Res.* **2019**, *272*, 139–157. [CrossRef]

18. Deb, P.P.; Bhattacharya, D.; Chatterjee, I.; Saha, A.; Mishra, A.R.; Ahammad, S.H. A decision-making model with intuitionistic fuzzy information for selection of enterprise resource planning systems. *IEEE Trans. Eng. Manag.* **2022**, 1–15. (Early Access). [CrossRef]
19. Albaity, M.; Mahmood, T. Medical diagnosis and pattern recognition based on generalized dice similarity measures for managing intuitionistic hesitant fuzzy information. *Mathematics* **2022**, *10*, 2815. [CrossRef]
20. Chang, K.H.; Cheng, C.H. A risk assessment methodology using intuitionistic fuzzy set in FMEA. *Int. J. Syst. Sci.* **2010**, *41*, 1457–1471. [CrossRef]
21. Riaz, M.; Akmal, K.; Almalki, Y.; Ahmad, D. Cubic intuitionistic fuzzy topology with application to uncertain supply chain management. *Math. Probl. Eng.* **2022**, *2022*, 9631579. [CrossRef]
22. Hussain, A.; Ullah, K.; Ahmad, J.; Karamti, H.; Pamucar, D.; Wang, H.L. Applications of the multiattribute decision-making for the development of the tourism industry using complex intuitionistic fuzzy Hamy mean operators. *Comput. Intell. Neurosci.* **2022**, *2022*, 8562390. [CrossRef] [PubMed]
23. Ullah, K. Picture fuzzy maclaurin symmetric mean operators and their applications in solving multiattribute decision-making problems. *Math. Probl. Eng.* **2021**, *2021*, 1098631. [CrossRef]
24. Yager, R.R. Pythagorean membership grades in multicriteria decision making. *IEEE Trans. Fuzzy Syst.* **2014**, *22*, 958–965. [CrossRef]
25. Mahmood, T.; Ullah, K.; Khan, Q.; Jan, N. An approach toward decision-making and medical diagnosis problems using the concept of spherical fuzzy sets. *Neural Comput. Appl.* **2019**, *31*, 7041–7053. [CrossRef]
26. Ghoushchi, S.J.; Bonab, S.R.; Ghiaci, A.M.; Haseli, G.; Tomaskova, H.; Hajiaghaei-Keshteli, M. Landfill site selection for medical waste using an integrated SWARA-WASPAS framework based on spherical fuzzy set. *Sustainability* **2021**, *13*, 13950. [CrossRef]
27. Ali, J.; Naeem, M. Multi-criteria decision-making method based on complex t-spherical fuzzy Aczel-Alsina aggregation operators and their application. *Symmetry* **2023**, *15*, 85. [CrossRef]
28. Akram, M.; Zahid, K.; Kahraman, C. A PROMETHEE based outranking approach for the construction of Fangcang shelter hospital using spherical fuzzy sets. *Artif. Intell. Med.* **2023**, *135*, 102456. [CrossRef]
29. Li, Z.X.; Liu, A.J.; Miao, J.; Yang, Y. A three-phase method for spherical fuzzy environment and application to community epidemic prevention management. *Expert Syst. Appl.* **2023**, *211*, 118601. [CrossRef]
30. Jin, Y.; Hussain, M.; Ullah, K.; Hussain, A. A new correlation coefficient based on T-spherical fuzzy information with its applications in medical diagnosis and pattern recognition. *Symmetry* **2022**, *14*, 2317. [CrossRef]
31. Haseli, G.; Ghoushchi, S.J. Extended base-criterion method based on the spherical fuzzy sets to evaluate waste management. *Soft Comput.* **2022**, *26*, 9979–9992. [CrossRef]
32. Alshammari, I.; Parimala, M.; Ozel, C.; Riaz, M. Spherical linear Diophantine fuzzy TOPSIS algorithm for green supply chain management system. *J. Funct. Space* **2022**, *2022*, 3136462. [CrossRef]
33. Hussain, A.; Ullah, K.; Yang, M.S.; Pamucar, D. Aczel-Alsina aggregation operators on T-spherical fuzzy (TSF) information with application to TSF multi-attribute decision making. *IEEE Access* **2022**, *10*, 26011–26023. [CrossRef]
34. Akram, M.; Ullah, K.; Pamucar, D. Performance evaluation of solar energy cells using the interval-valued T-spherical fuzzy Bonferroni mean operators. *Energies* **2022**, *15*, 292. [CrossRef]
35. Chen, W.; Yang, B.; Liu, Y. An integrated QFD and FMEA approach to identify risky components of products. *Adv. Eng. Inform.* **2022**, *54*, 101808. [CrossRef]
36. Liang, X.B.; Ma, W.F.; Ren, J.J.; Dang, W.; Wang, K.; Nie, H.L.; Cao, J.; Yao, T. An integrated risk assessment methodology based on fuzzy TOPSIS and cloud inference for urban polyethylene gas pipelines. *J. Clean. Prod.* **2022**, *376*, 134332. [CrossRef]
37. Paramanik, A.R.; Sarkar, S.; Sarkar, B. OSWMI: An objective-subjective weighted method for minimizing inconsistency in multi-criteria decision making. *Comput. Ind. Eng.* **2022**, *169*, 108138. [CrossRef]
38. Barukab, O.; Abdullah, S.; Ashraf, S.; Arif, M.; Khan, S.A. A new approach to fuzzy TOPSIS method based on entropy measure under spherical fuzzy information. *Entropy* **2019**, *21*, 1231. [CrossRef]
39. Chang, K.H. Integrating subjective-objective weights consideration and a combined compromise solution method for handling supplier selection issues. *Systems* **2023**, *11*, 74. [CrossRef]
40. Zhang, H.H.; Xu, Z.H.; Qian, H.; Su, X.Y. Failure mode and effects analysis based on Z-numbers and the graded mean integration representation. *CMES-Comp. Model. Eng. Sci.* **2023**, *134*, 1005–1019. [CrossRef]
41. Liu, S.; Yu, W.; Liu, L.; Hu, Y.A. Variable weights theory and its application to multi-attribute group decision making with intuitionistic fuzzy numbers on determining decision maker's weights. *PLoS ONE* **2019**, *14*, e0212636. [CrossRef] [PubMed]
42. Mathew, M.; Chakrabortty, R.K.; Ryan, M.J. A novel approach integrating AHP and TOPSIS under spherical fuzzy sets for advanced manufacturing system selection. *Eng. Appl. Artif. Intell.* **2020**, *96*, 103988. [CrossRef]
43. Gundogdu, F.K.; Kahraman, C. Spherical fuzzy sets and spherical fuzzy TOPSIS method. *J. Intell. Fuzzy Syst.* **2019**, *36*, 337–352. [CrossRef]
44. Maniya, K.; Bhatt, M.G. A selection of material using a novel type decision-making method: Preference selection index method. *Mater. Des.* **2010**, *31*, 1785–1789. [CrossRef]

45. Aguirre, P.A.G.; Perez-Dominguez, L.; Luviano-Cruz, D.; Noriega, J.J.S.; Gomez, E.M.; Callejas-Cuervo, M. PFDA-FMEA, an integrated method improving FMEA assessment in product design. *Appl. Sci.* **2021**, *11*, 1406. [CrossRef]
46. Ciani, L.; Guidi, G.; Patrizi, G. A critical comparison of alternative risk priority numbers in failure modes, effects, and criticality analysis. *IEEE Access* **2019**, *7*, 92398–92409. [CrossRef]

applied sciences

MDPI

Review

Comparing Vision Transformers and Convolutional Neural Networks for Image Classification: A Literature Review

José Maurício , Inês Domingues and Jorge Bernardino *

Polytechnic of Coimbra, Coimbra Institute of Engineering (ISEC), Rua Pedro Nunes, 3030-199 Coimbra, Portugal; a2018056151@isec.pt (J.M.); ines.domingues@isec.pt (I.D.)
* Correspondence: jorge@isec.pt

Abstract: Transformers are models that implement a mechanism of self-attention, individually weighting the importance of each part of the input data. Their use in image classification tasks is still somewhat limited since researchers have so far chosen Convolutional Neural Networks for image classification and transformers were more targeted to Natural Language Processing (NLP) tasks. Therefore, this paper presents a literature review that shows the differences between Vision Transformers (ViT) and Convolutional Neural Networks. The state of the art that used the two architectures for image classification was reviewed and an attempt was made to understand what factors may influence the performance of the two deep learning architectures based on the datasets used, image size, number of target classes (for the classification problems), hardware, and evaluated architectures and top results. The objective of this work is to identify which of the architectures is the best for image classification and under what conditions. This paper also describes the importance of the Multi-Head Attention mechanism for improving the performance of ViT in image classification.

Keywords: transformers; Vision Transformers (ViT); convolutional neural networks; multi-head attention; image classification

Citation: Maurício, J.; Domingues, I.; Bernardino, J. Comparing Vision Transformers and Convolutional Neural Networks for Image Classification: A Literature Review. *Appl. Sci.* **2023**, *13*, 5521. https://doi.org/10.3390/app13095521

Academic Editor: Yu-Dong Zhang

Received: 20 March 2023
Revised: 19 April 2023
Accepted: 26 April 2023
Published: 28 April 2023

1. Introduction

Nowadays, transformers have become the preferred models for performing Natural Language Processing (NLP) tasks. They offer scalability and computational efficiency, allowing models to be trained with more than a hundred billion parameters without saturating model performance. Inspired by the success of the transformers applied to NLP and assuming that the self-attention mechanism could also be beneficial for image classification tasks, it was proposed to use the same architecture, with few modifications, to perform image classification [1]. The author's proposal was an architecture, called Vision Transformers (ViT), which consists of breaking the image into 2D patches and providing this linear sequence of patches as input to the model. Figure 1 presents the architecture proposed by the authors.

In contrast to this deep learning architecture, there is another very popular tool for processing large volumes of data called Convolutional Neural Networks (CNN). The CNN is an architecture that consists of multiple layers and has demonstrated good performance in various computer vision tasks such as object detection or image segmentation, as well as NLP problems [2]. The typical CNN architecture starts with convolutional layers that pass through the kernels or filters, from left to right of the image, extracting computationally interpretable features. The first layer extracts low-level features (e.g., colours, gradient orientation, edges, etc.), and subsequent layers extract high-level features. Next, the pooling layers reduce the information extracted by the convolutional layers, preserving the most important features. Finally, the fully-connected layers are fed with the flattened output of the convolutional and pooling layers and perform the classification. Its architecture is shown in Figure 2.

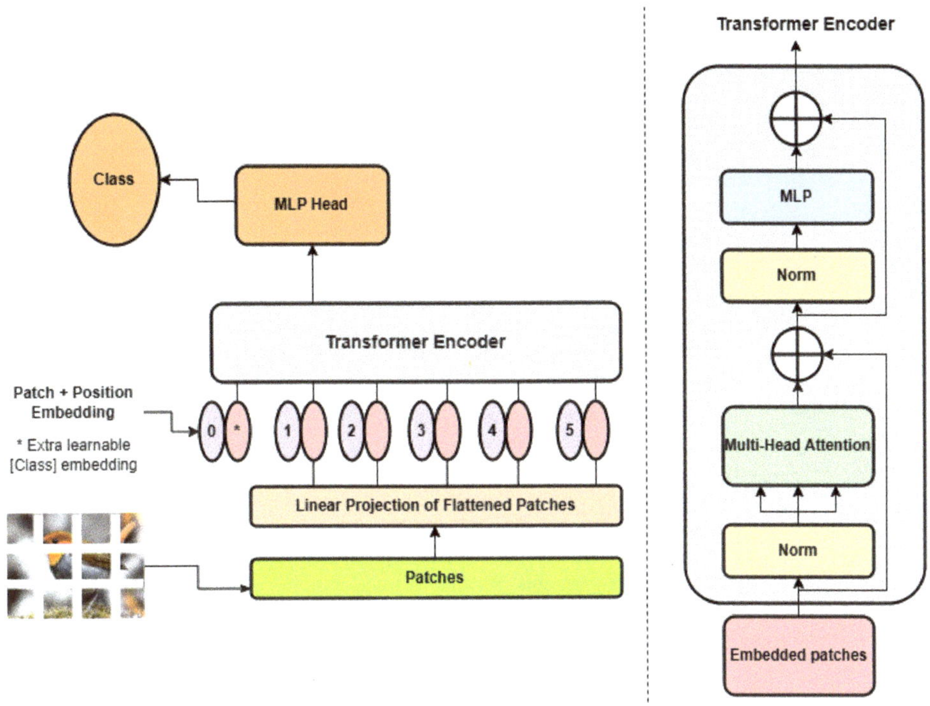

Figure 1. Example of an architecture of the ViT, based on [1].

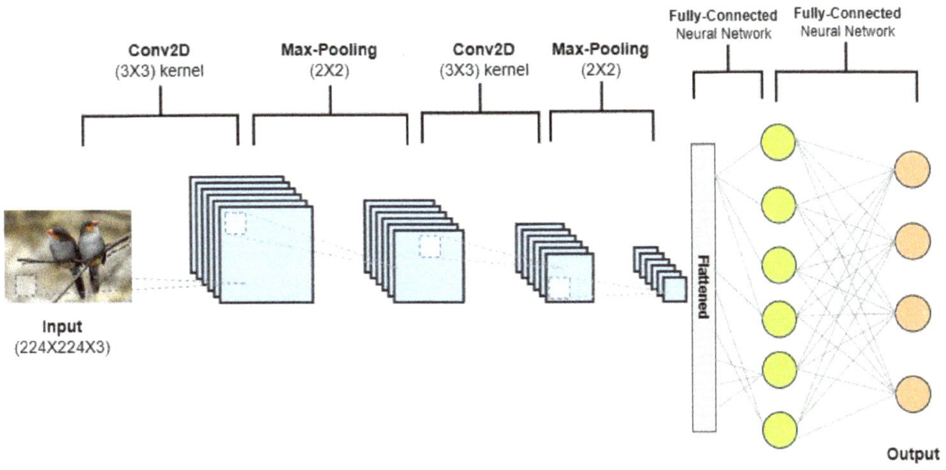

Figure 2. Example of an architecture of a CNN, based on [2].

With the increasing interest in Vision Transformers as a novel architecture for image recognition tasks, and the established success of CNNs in image classification, this work aims to review the state of the art in comparing Vision Transformers (ViT) and Convolutional Neural Networks (CNN) for image classification. Transformers offer advantages such as the ability to model long-range dependencies, adapt to different input sizes, and the potential for parallel processing, making them suitable for image tasks. However, Vision Transformers also face challenges such as computational complexity, model size, scalability to large datasets, interpretability, robustness to adversarial attacks, and generalization performance. These points highlight the importance of comparing ViTs with older and established CNN models.

The overall goal of this work is to understand what conditions have the most influence on the performance of the two Deep Learning architectures, and what characteristics differ between the two architectures, that allow them to perform differently for the same objective. Some of the aspects that will be compared include datasets considerations, robustness, performance, evaluation, interpretability, and architecture. Specifically, we aim to answer the following research questions:

RQ1—Can the ViT architecture have a better performance than the CNN architecture, regardless of the characteristics of the dataset?
RQ2—What influences CNNs that do not to perform as well as ViTs?
RQ3—How does the Multi-Head Attention mechanism, which is a key component of ViTs, influence the performance of these models in image classification?

In order to address these research questions, a literature review was conducted by searching various databases such as Google Scholar, Scopus, Web of Science, ACM Digital Library, and Science Direct using specific search terms. This paper presents the results of this review and analyses the methodologies and findings from the selected papers.

The rest of this paper is structured as follows. Section 2 describes the research methodology and search results. Section 3 presents the knowledge, methodology, and results found in the selected documents. Section 4 provides a brief overview of the reviewed papers and attempts to answer the three research questions. Section 5 discusses threats to the validity of the research. Section 6 overviews the strengths and weaknesses of each architecture and suggests future research directions, and Section 7 presents the main conclusions of this work.

2. Research Methodology

The purpose of a literature review is to evaluate, analyse and summarize the existing literature on a specific research topic, in order to facilitate the emergence of theoretical frameworks [3]. In this literature review, the aim is to synthesize the knowledge base, critically evaluate the methods used and analyze the results obtained in order to identify the shortcomings and improve the two aforementioned deep learning architectures for image classification. The methodology for conducting this literature review is based on the guidelines presented in [3,4].

2.1. Data Sources

ACM Digital Library, Google Scholar, Science Direct, Scopus, and Web of Science, were chosen as the data sources to extract the primary studies. The number of results found after searching papers in each of the data sources is shown in Table 1.

Table 1. Data sources and the number of obtained results.

Data Source	Number of Results	Number of Selected Papers
ACM Digital Library	19,159	1
Google Scholar	10,700	10
Science Direct	1437	3
Scopus	55	2
Web of Science	90	1

2.2. Search String

The research questions developed for this paper served as the basis for the search strings utilized in each of the data sources. Table 2 provides a list of the search strings used in each electronic database.

Table 2. Data sources and used search string.

Data Source	Search String
ACM Digital Library	((Vision Transformers) AND (convolutional neural networks) AND (images classification) AND (comparing))
Google Scholar	((ViT) AND (CNN) AND (Images Classification) OR (Comparing) OR (Vision Transformers) OR (convolutional neural networks) OR (differences))
Science Direct	((Vision Transformers) AND (convolutional neural networks) AND (images classification) AND (comparing))
Scopus	((ViT) AND (CNN) AND (comparing))
Web of Science	((ViT) AND (CNN) AND (comparing))

2.3. Inclusion Criteria

The inclusion criteria set to select the papers were that the studies were recent, had been written in English, and were published between January 2021 and December 2022. This choice of publication dates is based on the fact that ViTs were not proposed until the end of 2020 [1]. In addition, the studies had to demonstrate a comparison between CNNs and ViTs for image classification and could use any pre-trained model of the two architectures. Studies that presented a proposal for a hybrid architecture, where they combined the two architectures into one, were also considered. The dataset used during the studies did not have to be a specific one, but it had to be a dataset of images that allowed classification using both deep learning architectures.

2.4. Exclusion Criteria

Studies that oriented their research on using only one of the two deep learning architectures (i.e., Vision Transforms, or Convolutional Neural Networks) were excluded. Additionally, papers that were discovered to be redundant when searches were conducted throughout the chosen databases were eliminated. It was also defined that one of the exclusion criteria will be that the papers would have more than seven citations.

In summary, with the application of these criteria, 10,690 papers were excluded from the Google Scholar database, 89 papers from Web of Science, 53 papers from Scopus, 19,158 papers from ACM Digital Library, and 1434 papers from Science Direct.

2.5. Results

After applying the inclusion and exclusion criteria to the papers obtained in each of the electronic databases, seventeen (17) papers were selected for the literature review. Table 3 lists all the papers selected for this work, the year of publication and the type of publication.

Table 3. List of selected studies.

Ref.	Title	Year	Type
[5]	Adversarial Robustness Comparison of Vision Transformer and MLP-Mixer to CNNs	2021	Conference
[6]	Are Transformers More Robust Than CNNs?	2021	Conference
[7]	Detecting Pneumonia using Vision Transformer and comparing with other techniques	2021	Conference
[8]	Do Vision Transformers See Like Convolutional Neural Networks?	2021	Conference
[9]	Vision Transformer for Classification of Breast Ultrasound Images	2021	Conference
[10]	ConvNets vs. Transformers: Whose Visual Representations are More Transferable?	2021	Conference
[11]	A vision transformer for emphysema classification using CT images	2021	Journal
[12]	Comparing Vision Transformers and Convolutional Nets for Safety Critical Systems	2022	Conference
[13]	Convolutional Nets Versus Vision Transformers for Diabetic Foot Ulcer Classification	2022	Conference
[14]	Convolutional Neural Network (CNN) vs Vision Transformer (ViT) for Digital Holography	2022	Conference
[15]	Cross-Forgery Analysis of Vision Transformers and CNNs for Deepfake Image Detection	2022	Conference
[16]	Traffic Sign Recognition with Vision Transformers	2022	Conference
[17]	An improved transformer network for skin cancer classification	2022	Journal
[18]	CNN and transformer framework for insect pest classification	2022	Journal
[19]	Single-layer Vision Transformers for more accurate early exits with less overhead	2022	Journal
[20]	Vision transformer-based autonomous crack detection on asphalt and concrete surfaces	2022	Journal
[21]	Vision Transformers for Weeds and Crops Classification of High-Resolution UAV Images	2022	Journal

Figure 3 shows the distribution of the selected papers by year of publication.

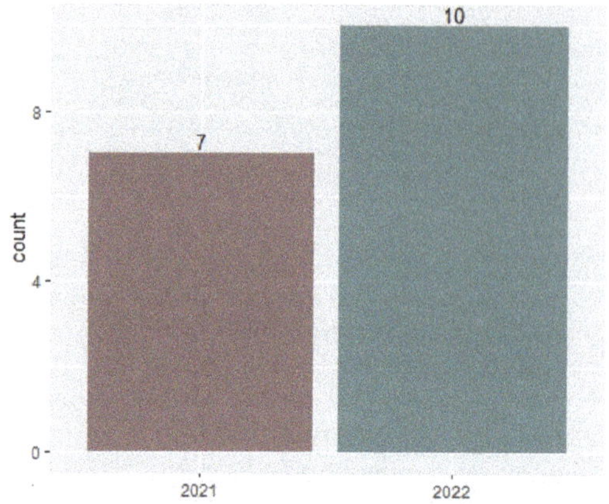

Figure 3. Distribution of the selected studies by years.

Figure 4 shows the distribution of the selected studies by application area. In the figure, most of the papers are generic in their application area. In these papers without a specific application area, the authors try to better understand the characteristics of the two architectures. For example, between CNNs and ViTs, the authors have tried to understand which of the architectures is more transferable. If architectures based on transformers are more robust than CNNs. And if the ViT will be able to see the same information as CNN with a different architecture. Within the health domain, some studies have been developed in different sub-areas, such as breast cancer, to show that ViT can be better than CNNs. The figure also shows that some work has been done, albeit to a lesser extent, in other application areas. Agriculture stands out with two papers comparing ViTs with CNNs.

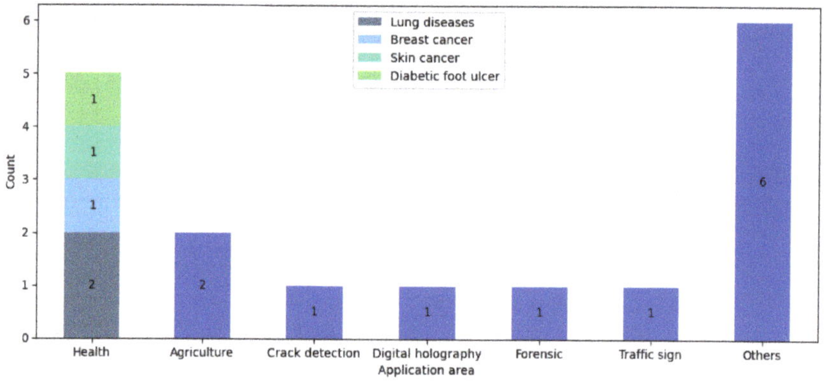

Figure 4. Distribution of the selected studies by application area.

3. Findings

An overview of the studies selected through the research methodology is shown in Table 4. This information summarizes the authors' approach, the findings, and other architectures that were used to build a comparative study. Therefore, to address the research questions, this section will offer an overview of the data found in the collected papers.

In the study developed in [12], the authors aimed to compare the two architectures (i.e., ViT and CNN), as well as the creation of a hybrid model that corresponded to the combination of the two. The experiment was conducted using the ImageNet dataset and perturbations were applied to the dataset images. It was concluded that ViT can perform better and be more resilient on images with natural or adverse disturbances than CNN. It was also found in this work that the combination of the two architectures results in a 10% improvement in accuracy (Acc).

The work done in [14] aimed to compare Vision Transformers (ViT) with Convolutional Neural Networks (CNN) for digital holography, where the goal was to reconstruct amplitude and phase by extracting the distance of the object from the hologram. In this work, DenseNet201, DenseNet169, EfficientNetB4, EfficientNetB7, ViT-B/16, ViT-B32 and ViT-L/16 architectures were compared with a total of 3400 images. They were divided into four datasets, original images with or without filters, and negative images with or without filters. The authors concluded that ViT despite having an accuracy like CNN, was more robust because, due to the self-attention mechanism, it can learn the entire hologram rather than a specific area.

The authors in [7] studied the performance of ViT in comparison with other architectures to detect pneumonia, through chest X-ray images. Therefore, a ViT model, a CNN network developed by the authors and the VGG-16 network were used for the study which focussed on a dataset with 5856 images. After the experiments performed, the authors concluded that ViT was better than CNN with 96.45% accuracy, 86.38% validation accuracy, 10.8% loss and 18.25% validation loss. In this work, it was highlighted that ViT has a self-attention mechanism that allows splitting the image into small patches that are trainable, and each part of the image can be given an importance. However, the attention mechanism as opposed to the convolutional layers makes ViT's performance saturate fast when the goal is scalability.

In the study [21] the goal was to compare ViT with state-of-art CNN networks to classify UAV images to monitor crops and weeds. The authors compared the influence of the size of the training dataset on the performance of the architectures and found that ViT performed better with fewer images than CNN networks in terms of F1-Score. They concluded that ViT-B/16 was the best model to do crop and weed monitoring. In comparison with CNN networks, ViT could better learn the patterns of images in small datasets due to the self-attention mechanism.

In the scope of lung diseases, the authors in [11] investigated the performance of ViT models to automatically classify emphysema subtypes through Computed Tomography (CT) images in comparison with CNN networks. In this study, they performed a comparative study between the two architectures using a dataset collected by the authors (3192 patches) and a public dataset of 168 patches taken from 115 HRCT slides. In addition to this, they also verified the importance of pre-trained models. They concluded that ViT failed to generalize when trained with fewer images, because when comparing the pre-training accuracy with 91.27% on the training and 70.59% on the test.

In the work in [9], a comparison between state-of-the-art CNNs and ViT models for Breast ultrasound image classification was developed. The study was performed with two different datasets: the first containing 780 images and the second containing 163 images. The following architectures were selected for the study: ResNet50, VGG-16, Inception, NASNET, ViT-S/32, ViT-B/32, ViT-Ti/16, R + ViT-Ti/16 and R26 + ViT-S/16. ViT models were found to perform better than CNN networks for image classification. The authors also highlighted that ViT models could perform better when they were trained with a small dataset, because via the attention mechanism, it was possible to collect more information from different patches, instead of collecting information from the image.

Benz et al., in [5], compared ViT models, with the MLP-Mixer architecture and with CNNs. The goal was to evaluate which architecture was more robust in image classification. The study consisted of generating perturbations and adverse examples in the images and understanding which of the architectures was most robust. However, this study did not aim to analyse the causes. Therefore, the authors concluded that ViT were more robust than CNNs to adversarial attacks and from a features perspective CNN networks were more sensitive to high-frequency features. It was also described that the shift-variance property of convolutional layers may be at the origin of the lack of robustness of the network in the classification of images that have been transformed.

The authors in [15] performed an analysis between ViT and CNN models aimed at detecting deepfake images. The experiment consisted in using the ForgeryNet dataset with 2.9 million images and 220 thousand video clips, together with three different image manipulation techniques, where they tried to train the models with real and manipulated images. By training the ViT-B model and the EfficientNetV2 network the authors demonstrated that the CNN network could generalize better and obtain higher training accuracy. However, ViT could have better generalization, reducing the bias in the identification of anomalies introduced by one or more different techniques to introduce anomalies.

Chao Xin et al. [17] aimed to compare their ViT model with CNN networks and with another ViT model to perform image classification to detect skin cancer. The experiment conducted by the authors used a public HAM10000 dataset with dermatoscopic skin cancer images and a clinical dataset collected through dermoscopy. In this study, a multi-scale image and the overlapping sliding window were used to serialize the images. They also used contrastive learning to improve the similarity of different labels and minimize the similarity in the same label. Thus, the ViT model developed was better for skin cancer classification using these techniques. However, the authors also demonstrated the effectiveness of balancing the dataset on the model performance, but they did not present the F1-Score values before the dataset is balanced to verify the improvement.

The authors in [19] aimed to study if ViT models could be an alternative to CNNs in time-critical applications. That is, for edge computing instances and IoT networks, applications using deep learning models consume multiple computational resources. The experiment used pre-trained networks such as ResNet152, DenseNet201, InceptionV3, and SL-ViT with three different datasets in the scope of images, audio, and video. They concluded that the ViT model introduced less overhead and performed better than the architectures used. It was also shown that increasing the kernel size of convolutional layers and using dilated convolutional caused a reduction in the accuracy of a CNN network.

In a study carried out in [20], the authors tried to find in ViTs an alternative solution to CNN networks for asphalt and concrete crack detection. The authors concluded that ViTs, due to the self-attention mechanism, had better performance in crack detection images with intense noise. CNN networks in the same images suffered from a high number of false negative rates, as well as the presence of biases in image classification.

Haolan Wang in [16] aimed to analyse eight different Vision Transformers and compare them with the performance of a pre-trained CNN network and without the pre-trained parameters to perform traffic signal recognition in autonomous driving systems. In this study, three different datasets with images of real-world traffic signals were used. This allowed the authors to conclude that the pre-trained DenseNet161 network had a higher accuracy than the ViT models to do traffic sign recognition. However, it was found in this work that ViT models performed better than the DenseNet161 network without the pre-trained parameters. From this work, it was also possible to conclude that the ViT models with a total number of parameters equal to or greater than the CNN networks, used during the experiment, had a shorter training time.

The work done in [13] compared CNN networks with Vision Transformers models for the classification of Diabetic Foot Ulcer images. For the study, the authors decided to use the following architectures: Big Image Transfer (BiT), EfficientNet, ViT-base and Data-efficient Image Transformers (DeIT) upon a dataset composed of 15,683 images. A further aim of this study to compare the performance of deep learning models using Stochastic Gradient Descent (SGD) [22] with Sharpness-Aware Optimization (SAM) [23,24]. These two tools are optimizers that seek to minimize the value of the loss function, improving the generalization ability of the model. However, SAM minimizes the value of the loss function and the sharpness loss, looking for parameters in the neighbourhood with a low loss. Therefore, this work concluded that the SAM optimizer originated an improvement in the values of F1-Score, AUC, Recall and Precision in all the architectures used. However, the authors did not present the training and test values that allow for evaluating the improvement in the generalization of the models. Therefore, the BiT-ResNetX50 model with the SAM optimizer obtained the best performance for the classification of Diabetic Foot Ulcer images with F1-Score = 57.71%, AUC = 87.68%, Recall = 61.88%, and Precision = 57.74%.

The authors in [18] performed a comparative study between ViT models and CNN networks used in the state of the art with a model developed by them, where they combined CNN and transformers to perform insect pest recognition to protect agriculture worldwide. This study involved three public datasets: the IP102 dataset, the D0 dataset and Li's dataset. The algorithm created by the authors consisted of using the sequence of inputs formed by the CNN feature maps to make the model more efficient, and a flexible attention-based classification head was implemented to use the spatial information. Comparing the results obtained, the proposed model obtained a better performance in insect pest recognition with an accuracy of 74.897%. This work demonstrated that fine-tuning worked better on Vision Transformers than CNN, but on the other hand, this caused the number of parameters, the size, and the inference time of the model to increase significantly with respect to CNN networks. Through their experiments, the authors also demonstrated the advantage of using decoder layers in the proposed model to perform image classification. The greater the number of decoder layers, the greater the accuracy value of the model. However, this increase in the number of decoder layers increased the number of parameters, the size, and the inference time of the model. In other words, the architecture to process the images consumes far greater computational resources, which may not compensate for the increase in accuracy value with few layers. In the case of this study, the increase from one layer to three decoder layers represented only an increase of 0.478% in the accuracy value.

Several authors in [6,8,10] went deeper into the investigation and aimed to understand how the learning process of Vision Transformers works if ViT could be more transferable and better understand if the transform-based architecture were more robust than CNNs. In this sense, the authors in [8] intended to analyse the internal representations of ViT and CNN structures in image classification benchmarks and found differences between them. One of the differences was that ViT has greater similarity between high and low layers, while CNN architecture needs more low layers to compute similar representations in smaller datasets. This is due to the self-attention layers implemented in ViT, which allows it to aggregate information from other spatial locations, vastly different from the fixed field sizes in CNNs. They also observed that ViTs in the lower, self-attention layers can access information from local heads (small distances) and global heads (large distances). Whereas CNNs have access to information locally in the lower layers. On the other hand, the authors in [10] systematically analysed the transfer learning capacity in the two architectures. The study was conducted by comparing the performance of the two architectures on single-task and multi-task learning problems, using the ImageNet dataset. Through this study, the authors concluded that the transform-based architecture contained more transferable representations compared to convolutional networks for fine-tuning, presenting better performance and robustness in multi-task learning problems.

In another study carried out in [6], the goal was to prove if ViT were more robust than CNN as the most recent studies have shown. The authors developed their work comparing the robustness of the two architectures using two different types of perturbations: adversarial samples, which consists in evaluating the robustness of deep learning architectures in images with human-caused perturbations (i.e., data augmentation) and out-of-distribution samples, which consists in evaluating the robustness of the architectures in benchmarks of classification images. Through this experiment, it was demonstrated that by replacing the activation function ReLU by the activation function of transformer-based architecture (i.e., GELU) the CNN network was more robust than ViT in adversarial samples. In this study, it was also demonstrated that CNN networks were more robust than ViT in patch-based attacks. However, the authors concluded that the self-attention mechanism was the key to the robustness of the transformer-based architecture in most of the experiments performed.

Table 4. Overview of selected studies.

Ref.	Datasets	Images Size	Number of Classes	Hardware	Evaluated Architectures	Best Architecture	Best Results
[5]	ImageNet-1K (more than 1.431 M images) for training and ImageNet-C for validation	224 × 224	2	N/A	ViT-B/16, ViT-L/16, Mixer-B/16, Mixer-B/16, RN18 (SWSL), RN50 (SWSL), RN18 and RN50	ViT-L/16	82.89% of Acc
[6]	ImageNet-A; ImageNet-C and Stylized ImageNet	224 × 224	N/A	N/A	ResNet50 and DeiT-S	N/A	N/A
[7]	5856 images collected by X-ray	250 × 250 to ViT 224 × 224 to CNN	2	Intel Core i5-8300H 2.30 GHz	ViT, CNN and VGG16	ViT	96.45% Acc, 86.38% val. Acc, 10.92% loss and 18.25% val. Loss
[8]	ImageNetILSVRC 2012 (1.78 M images)	224 × 224	1000	N/A	ViT-B/32, ViT-L/16, ViT-H/14, ResNet50 and ResNet152	N/A	N/A
[9]	Public dataset 1 [25] with 780 images; Public dataset 2 [26] with 163 images	224 × 224	3	N/A	ViT-S/32, ViT-B/32, ViT-Ti/16, R26 + S/16, R + Ti/16, VGG, Inception and NASNET	ViT-B/32	86.7% Acc and 95% AUC
[10]	Flower 102 (4080 to 11,016 images); CUB 200 (11,788 images); Indoor 67 (15,620 images); NY Depth V2 (1449 images); WikiArt; COVID-19 Image Data Collection (700 images); Caltech101 (9146 images); FG-NET (1002 images)	384 × 384; 224 × 224; 300 × 300	40 to 102	N/A	R-101 × 3, R-152 × 4, ViT-B/16, ViT-L/16, and Swim-B	N/A	N/A
[11]	3192 images collected by CT and 160 images of a public dataset with CT biomarkers	61 × 61	4	Intel Core i7-9700 3.0 GHz, 326 GB RAM; NVIDIA GeForce RTX 2080 Ti (116 GB DDR6)	AlexNet, VGG-16, InceptionV3, MobileNetV2, ResNet34, ResNet50 and ViT	ViT	95.95% Acc
[12]	ImageNet-C benchmark	224 × 224	2	NVIDIA Quadro A6000	ViT-L/16, CNN, hybrid model (BiT-M + ResNet152 × 4	Hybrid model	99.20% Acc
[13]	Dataset provided in DFUC 2021 challenge (15,683 images)	224 × 224	4	NVIDIA GeForce RTX 3080, 10 GB memory	EfficientNetB3, BiT-ResNeXt50, ViT-B/16 and DeiT-S/16	BiT-ResNeXt50	88.49% AUC, 61.53% F1-Score, 65.59% recall and 60.53% precision
[14]	3400 images collected by holographic camera	512 × 512	10	NVIDIA V100	EfficientnetB7, Densenet169 and ViT-B/16	ViT-B/16	99% Acc
[15]	ForgeryNet with 2.9 M images	N/A	2	N/A	EfficientNetV2 and ViT-B	N/A	N/A
[16]	German dataset (51,830 images); Indian dataset (1976 images); Chinese dataset (18,168 images)	128 × 128	15, 43 and 103	AMD Ryzen 7 5800H; NVIDIA GeForce RTX 3070	DenseNet161, ViT, DeepViT, MLP-Mixer, CvT, PiT, CaiT, CCT, CrossViT and Twins-SVT	CCT	99.04% Acc

Table 4. *Cont.*

Ref.	Datasets	Images Size	Number of Classes	Hardware	Evaluated Architectures	Best Architecture	Best Results
[17]	HAM10000 dataset (10,015 images); 1016 images collected by dermoscopy	224 × 224	3	Intel i7; 2x NVIDIA RTX 3060, 12 GB	MobileNetV2, ResNet50, InceptionV2, ViT and Proposed ViT model	Proposed ViT model	94.10% Acc, 94.10% precision and 94.10% F1-Score
[18]	IP102 dataset (75,222 images); D0 dataset (4508 images; Li's dataset (5629 images)	224 × 224; 480 × 480	10 to 102	Train: Intel Xeon; 8x NVIDIA Tesla V100, 256 GB. Test: Intel Core; NVIDIA GTX 1060 Ti, 16 GB	ResNet, EfficientNetB0, EfficientB1, RepVGG, VGG-16, ViT-L/16 and hybrid model proposed	Proposed hybrid model	99.47% Acc on the D0 dataset and 97.94% Acc on the Li's dataset
[19]	CIFAR-10 and CIFAR-100 (6000 images); Speech commands (100,503 1-second audio clips); GTZAM (1,00,030-second audio clips); DISCO (1935 images)	224 × 224 px; 1024 × 576 px. Spectrograms: 229 × 229 samples; 512 × 256 samples	10 to 100	4x NVIDIA 2080 Ti	SL-ViT, ResNet152, DenseNet201 and InceptionV3	SL-ViT	71.89% Acc
[20]	CrackTree260 (260 images); Ozegenel (458 images); Lab's on dataset (80,000 images)	256 × 256; 448 × 448	2	N/A	TransUNet, U-Net, DeepLabv3+ and CNN + ViT	CNN + ViT	99.55% Acc and 99.57% precision
[21]	10,265 images collected by Pilgrim technologies UAV with Sony ILCE-7R-36 mega pixels	64 × 64	5	Intel Xeon E5-1620 V4 3.50 GHz with 8 processor, 16 GB RAM; NVIDIA Quadro M2000	ViT-B/16, ViT-B/32, EfficientNetB, EfficientNetB1 and ResNet50	ViT-B/16	99.8% Acc

4. Discussion

The results can be summarized as follows. In [12], ViTs were found to perform better and be more resilient to images with natural or adverse disturbances compared to CNNs. Another study [14] concluded that ViTs are more robust in digital holography because they can access the entire hologram rather than just a specific area, giving them an advantage. ViTs have also been found to outperform CNNs in detecting pneumonia in chest X-ray images [7] and in classifying UAV images for crop and weed monitoring with small datasets [21]. However, it has been noted that ViT performance may saturate if scalability is the goal [7]. In a study on classifying emphysema subtypes in CT images [11], ViTs were found to struggle with generalization when trained on fewer images. Nevertheless, ViTs were found to outperform CNNs in breast ultrasound image classification, especially with small datasets [9]. Another study [5] found that ViTs are more robust to adversarial attacks and that CNNs are more sensitive to high-frequency features. The authors in [15] found that CNNs had higher training accuracy and better generalization, but ViTs showed potential to reduce bias in anomaly detection. In [17], the authors claimed that the ViT model showed better performance for skin cancer classification. ViTs have also been shown to introduce less overhead and perform better for time-critical applications in edge computing and IoT networks [19]. In [20], the authors investigated the use of ViTs for asphalt and concrete crack detection and found that ViTs performed better due to the self-attention mechanism, especially in images with intense noise and biases. Wang [16] found that a pre-trained CNN network had higher accuracy, but the ViT models performed better than the non-pre-trained CNN network and had a shorter training time. The authors in [13] used several models for diabetic foot ulcer image classification and compared SGD and SAM optimizers, concluding that the SAM optimizer improved several evaluation metrics. In [18], the authors showed that fine-tuning performed better on ViT models than CNNs for insect pest recognition.

Therefore, based on the information gathered from the selected papers, we attempt to answer the research questions posed in Section 1:

RQ1—Can the ViT architecture have a better performance than the CNN architecture, regardless of the characteristics of the dataset?

The literature review shows that ViT in image processing can be more efficient in smaller datasets due to the increase of relations created between images through the self-attention mechanism. However, it is also shown that if ViT trained with little data will have less generalization ability and worse performance compared to CNN's.

RQ2—What influences the CNNs that do not allow them to perform as well as the ViTs?

Shift-invariance is a limitation of CNN that makes the same architecture not have a satisfactory performance because the introduction of noise in the input images makes the same architecture unable to get the maximum information from the central pixels. However, the authors in [27] propose the addition of an anti-aliasing filter which combines blurring with subsampling in the Convolutional, MaxPooling and AveragePooling layers. Demonstrating through the experiment carried out that the application of this filter originates a greater generalization capacity and an increase in the accuracy of CNN. Furthermore, increasing the kernel size in convolutional layers and using dilated convolution have been shown as limitations that deteriorate the performance of CNNs against ViTs.

RQ3—How does the Multi-Head Attention mechanism, which is a key component of ViTs, influence the performance of these models in image classification?

The Attention mechanism is described as the mapping of a query and a set of key-value pairs to an output, the output being the result of a weighted sum of the values, in which the weight given is calculated, through the query with the corresponding key by a compatibility function. Multi-head Attention mechanism instead of performing a single attention function will perform multiple projections of attention functions [28]. This mechanism improves the ViT architecture because it allows it to extract more information from each pixel of the images that have been placed inside the embedding. In addition, this mechanism can have better performance if the images have more secondary elements that illustrate the central element. And since this mechanism performs several computations in parallel, it reduces the computational cost [29].

Overall, ViTs have shown promising performance compared to CNNs in various applications, but there are limitations and factors that can affect their performance, such as dataset size, scalability, and pre-training accuracy.

5. Threats to Validity

This section discusses internal and external validity threats. The validity of the entire process performed in this study is demonstrated and how the results of this study can be replicated in other future experiments.

In this literature review, different search strings were used in each of the selected data sources, resulting in different results from each source. This approach may introduce a bias into the validation of the study, as it makes it difficult to draw conclusions about the diversity of studies obtained by replicating the same search. In addition, the maturity of the work was identified as an internal threat to validity, as the ViT architecture is relatively new and only a limited number of research projects have been conducted using it. In order to draw more comprehensive conclusions about the robustness of ViT compared to CNN, it is imperative that this architecture is further disseminated and deployed, thereby making more research available for analysis.

In addition to these threats, this study did not use methods that would allow to quantitatively and qualitatively analyse the results obtained in the selected papers. This may bias the validity of this review in demonstrating which of the deep learning architectures is more efficient in image processing.

The findings obtained in this study could be replicated in other future research in image classification. However, the results obtained may not be the same as those described by the selected papers because it has been proven that for different problems and different methodologies used, the results are different. In addition, the authors do not describe in sufficient detail all the methodologies they used, nor the conditions under which the experiment was performed.

6. Strengths, Limitations, and Future Research Directions

The review made it possible to identify not only the strengths of each architecture (outlined in Section 6.1), but also their potential for improvement (described in Section 6.2). Future research directions were also derived from this and are presented in Section 6.3.

6.1. Strengths

Both CNNs and ViTs have their own advantages, and some common ones. This section will explore these in more detail, including considerations on the Datasets, Robustness, Performance optimization, Evaluation, Explainability and Interpretability, and Architectures.

6.1.1. Dataset Considerations

CNNs have been widely used and extensively studied for image-related tasks, resulting in a rich literature, established architectures, and pre-trained models, making them accessible and convenient for many datasets. On the other hand, ViTs can process patches in parallel, which can lead to efficient computation, especially for large-scale datasets, and allow faster training and inference. ViTs can also handle images of different sizes and aspect ratios without losing resolution, making them more scalable and adaptable to different datasets and applications.

6.1.2. Robustness

CNNs are inherently translation-invariant, making them robust to small changes in object position or orientation within an image. The main advantage of ViTs is their ability to effectively capture global contextual information through the self-attention mechanism, enabling them to model long-range dependencies and contextual relationships, which can improve robustness in tasks that require understanding global context. ViTs can also adaptively adjust the receptive fields of the self-attention mechanism based on input data, allowing them to better capture both local and global features, making them more robust to changes in scale, rotation, or perspective of objects.

Both architectures can be trained using data augmentation techniques, such as random cropping, flipping, and rotation, which can help improve robustness to changes in input data and reduce overfitting. Another technique is known as adversarial training, where they are trained on adversarial examples: perturbed images designed to confuse the model, to improve its ability to handle input data with adversarial perturbations. Combining models, using ensemble methods, such as bagging or boosting, can also improve robustness by exploiting the diversity of multiple models, which can help mitigate the effects of individual model weaknesses.

6.1.3. Performance Optimization

CNNs can be effectively compressed using techniques such as pruning, quantization, and knowledge distillation, reducing model size and improving inference efficiency without significant loss of performance. They are also well-suited for hardware optimization, with specialized hardware accelerators (e.g., GPUs, TPUs) designed to perform convolutional operations efficiently, leading to optimized performance in terms of speed and energy consumption.

ViTs can efficiently scale to handle high-resolution images or large-scale datasets because they operate on the entire image at once and do not require processing of local receptive fields at multiple spatial scales, potentially resulting in improved performance in terms of scalability.

Transfer Learning for pre-train on large-scale datasets and fine-tune on smaller datasets can potentially lead to improved performance with limited available data and can be used with both architectures.

6.1.4. Evaluation

CNNs have been widely used in image classification tasks for many years, resulting in well-established benchmarks and evaluation metrics that allow meaningful comparison and evaluation of model performance. The standardized evaluation protocols, such as cross-validation or hold-out validation, which provide a consistent framework for evaluating and comparing model performance across different datasets and tasks, are applicable for both architectures.

6.1.5. Explainability and Interpretability

CNNs produce feature maps that can be visualized, making it possible to interpret the behaviour of the model by visualizing the learned features or activations in different layers. They capture local features in images, such as edges or textures, which can lead to interpretable features that are visually meaningful and can provide insight into how the model is processing the input images. ViTs, on the other hand, are designed to capture global contextual information, making them potentially more interpretable in tasks that require an understanding of long-range dependencies or global context. They have a hierarchical structure with self-attention heads that can be visualized and interpreted individually, providing insights into how different heads attend to different features or regions in the input images.

6.1.6. Architecture

CNNs have a wide range of established architecture variants, such as VGG, ResNet, and Inception, with proven effectiveness in various image classification tasks. These architectures are well-tested and widely used in the deep learning community. ViTs can be easily modified to accommodate different input sizes, patch sizes, and depth, providing flexibility in architecture design and optimization.

6.2. Limitations

Despite their many advantages and the breakthroughs made over the years. There are still some drawbacks to the architectures studied. This section focuses on these.

6.2.1. Dataset Considerations

CNNs can be susceptible to biases present in training datasets, such as biased sampling or label noise, which can affect the validity of training results. They typically operate on fixed input spatial resolutions, which may not be optimal for images of varying size or aspect ratio, resulting in information loss or distortion. While pre-trained models for CNNs are well-established, pre-trained models for ViT are (still) less common for some datasets, which may affect the ease of use in some situations.

6.2.2. Robustness

CNNs may struggle to capture long-range contextual information, as they focus primarily on local feature extraction, which may limit the ability to understand global context, leading to reduced robustness in tasks that require global context, such as scene understanding, image captioning or fine-grained recognition.

Both architectures can be prone to overfitting, especially when the training data is limited or noisy, which can lead to reduced robustness to input data outside the training distribution. Adversarial attacks can also pose a challenge to the robustness of both architectures. In particular, ViTs do not have an inherent spatial inductive bias like CNNs, which are specifically designed to exploit the spatial locality of images. This can make them more vulnerable to certain types of adversarial attacks that rely on spatial information, such as spatially transformed adversarial examples.

6.2.3. Performance Optimization

CNNs can suffer from reduced performance and increased memory consumption when applied to high-resolution images or large-scale datasets, as they require processing of local receptive fields at multiple spatial scales, leading to increased computational requirements. Compared to CNNs, ViTs are computationally expensive, especially as the image size increases or model depth increases, which may limit their use in certain resource-constrained environments. Reduced computational complexity can sometimes result in decreased robustness, as models may not have the ability to learn complex features that can help generalize well to adversarial examples.

6.2.4. Evaluation

As mentioned above, CNNs are primarily designed for local feature extraction and may struggle to capture long-range contextual dependencies, which can limit the evaluation performance in tasks that require understanding of global context or long-term dependencies. ViTs are relatively newer than CNNs, and as such, may lack well-established benchmarks or evaluation metrics for specific tasks or datasets, which can make performance evaluation difficult and less standardized.

6.2.5. Explainability and Interpretability

Despite well-established methods for model interpretation, CNNs still lack full interpretability because the complex interactions between layers and neurons can make it difficult to fully understand the model's decision-making process, particularly in deeper layers of the network.

While ViTs produce attention maps for interpretability, the complex interactions between self-attention heads can still present challenges in accurately interpreting the model's behaviour. ViTs can have multiple heads attending to different regions, which can make it difficult to interpret the interactions between different attention heads and to understand the reasoning behind model predictions.

6.2.6. Architecture

CNNs typically have fixed, predefined model architectures, which may limit the flexibility to adapt to specific task requirements or incorporate domain-specific knowledge, potentially affecting performance optimization. For ViTs, the availability of established architecture variants is still limited, which may require more experimentation and exploration to find optimal architectures for specific tasks.

6.3. Future Research Directions

As future research, meta-analysis or systematic reviews should be conducted within the scope of this review to provide the scientific community with more detail on which of the architectures is more effective at image classification, in addition to specifying under what conditions a particular architecture stands out from the others. It is therefore necessary to facilitate the choice of the deep learning architecture to be used in future image classification problems. This section aims to provide guidelines for future research in this area.

6.3.1. Dataset Considerations

The datasets used in most studies may not be representative of real-world scenarios. Future research should consider using more diverse datasets that better reflect the complexity and variability of real-world images. As an example, it would be interesting to study the impact that image resolution might have on the performance of deep learning architectures. That is, it would be important to find out in which of the architectures (i.e., ViT, CNN, and MLP-Mixer) the resolution of the images will influence their performance, as well as what impact it will have on the processing time of the deep learning architectures.

6.3.2. Robustness

As documented in [5], deep learning models are typically vulnerable to adversarial attacks, where small perturbations to an input image can cause the model to misclassify it. Future research should focus on developing architectures that are more robust to adversarial attacks (for example by further augmenting the robustness of ViTs), as well as exploring ways to detect and defend against these attacks.

Beyond that, most studies (as the ones reviewed in this work) have focused on the performance of deep learning architectures on image classification tasks, but there are many other image processing tasks (such as object detection, segmentation, and captioning) that could benefit from the use of these architectures. Future research should further explore the effectiveness of these architectures on these tasks.

6.3.3. Performance Optimization

Deep learning architectures require substantial amounts of labelled data to achieve high performance. However, labelling data is time-consuming and expensive. Future research should explore ways to improve the efficiency of deep learning models, such as developing semi-supervised learning methods or transfer learning (following up on the finding in [10]) that can leverage pre-trained models.

In addition, the necessity of large amounts of labelled data requires significant computational resources, which limits the deployment on resource-constrained devices. Future research should focus on developing architectures that are optimized for deployment on these devices, as well as exploring ways to reduce the computational cost of existing architectures. It should explore the advantages of the implementation of the knowledge distillation of deep learning architectures to reduce computational resources.

6.3.4. Evaluation

The adequacy of the metrics to the task and problem at hand is also another suggested line of future research. Most studies have used standard performance metrics (such as accuracy and F1-score) to evaluate the performance of deep learning architectures. Future research should consider using more diverse metrics that better capture the strengths and weaknesses of different architectures.

6.3.5. Explainability and Interpretability

Deep learning models are often considered as black boxes because they do not provide insight into the decision-making process. This may prevent the usage of the models in certain areas such as justice and healthcare [30], among others. Future research should focus on making these models more interpretable and explainable. For example, by designing transformer architectures that provide visual explanations of their decisions or by developing methods for extracting features that are easily interpretable.

6.3.6. Architecture

In future investigations, it will be necessary to study the impact of the MLP-Mixer deep learning architecture in image processing, what are the characteristics that allow it to have a performance superior to CNNs, but inferior to the performance obtained by the ViT architecture [5]. Future research should also focus on developing novel architectures that can achieve high performance with fewer parameters or that are more efficient in terms of computation and memory usage.

7. Conclusions

This work has reviewed recent studies done in image processing to give more information about the performance of the two architectures and what distinguishes them. A common feature across all papers is that transformer-based architecture or the combination of ViTs with CNN allows for better accuracy compared to CNN networks. It has also been shown that this new architecture, even with hyperparameters fine-tuning, can be lighter than the CNN, consuming fewer computational resources and taking less training time as demonstrated in the works [16,19].

In summary, the ViT architecture is more robust than CNN networks for images that have noise or are augmented. It manages to perform better compared to CNN due to the self-attention mechanism because it makes the overall image information accessible from the highest to the lowest layers [12]. On the other hand, CNN's can generalize better with smaller datasets and get better accuracy than ViTs, but in contrast, ViTs have the advantage of learning information better with fewer images. This is because the images are divided into small patches, so there is a greater diversity of relationships between them.

Author Contributions: Conceptualization, J.B.; Methodology, J.M. and J.B.; Software, J.M; Validation, J.M., I.D. and J.B.; Formal analysis, J.M., I.D. and J.B.; Investigation, J.M.; Resources, J.M.; Data curation, J.M.; Writing—original draft preparation, J.M.; Writing—review and editing, J.M., I.D. and J.B.; Supervision, J.B. and I.D.; Project administration, J.B. and I.D.; Funding acquisition, J.B. All authors have read and agreed to the published version of the manuscript.

Funding: This research received no external funding.

Data Availability Statement: Data sharing is not applicable to this article.

Conflicts of Interest: The authors declare no conflict of interest.

References

1. Dosovitskiy, A.; Beyer, L.; Kolesnikov, A.; Weissenborn, D.; Zhai, X.; Unterthiner, T.; Dehghani, M.; Minderer, M.; Heigold, G.; Gelly, S.; et al. An Image Is Worth 16x16 Words: Transformers for Image Recognition at Scale. *arXiv* **2020**, arXiv:2010.11929. [CrossRef]
2. Saha, S. A Comprehensive Guide to Convolutional Neural Networks—The ELI5 Way. Available online: https://towardsdatascience.com/a-comprehensive-guide-to-convolutional-neural-networks-the-eli5-way-3bd2b1164a53 (accessed on 8 January 2023).

3. Snyder, H. Literature Review as a Research Methodology: An Overview and Guidelines. *J. Bus. Res.* **2019**, *104*, 333–339. [CrossRef]
4. Matloob, F.; Ghazal, T.M.; Taleb, N.; Aftab, S.; Ahmad, M.; Khan, M.A.; Abbas, S.; Soomro, T.R. Software Defect Prediction Using Ensemble Learning: A Systematic Literature Review. *IEEE Access* **2021**, *9*, 98754–98771. [CrossRef]
5. Benz, P.; Ham, S.; Zhang, C.; Karjauv, A.; Kweon, I.S. Adversarial Robustness Comparison of Vision Transformer and MLP-Mixer to CNNs. *arXiv* **2021**, arXiv:2110.02797. [CrossRef]
6. Bai, Y.; Mei, J.; Yuille, A.; Xie, C. Are Transformers More Robust Than CNNs? *arXiv* **2021**, arXiv:2111.05464. [CrossRef]
7. Tyagi, K.; Pathak, G.; Nijhawan, R.; Mittal, A. Detecting Pneumonia Using Vision Transformer and Comparing with Other Techniques. In Proceedings of the 2021 5th International Conference on Electronics, Communication and Aerospace Technology (ICECA), IEEE, Coimbatore, India, 2 December 2021; pp. 12–16.
8. Raghu, M.; Unterthiner, T.; Kornblith, S.; Zhang, C.; Dosovitskiy, A. Do Vision Transformers See Like Convolutional Neural Networks? *arXiv* **2021**, arXiv:2108.08810. [CrossRef]
9. Gheflati, B.; Rivaz, H. Vision Transformer for Classification of Breast Ultrasound Images. *arXiv* **2021**, arXiv:2110.14731. [CrossRef]
10. Zhou, H.-Y.; Lu, C.; Yang, S.; Yu, Y. ConvNets vs. Transformers: Whose Visual Representations Are More Transferable? In Proceedings of the 2021 IEEE/CVF International Conference on Computer Vision Workshops (ICCVW), IEEE, Montreal, BC, Canada, 17 October 2021; pp. 2230–2238.
11. Wu, Y.; Qi, S.; Sun, Y.; Xia, S.; Yao, Y.; Qian, W. A Vision Transformer for Emphysema Classification Using CT Images. *Phys. Med. Biol.* **2021**, *66*, 245016. [CrossRef]
12. Filipiuk, M.; Singh, V. Comparing Vision Transformers and Convolutional Nets for Safety Critical Systems. *AAAI Workshop Artif. Intell. Saf.* **2022**, *3087*, 1–5.
13. Galdran, A.; Carneiro, G.; Ballester, M.A.G. Convolutional Nets Versus Vision Transformers for Diabetic Foot Ulcer Classification. *arXiv* **2022**, arXiv:2111.06894. [CrossRef]
14. Cuenat, S.; Couturier, R. Convolutional Neural Network (CNN) vs Vision Transformer (ViT) for Digital Holography. In Proceedings of the 2022 2nd International Conference on Computer, Control and Robotics (ICCCR), IEEE, Shanghai, China, 18 March 2022; pp. 235–240.
15. Coccomini, D.A.; Caldelli, R.; Falchi, F.; Gennaro, C.; Amato, G. Cross-Forgery Analysis of Vision Transformers and CNNs for Deepfake Image Detection. In Proceedings of the 1st International Workshop on Multimedia AI against Disinformation, Newark, NJ, USA, 27–30 June 2022; Association for Computing Machinery: New York, NY, USA, 2022; pp. 52–58.
16. Wang, H. Traffic Sign Recognition with Vision Transformers. In Proceedings of the 6th International Conference on Information System and Data Mining, Silicon Valley, CA, USA, 27–29 May 2022; Association for Computing Machinery: New York, NY, USA, 2022; pp. 55–61.
17. Xin, C.; Liu, Z.; Zhao, K.; Miao, L.; Ma, Y.; Zhu, X.; Zhou, Q.; Wang, S.; Li, L.; Yang, F.; et al. An Improved Transformer Network for Skin Cancer Classification. *Comput. Biol. Med.* **2022**, *149*, 105939. [CrossRef]
18. Peng, Y.; Wang, Y. CNN and Transformer Framework for Insect Pest Classification. *Ecol. Inform.* **2022**, *72*, 101846. [CrossRef]
19. Bakhtiarnia, A.; Zhang, Q.; Iosifidis, A. Single-Layer Vision Transformers for More Accurate Early Exits with Less Overhead. *Neural Netw.* **2022**, *153*, 461–473. [CrossRef]
20. Asadi Shamsabadi, E.; Xu, C.; Rao, A.S.; Nguyen, T.; Ngo, T.; Dias-da-Costa, D. Vision Transformer-Based Autonomous Crack Detection on Asphalt and Concrete Surfaces. *Autom. Constr.* **2022**, *140*, 104316. [CrossRef]
21. Reedha, R.; Dericquebourg, E.; Canals, R.; Hafiane, A. Vision Transformers for Weeds and Crops Classification of High Resolution UAV Images. *Remote Sens.* **2022**, *14*, 592. [CrossRef]
22. Bottou, L.; Bousquet, O. The Tradeoffs of Large Scale Learning. In *Advances in Neural Information Processing Systems*; Platt, J., Koller, D., Singer, Y., Roweis, S., Eds.; Curran Associates, Inc.: Vancouver, BC, Canada, 2007; Volume 20.
23. Foret, P.; Kleiner, A.; Mobahi, H.; Neyshabur, B. Sharpness-Aware Minimization for Efficiently Improving Generalization. *arXiv* **2020**, arXiv:2010.01412. [CrossRef]
24. Korpelevich, G.M. The Extragradient Method for Finding Saddle Points and Other Problems. *Ekon. Mat. Metod.* **1976**, *12*, 747–756.
25. Al-Dhabyani, W.; Gomaa, M.; Khaled, H.; Fahmy, A. Dataset of Breast Ultrasound Images. *Data Brief* **2020**, *28*, 104863. [CrossRef]
26. Yap, M.H.; Pons, G.; Marti, J.; Ganau, S.; Sentis, M.; Zwiggelaar, R.; Davison, A.K.; Marti, R. Automated Breast Ultrasound Lesions Detection Using Convolutional Neural Networks. *IEEE J. Biomed. Health Inform.* **2018**, *22*, 1218–1226. [CrossRef]
27. Zhang, R. Making Convolutional Networks Shift-Invariant Again. *arXiv* **2019**, arXiv:1904.11486. [CrossRef]
28. Vaswani, A.; Shazeer, N.M.; Parmar, N.; Uszkoreit, J.; Jones, L.; Gomez, A.N.; Kaiser, L.; Polosukhin, I. Attention Is All You Need. *Neural Inf. Process. Syst.* **2017**, *30*, 3762. [CrossRef]
29. Zhou, D.; Kang, B.; Jin, X.; Yang, L.; Lian, X.; Jiang, Z.; Hou, Q.; Feng, J. DeepViT: Towards Deeper Vision Transformer. *arXiv* **2021**, arXiv:2103.11886. [CrossRef]
30. Amorim, J.P.; Domingues, I.; Abreu, P.H.; Santos, J.A.M. Interpreting Deep Learning Models for Ordinal Problems. In Proceedings of the European Symposium on Artificial Neural Networks, Bruges, Belgium, 25–27 April 2018.

applied sciences

MDPI

Article

Efficient Data Transfer by Evaluating Closeness Centrality for Dynamic Social Complex Network-Inspired Routing

Manuel A. López-Rourich [1,*] and Francisco J. Rodríguez-Pérez [2]

[1] Department of Crowdsourcing Mobile Data Intelligence, Trecone Solutions, 10003 Cáceres, Spain
[2] Department of Computing and Telematics Systems Engineering, University of Extremadura, 10003 Cáceres, Spain; fjrodri@unex.es
* Correspondence: alopez@trecone.com; Tel.: +34-927251657

Abstract: Social Complex Networks in communication networks are pivotal for comprehending the impact of human-like interactions on information flow and communication efficiency. These networks replicate social behavior patterns in the digital realm by modeling device interactions, considering friendship, influence, and information-sharing frequency. A key challenge in communication networks is their dynamic topologies, driven by dynamic user behaviors, fluctuating traffic patterns, and scalability needs. Analyzing these changes is essential for optimizing routing and enhancing the user experience. This paper introduces a network model tailored for Opportunistic Networks, characterized by intermittent device connections and disconnections, resulting in sporadic connectivity. The model analyzes node behavior, extracts vital properties, and ranks nodes by influence. Furthermore, it explores the evolution of node connections over time, gaining insights into changing roles and their impact on data exchange. Real-world datasets validate the model's effectiveness. Applying it enables the development of refined routing protocols based on dynamic influence rankings. This approach fosters more efficient, adaptive communication systems that dynamically respond to evolving network conditions and user behaviors.

Keywords: dynamic complex networks; opportunistic social mobility patterns; device-to-device data routing; spray and wait routing; quality of service

Citation: López-Rourich, M.A.; Rodríguez-Pérez, F.J. Efficient Data Transfer by Evaluating Closeness Centrality for Dynamic Social Complex Network-Inspired Routing. *Appl. Sci.* **2023**, *13*, 10766. https://doi.org/10.3390/app131910766

Academic Editor: Christos Bouras

Received: 12 August 2023
Revised: 21 September 2023
Accepted: 25 September 2023
Published: 27 September 2023

1. Introduction

The vast increase in autonomous and heterogeneous wireless devices poses challenges for the future of communication systems due to the complexity of their interconnection, which involves multiple networking technologies and a wide range of device capabilities. According to statistics provided by Statista [1], the global number of smartphones reached nearly 6.6 billion in 2022 and is projected to surpass 7.8 billion by 2028. In other words, a world of pervasive mobile devices is being built that has vast processing capabilities and allows for smooth communication among them, enabling greater connectedness. Each node or gadget in a mobile communication network is a connecting point that is innately connected to a person who is moving and takes part in the network's data exchange. Moreover, Mobile Social Networks (MSNs) leverage wireless devices and function as a communication infrastructure designed for point-to-point and short-range communications, seeking increased data exchange efficiency by adapting to the typical movements and behaviors of individuals using mobile devices [2–5]. In this regard, opportunistic networks emerge as a classification of wireless networks characterized by sporadic, unreliable, or constrained user-to-user ad hoc connections [6]. In such networks, conventional routing algorithms often depend on the "storage-carry-and-forward" approach, whereby a node forwards messages to a varying number of neighboring nodes it encounters based on the specific routing algorithm employed. Nevertheless, this flooding strategy can result in a proliferation of message duplicates, potentially leading to network and device congestion [7].

Within such a context, a Dynamic Social Complex Network (DSCN) can be defined as a network that leverages human social behavior, such as daily routines, mobility patterns, and interests, to facilitate message routing and data sharing over time. In these networks, nodes (users with mobile devices) can form on-the-fly social networks to communicate with each other. Considering users' social routines when determining whether a node should retransmit a message to another node can reduce transmission delay and routing overhead [8]. Consequently, minimizing routing overhead will decrease the average number of hops that message routes traverse before reaching their destination.

The search for the most influential or important nodes is a critical component of analyzing and comprehending the network topology dynamics due to its intrinsic role in determining the network's overall structure or efficiency [9–11]. Influential devices often act as key hubs for data exchange and efficient communication pathways and, depending on the objective, the significance of a node can vary. There are several metrics of node importance, such as degree centrality, closeness centrality, or social centrality. The first refers to the number of its direct connections to other nodes and analyzes its level of activity in the network's topology compared to others. Closeness centrality measures a node's proximity to other nodes and indicates how efficiently it can access or distribute information within the network. Finally, social centrality aims to capture the extent to which certain nodes in the network hold influential positions in terms of data forwarding or control over the network's social dynamics. Social centrality metrics take into account factors such as connectivity, interactions, and relationships among nodes to assess their relative significance in the social behavior of the network.

However, conventional centrality metrics used in traditional networks are not useful in DSCNs, as they rely on a static network model where there are multiple connections and disconnections over time that are usually aggregated into a single binary network. As a result, traditional metrics have been extended to work with weighted or dynamic topologies [12–15]. The authors of [16] considered the number of connections as link weights and redefined the centrality metrics to consider both the number of links and their weights in the graphs. On the other hand, the authors of [17] proposed time-based measures that leverage the temporal patterns of changing topologies. Furthermore, individuals have inherent social tendencies, and their behavioral patterns, which are substantially influenced by the patterns of interaction among individuals, are not random [18,19]. Thus, when identifying hub nodes in the network, mobility patterns, spatiotemporal connections, and social behavior must also be considered.

In examining the current landscape of detecting influential nodes, several notable works have delved into incorporating both non-social and social attributes associated with network nodes. The research conducted by the authors of [20] focuses on centrality metrics, which help identify important nodes in a network, crucial for understanding network structures and behaviors. Static and dynamic centrality metrics are discussed, including their relevance in weighted networks. The study highlights challenges, proposes new centrality metrics, and emphasizes the importance of considering temporal aspects in network analysis. It also explores network resilience and the impact of centrality on fault tolerance. However, the authors do not explore a broader range of centrality measures or compare this measure with other existing centrality measures comprehensively. The research conducted by the authors of [21] delves into online information propagation within complex networks, emphasizing the critical role of influential nodes in network structure and operation. The paper classifies centrality measures into global, local, and semi-local types, exploring their effectiveness in identifying influential nodes. It introduces a novel centrality measure, 'centripetal centrality,' and presents an algorithm, 'seeds exclusion,' to enhance information propagation. The work demonstrates the effectiveness of 'centripetal centrality' in identifying key nodes and improving propagation effects. The authors assume that identifying influential spreaders is essential for maximizing information coverage. However, this assumption might not always hold true, especially in scenarios where the objective differs from maximizing information spread. For instance, in communication networks, it is crucial to minimize overhead rather than increase it. The authors of [22] address the prediction of social network dynamics and evolution, distinguishing between short-term dynamics and long-term changes. The proposed methodology, MONDE, utilizes hidden Markov models and a genetic algorithm to predict individual, group, and network dynamics. The approach aims to provide a comprehensive view of network evolution and dynamics, benefiting fields such as marketing and public security by aiding decision-making and strategy planning. However, the accuracy and effectiveness of MONDE heavily depend on the quality and availability of data, especially the posting activities and comments used for feature extraction. Incomplete or inaccurate data could lead to less reliable predictions, particularly in the case of low-density networks where empty or discontinuous samples may exist in the data. The study carried out by the author of [23] explores critical node detection and introduces a novel centrality measure, known as isolating centrality, to identify nodes that significantly impact network connectedness. The paper emphasizes the importance of accurately identifying critical nodes for ensuring network reliability and provides a comparative analysis of centrality measures' performance. It also investigates the correlation between leverage centrality and critical nodes, showcasing the effectiveness of the proposed centrality measure. However, it is worth noting that the effectiveness of this proposed measure is influenced by the structure of the nodes' neighborhood, especially in detecting critical nodes that segregate the network into connected components. This dependency might limit its effectiveness in certain low-density topologies. The authors of [24] focus on seed node selection in online social networks (OSNs) for information propagation and influence maximization. The study explores various centrality measures, such as clustering coefficients and node degree, to identify influential seed nodes. It considers Twitter as a platform for opinion generation and discusses the relevance of centrality measures as seed nodes in large-scale networks. The study also conducts a comparative analysis using benchmark similarity measures to assess the effectiveness of different centrality measures in seed node selection. The study acknowledges that the effectiveness of seed node selection is influenced by the network's structure. Certain propagation approaches, like Random Walk, are affected by local clustering. This sensitivity to network structure implies that the effectiveness of the proposed approach could vary significantly in network topologies with insufficient connections.

In summary, these works propose or utilize specific centrality measures to assess the importance of nodes in a network, hence their focus on identifying influential or critical nodes within the network as shown in Table 1. These nodes are deemed essential for information propagation and can play a crucial role in maximizing information coverage within the network. However, the objective is not always to maximize the pathways through which information circulates. In communication networks, it is preferable to maintain low overhead values to avoid unnecessary consumption of memory resources in intermediary devices forwarding data to their destinations. Furthermore, these works exhibit a certain dependence on network topology, implying that effectiveness could vary in cases of low connection density, as observed in opportunistic networks.

Based on the aforementioned concerns, this research paper aims to identify and rank hub nodes using a dynamic network model to analyze how device connections evolve over time. For that, the behavior of a DSCN is gathered by a progression of graphs as the devices connect and disconnect throughout the network operation. First, we introduce a novel local centrality metric, Dynamic Degree centrality, as we believe that both the number of neighbors and the frequency of connections with them serve as valuable cues of a node's importance in the network.

This metric seamlessly integrates both factors, effectively gauging the node's centrality based on the progression of its connections and contact frequency with neighboring nodes. Furthermore, we have developed a closeness centrality measure to address the potential impact of longer forwarding delays on storage capacity utilization (network overhead) and its subsequent influence on data forwarding likelihood. To quantify a device's global centrality, we propose the Dynamic Closeness centrality based on the temporal evolution network model, which considers forwarding overhead. We also propose the Social-based Closeness Centrality Metric, which considers social relations to provide an effective centrality metric to ensure that data are carried and forwarded by relay devices with a high likelihood of reaching the destination host. This is because social relations and behaviors among wireless users are typically long-term characteristics and less fluctuating than device mobility. Thus, to assess the usefulness of our suggested centrality metrics and to examine the properties of the centrality distribution applied to various Quality of Service (QoS) measurements, we evaluate the results of experiments run on real-world datasets.

Table 1. Comparison of related works on the detection of the most influential nodes in a network.

Study	Objective	Methodology	Key Findings
[20]	Detection of influential nodes in dynamic weighted networks.	Time-ordered weighted graph models with Opshal's algorithms, considering temporal aspects.	New hybrid centrality measure: Temporal Closeness-Closeness measure.
[21]	Identification of influential spreaders.	Integrate degree, constraint coefficient, and k-shell for a comprehensive assessment of node importance.	Centripetal centrality as an effective measure to identify influential nodes.
[22]	Prediction of the dynamics and evolution of a social network.	Two-layer HMM to model individual and group dynamics.	MONDE, demonstrating prediction accuracy rates for dynamics and evolution in social networks.
[23]	Detection of critical nodes of networks.	Compare centrality measures' effectiveness.	Isolating centrality as an effective measure for identifying critical nodes.
[24]	Correlation between seed node detection and information flow.	Investigate different centrality measures for seed node detection.	Emphasize the impact of network structure on seed node selection.

The remainder of this paper is organized as follows: Section 2 provides an overview of the dynamic network model used. The proposed local and global influence metrics are described in detail in Section 3. In Section 4, we present the experimental results of all the proposed centrality metrics, including a comparative analysis of different QoS metrics. Finally, we offer conclusions in Section 5.

2. Model and Method

In this part, we first use the dynamic network model to show how the topological structures of DSCNs are constantly evolving. Using this model as a guide, we look at people's social connections and movement patterns to create new interpretations of traditional influence measurements that are based on the network's dynamic.

A graph G is made up of a limited number of nodes (V) and edges (E). Since there cannot be an empty set of nodes, $V \neq \varnothing$, and thus $V = \{v_1, v_2, \ldots, v_n\}$. Pairs of nodes (v_i, v_j) that represent some sort of connection pattern between nodes make up the collection of edges E. The terms nearby and neighboring are used to describe two nodes connected by an edge. The network is referred to as undirected if the edges are unordered, where $(v_i, v_j) = (v_j, v_i)$.

If there is a relationship between the nodes v_i, v_j, then an adjacency matrix M with elements $m_{ij} = 1$ and 0 otherwise can be used to fully describe network G. Unweighted or binary networks are examples of this. In general, G is characterized using an adjacency matrix, where $m_{ij} \geq 1$ if there is an edge between nodes v_i, v_j, and 0 otherwise, where the edges contain a numerical value measuring a feature of the edge. A network G is also considered to be connected if, for every pair of distinct nodes v_i, v_j, there is a route from v_i to v_j; otherwise, it is said to be unconnected [25]. However, complex networks are a special kind of graph in which the nodes and edges have complicated organizational structures and non-trivial topological properties. Since these networks contain complicated patterns and features, the interactions between the pieces in this situation are not clear-cut or simple.

Dynamic Model of a Complex Network

A Dynamic Complex Network (DCN) is made up of several nodes, which stand in for individual devices, and edges, which represent the connections or temporal interactions between them. Traditional static complex network models cannot adequately capture such dynamic evolution since the topology and device placements coevolve over time. The time-ordered network was suggested by the authors of [17] to transform a dynamic network into a static network with directed flows. The authors of [26] analyzed the uniformity of device behavior over time. We build upon their work and propose a dynamic network model that captures the evolving nature of a DCN. Our objective is to predict the behavioral trends of devices in the network and measure their QoS parameters. The authors of [26] rely on Shannon entropy to verify the uniformity of device behavior, whereas we go further by conducting regression studies to analyze not only uniformity but also the trend of node behavior and directly apply it to QoS metrics in opportunistic networks with low connection density.

By considering the temporal sequence, length, and correlations between connections or devices happening at various moments in time, our model seeks to give a clear and thorough framework for understanding the evolutionary patterns of the network. Our model illustrates the evolution of interactions between devices in a DCN over a certain time period by using a series of snapshots. We shall outline the basic concepts of the dynamic network model in the parts below:

Define a finite set of devices V (nodes) and a set of connections E (edges) between these devices. The connections between devices are assumed to take place over a time span T. We use L to denote the duration of each spatial snapshot (or time window size), and $F_T = \frac{T}{L}$ represents the number of spatial snapshots during the time span T. The dynamics of the network can be subsequently described by $G_T = (V_T, E_T, M_T)$,

where

$V_T \subseteq V$ is the collection of all networked devices throughout the duration of T.

$E_T \subseteq E$ is the set of edges that stands for connections between devices throughout the course of time T.

M_T is a sequence of connectivity matrices that record contact events of devices during the time span T.

A discretized collection of static complex networks, $G_T = \{G_1, G_2, \ldots, G_{F_T}\}$, can be used to simulate a DCN. In this model, the edges in each connection matrix are not binary as they are in the adjacency matrix of an unweighted graph. The connectivity matrix's edge weights, which range from 0 to F_T, indicate how frequently points of contact occur.

For the sake of clarity, we provide the following example in Figure 1, where the network's aggregated view is represented by G_T. In this scenario, two devices are said to be connected if they have made contact within a time interval t_i. The network snapshots are denoted as $\{G_1, G_2, \ldots, G_{10}\}$. All information from both geographical and temporal data is included in this network's representation. Figure 2 displays the connection matrixes in order.

The connectivity matrix is symmetric since each snapshot is an undirected graph with a connection between devices denoting the presence of a contact link in both directions. For instance, the weight of M_{AC} in M_1 is 0, indicating that device A and device C did not make contact during time unit t_1. One link between device A and device C was represented by the weight of M_{AC} in M_8, which is 1, during the time unit t_8.

The costs of the routes between a source and a destination in the domain G_T can be represented as a function of delay, $\{\delta_{ij}x_{ij}\}_{t_{ij}}$, if the objective is to minimize the message delivery delay, or as a function of load, $\{\lambda x_{ij}\}_{t_{ij}}$, if the objective is to minimize the overhead.

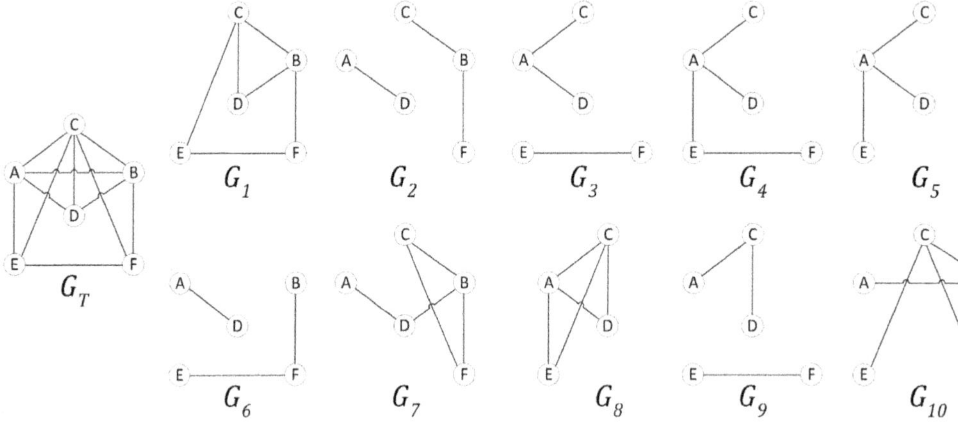

Figure 1. Example of a dynamic network model.

	A	B	C	D	E	F
A	-	0	0	0	0	0
B	0	-	1	1	0	1
C	0	1	-	1	1	0
D	0	1	1	-	0	0
E	0	0	1	0	-	1
F	0	1	0	0	1	-

M_1

	A	B	C	D	E	F
A	-	0	0	1	0	0
B	0	-	1	0	0	1
C	0	1	-	0	0	0
D	1	0	0	-	0	0
E	0	0	0	0	-	0
F	0	1	0	0	0	-

M_2

	A	B	C	D	E	F
A	-	0	1	1	0	0
B	0	-	0	0	0	0
C	1	0	-	0	0	0
D	1	0	0	-	0	0
E	0	0	0	0	-	1
F	0	0	0	0	1	-

M_3

	A	B	C	D	E	F
A	-	0	1	1	1	0
B	0	-	0	0	0	0
C	1	0	-	0	0	0
D	1	0	0	-	0	0
E	1	0	0	0	-	1
F	0	0	0	0	1	-

M_4

	A	B	C	D	E	F
A	-	0	1	1	1	0
B	0	-	0	0	0	0
C	1	0	-	0	0	0
D	1	0	0	-	0	0
E	1	0	0	0	-	0
F	0	0	0	0	0	-

M_5

	A	B	C	D	E	F
A	-	0	0	1	0	0
B	0	-	0	0	0	1
C	0	0	-	0	0	0
D	1	0	0	-	0	0
E	0	0	0	0	-	1
F	0	1	0	0	1	-

M_6

	A	B	C	D	E	F
A	-	0	0	1	0	0
B	0	-	1	1	0	1
C	0	1	-	0	0	1
D	1	1	0	-	0	0
E	0	0	0	0	-	0
F	0	1	1	0	0	-

M_7

	A	B	C	D	E	F
A	-	0	1	1	1	0
B	0	-	0	0	0	0
C	1	0	-	1	1	0
D	1	0	1	-	0	0
E	1	0	1	0	-	0
F	0	0	0	0	0	-

M_8

	A	B	C	D	E	F
A	-	0	1	0	0	0
B	0	-	0	0	0	0
C	1	0	-	1	0	0
D	0	0	1	-	0	0
E	0	0	0	0	-	1
F	0	0	0	0	1	-

M_9

	A	B	C	D	E	F
A	-	1	0	0	0	0
B	1	-	1	0	0	0
C	0	1	-	0	1	1
D	0	0	0	-	0	0
E	0	0	1	0	-	0
F	0	0	1	0	0	-

M_{10}

Figure 2. Sequence of connectivity matrixes of the dynamic network model in Figure 1.

The Delay matrix (D) of existing routes between a source device s and a destination device d in the domain G_T can be obtained as follows:

$$D(s,d) = \sum_{i=1}^{|V|-1} \sum_{j=2}^{|V|} \{\delta_{ij} x_{ij}\}_{t_{ij}} \tag{1}$$

The Load matrix (L) of all existing routes between a source device s and a destination device d in the domain G_T can be obtained as follows:

$$L(s,d) = \sum_{i=1}^{|V|-1} \sum_{j=2}^{|V|} \{\lambda x_{ij}\}_{t_{ij}} + \lambda \tag{2}$$

where

δ_{ij} represents the forwarding delay from device i to device j.
λ represents the message size.

$x_{ij} = 1$ if devices $\{i, j\}$ contact at any time and that link is used in a route between the source device s and the destination d (i.e., device i decides to forward a copy of the message to device j).
$x_{ij} = 0$ if devices $\{i, j\}$ do not connect or if the link is not used in any route between the source device s and destination d.
t_{ij} represents the encounter time of device i and device j.

We suggest the Dynamic Shortest Path Method, drawing on the aforementioned factors. This approach aims to achieve a compromise between decreasing communication costs, guaranteeing equitable load distribution, and obtaining the ideal delivery delay. To do this, we employ a tuning parameter α that enables the three factors to be considered when determining the optimum route between source and destination nodes. The variables $\{\alpha_1, \alpha_2, \alpha_3\}$ can also be changed depending on the analysis to give the three variables different relative weights. Next, we list the equations that describe the dynamic shortest path approach that is suggested:

- Delivery Delay:

$$\alpha_1 min(D(s,d)) = \alpha_1 min\left(\sum_{i=1}^{|V|-1}\sum_{j=2}^{|V|}\{\delta_{ij}x_{ij}\}_{t_{ij}}\right) \quad (3)$$

- Load Balancing:

$$\alpha_2 min\left(\frac{1}{\sum_{i=1}^{|V|-1}\sum_{j=2}^{|V|}x_{ij}}\sum_{i=1}^{|V|-1}\sum_{j=2}^{|V|}\left(\{\delta_{ij}x_{ij}\}_{t_{ij}} - \frac{1}{\sum_{i=1}^{|V|-1}\sum_{j=2}^{|V|}x_{ij}}\sum_{i=1}^{|V|-1}\sum_{j=2}^{|V|}\{\delta_{ij}x_{ij}\}_{t_{ij}}\right)^2\right)^{1/2} \quad (4)$$

- Communication Overhead:

$$\alpha_3 min(L(s,d)) = \alpha_3 min\left(\sum_{i=1}^{|V|-1}\sum_{j=2}^{|V|}\left(\{\lambda x_{ij}\}_{t_{ij}}\right) + \lambda\right) \quad (5)$$

subject to the following restrictions:

$\delta_{ij} > 0, \forall i, j \in V$
$\lambda > 0$
$x_{ij} \in \{0, 1\}, \forall i, j \in V$
$\sum_{j=2}^{|V|} x_{1j} = 1$ (the shortest path only uses one link from the source device).
$\sum_{i=1}^{|V|-1} x_{i|V|} = 1$ (the shortest path only uses one link to the destination device).
$\sum_{i=1}^{|V|-1} x_{ik} = \sum_{j=2}^{|V|} x_{kj}$, $\forall\ k \in \{2, 3, \ldots |V| - 1\}$ (in the shortest path, if a link arriving at device k is used, then a single link leaving k will be used).
$t_{1,2} \leq t_{2,3} \leq \ldots \leq t_{(T-1),T} \leq$ (represents the connections of intermediary devices based on the time order).

The example of Figure 3 lists the values of the three variables for the four pathways (in different colors) from device A to device B depending on the time order to demonstrate the efficacy of the suggested strategy.

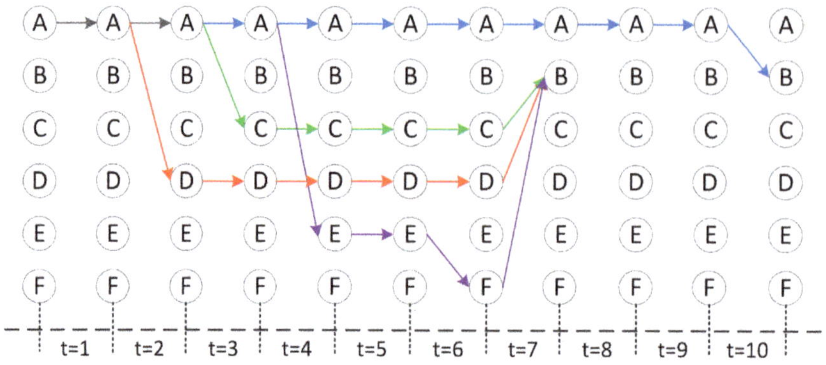

Figure 3. Illustration of the paths from device A to device B based on the time order in Figure 2.

- Number of Paths between devices A and B is $p(A,B) = \sum_{i=1}^{|V|-1} M_T^i(A,B) = 4$.
- Delay matrix: $D(A,B) = \sum_{i=1}^{|V|-1} \sum_{j=2}^{|V|} \{\delta_{ij}x_{ij}\}_{t_{ij}} =$

$$\{A,B\} \Rightarrow \delta_{AB}x_{AB} = 10t \text{ seconds}$$
$$\{A,C,B\} \Rightarrow \delta_{AC}x_{AC} + \delta_{CB}x_{CB} = 3t + 4t = 7t \text{ seconds}$$
$$\{A,D,B\} \Rightarrow \delta_{AD}x_{AD} + \delta_{DB}x_{DB} = 2t + 5t = 7t \text{ seconds}$$
$$\{A,E,F,B\} \Rightarrow \delta_{AE}x_{AE} + \delta_{EF}x_{EF} + \delta_{FB}x_{FB} = 4t + 2t + t = 7t \text{ seconds}$$

- Load balancing (LB):

$$\{A,B\} \Rightarrow LB = \left(\tfrac{1}{1}\left((10t - 10t)^2\right)\right)^{1/2} = 0t \text{ seconds}$$
$$\{A,C,B\} \Rightarrow LB = \left(\tfrac{1}{2}\left((3t - 3.5t)^2 + (4t - 3.5t)^2\right)\right)^{1/2} = 0.5t \text{ seconds}$$
$$\{A,D,B\} \Rightarrow LB = \left(\tfrac{1}{2}\left((2t - 3.5t)^2 + (5t - 3.5t)^2\right)\right)^{1/2} = 1.5t \text{ seconds}$$
$$\{A,E,F,B\} \Rightarrow LB = \left(\tfrac{1}{3}\left((4t - 2.33t)^2 + (2t - 2.33t)^2 + (1t - 2.33t)^2\right)\right)^{1/2} = $$
$$= 1.5764t \text{ seconds}$$

- Load matrix: $L(A,B) = \sum_{i=1}^{|V|-1} \sum_{j=2}^{|V|} \left(\{\lambda x_{ij}\}_{t_{ij}}\right) + \lambda =$

$$\{A,B\} \Rightarrow load = \lambda x_{AB} + \lambda = 2\lambda \; bytes$$

$$\{A,C,B\} \Rightarrow load = \lambda x_{AC} + \lambda x_{CB} + \lambda = 3\lambda \; bytes$$

$$\{A,D,B\} \Rightarrow load = \lambda x_{AD} + \lambda x_{DB} + \lambda = 3\lambda \; bytes$$

$$\{A,E,F,B\} \Rightarrow load = \lambda x_{AE} + \lambda x_{EF} + \lambda x_{FB} + \lambda = 4\lambda \; bytes$$

As shown in Table 2 considering the influence of delivery delay, communication overhead, and load balancing on routing performances, we observe that there are three paths with the shortest delivery delay (7t) from device A to device B: {A, C, B}, {A, D, B} and {A, E, F, B}. However, the length of paths {A, C, B} and {A, D, B} is shorter than that of path {A, E, F, B}. Moreover, the longest forwarding delay of the path {A, C, B} is three time periods, from t = 4 to t = 6 (load balancing = 0.5t), which is shorter than that of path {A, D, B}, with four time periods, from t = 3 to t = 6 (load balancing = 1.5t), so {A, C, B} have better load balancing. Finally, we observe that path $\{A, B\}$ has the shortest hop number but the longest delivery delay ($10t$). In summary, path $\{A, C, B\}$ is the best path from device A to device B considering the trade-off between delivery delay, communication overhead, and load balancing.

Table 2. Dynamic shortest path identification from device *A* to device *B*.

Title 1	Latency	Load Balancing	Overhead
$\{A, B\}$	$10t$	$0t$	2λ
$\{A, C, B\}$	$7t$	$0.5t$	3λ
$\{A, D, B\}$	$7t$	$1.5t$	3λ
$\{A, E, F, B\}$	$7t$	$1.5764t$	4λ

Using flooding or epidemic routing, device *B* should receive 4 copies of the message. One from device *A* at time $t = 10$ and three copies from C, *D* and F at $t = 7$, assuming a total load of $CO - p_{AB}\lambda + \lambda = \sum_{i=1}^{|V|-1}\sum_{j=2}^{|V|}\left(\{\lambda x_{ij}\}_{t_{ij}} + \lambda\right) - p_{AB}\lambda + \lambda = 12\lambda - 4\lambda + \lambda = 9\lambda$ bytes.

However, by selecting the best route, overhead improvement can be obtained as $\sum_{i=1}^{|V|-1}\sum_{j=2}^{|V|}\left(\{\lambda x_{ij}\}_{t_{ij}} + \lambda\right) - p_{AB}\lambda + \lambda - min\left(\sum_{i=1}^{|V|-1}\sum_{j=2}^{|V|}\{\lambda x_{ij}\}_{t_{ij}} + \lambda\right) = 6\lambda$ bytes, which is an improvement of 33%.

3. Influence Metrics

Node centrality or impact is the process of classifying nodes or devices in a network according to their importance or effect. This statistic evaluates a device's significance or effect on the network as a whole. There are several ways to determine device centrality, and each one takes a different strategy to pinpoint the most important nodes. These methods are intended to identify the importance and function of each device inside the network. Our goal is to order the network nodes according to their impact, allowing for the creation of new routing protocols that improve the QoS of the network. We can increase QoS levels by using just the most powerful nodes for data forwarding.

Degree and Closeness are two conventionally used common centrality measurements. While the Closeness metric considers the global topological information, Degree centrality is based on local topological information and assesses the node's local importance in the network. These influence measures are defined as follows for a network $G = (V, E)$:

3.1. Local Influence

To determine a node's local influence, it is straightforward to assess the centrality of the node within the network. The quantity of direct connections a node has to other nodes determines its degree of centrality. The following is the mathematical formula for determining the degree centrality of a given node *j*:

1. If it is an unweighted and undirected network,

$$D(j) = \sum_{i}^{|V|} x_{ij} \tag{6}$$

where $x_{ij} = x_{ji} = 1$ if and only if nodes *i* and *j* are connected; $x_{ij} = x_{ji} = 0$ otherwise.

2. If it is a weighted and undirected network,

$$D(j) = \sum_{i}^{|V|} \omega_{ij} x_{ij} \tag{7}$$

where

$\omega_{ij} = \omega_{ji} = 1$ is the cost of the link (i, j).
$x_{ij} = x_{ji} = 1$ if and only if nodes i and j are connected; $x_{ij} = x_{ji} = 0$ otherwise.

Dynamic Degree Metric

Since nodes in static binary networks may only be either linked or disconnected, the traditional degree centrality statistic was initially created for those types of networks. The interactions between nodes in DSCN networks are not binary, though, and the topology of these networks is continually changing. Individuals in DSCNs often have a small number of regular connections in addition to sporadic encounters. Connections made often have a tendency to be stronger than those made infrequently. As a result, if a person interacts with their neighbors regularly and has a larger number of neighbors, they are more likely to be able to engage with new individuals. Therefore, it is crucial to consider the following three behavioral traits in DSCNs in order to effectively evaluate a device's local influence: a large number of neighbors, a high number of neighbor contact instances, and a positive evolution in the frequency of neighbor contact over time.

Regression analysis is a commonly used method for investigating data distribution patterns in the fields of information science and statistical modeling [27,28]. This theory has been used by us to investigate the connections between the temporal changes in the connection time distributions among devices.

Considering a device v which has $\sum_{k=1,k\neq v}^{|V|}\{x_{vk}\}_T$ connections with neighbors during the time span T, $\sum_{k=1,k\neq v}^{|V|}\{x_{vk}\}_{t_i}$ is the number of neighbors of v during the snapshot t_i and $\sum_{k=1,k\neq v}^{|V|}\{\varphi_{vk}x_{vk}\}_T$ is the frequency of contact times between device v and its neighbors during the time span T. Then, the evolution trend of connections of device v during the time span T is defined as follows:

$$\tau_D(v) = \frac{F_T\sum_{i=1}^{T}\left(t_i\sum_{k=1,k\neq v}^{|V|}\{x_{vk}\}_{t_i}\right) - \sum_{i=1}^{T}t_i\sum_{i=1}^{T}\left(\sum_{k=1,k\neq v}^{|V|}\{x_{vk}\}_{t_i}\right)}{F_T\sum_{i=1}^{T}t_i^2 - \left(\sum_{i=1}^{T}t_i\right)^2} \tag{8}$$

The device tends to increase the frequency of interaction with its neighbors over the time period T if the value of $\tau_T(v)$ is positive. An equitable distribution of contact frequency with a specific node's neighbors is indicated by a value of 0, while a negative value denotes a reduction in contact frequency over time. Therefore, devices will have a more favorable contact dynamic if they have more neighbors and increased contact probabilities with those neighbors. For that, we propose the Dynamic Degree metric of a device v ($DTE(v)$), which takes into account the following properties, as more interactions with neighbors lead to stronger links with them:

$$DTE(v) = \alpha\tau_D(v) + (1-\alpha)\sum_{i=1}^{F_T}\frac{\sum_{i=1}^{F_T}\{x_{vk}\}_{t_i} - F_T^{-1}\sum_{k=1,k\neq v}^{F_T}\{\varphi_{vk}x_{vk}\}_T}{\left(F_T^{-1}\sum_{i=1}^{F_T}\left(\sum_{k=1,k\neq v}^{|V|}\{x_{vk}\}_{t_j} - F_T^{-1}\sum_{k=1,k\neq v}^{|V|}\{\varphi_{vk}x_{vk}\}_T\right)^2\right)^{1/2}} \tag{9}$$

The user-defined parameter value $\alpha \in [0,1]$ regulates the importance or weight of connection evolution and the frequency of contact moments. Please note that this value adheres to a zero-sum condition, meaning that increasing the weight of one element would inherently decrease the weight of the other factor. If significant patterns of growth or decline in the metric's values are observed, then $\alpha \to 1$. If the data show uniform distributions across time, $\alpha \to 0$.

3.2. Global Influence

A typical global centrality metric known as 'Closeness' uses the shortest routes to calculate distances between each node and every other node in the network. However, due to the specific characteristics of DSCNs, this statistic often results in inaccurate estimates. To address these issues, we have developed a unique approach to calculating the shortest paths that more accurately represents the information propagation patterns within DSCNs over time. With this approach, we subsequently formulated a refined definition of the global influence measure, accounting for these distinctive qualities of DSCNs.

The frequent partitioning of topologies and intermittent connections that characterize DSCNs often lead to higher storage capacity utilization. The constrained storage space on a device can present a hurdle for efficient routing, particularly if it receives messages more quickly than it can transmit them to the next relay device. This situation can result in uneven load balancing, which significantly impacts the overall routing efficiency within DSCNs. Furthermore, employing an excessive number of devices as relays for a message can introduce unnecessary communication overhead, exacerbating routing performance issues. Hence, achieving a balance between delivery delay, load distribution, and communication overhead becomes imperative when making routing decisions in DSCNs.

3.2.1. Dynamic Closeness Metric

One centrality measure that relies on distance is Closeness. It is determined by averaging the shortest distances (involving the fewest nodes, thus minimizing overhead) from a specific node to all other nodes in the network. This is equivalent to summing the shortest distances (d_{short}) and dividing by the number of nodes (referred to as the network order, denoted as $|V|$), minus one, as node j itself is excluded from this calculation:

$$Average\ path\ lengths(v) = \frac{\sum_{k=1,k\neq v}^{|V|-1} min(D(v,k))}{|V|-1} = \frac{\sum_{k=1,k\neq v}^{|V|-1} min\left(\sum_{i=1}^{|V|-1}\sum_{j=2}^{|V|}\left(\{\lambda x_{ij}\}_{t_{ij}}\right)+\lambda\right)}{|V|-1} \quad (10)$$

where $x_{ij} = 1$ if the link $\{i, j\}$ is used in a route between the source node v and the destination node k and $x_{ij} = 0$ otherwise.

The lower the above value, the closer a node is to the center of the network. For this reason, closeness is defined as the reciprocal of Equation (10), so that the more centered a node v is in the network, the higher its closeness metric is:

$$C_{CLO}(v) = \frac{|V|-1}{\sum_{k=1,k\neq v}^{|V|-1} min\left(\sum_{i=1}^{|V|-1}\sum_{j=2}^{|V|}\left(\{\lambda x_{ij}\}_{t_{ij}}\right)+\lambda\right)} \quad (11)$$

The Closeness measure in the network is strongly related to the rate of information propagation between devices as well as the timeframes at which messages are transmitted over the network. This metric offers a means to evaluate how accessible a device is within the network.

Let us examine an example calculation for the connected devices $\{A, B, C, D, F\}$ during G_7, as shown in Figure 4.

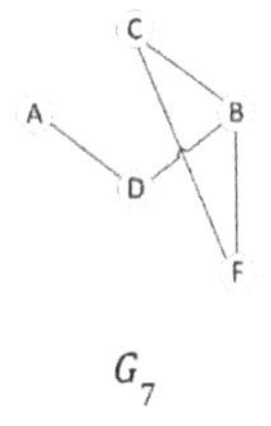

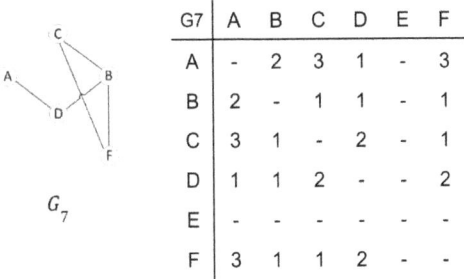

G7	A	B	C	D	E	F
A	-	2	3	1	-	3
B	2	-	1	1	-	1
C	3	1	-	2	-	1
D	1	1	2	-	-	2
E	-	-	-	-	-	-
F	3	1	1	2	-	-

Figure 4. Network connections during instant G_7, based on the example of Figure 1.

$$C_{CLO}(A) = \frac{6-1}{\lambda(2+3+1+3)+\lambda} = \frac{5}{10\lambda}$$

$$C_{CLO}(B) = \frac{6-1}{\lambda(2+1+1+1)+\lambda} = \frac{5}{6\lambda}$$

$$C_{CLO}(C) = \frac{6-1}{\lambda(3+1+2+1)+\lambda} = \frac{5}{8\lambda}$$

$$C_{CLO}(D) = \frac{6-1}{\lambda(1+1+2+2)+\lambda} = \frac{5}{7\lambda}$$

$$C_{CLO}(F) = \frac{6-1}{\lambda(3+1+1+2)+\lambda} = \frac{5}{8\lambda}$$

Therefore, the ranking of nodes according to their closeness index is as follows:

$$Ranking = \{B, D, C, F, A\} \tag{12}$$

Regrettably, calculating the Closeness metric (as indicated in Equation (11)) for each node in a network requires knowledge of the distances between all pairs of vertices. In disconnected networks, where nodes belong to distinct components or subnetworks that lack any larger linked subnetwork, the distance between two nodes is traditionally considered infinite, as depicted in the example shown in Figure 1, rendering Closeness inapplicable. As a result, the reciprocal becomes 0, and the sum in the equation (Equation (11)) diverges. Devices often belong to various components, rendering Closeness values irrelevant for all devices in the network except those within the largest component. Consequently, the computation of the Closeness metric must exclude devices that are part of smaller components.

Through the utilization of dynamic shortest paths, we can overcome the limitation of the conventional Closeness metric in disconnected networks. By employing a method that accumulates the reciprocal of path costs instead of the reciprocal of the total path cost, we can redefine the Closeness metric. This approach takes into consideration communication costs or overhead. As a result, the Dynamic Closeness metric is defined as the sum of the reciprocals of distances, rather than the reciprocal of the sum of distances:

$$C_{CLO}(v) = \sum_{k=1,k\neq v}^{|V|} \left(\frac{1}{min\left(\sum_{i=1}^{|V|-1} \sum_{j=2}^{|V|} \left(\{\lambda x_{ij}\}_{t_{ij}} \right) + \lambda \right)} \right) \tag{13}$$

The adoption of the Dynamic Closeness measure prevents situations where an infinite distance dominates over other distances. Additionally, this measure can be standardized by considering that in a network with a star topology, the maximum value is achieved by the central node, which is equal to $|V| - 1$ (the longest distance possible in a network with $|V|$ nodes is $|V| - 1$, i.e., in a chain-connected network). The standardized value of the central node in a star network is 1, while the value for the leaf nodes is

$$\frac{1}{|V|-1}\left(\frac{1}{1} + (|V|-2)\frac{1}{2}\right) = \frac{|V|}{2(|V|-1)} \tag{14}$$

Thus, the centrality index is now defined by

$$C_{CLO}(v) = \frac{1}{|V|-1} \sum_{k=1,k\neq v}^{|V|} \left(\frac{1}{min\left(\sum_{i=1}^{|V|-1} \sum_{j=2}^{|V|} \left(\{\lambda x_{ij}\}_{t_{ij}} \right) + \lambda \right)} \right), \tag{15}$$

subject to $t_{k,k+1} \leq t_{k+1,k+2} \leq \ldots \leq t_{|V|-1,|V|}$.

Thus, considering $(|V|-1)^{-1}\sum_{k=1,k\neq v}^{|V|}\left(min\left(\sum_{i=1}^{|V|-1}\sum_{j=2}^{|V|}\left(\{\lambda x_{ij}\}_{t_{ij}}\right)+\lambda\right)\right)^{-1}$, the closeness evolution of device v during the time span T is defined as

$$\tau_C(v) = \frac{F_T\sum_{l=1}^{T}\left(t_l(|V|-1)^{-1}\sum_{k=1,k\neq v}^{|V|}\left(\min\left(\sum_{i=1}^{|V|-1}\sum_{j=2}^{|V|}\left(\{\lambda x_{ij}\}_{t_l}\right)+\lambda\right)\right)^{-1}\right)}{F_T\sum_{l=1}^{T}t_l^2-\left(\sum_{l=1}^{T}t_l\right)^2} - \frac{\sum_{l=1}^{T}t_l\sum_{l=1}^{T}\left((|V|-1)^{-1}\sum_{k=1,k\neq v}^{|V|}\left(\min\left(\sum_{i=1}^{|V|-1}\sum_{j=2}^{|V|}\left(\{\lambda x_{ij}\}_{t_l}\right)+\lambda\right)\right)^{-1}\right)}{F_T\sum_{l=1}^{T}t_l^2-\left(\sum_{l=1}^{T}t_l\right)^2} \tag{16}$$

In conclusion, we put forward the Dynamic Closeness for device v ($CTE(v)$), which takes into account the fluctuations in the Closeness measure, as defined:

$$CTE(v) = \alpha\tau_C(v) + (1-\alpha)\frac{1}{F_T}\sum_{l=1}^{F_T}\left(\frac{1}{|V|-1}\sum_{k=1,k\neq v}^{|V|}\left(\frac{1}{min\left(\sum_{i=1}^{|V|-1}\sum_{j=2}^{|V|}\left(\{\lambda x_{ij}\}_{t_l}\right)+\lambda\right)}\right)\right) \tag{17}$$

where $x_{ij} = 1$ if the link $\{i,j\}$ is used in any route between nodes v and k and $x_{ij} = 0$ otherwise.

3.2.2. Social Closeness Metric

Human social relationships typically display greater stability than transmission links between mobile devices due to the complex network conditions present, for instance, in Opportunistic Mobile Social Networks (OppMSNs), characterized by intermittent connectivity that results in unstable end-to-end paths between devices. As a result, OppMSN routing decisions may be made more efficient using social indicators.

It is noticed that people keep both regular and sporadic interactions within their social surroundings. Information propagation greatly depends on the degree of contact between nodes. If the sender often communicates with the destination device, the sender may be aware of the times when they are most likely to run across the destination or nodes that are very likely to cross paths with the destination in the future [29,30]. Conversely, the likelihood of two devices knowing one another improves if they have a greater number of friends in common.

We examine the devices' past contacts in order to develop the Opportunistic Relationship Index (ORI), a social metric that is derived from important structural characteristics of a complex network, specifically the contact durations between devices, their shared neighbors, and distances [31]. In order to reflect the possibility of establishing a connection between devices v and k, the score is calculated as shown in Equation (17) for each pair of unconnected devices v and k. In this equation, $\sum_{i=1}^{T}\{\varphi_{vk}x_{vk}\}_{t_i}$ represents the frequency of contact occurrences between devices v and k within the time span T, and $\sum_{i=1}^{|V|-1}\sum_{j=2}^{|V|}x_{ij}$ denotes the distance matrix of existing paths between the two devices:

$$ORI_T(v,k) = \begin{cases} \left(\sum_{i=1}^{T}\{\varphi_{vk}x_{vk}\}_{t_i}\right)^{\frac{|\Gamma(v)\cap\Gamma(k)|+1}{2}} & \text{if } \Gamma(v)\cap\Gamma(K)\neq 0 \\ \left(\sum_{i=1}^{T}\{\varphi_{vk}x_{vk}\}_{t_i}\right)^{\frac{1}{min(\sum_{i=1}^{|V|-1}\sum_{j=2}^{|V|}x_{ij})}} & \text{otherwise,} \end{cases} \tag{18}$$

where $\Gamma(v) = \sum_{i=1}^{T}\sum_{l=1,l\neq v}^{|V|}\{x_{vl}\}_{t_i}$ and $\Gamma(k) = \sum_{i=1}^{T}\sum_{l=1,l\neq k}^{|V|}\{x_{kl}\}_{t_i}$ represent the respective sets of neighbors for devices v and k, respectively, over the time span T.

Figure 5 in this context shows the subnetwork created from Appendix A Figure A1, only showing the four current routes connecting device A and device B, with the weights denoting the calculated ORI.

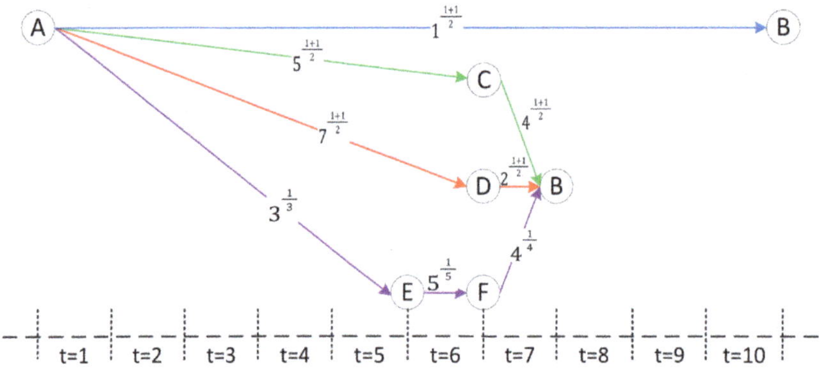

Figure 5. Illustration of the paths from device *A* to device *B* of example in Figure 1, where the weights represent the ORI.

Then, we present the shortest path based on opportunistic relationships, which include both opportunistic relationships and communication costs. We invert the weights to find the path with the lowest weight since the weight denotes the *ORI*. We use a tuning parameter to ensure that the ORI and the number of intermediary devices affect the choice of the best path, incorporating both communication cost and *ORI*. The following is the social measure used in Equation (18) to determine the opportunistic cost ($Cost_{opp}$) of a path between a source device (s) and a destination device (d):

$$Cost_{opp}(s, s+1, \ldots, d) = \sum_{i=s}^{d-1} \left(\frac{1}{ORI_T(i, i+1)} \right)^{\alpha}, \tag{19}$$

subject to $t_{s,s+1} \leq t_{s+1,s+2} \leq \cdots \leq t_{d-1,d}$.

The *ORI* between devices i and j before their interaction with the next relay device is indicated by the symbol ORI_T in Equation (18). The moment at which device i and device j first made contact is represented by $t_{i,j}$, and the sequence in which connections between intermediary devices are made is indicated by $t_{s,s+1} \leq t_{s+1,s+2} \leq \cdots \leq t_{d-1,d}$.

We further normalize the series by dividing the geometric mean of ORI_T by the highest ORI_T, as the routing choice may not function effectively with a low ORI_T value across devices. The tuning parameter is this normalized value, which enables us to gauge the degree of dispersion between low and high ORI_T values. As a result, the following is how the definition of α is stated:

$$\alpha_{s,s+1,\ldots d} = \frac{\left(\prod_{i=s}^{d-1} ORI_T(i, i+1) \right)^{\frac{1}{|\{s,d\}|}}}{max(ORI_T(s, s+1), ORI_T(s+1, s+2), \ldots, ORI_T(d-1, d))}, \tag{20}$$

where $|\{s, d\}|$ represents the number of hops in the path from the source device s to the destination device d.

By adding considerations of the shortest pathways, social interactions between nodes, and communication cost, we use the opportunistic-based shortest path approach to expand the Closeness measure in this way. The Social Closeness metric (CLO_{opp}) is calculated in the manner described in the following:

$$\begin{aligned} CLO_{opp}(v) &= \frac{1}{|V|-1} \sum_{k=1, k \neq v}^{|V|} \left(\frac{1}{\min\left(Cost_{opp}(v,k)\right)} \right) = \\ &= \frac{1}{|V|-1} \sum_{k=1, k \neq v}^{|V|} \left(\frac{1}{min\left(\sum_{i=v}^{k-1} \left(\frac{1}{ORI_T(i,i+1)} \right)^{\alpha} \right)} \right), \end{aligned} \tag{21}$$

subject to $t_{v,v+1} \leq t_{v+1,v+2} \leq \ldots \leq t_{k-1,k}$, in which $\min(Cost_{opp}(v,k))$ first computes all the paths between v and k, then calculates the $Cost_{opp}$ of each path, and finally selects the minimum $Cost_{opp}$ among all.

Hence, taking into consideration the expression $(|V|-1)^{-1}\sum_{k=1,k\neq v}^{|V|}\left(min\left(\sum_{i=v}^{k-1}\left(\frac{1}{\{ORI_T(i,i+1)\}_{t_i}}\right)^{\alpha}\right)\right)^{-1}$, we can evaluate a device v's social closeness at the specified snapshot t_l. As a result, Equation (22) describes how the social closeness of device v changed throughout time T:

$$\tau_{opp}(v) = \frac{F_T\sum_{l=1}^{T}\left(t_l(|V|-1)^{-1}\sum_{k=1,k\neq v}^{|V|}\left(\min\left(\sum_{i=v}^{k-1}\left(\frac{1}{\{ORI(i,i+1)\}_{t_l}}\right)^{\alpha}\right)^{-1}\right)\right)}{F_T\sum_{l=1}^{T}t_l^2-\left(\sum_{l=1}^{T}t_l\right)^2} - \frac{\sum_{l=1}^{T}t_l\sum_{l=1}^{T}\left((|V|-1)^{-1}\sum_{k=1,k\neq v}^{|V|}\left(\min\left(\sum_{i=v}^{k-1}\left(\frac{1}{\{ORI(i,i+1)\}_{t_l}}\right)^{\alpha}\right)^{-1}\right)\right)}{F_T\sum_{l=1}^{T}t_l^2-\left(\sum_{l=1}^{T}t_l\right)^2} \tag{22}$$

In conclusion, we provide the definition of the Social Closeness metric of a node v ($CTE_{opp}(v)$), which combines the dynamics of the Social Closeness metric in:

$$CTE_{opp}(v) = \tau_{opp}(v)^{\alpha} + \frac{1}{F_T}\sum_{l=1}^{F_T}\left(\frac{1}{|V|-1}\sum_{k=1,k\neq v}^{|V|}\left(\frac{1}{\min\left(\sum_{i=v}^{k-1}\left(\frac{1}{\{ORI(i,i+1)\}_{t_l}}\right)^{\alpha}\right)}\right)\right), \tag{23}$$

subject to $t_{v,v+1} \leq t_{v+1,v+2} \leq \cdots \leq t_{k-1,k}$.

4. Results and Discussion

Our goal is primarily to discuss the dataset used to assess various QoS measures and the analytical process once we have shown the production of rankings with the most significant nodes based on different centrality metrics. We will next go through how we integrated these rankings into message routing.

Using a collection of Reality Mining datasets [32], the recommended algorithm's efficacy has been evaluated. These datasets embody a complex social system by capturing data from 100 mobile phones over a span of 9 months. The authors demonstrate how common Bluetooth-enabled mobile phones can be used to measure information access and utilization in a variety of settings, detect social patterns in users' daily activities, infer relationships, identify socially significant locations, and model organizational patterns.

However, it is worth noting that we utilize a modified version of the Reality dataset provided by the authors of [33]. As stated in the same reference, there is no significant activity before and after the timestamp ranges 1,094,545,041 and 1,111,526,856. Therefore, the simulations presented in this paper exclusively employ the data within that time interval, as shown in Table 3.

Table 3. Characteristics of the dataset.

Feature	Value
Number of devices	97
Environment	Campus
Dataset duration	246 days
Dataset duration used	196 days
Encounter prob. 1st 1/4 day	0.0003
Encounter prob. 2nd 1/4 day	0.0011
Encounter prob. 3rd 1/4 day	0.0019
Encounter prob. 4th 1/4 day	0.0012
Percentage of dataset duration for the Training Graph (GT)	75%
Percentage of dataset duration for the Probe Graph (GP)	25%
Network density	<0.5%
Number of contacts of the top 20 devices	4–9

Regarding the methodology for analyzing the dataset, it has undergone processing using a similar approach employed when implementing Machine Learning models. In this manner, the dataset is partitioned into two distinct non-overlapping graphs known as the training (GT) and probe (GP) graphs. The Training Graph (GT) is constructed by selecting a subset that represents the initial 75% of node interactions within the dataset. The remaining edges, not included in GT, constitute the Probe Graph (GP). Likewise, the edges included in GT are denoted as ET, while those in GP are referred to as EP, i.e., E = ET + EP. It should be noted that ET and EP are mutually exclusive; however, there may be overlapping nodes between GT and GP. For our experiments, we have allocated 75% of the edges to ET and the remaining 25% to EP.

The simulations have been conducted by simulating the GP (Probe Graph) with an implementation of a total of five routing algorithms. On one hand, we include the conventional ones typically used to evaluate QoS in OppNets, that is Spray and Wait (S&W), Prophet versions 1 and 2, and Epidemic (four algorithms). On the other hand, we incorporate a modified version of the S&W algorithm, which is evaluated three times based on a parameterized ranking of the most significant nodes according to the metrics described in the previous sections (Dynamic Degree, Dynamic Closeness, and Social Closeness), which sums a total of five routing algorithms. The results of the simulations will be presented in Sections 4.3.4–4.3.6.

The algorithms have been developed using The ONE (Opportunistic Network Environment) simulator [34] and can be accessed from a public repository located at https://github.com/sito25/pubtesis.git, (accessed on 21 September 2023) under the GNU Lesser General Public License v.3.0. This simulator is specifically designed for opportunistic networks and was initially developed at Aalto University in 2009. The ONE provides a wide range of capabilities, including the generation of node movements using various models, replication of message traffic and routing, cache management, and visualization of both mobility and message transmission through its graphical user interface. Additionally, it offers diverse reporting options, such as node movements, message transmission, and general statistics. Currently, it is collaboratively maintained by Aalto University and Technische Universität München, boasting a robust user community. The version utilized in our research is 1.6.0, implemented in Java.

4.1. Network Density

As mentioned earlier, opportunistic networks are a type of low-density networks that traditionally focus on self-organized and ad hoc mobile networks. These networks often experience frequent disruptions, delays, and intermittent connectivity, leading to a lack of end-to-end connections within the environment. In such scenarios, wireless devices can temporarily store information and forward it to other devices that are more likely to be within communication range of the intended destination when an opportunity for connection arises.

The density of an opportunistic network is determined by the ratio of edges present in a graph to the maximum number of edges the graph can contain. This ratio provides a conceptual idea of the network's connectivity in terms of link density. Specifically, network density is defined as the ratio of the number of connections to the maximum possible connections.

A network is considered dense when the number of links is close to the maximum possible, where every pair of devices is connected by a single link. Conversely, a network with few links is considered sparse. This concept provides an understanding of the level of connectivity and density within the network [35,36]. Therefore, to determine the maximum number of connections in the network, we can derive it as follows:

$$MaxConn = \frac{|V|(|V|-1)}{2} \tag{24}$$

Let us now introduce the formula for calculating network density. The network density is calculated by dividing the total number of connections existing in the network $G(V, E)$ by the maximum possible number of connections that could potentially exist within the network. Let us examine the formula in detail:

$$Density = \frac{\frac{|E|}{|V|(|V|-1)}}{2} = \frac{2|E|}{|V|(|V|-1)} \tag{25}$$

In our study, Figure 6 displays the likelihood of the presence of complete pathways, which is linked to the density of connections within the Reality Mining datasets. This likelihood, denoted as $P(EE)$, can be defined as the ratio of the number of established end-to-end routes to the total number of possible connections.

$$P(EE) = \left(\sum_{s=1}^{|V|-1} \sum_{d=1}^{|V|-1} \sum_{i=1}^{|V|-1} M_T^i(s,d) \right) \left(\frac{|V|(|V|-1)}{2} \right)^{-1} \tag{26}$$

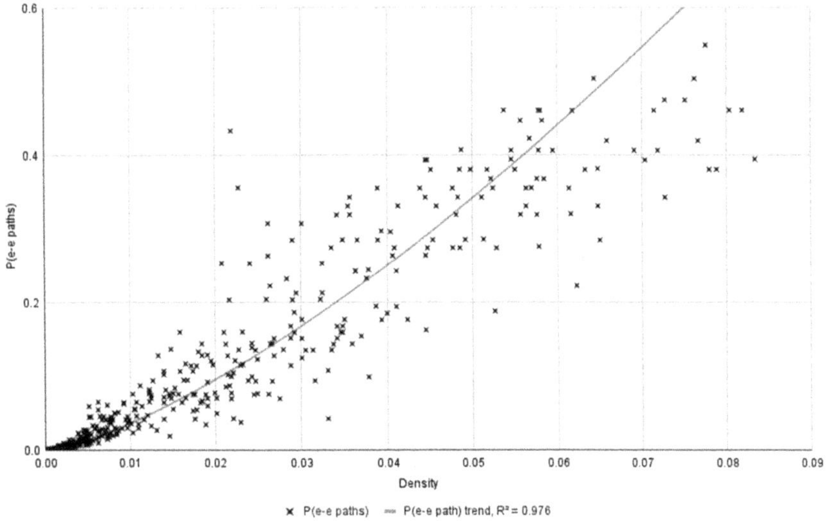

Figure 6. End-to-end paths probability in the datasets as a function of the connection density.

Figure 6 illustrates the network structure of the used dataset, which is characterized as a sparsely connected network with a consistent connection density among devices that does not exceed 8.5% throughout its dynamic nature. The majority of density values are below 1%. As a result, the upper limit for the probability of end-to-end connections remains below 55%, with a significant concentration below 0.5%. Therefore, it is imperative to consider that the utilized datasets define an opportunistic network with a very low density of connections.

4.2. Effectiveness Analysis of the Proposed Metrics

We may examine how successfully suggested metrics capture and quantify the intended attributes or characteristics of network devices by evaluating their efficacy. It enables us to assess how well these metrics capture the underlying ideas of ranking, connection, or social impact. As a result, we can evaluate their effectiveness and decide which metrics are more appropriate for achieving our study goals. Based on their capacity to capture the necessary elements of device centrality, these analyses aid in the selection of the most suitable centrality metrics. We can also establish whether these metrics provide useful information and can be relied upon when making network design decisions. For performance evaluation, comparative analysis, hypothesis validation, and determining their practical relevance, evaluating the efficacy of local and global centrality measures in the network is essential.

4.2.1. Local Metrics

We compared Dynamic Degree during the studies to two benchmark measures, namely, Degree and Weighted Degree. Figure 7 uses the Reality Mining datasets to show the outcomes of these studies. The top-N devices are sorted according to their Weighted Degree and Dynamic Degree, and Figure 7a shows the plotted curves reflecting the average number of nearby devices among those devices.

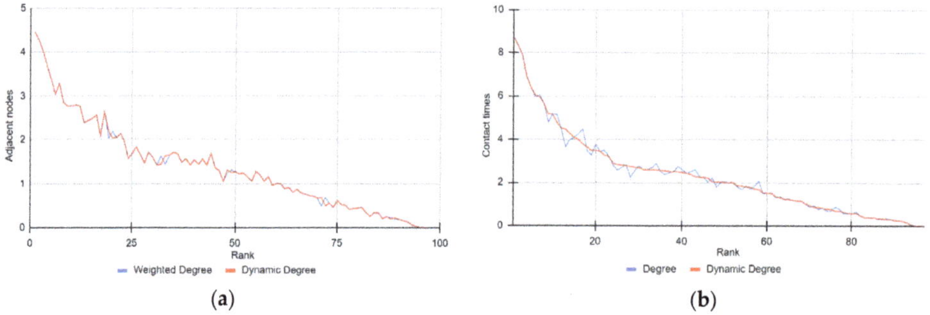

Figure 7. Effectiveness of local metrics: (**a**) Number of neighboring devices among the top-N devices; (**b**) average number of contacts among the top-N devices.

An effective local impact metric in Figure 7a should show a diminishing trend since a node with strong local influence often shows a high number of nearby nodes. The curves for the two measures shown in the picture, however, are rather near to one another. This resemblance could result from certain traits or distinctive qualities that the dataset itself possesses.

Figure 7b shows the plotted curves for the top-N devices ordered by their Degree and Dynamic Degree, which reflect the average number of connections. In light of the fact that a device with a strong local effect should interact with its neighbors often, the Dynamic Degree curve has a smoother downward slope and performs better than the Degree metric. Therefore, of the three influence measures, the Dynamic Degree meter performs the best since it establishes a balance between the number of nearby devices and the frequency of encounters.

4.2.2. Global Metrics

Using real-world datasets, we analyzed the distribution properties of suggested global influence indicators. Figure 8 displays the Complementary Cumulative Distribution Functions (CCDF) for the suggested global centralities, with the horizontal axis denoting the order of device effect. Here, it is clear that the distributions of the Closeness measure are not uniform when taking into account various centrality techniques. This suggests that the selection of the centrality approach affects the measures' distributions. These results help us determine the best tools to increase the effectiveness of data transmission in OppMSNs. These features can be used by routing algorithms to choose the most reliable device as a relay. As a result, the impact measures suggest various possible contributions of the same device to information propagation when combined with various centrality approaches. This connection between device impact and route design, in our opinion, is a key element in enabling effective information propagation inside OppMSNs.

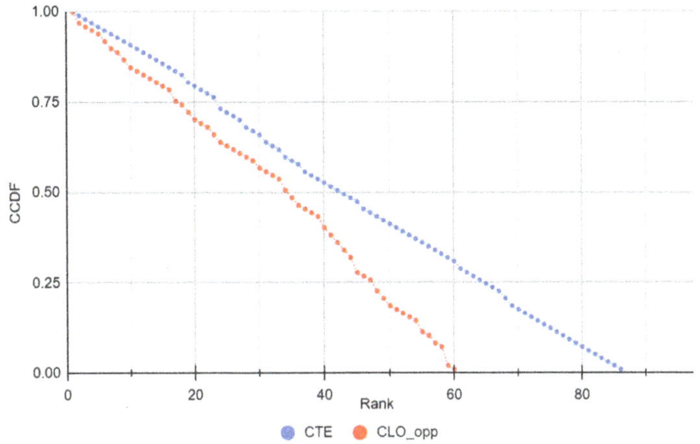

Figure 8. Illustration of Complementary Cumulative Distribution Function of proposed global centrality metrics.

4.2.3. Correlation Analysis between Local and Global Metrics

Our goal is to study the probable association between the ranking of global influence and the ranking of local impact through time and to address the consequences of anticipating global influence by looking at the consistency and predictability of human social qualities.

A statistical metric used to quantify the strength of the association between the relative changes of two variables is the correlation coefficient. It includes values between −1 and 1. A complete negative correlation is represented by a correlation coefficient of −1, and a perfect positive correlation is represented by a correlation coefficient of 1. The absence of a linear link between the changes in the two variables is shown by a correlation value of 0.

A statistical metric used to quantify the degree of linear association between two variables is the Pearson correlation coefficient. This coefficient reveals the nature and strength of the relationship, given that a change in one variable leads to a proportionate change in the other. When there is no apparent association, the Pearson coefficient returns a value of 0.

Another correlation statistic used to assess rank correlation, which indicates the statistical dependence between the ranks of two variables, is Spearman's rank correlation coefficient. The Spearman coefficient measures the extent to which the relationship between variables can be characterized by a monotonic function, in contrast to a linear relationship where the rate of increase or decrease is constant. Without necessarily adhering to a consistent rate of change, it assesses how well the data align monotonically. The Spearman correlation evaluates the monotonic relationship between two variables, whether they are continuous or ordinal in nature, by considering the ranked values of each variable rather than the raw data. In a monotonic relationship, one of the following is true:

- As one variable increases, the value of the other variable decreases; or
- Conversely, as one variable increases, the value of the other variable also increases.

The methods employed by the two correlation coefficients are fundamentally distinct. While the Spearman coefficient considers both linear and monotonic correlations, the Pearson coefficient focuses exclusively on linear relationships between variables. Furthermore, Spearman utilizes rank-ordered variables, whereas Pearson employs the raw data values of the variables.

It is advisable to employ the Spearman coefficient instead of the Pearson coefficient when a scatterplot reveals a potential link that could be either monotonic or linear. Using the Spearman coefficient does not cause any harm, even if the data eventually demonstrate a perfect linear relationship. Nevertheless, choosing Pearson's coefficient might lead to missing crucial insights that Spearman could provide in cases where the connection is not exactly linear. Therefore, as illustrated in Figure 9, we utilize the Spearman correlation coefficient to analyze the relationship between the ranks of Closeness and Degree throughout their dynamics. By establishing a consistent relative order of observations within each variable (e.g., first, second, third, and so on), it is intuitively understandable that the Spearman correlation between two variables becomes strong when observations possess similar (or identical, resulting in a correlation of 1) ranks. Conversely, the correlation is low when observations exhibit disparate ranks across the two variables (or entirely opposite rankings, leading to a correlation of −1).

Let $(x_1, y_1), (x_2, y_2), \ldots, (x_n, y_n)$ represent a collection of composite rankings from two distinct ranking lists, X and Y. The n raw scores (x_i, y_i are converted to ranks ($R(x_i)$, $R(y_i)$), and r_s is derived as follows:

$$r_s = \rho(R(X), R(Y)) = \frac{cov\left(R(X), R(Y)\right)}{\sigma\left(R(X)\right)\sigma\left(R(Y)\right)} = \\ = \frac{\left(2n^2\right)^{-1}\sum_{i=2}^{n}\sum_{j=1}^{n}\left(R(x_i)-R(x_j)\right)\left(R(y_i)-R(y_j)\right)}{\left(\frac{1}{n}\sum_{i=1}^{n}\left(R(x_i)-\frac{1}{n}\sum_{i=1}^{n}R(x_i)\right)^2\right)^{\frac{1}{2}}\left(\frac{1}{n}\sum_{i=1}^{n}\left(R(y_i)-\frac{1}{n}\sum_{i=1}^{n}R(y_i)\right)^2\right)^{\frac{1}{2}}}, \tag{27}$$

where

ρ represents the application of the Pearson correlation coefficient to the rank-transformed variables.
$cov(R(X), R(Y))$ is the covariance of the rank variables.
$\sigma(R(X))$ and $\sigma(R(Y))$ represent the standard deviations of the rank-transformed variables.

With mean and standard deviation statistical significances (p-values) of 0.0035 and 0.0381, respectively, the majority of Spearman's rank correlation coefficients for the proximity measure in Figure 9 exhibit values near 0.75. This observation underscores the strong relationship between this measure and the Dynamic Degree. As the window period lengthens, the curve remains largely stable despite minor fluctuations in the correlation coefficients. Given that the Degree measure influences the Closeness metric in the dynamic context, it becomes conceivable to formulate an approximation of Closeness metric values through a combination of social-based and dynamic-based methods. These characteristics offer opportunities to enhance the accuracy of our closeness prediction strategies and to devise more effective forwarding algorithms in OppMSNs.

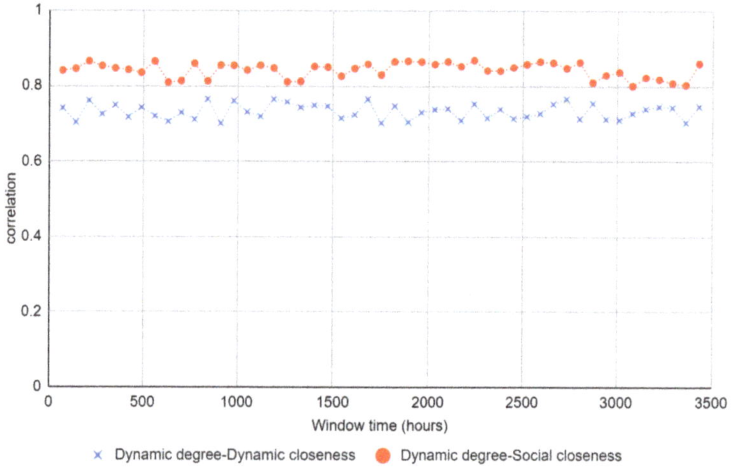

Figure 9. Ranking of Spearman correlations by varying snapshots during the time span.

4.3. *Quality of Service Metrics Analysis*

Latency, overhead, or hop count are QoS metrics that allow us to assess the performance and efficiency of the network. By quantifying the latency, we can determine how quickly data or messages are transmitted from the source to the destination. An overhead calculation helps assess the additional resources or data required to support the communication process. By quantifying overhead, we can identify potential inefficiencies or resource-intensive aspects of the network. This information is valuable for optimizing network performance and ensuring efficient resource utilization.

On the other hand, a hop count calculation helps evaluate the number of network devices or "hops" required for data to travel from the source to the destination. Lower hop counts generally imply a more direct and efficient routing path, resulting in reduced latency and improved overall network performance.

Overall, accurate calculation and analysis of latency, overhead, and hop count are crucial for performance evaluation, resource optimization, reliability assessment, routing efficiency analysis, identifying areas for improvement, and making informed decisions regarding network design and configuration.

In summary, precise calculations and analyses of latency, overhead, and hop count are pivotal for performance evaluation, resource optimization, reliability assessment, and routing efficiency analysis. To fulfill these requirements, we have implemented a set of algorithms to evaluate these QoS metrics and incorporate our proposals into The ONE—an opportunistic network simulator that is well-suited for simulating and studying such networks, as mentioned earlier.

4.3.1. Updating Training Matrix on Contact

The analysis for the used dataset involves simulating data analysis with a similar methodology to the one that is used to train Machine Learning models. This simulation includes dividing the data into a training subset and a test subset. As mentioned earlier, we allocate 75% of the simulation data to train a set of adjacency matrices. In these matrices, we simply increment the counter for each connection between pairs of nodes (device, another host). Algorithm 1 generates adjacency matrices that not only indicate connected nodes but also capture the connectivity capacity of nodes, making them likely to be chosen as message transporters to other nodes.

It is important to note that matrices are calculated with a specific frequency to ensure that different adjacency matrices are collected, reflecting the evolving connectivity of the nodes.

Algorithm 1. Updating training matrix on contact

Input: *TM* (Training Matrix), *N*1 and *N*2 (new contact between two nodes)
Output: *TM* (a new version of Training Matrix)

```
begin
    if TM[N1, N2] == null then
        TM[N1, N2] = 0;
    if TM[N2, N1] == null then
        TM[N2, N1] = 0; TM[N1, N2] +=1;
    TM[N1, N2] +=1;
    TM[N1, N2] +=1;
return TM
```

Computational complexity analysis allows us to evaluate the efficiency of an algorithm in terms of resource usage, such as time and memory [37]. By understanding the complexity, we can estimate how the algorithm will respond to varying input sizes. In the case of Algorithm 1, the complexity is simply O(1) because when two nodes meet each other, the corresponding cells of the adjacency matrix (TM[*N*1, *N*2] and TM[*N*2, *N*1]) are updated, indicating how many times those nodes have been found throughout the considered period. The spatial complexity, on the other hand, is determined by the need to store contact information in an N × N matrix, where N is the number of nodes in the topology. However, since we also store M matrices to reflect the evolution of the adjacency matrix between nodes, the spatial complexity turns out to be O(M × N2).

For each matrix, we generate an intermediate ranking of nodes with good connectivity. These rankings are then combined to obtain a final ranking, as described in Section 3, Influence Metrics, depending on the metric used. The rankings assist the routing algorithm in determining whether to forward a new copy of the package to the identified node.

4.3.2. Calculation of Friends Nodes upon Contact

In our proposal, the concept of a friend refers to a node that has connected with another node and is likely to reconnect within a relatively short time, based on the principle of temporal locality. To implement this idea of temporal locality, we introduce the friend concept, which involves setting a timer for each pair of connecting devices. This timer is activated after the nodes establish a connection. If they reconnect before the timer expires, we consider them friends who frequently connect. On the other hand, if the timer has expired by the time they encounter each other again, they are still friends but do not connect frequently.

The concept of a friend represents a list of nodes to which the given node has previously connected. This list is closely related to the adjacency matrices described in Algorithm 1, as both implementations rely on node connections. However, unlike the adjacency matrices, the friends list is not reset with each new matrix. Instead, it is continually updated throughout the simulation as the node forms new connections.

It is important to note that the friends list is not utilized during the training phase (when running Algorithm 2). Instead, it is used during testing, which will be explained further in Algorithm 3.

Moreover, the concept of a friend can be utilized to assess the extent to which these friends adhere to the notion of temporal locality. For instance, a connection counter between them can be employed within the context of their unexpired timer. In these simulations, we use the concept of friends as a list of nodes to which a specific node has connected throughout the entire simulation.

Algorithm 2. Calculation of Friend Nodes on contact

```
Input:   N1 LNF (List of N1 friends), N1 and N2 (new contact between two nodes)
Output:  N1 LNF (a new version of N1 friends), T(N1,N2) (timeout between N1 and N2)
and TL(N1,N2) (temporal locality between L1 and L2)
begin
    MAX_TIMEOUT = 20,000;
    if N2 in N1 LNF then
        elapsed_time = (current_time - T(N1,N2));
        if elapsed_time < MAX_TIMEOUT then
            TL(N1, N2) +=1;
    else
        add N2 to N1 LNF;
    T(N1, N2) = current_time;
return N1 LNF, T(N1,N2), TL(N1,N2)
repeat with the input: N2 LNF, N1 and N2
```

The complexity is also O(1) because this algorithm simply updates the list of friends (LNF) of a node and vice versa when it connects to another. The space complexity is O(N2) because the friend list of each node is not renewed during the simulation, as is the case with TM adjacency matrices during training. Instead, a friend is added to each contact in case they both meet and connect for the first time.

4.3.3. Routing Decision on Contact

The main objective of Algorithm 3 is to determine whether node *A*, given the connection between nodes *A* and *B*, should send a copy of the messages it carries to node *B*. This decision aims to minimize overhead. Instead of sending copies to every encountered node *B*, it is preferable to choose a node with better connectivity. Such a node is more likely to have greater access to a larger network.

Algorithm 3. Routing decision on contact

```
Input:   N1 and N2 (new contact between two nodes), R (best nodes ranking), N1 LNF
Output:  N2 messages queue
begin
  | for each message in N1 queue do
  |   | N3 = obtain message destination;
  |   | is_one_of_best_nodes = (N2 in R);
  |   | are_friends = (N3 in N1 LNF);
  |   | arrived_to_destination = (N2 == N3);
  |   | if is_one_of_best_nodes or are_friends or arrived_to_destination then
  |   |     forward message to N2 queue;
  | end for
return N2 messages queue
```

This algorithm requires the use of different data structures to determine whether node N1 forwards the messages in its queue to *N*2. Therefore, these data structures are accessed in a single loop, where the processing of each one is decided. The time complexity is O(N). In terms of spatial complexity, we need a list M for each node N to act as a message buffer, and a unique list to store the ranking of nodes. Consequently, the spatial complexity is $O(N \times M)$.

This approach restricts the generation of message copies by node *A* (a finite number in the case of S&W or unlimited in the case of Epidemic) to specific connections where node *B* exhibits one of the following three characteristics:

- Node *B* ranks among the top positions in the ranking obtained through Algorithm 1. Being a node with good connectivity, it is more likely to successfully deliver the packet to the intended recipient or another node that can assist in reaching the message's destination;
- The source and destination nodes of the message are friends. This indicates that they have previously connected and are likely to reconnect. Therefore, node *A*, carrying the message, is allowed to deliver a copy to node *B*;
- Node *B* is the intended destination of the message. In this scenario, it is logical for node *A* to deliver the message to node *B*.

4.3.4. Packet Latency

Based on Figure 10, it becomes apparent that our algorithms achieve a decreased average packet delay in comparison to the original Spray and Wait protocol, with an average reduction of approximately 2% with respect to S&W and more than 10% with respect to Epidemic or Prophets v1 and v2 protocols. By integrating equivalent buffer sizes, minimizing overhead, and intelligent packet forwarding selection (unlike S&W, which disseminates packets to all encountered nodes), our approach facilitates expedited packet delivery to their intended destination.

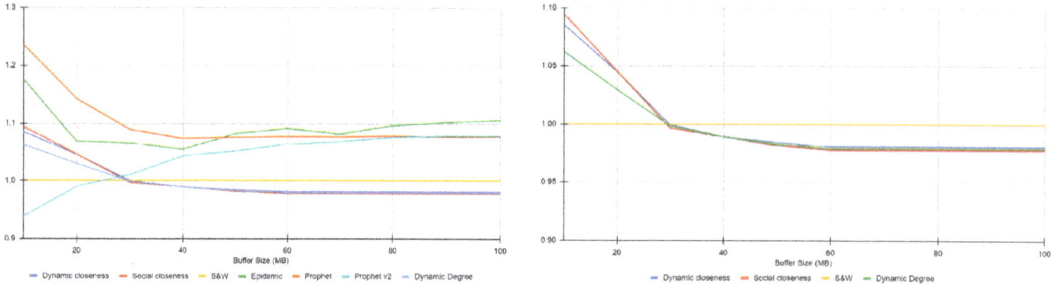

(**a**) (**b**)

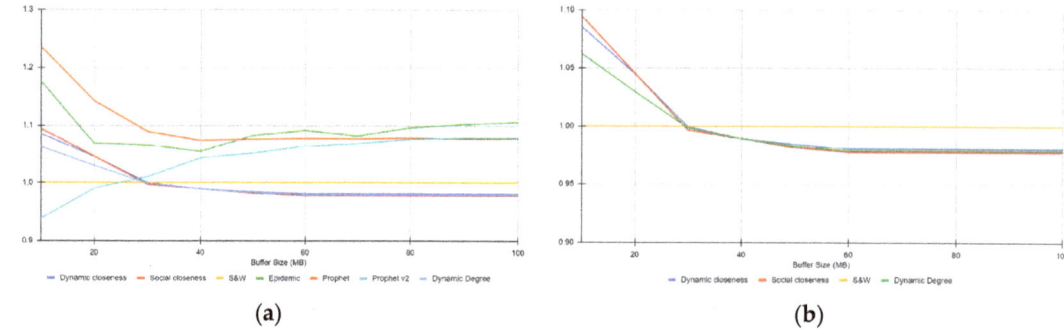

Figure 10. Average latency for different routing protocols, depending on the buffer size, normalized with respect to the results of the S&W routing protocol: (**a**) comparison among all the analyzed routing protocols; (**b**) comparison of the proposed metrics and S&W.

4.3.5. Path Length

As illustrated in Figure 11, the average number of hops taken by packets is influenced by the node selection procedure. While this effect may not be readily apparent from the graph, our algorithms have exhibited a slight decrease in the number of hops, surpassing S&W by more than 3%, and significantly outperforming the Epidemic or Prophet v1 and v2 routing protocols. This reduction in hops is accomplished through our meticulous node selection process and the principles we employ to determine packet forwarding. Consequently, the packets take fewer diversions along their routing path.

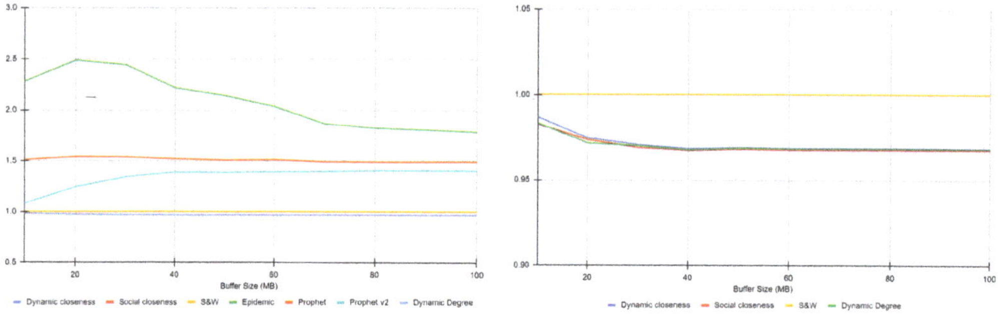

(**a**) (**b**)

Figure 11. Hop count of delivered packets for different routing protocols, depending on the buffer size, normalized with respect to the results of the S&W routing protocol: (**a**) comparison among all the analyzed routing protocols; (**b**) comparison of the proposed metrics and S&W.

4.3.6. Route Overhead

The execution of Algorithm 3 within the proposed routing protocol determines the quality of the connection between the source device and destination node. This leads to a more restricted packet transmission approach compared to S&W, Epidemic, and Prophet v1 and v2 protocols, resulting in a great reduction in overhead (more than 32% less), as shown in Figure 12. Although the proposed algorithm retains the same number of message copies, they are no longer forwarded to all nodes but only to those that satisfy the conditions specified in Algorithm 3.

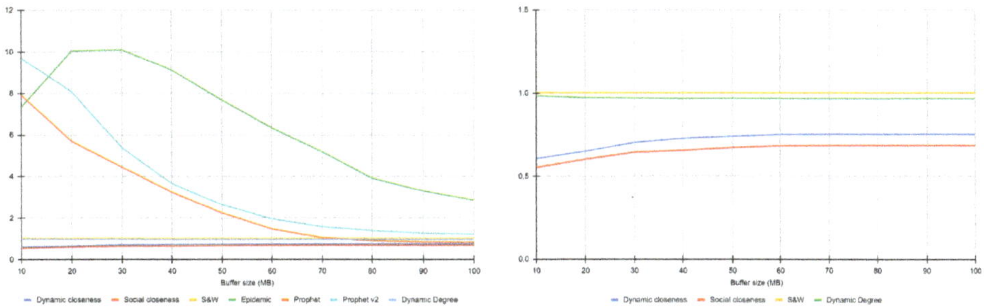

(**a**) (**b**)

Figure 12. Overhead comparison of different routing protocols based on buffer size relative to the results of the S&W routing protocol: (**a**) comparison among all the analyzed routing protocols; (**b**) comparison of the proposed metrics and S&W.

4.3.7. Discussion of the Results

In the realm of networking protocols and QoS optimization, effective data transmission and low-latency routing are of utmost importance. This study deeply delves into the evaluation and refinement of routing algorithms to address these critical aspects, specifically comparing the performance of our proposed algorithms against well-established ones such as Spray and Wait, Epidemic, Prophet v1, and Prophet v2. The comparative analysis encompasses essential metrics, including packet latency, the number of hops, and overhead. It sheds light on the superior efficiency and effectiveness achieved by our meticulously designed algorithms. Subsequently, the discussion encapsulates noteworthy findings and implications of our research, illuminating the promising advancements in network optimization and reliability.

Concerning packet latency, our algorithms significantly reduce average packet delay compared to the original Spray and Wait protocol, with an average reduction of approximately 2% compared to Spray and Wait, and more than 10% compared to Epidemic or Prophets v1 and v2 protocols. Additionally, our algorithms demonstrate a reduction in the number of hops, surpassing Spray and Wait by more than 3%, and significantly outperforming the Epidemic or Prophet v1 and v2 routing protocols. This reduction in hops is achieved through our meticulous node selection process and the principles we employ to determine packet forwarding. Regarding overhead, our algorithms adopt a more restricted approach to packet transmission compared to Spray and Wait, Epidemic, and Prophet v1 and v2 protocols, resulting in a substantial reduction in overhead (more than 32% less).

In summary, our study highlights the remarkable performance enhancements achieved by our proposed algorithms. The reductions in packet latency, number of hops, and overhead represent significant advancements over established protocols like Spray and Wait, Epidemic, Prophet v1, and Prophet v2. These improvements are attributed to our meticulous node selection process and refined packet forwarding principles. The results underscore the potential impact of our algorithms in optimizing QoS for routing in various network scenarios, emphasizing their significance in advancing network efficiency and reliability.

5. Conclusions and Future Work

The extraction of the most influential nodes in a Complex Network is crucial for seeking more efficient data transmission within the network, evaluating its resilience, making better routing decisions, and gaining a deeper understanding of its dynamics. Our study arises from the need to find metrics that measure the influence of nodes in a DSCN, aiming to enhance the network's QoS metrics.

Initially, we employed a network model that transformed the operation of OppMSN over time into a discretized time series of CSNs, to analyze the network's dynamic topology and the pattern of connections among devices. This approach provides a more accurate framework to analyze the evolution of these patterns, based on regression analysis, rather than using a single static aggregated network. In fact, the connections between devices have been analyzed from the perspective of dynamic centrality metrics, as well as from the perspective of a social complex network, extracting relationship patterns to detect the most influential devices over others throughout the network's operation.

In this study, real datasets have been used for validation to showcase the effectiveness of the conducted experiments. The efficacy of different metrics employed on the datasets and potential correlations between them have been verified. Finally, based on influence dynamic rankings, our algorithms have facilitated better decision-making regarding the selection of nodes most suitable for routing data toward their destination in the datasets, leading to enhancements in standard QoS metrics.

Moving forward, our future work involves analyzing the evolutionary characteristics of influence distribution using additional real datasets with more devices and enhanced connectivity among them. Furthermore, we will explore other combinations of centrality metrics and similarity indices to enhance the accuracy of classifying devices in an importance ranking. Additionally, we aim to investigate the concept of "friend" as a measure of temporal locality between each pair of nodes, evaluating its relationship with the connection capacity of a node with others.

Author Contributions: Conceptualization, F.J.R.-P.; Methodology, F.J.R.-P.; Formal analysis, F.J.R.-P.; Investigation, M.A.L.-R.; Resources, M.A.L.-R.; Data curation, M.A.L.-R.; Writing—review & editing, M.A.L.-R. and F.J.R.-P. All authors have read and agreed to the published version of the manuscript.

Funding: This research was funded in part by TED2021-131699B-I00/ AEI/10.13039/501100011033/ European Union NextGenerationEU/PRTR.

Institutional Review Board Statement: Not applicable.

Informed Consent Statement: Not applicable.

Data Availability Statement: The data presented in this study are available in the article.

Conflicts of Interest: The authors declare no conflict of interest.

Appendix A

We quantify the Dynamic Degree of the devices in the example presented in Figure 1 for clarity. The conclusions are drawn from Figure A1, where edge weights represent contact times, and $\alpha = 0.5$ is employed to give equal relative weight to the evolutionary trend of connections and the frequency of contact times.

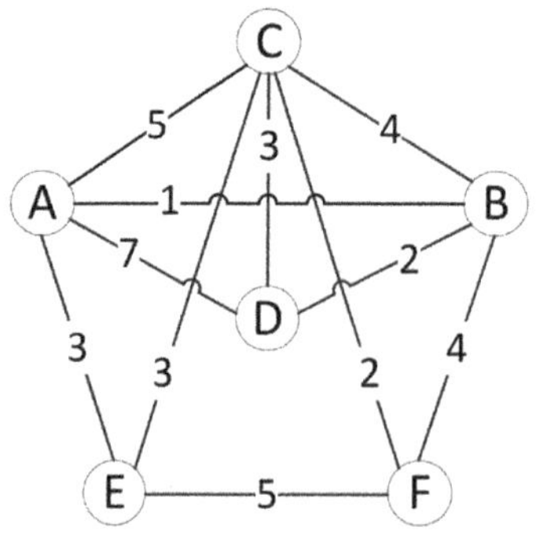

Figure A1. Aggregated network derived from contact frequency of Figure 1.

$$DTE_T(A) = 0.5{\cdot}0.0364 + (1-0.5)\frac{(5+1+7+3)}{10} = 0.0182 + 0.8 = 0.8182$$

$$DTE_T(B) = 0.5\cdot(-0.0788) + (1-0.5)\frac{(4+1+2+4)}{10} = (-0.0394) + 0.55 = 0.5106$$

$$DTE_T(C) = 0.5\cdot0.1151 + (1-0.5)\frac{(5+3+3+2+4)}{10} = 0.05755 + 0.85 = 0.90755$$

$$DTE_T(D) = 0.5\cdot(-0.0788) + (1-0.5)\frac{(7+3+2)}{10} = (-0.0394) + 0.6 = 0.5606$$

$$DTE_T(E) = 0.5\cdot(-0.0182) + (1-0.5)\frac{(3+3+5)}{10} = (-0.0091) + 0.55 = 0.5409$$

$$DTE_T(F) = 0.5\cdot(-0.0545) + (1-0.5)\frac{(5+2+4)}{10} = (-0.02725) + 0.55 = 0.52275$$

For instance, based on the following details, a preliminary ranking of nodes can be established: Device *A* interacts with its neighbors more frequently than device *B*, despite both having the same three neighbors. As a result, device *A* displays a higher Dynamic Degree than device *B*. Similar to the previous example, device *E* receives a better score due to its greater trend of contact development, even if device *D* contacts its neighbors more frequently. Device *C* achieves a higher Dynamic Degree than Device *A* due to having more neighbors and a faster rate of contact development. Consequently, the nodes are ranked as follows according to their Dynamic Degree index: $\text{Ranking}_{\text{DTE}}$ = {*C*, *A*, *D*, *E*, *F*, *B*}.

Let us examine now an example calculation in Figure A2, showcasing an unconnected network during G_2.

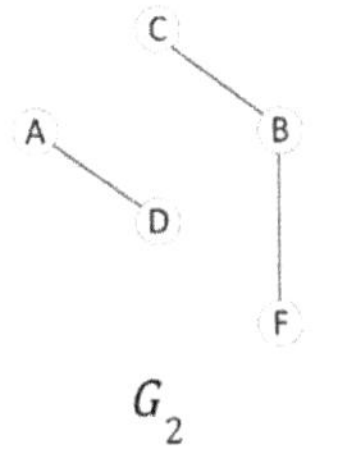

G_2	A	B	C	D	E	F
A	-	∞	∞	1	∞	∞
B	∞	-	1	∞	∞	1
C	∞	1	-	∞	∞	2
D	1	∞	∞	-	∞	∞
E	∞	∞	∞	∞	-	∞
F	∞	1	2	∞	∞	-

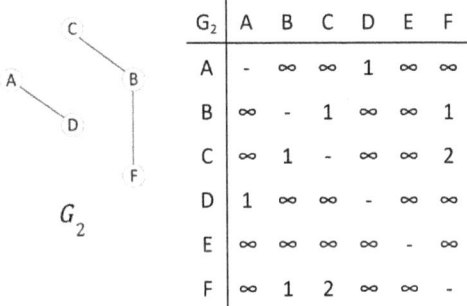

Figure A2. Network connections during instant G_2, based on the example of Figure 1.

$$C_{CLO}(A) = \lim_{x\to\infty}\left(\frac{1}{6-1}\left(\frac{1}{x}+\frac{1}{x}+\frac{1}{2\lambda}+\frac{1}{x}+\frac{1}{x}\right)\right) = \frac{1}{10\lambda}$$

$$C_{CLO}(B) = \lim_{x\to\infty}\left(\frac{1}{6-1}\left(\frac{1}{x}+\frac{1}{2\lambda}+\frac{1}{x}+\frac{1}{x}+\frac{1}{2\lambda}\right)\right) = \frac{1}{5\lambda}$$

$$C_{CLO}(C) = \lim_{x\to\infty}\left(\frac{1}{6-1}\left(\frac{1}{x}+\frac{1}{2\lambda}+\frac{1}{x}+\frac{1}{x}+\frac{1}{3\lambda}\right)\right) = \frac{1}{6\lambda}$$

$$C_{CLO}(D) = \lim_{x\to\infty}\left(\frac{1}{6-1}\left(\frac{1}{2\lambda}+\frac{1}{x}+\frac{1}{x}+\frac{1}{x}+\frac{1}{x}\right)\right) = \frac{1}{10\lambda}$$

$$C_{CLO}(E) = \lim_{x\to\infty}\left(\frac{1}{6-1}\left(\frac{1}{x}+\frac{1}{x}+\frac{1}{x}+\frac{1}{x}+\frac{1}{x}\right)\right) = 0$$

$$C_{CLO}(F) = \lim_{x\to\infty}\left(\frac{1}{6-1}\left(\frac{1}{x}+\frac{1}{2\lambda}+\frac{1}{3\lambda}+\frac{1}{x}+\frac{1}{x}\right)\right) = \frac{1}{6\lambda}$$

Upon analyzing this small example, it becomes apparent that when computed on an unconnected network, the Closeness metric tends to result in lower values, indicating the difficulty of communication between devices belonging to different components. Additionally, the devices within the same component experience an increase in their centrality, as all values are non-zero in the calculation. Consequently, the metric places greater emphasis on nodes that are well-connected.

Let us see the same example of calculation as before for connected devices $\{A, B, C, D, F\}$, from G_7 as shown in Figure 4, but now using the new closeness equation (Equation (15)):

$$C_{CLO}(A) = \lim_{x\to\infty}\left(\frac{1}{6-1}\left(\frac{1}{3\lambda}+\frac{1}{4\lambda}+\frac{1}{2\lambda}+\frac{1}{x}+\frac{1}{4\lambda}\right)\right) = \frac{4}{15\lambda}$$

$$C_{CLO}(B) = \lim_{x\to\infty}\left(\frac{1}{6-1}\left(\frac{1}{3\lambda}+\frac{1}{2\lambda}+\frac{1}{2\lambda}+\frac{1}{x}+\frac{1}{2\lambda}\right)\right) = \frac{11}{30\lambda}$$

$$C_{CLO}(C) = \lim_{x\to\infty}\left(\frac{1}{6-1}\left(\frac{1}{4\lambda}+\frac{1}{2\lambda}+\frac{1}{3\lambda}+\frac{1}{x}+\frac{1}{2\lambda}\right)\right) = \frac{19}{60\lambda}$$

$$C_{CLO}(D) = \lim_{x\to\infty}\left(\frac{1}{6-1}\left(\frac{1}{2\lambda}+\frac{1}{2\lambda}+\frac{1}{3\lambda}+\frac{1}{x}+\frac{1}{3\lambda}\right)\right) = \frac{1}{3\lambda}$$

$$C_{CLO}(E) = \lim_{x\to\infty}\left(\frac{1}{6-1}\left(\frac{1}{x}+\frac{1}{x}+\frac{1}{x}+\frac{1}{x}+\frac{1}{x}\right)\right) = 0$$

$$C_{CLO}(F) = \lim_{x\to\infty}\left(\frac{1}{6-1}\left(\frac{1}{4\lambda}+\frac{1}{2\lambda}+\frac{1}{2\lambda}+\frac{1}{3\lambda}+\frac{1}{x}\right)\right) = \frac{19}{60\lambda}$$

The new ranking of devices based on their closeness index, which is determined by using the sum of reciprocal distances instead of the reciprocal sum of distances, is in line with the ranking obtained by using the traditional equation (Equation (11)) on a network that is well-connected (Equation (12)):

$$Ranking = \{B, D, C, F, A\}$$

To maintain clarity in the example illustrated in Figure 1, we opt for a value of 0.5 when assessing the closeness of the devices. This choice ensures that the Closeness measure and its dynamics carry equal relative weight.

$$CTE_T(A) = 0.5{\cdot}0.0174 + (1-0.5)\frac{2.0667(\lambda^{-1})}{10} = 0.1167\lambda^{-1}$$

$$CTE_T(B) = 0.5{\cdot}(-0.0026) + (1-0.5)\frac{1.43344(\lambda^{-1})}{10} = 0.0704\lambda^{-1}$$

$$CTE_T(C) = 0.5{\cdot}0.0056 + (1-0.5)\frac{2.3501(\lambda^{-1})}{10} = 0.1203\lambda^{-1}$$

$$CTE_T(D) = 0.5{\cdot}(-0.0122) + (1-0.5)\frac{1.9833(\lambda^{-1})}{10} = 0.0931\lambda^{-1}$$

$$CTE_T(E) = 0.5{\cdot}(-0.0001) + (1-0.5)\frac{1.8167(\lambda^{-1})}{10} = 0.0908\lambda^{-1}$$

$$CTE_T(F) = 0.5{\cdot}(-0.0065) + (1-0.5)\frac{1.7667(\lambda^{-1})}{10} = 0.0851\lambda^{-1}$$

The new ranking of nodes according to their Dynamic Closeness index is as follows:

$$Ranking = \{C, A, D, E, F\}$$

The best path from device A to device B in Figure 5 is then determined using the opportunistic-based shortest path method. Table A1 presents the findings of this investigation.

$$ORI_T(A,B) = \left(\sum_{i=1}^{T}\{\varphi_{AB}x_{AB}\}_{t_i}\right)^{\frac{|\Gamma(A)\cap\Gamma(B)|+1}{2}} = 1^{\frac{1+1}{2}}$$

$$ORI_T(A,C) = \left(\sum_{i=1}^{T}\{\varphi_{AC}x_{AC}\}_{t_i}\right)^{\frac{|\Gamma(A)\cap\Gamma(C)|+1}{2}} = 5^{\frac{1+1}{2}}$$

$$ORI_T(C,B) = \left(\sum_{i=1}^{T}\{\varphi_{CB}x_{CB}\}_{t_i}\right)^{\frac{|\Gamma(C)\cap\Gamma(B)|+1}{2}} = 4^{\frac{1+1}{2}}$$

$$ORI_T(A,D) = \left(\sum_{i=1}^{T}\{\varphi_{AD}x_{AD}\}_{t_i}\right)^{\frac{|\Gamma(A)\cap\Gamma(D)|+1}{2}} = 7^{\frac{1+1}{2}}$$

$$ORI_T(D,B) = \left(\sum_{i=1}^{T}\{\varphi_{DB}x_{DB}\}_{t_i}\right)^{\frac{|\Gamma(D)\cap\Gamma(B)|+1}{2}} = 2^{\frac{1+1}{2}}$$

$$ORI_T(A,E) = \left(\sum_{i=1}^{T}\{\varphi_{AE}x_{AE}\}_{t_i}\right)^{\frac{|\Gamma(A)\cap\Gamma(E)|+1}{2}} = 3^{1/3}$$

$$ORI_T(E,F) = \left(\sum_{i=1}^{T}\{\varphi_{EF}x_{EF}\}_{t_i}\right)^{\frac{|\Gamma(E)\cap\Gamma(F)|+1}{2}} = 5^{1/5}$$

$$ORI_T(F,B) = \left(\sum_{i=1}^{T}\{\varphi_{FB}x_{FB}\}_{t_i}\right)^{\frac{|\Gamma(F)\cap\Gamma(B)|+1}{2}} = 4^{1/4}$$

$$\alpha_{A,B} = \frac{(ORI_T(A,B))^{\frac{1}{|\{A,B\}|}}}{max(ORI_T(A,B))} = \frac{\left(1^{\frac{1+1}{2}}\right)^{\frac{1}{1}}}{1^{\frac{1+1}{2}}} = 1$$

$$\alpha_{A,C,B} = \frac{(ORI_T(A,C)\cdot ORI_T(C,B))^{\frac{1}{2}}}{max(ORI_T(A,C),\ ORI_T(C,B))} = \frac{\left(5^{\frac{1+1}{2}}\cdot 4^{\frac{1+1}{2}}\right)^{\frac{1}{2}}}{5^{\frac{1+1}{2}}} = 0.8944$$

$$\alpha_{A,D,B} = \frac{(ORI_T(A,D)\cdot ORI_T(D,B))^{\frac{1}{2}}}{max(ORI_T(A,D),\ ORI_T(D,B))} = \frac{\left(7^{\frac{1+1}{2}}\cdot 2^{\frac{1+1}{2}}\right)^{\frac{1}{2}}}{7^{\frac{1+1}{2}}} = 0.5345$$

$$\alpha_{A,E,F,B} = \frac{(ORI_T(A,E)\cdot ORI_T(E,F)\cdot ORI_T(F,B))^{\frac{1}{3}}}{max(ORI_T(A,E),\ ORI_T(E,F),\ ORI_T(F,B))} = \frac{\left(3^{\frac{1}{3}}\cdot 5^{\frac{1}{5}}\cdot 4^{\frac{1}{4}}\right)^{\frac{1}{3}}}{3^{\frac{1}{3}}} = 0.9789$$

$$Cost_{opp}(A,B) = \left(\frac{1}{1^{\frac{1+1}{2}}}\right)^{1} = 1$$

$$Cost_{opp}(A,C,B) = \left(\frac{1}{5^{\frac{1+1}{2}}}\right)^{0.8944} + \left(\frac{1}{4^{\frac{1+1}{2}}}\right)^{0.8944} = 0.5226$$

$$Cost_{opp}(A,D,B) = \left(\frac{1}{7^{\frac{1+1}{2}}}\right)^{0.5345} + \left(\frac{1}{2^{\frac{1+1}{2}}}\right)^{0.5345} = 1.0437$$

$$Cost_{opp}(A,E,F,B) = \left(\frac{1}{3^{\frac{1}{3}}}\right)^{0.9789} + \left(\frac{1}{5^{\frac{1}{5}}}\right)^{0.9789} + \left(\frac{1}{4^{\frac{1}{4}}}\right)^{0.9789} = 2.1408$$

Table A1. Results of the opportunistic social shortest path method to identify the best path from device A to device B.

Path	Latency	Load Balancing	Overhead	ORI_T	ORI_T Balancing	$Cost_{opp}$
$\{A, B\}$	$10t$	$0t$	2λ	$1^{2/2}$	0	1
$\{A, C, B\}$	$7t$	$0.5t$	3λ	$5^{2/2}, 4^{2/2}$	4.5	0.5226
$\{A, D, B\}$	$7t$	$1.t$	3λ	$7^{2/2}, 2^{2/2}$	4.5	1.0437
$\{A, E, F, B\}$	$7t$	$1.5764t$	4λ	$3^{1/3}, 5^{1/5}, 4^{1/4}$	1.5087	2.1408

The path $\{A, B\}$ represents the shortest distance, but it possesses the lowest $ORI_T(A, B)$ value among all routes. On the other hand, the path $\{A, E, F, B\}$ has the longest path distance with a $ORI_T(A, E, F, B)$ value of 4.2362 (the sum of the three ORI_T), which is the second lowest. Comparatively, both the paths $\{A, C, B\}$ and $\{A, D, B\}$ share the same path distance, with $ORI_T(A, C, B) = ORI_T(A, D, B) = 9$. However, $Cost_{opp}(A, C, B)$ is lower than $Cost_{opp}(A, D, B)$. As a consequence, when both social relationships (ORI_T) and communication costs are taken into consideration, the path $\{A, C, B\}$ is shown to be the best option for moving from device A to device B.

Therefore, to obtain the ranking based on CLO_{opp}, one must compute the CLO_{opp} of each node with respect to the rest:

$$
\begin{aligned}
CLO_{opp}(A) &= \frac{1}{6-1}\left(\frac{1}{\min(Cost_{opp}(A,B))} + \frac{1}{\min(Cost_{opp}(A,C))} + \frac{1}{\min(Cost_{opp}(A,D))}\right.\\
&\left.+\frac{1}{\min(Cost_{opp}(A,E))} + \frac{1}{\min(Cost_{opp}(A,F))}\right)\\
&= \frac{1}{6-1}\left(\frac{1}{0.5264} + \frac{1}{0.2} + \frac{1}{0.1429} + \frac{1}{0.6934} + \frac{1}{1.4289}\right) = 3.5343\\
CLO_{opp}(B) &= \frac{1}{6-1}\left(\frac{1}{\min(Cost_{opp}(B,A))} + \frac{1}{\min(Cost_{opp}(B,C))} + \frac{1}{\min(Cost_{opp}(B,D))}\right.\\
&\left.+\frac{1}{\min(Cost_{opp}(B,E))} + \frac{1}{\min(Cost_{opp}(B,F))}\right)\\
&= \frac{1}{6-1}\left(\frac{1}{0.5264} + \frac{1}{0.25} + \frac{1}{0.5} + \frac{1}{0.4378} + \frac{1}{0.7071}\right) = 2.0019\\
CLO_{opp}(C) &= \frac{1}{6-1}\left(\frac{1}{\min(Cost_{opp}(C,A))} + \frac{1}{\min(Cost_{opp}(C,B))} + \frac{1}{\min(Cost_{opp}(C,D))}\right.\\
&\left.+\frac{1}{\min(Cost_{opp}(C,E))} + \frac{1}{\min(Cost_{opp}(C,F))}\right)\\
&= \frac{1}{6-1}\left(\frac{1}{0.2} + \frac{1}{0.25} + \frac{1}{0.4497} + \frac{1}{1.2428} + \frac{1}{1.2523}\right) = 2.5654\\
CLO_{opp}(D) &= \frac{1}{6-1}\left(\frac{1}{\min(Cost_{opp}(D,A))} + \frac{1}{\min(Cost_{opp}(D,B))} + \frac{1}{\min(Cost_{opp}(D,C))}\right.\\
&\left.+\frac{1}{\min(Cost_{opp}(D,E))} + \frac{1}{\min(Cost_{opp}(D,F))}\right)\\
&= \frac{1}{6-1}\left(\frac{1}{0.1429} + \frac{1}{0.5} + \frac{1}{0.4497} + \frac{1}{1.2603} + \frac{1}{1.3055}\right) = 2.5562\\
CLO_{opp}(E) &= \frac{1}{6-1}\left(\frac{1}{\min(Cost_{opp}(E,A))} + \frac{1}{\min(Cost_{opp}(E,B))} + \frac{1}{\min(Cost_{opp}(E,C))}\right.\\
&\left.+\frac{1}{\min(Cost_{opp}(E,D))} + \frac{1}{\min(Cost_{opp}(E,F))}\right)\\
&= \frac{1}{6-1}\left(\frac{1}{0.6394} + \frac{1}{1.4378} + \frac{1}{1.2428} + \frac{1}{1.2603} + \frac{1}{0.7248}\right) = 1.0474\\
CLO_{opp}(F) &= \frac{1}{6-1}\left(\frac{1}{\min(Cost_{opp}(F,A))} + \frac{1}{\min(Cost_{opp}(F,B))} + \frac{1}{\min(Cost_{opp}(F,C))}\right.\\
&\left.+\frac{1}{\min(Cost_{opp}(F,D))} + \frac{1}{\min(Cost_{opp}(F,E))}\right)\\
&= \frac{1}{6-1}\left(\frac{1}{1.4289} + \frac{1}{0.7071} + \frac{1}{1.2523} + \frac{1}{1.3055} + \frac{1}{0.7248}\right) = 1.0116
\end{aligned}
$$

So, the new ordered ranking of nodes based on their Social Closeness index is as follows:

$$Ranking_{CLO\_opp} = \{A, C, D, B, F, E\}$$

References

1. Number of Smartphone Mobile Network Subscriptions Worldwide from 2016 to 2022, with Forecasts from 2023 to 2028. Available online: https://www.statista.com/statistics/330695/number-of-smartphone-users-worldwide/ (accessed on 30 June 2023).
2. Zhang, H.; Chen, Z.; Wu, J.; Liu, K. FRRF: A Fuzzy Reasoning Routing-Forwarding Algorithm Using Mobile Device Similarity in Mobile Edge Computing-Based Opportunistic Mobile Social Networks. *IEEE Access* **2019**, *7*, 35874–35889. [CrossRef]
3. Gantha, S.S.; Jaiswal, S.; Ppallan, J.M.; Arunachalam, K. Path Aware Transport Layer Solution for Mobile Networks. *IEEE Access* **2020**, *8*, 174605–174613. [CrossRef]
4. Yuan, X.; Yao, H.; Wang, J.; Mai, T.; Guizani, M. Artificial Intelligence Empowered QoS-Oriented Network Association for Next-Generation Mobile Networks. *IEEE Trans. Cogn. Commun. Netw.* **2021**, *7*, 856–870. [CrossRef]
5. He, B.; Wang, J.; Qi, Q.; Sun, H.; Liao, J. RTHop: Real-Time Hop-by-Hop Mobile Network Routing by Decentralized Learning with Semantic Attention. *IEEE Trans. Mob. Comput.* **2023**, *22*, 1731–1747. [CrossRef]
6. Soelistijanto, B.; Howarth, M.P.; Qi, Q.; Sun, H.; Liao, J. Transfer Reliability and Congestion Control Strategies in Opportunistic Networks: A Survey. *IEEE Commun. Surv. Tutor.* **2014**, *16*, 538–555. [CrossRef]
7. Xiong, F.; Xia, L.; Xie, J.; Sun, H.; Wang, H.; Li, A.; Yu, Y. Is Hop-by-Hop Always Better Than Store-Carry-Forward for UAV Network? *IEEE Access* **2019**, *7*, 154209–154223. [CrossRef]
8. Anh Duong, D.V.; Kim, D.Y.; Yoon, S. TSIRP: A Temporal Social Interactions-Based Routing Protocol in Opportunistic Mobile Social Networks. *IEEE Access* **2021**, *9*, 72712–72729. [CrossRef]
9. Hajarathaiah, K.; Enduri, M.K.; Dhuli, S.; Anamalamudi, S.; Cenkeramaddi, L.R. Generalization of Relative Change in a Centrality Measure to Identify Vital Nodes in Complex Networks. *IEEE Access* **2023**, *11*, 808–824. [CrossRef]
10. Qiu, L.; Zhang, J.; Tian, X.; Zhang, S. Identifying Influential Nodes in Complex Networks Based on Neighborhood Entropy Centrality. *Comput. J.* **2021**, *64*, 1465–1476. [CrossRef]
11. Ibrahim, M.H.; Missaoui, R.; Vaillancourt, J. Cross-Face Centrality: A New Measure for Identifying Key Nodes in Networks Based on Formal Concept Analysis. *IEEE Access* **2020**, *8*, 206901–206913. [CrossRef]
12. Zhu, Y.; Ma, H. Ranking Hubs in Weighted Networks with Node Centrality and Statistics. In Proceedings of the Fifth International Conference on Instrumentation and Measurement, Computer, Communication and Control (IMCCC), Qinhuangdao, China, 18–20 September 2015.
13. Liu, M.; Zeng, Y.; Jiang, Z.; Liu, Z.; Ma, J. Centrality Based Privacy Preserving for Weighted Social Networks. In Proceedings of the 13th International Conference on Computational Intelligence and Security (CIS), Hong Kong, China, 15–18 December 2017.
14. Niu, J.; Fan, J.; Wang, L.; Stojinenovic, M. K-hop centrality metric for identifying influential spreaders in dynamic large-scale social networks. In Proceedings of the IEEE Global Communications Conference, Austin, TX, USA, 8–12 December 2014.
15. Rogers, T. Null models for dynamic centrality in temporal networks. *J. Complex Netw.* **2015**, *3*, 113–125. [CrossRef]
16. Opsahl, T.; Agneessens, F.; Skvoretz, J. Node centrality in weighted networks: Generalizing degree and shortest paths. *Soc. Netw.* **2010**, *32*, 245–251. [CrossRef]
17. Kim, H.; Anderson, R. Temporal node centrality in complex networks. *Phys. Rev. E* **2012**, *85*, 026107. [CrossRef] [PubMed]
18. Trajkovic, L. Complex Networks. In Proceedings of the IEEE 19th International Conference on Cognitive Informatics & Cognitive Computing (ICCI*CC), Beijing, China, 26–28 September 2020.
19. Zhang, S.-S.; Liang, X.; Wei, Y.-D.; Zhang, X. On Structural Features, User Social Behavior, and Kinship Discrimination in Communication Social Networks. *IEEE Trans. Comput. Soc. Syst.* **2020**, *7*, 425–436. [CrossRef]
20. Elmezain, M.; Othman, E.A.; Ibrahim, H.M. Temporal Degree-Degree and Closeness-Closeness: A New Centrality Metrics for Social Network Analysis. *Mathematics* **2021**, *9*, 2850. [CrossRef]
21. Wang, Y.; Li, H.; Zhang, L.; Zhao, L.; Li, W. Identifying influential nodes in social networks: Centripetal centrality and seed exclusion approach. *Chaos Solitons Fractals* **2022**, *162*, 112513. [CrossRef]
22. Caschera, M.C.; D'Ulizia, A.; Ferri, F.; Grifoni, P. MONDE: A method for predicting social network dynamics and evolution. *Evol. Syst.* **2019**, *10*, 363–379. [CrossRef]
23. Ugurlu, O. Comparative analysis of centrality measures for identifying critical nodes in complex networks. *J. Comput. Sci.* **2022**, *62*, 101738. [CrossRef]
24. Dey, P.; Bhattacharya, S.; Roy, S. A Survey on the Role of Centrality as Seed Nodes for Information Propagation in Large Scale Network. *ACM/IMS Trans. Data Sci.* **2021**, *2*, 1–25. [CrossRef]
25. Omar, Y.M.; Plapper, P. A Survey of Information Entropy Metrics for Complex Networks. *Entropy* **2020**, *22*, 1417. [CrossRef]
26. Gao, Z.; Shi, Y.; Chen, S. Measures of node centrality in mobile social networks. *Int. J. Mod. Phys. C* **2015**, *26*, 1550107. [CrossRef]
27. Wei, B.; Kawakami, W.; Kanai, K.; Katto, J. A History-Based TCP Throughput Prediction Incorporating Communication Quality Features by Support Vector Regression for Mobile Network. In Proceedings of the 13th International Conference on Information and Communication Technology Convergence (ICTC), Taichung, Taiwan, 11–13 December 2017.
28. Awane, H.; Ito, Y.; Koizumi, M. Study on QoS Estimation of In-vehicle Ethernet with CBS by Multiple Regression Analysis. In Proceedings of the IEEE International Symposium on Multimedia (ISM), Jeju Island, Republic of Korea, 19–21 October 2022.
29. Chen, W.; Zhou, Y. A Link Prediction Similarity Index Based on Enhanced Local Path Method. In Proceedings of the 40th Chinese Control Conference (CCC), Shanghai, China, 26–28 July 2021.
30. Varma, S.; Shivam, S.; Thumu, A.; Bhushanam, A.; Sarkar, D. Jaccard Based Similarity Index in Graphs: A Multi-Hop Approach. In Proceedings of the IEEE Delhi Section Conference (DELCON), New Delhi, India, 11–13 February 2022.

31. Ahmad, I.; Akhtar, M.; Noor, S.; Shahnaz, A. Missing Link Prediction using Common Neighbor and Centrality based Parameterized Algorithm. *Nat. Sci. Rep.* **2020**, *10*, 364. [CrossRef] [PubMed]
32. Eagle, N.; Pentland, A. Reality mining: Sensing complex social systems. *Pers. Ubiquitous Comput.* **2006**, *10*, 255–268. [CrossRef]
33. Orlinski, M.; Filer, N. The rise and fall of spatio-temporal clusters in mobile ad hoc networks. *Ad Hoc Netw.* **2013**, *11*, 1641–1654. [CrossRef]
34. Anulakshmi, S.; Anand, S.; Ramesh, M.V. Impact of Network Density on the Performance of Delay Tolerant Protocols in Heterogeneous Vehicular Network. In Proceedings of the International Conference on Wireless Communications Signal Processing and Networking (WiSPNET), Chennai, India, 21–23 March 2019.
35. Sati, M.; Shanab, S.; Elshawesh, A.; Sati, S.O. Density and Degree Impact on Opportunistic Network Communications. In Proceedings of the IEEE 1st International Maghreb Meeting of the Conference on Sciences and Techniques of Automatic Control and Computer Engineering MI-STA, Tripoli, Libya, 25–27 May 2021.
36. Dede, J.; Förster, A.; Hernández-Orallo, E.; Herrera-Tapia, J.; Kuladinithi, K.; Kuppusamy, V.; Vatandas, Z. Simulating opportunistic networks: Survey and future directions. *IEEE Commun. Surv. Tutor.* **2017**, *20*, 1547–1573. [CrossRef]
37. Khan, M.; Liu, M.; Dou, W.; Yu, S. vGraph: Graph Virtualization towards Big Data. In Proceedings of the Third IEEE International Conference on Advanced Cloud and Big Data (CBD), Yangzhou, China, 30 October–1 November 2015.

MDPI
St. Alban-Anlage 66
4052 Basel
Switzerland
www.mdpi.com

Applied Sciences Editorial Office
E-mail: applsci@mdpi.com
www.mdpi.com/journal/applsci

www.ingramcontent.com/pod-product-compliance
Lightning Source LLC
LaVergne TN
LVHW070235100526
838202LV00015B/2131